Agricultural Marketing

Author

Dr. K. Nirmal Ravi Kumar is presently working as an Associate Professor and Head (Agricultural Economics) in Agricultural College, Mahanandi in Acharya N.G. Ranga Agricultural University. He had a brilliant academic career and he specialized in 'Agricultural Marketing' both in his post-graduate and doctoral programmes. He is actively involved both in agricultural research and teaching activities during the past eleven years in the University. He published 44 articles in popular agricultural journals. He also contributed two technical bulletins on economic aspects of irrigation water management highlighting the research priorities in major irrigation commands of Andhra Pradesh and need based technological interventions to address the same during his active stint in "Andhra Pradesh Water Management Project", an international project funded by The Royal Netherlands Embassy. His interested areas include International trade of Indian agriculture, Farming systems approach, Irrigation water management etc. The following are the other major contributions (books) by the same author:

- ☆ Indian Agriculture in the 21st Century: Challenges and Opportunities
- ☆ World Trade Agreement and Indian Agriculture: Implementation Experience
- ☆ Microeconomic Analysis in Agriculture in 2 Vols (Set)
- ☆ Objective Agricultural Economics 2nd Revised and Enlarged edn
- ☆ Farmers Indebtedness in India: An Economic Analysis, and
- ☆ Research Methodology for Agricultural Economics

Agricultural Marketing

— Volume 1 —

K. Nirmal Ravi Kumar
Associate Professor and Head (Agricultural Economics)
Agricultural College,
Mahanandi

2014
Daya Publishing House®
A Division of
Astral International Pvt. Ltd.
New Delhi – 110 002

Cataloging in Publication Data—DK
Courtesy: D.K. Agencies (P) Ltd. <docinfo@dkagencies.com>

Ravi Kumar, K. N. (Kotamraju Nirmal), 1969-
Agricultural marketing / K. Nirmal Ravi Kumar.
2 v. cm.
Includes bibliographical references (p.).
ISBN 9789351242390 (set) (Hardbound)
ISBN 9789351242512 (v. 1) (Hardbound)
ISBN 9789351242529 (v. 2) (Hardbound)
ISBN 9789351300885 (set) (International edition)
ISBN 9789351300892 (v. 1) (International edition)
ISBN 9789351300908 (v. 2) (International edition)

1. Farm produce—India—Marketing. 2. Produce trade—Government policy—India. 3. Agriculture—Economic aspects—India. I. Title.

DDC 381.410954 23

Published by : **Daya Publishing House®**
A Division of
Astral International Pvt. Ltd.
– ISO 9001:2008 Certified Company –
81, Darya Ganj, Near Hindi Park,
Delhi Medical Association Road,
New Delhi - 110 002
Phone: 011-4354 9197, 2327 8134
Fax: +91-11-2324 3060
E-mail: info@astralint.com
Website: www.astralint.com

Laser Typesetting : **Classic Computer Services**, Delhi - 110 035

Printed at : **Salasar Imaging Systems**, Delhi - 110 035

PRINTED IN INDIA

– Dedicated to –

Late K. RAMA CHANDRA RAO

ACHARYA N.G. RANGA AGRICULTURAL UNIVERSITY
ADMINISTRATIVE OFFICE, RAJENDRANAGAR, HYDERABAD-500030

V. NAGI REDDY, I.A.S
Vice-Chancellor

Phone : +91-40-24015035(O)
+91-40-24015031(F)
Email: angrau_vc@yahoo.com
GRAMS: "AGRIVARSITY"

Foreword

Even today, 'Agriculture' has often been considered as production, rather than market-oriented, whether in a developed or a developing country context. But, in the era of liberalized trade regime, the subject of 'Agricultural Marketing' has assumed crucial significance in boosting the economic development of a country. Further, down the supply chain, the production and marketing scenarios are totally demand driven with the main objectives of increasing focus on consumer needs, exploit the market opportunities and ensure maximization of consumer's satisfaction. The revamping of Indian agriculture suiting to the trends in global market provides tremendous market opportunities and offer challenges for gaining due share in the international market. To have a thorough analysis on these aspects, it demands basic and fundamental knowledge on the various concepts of the subject of 'Agricultural Marketing' and this book deserves a special mention in this regard. The overall objective of this book is to provide a comprehensive knowledge on the basic concepts in Agricultural Marketing and analyzing the same with reference to the Indian agricultural marketing system in the context of WTO. I complement the sincere efforts of Dr. K. Nirmal Ravi Kumar in bringing out this text book duly considering the latest syllabi, as prescribed by the IV Deans' Committee recommendations appointed by the Indian Council of Agricultural Research, New Delhi. This text book is written in a style intended to provide basic, practical knowledge and techniques for the benefit of the students and also helps the marketing personnel, who are untrained in the discipline of Agricultural Marketing. I hope this text book will be very useful and instructive for the graduates, post-graduates and doctoral students, who wish to pursue their career in the field of Agricultural Economics. It will also be of immense use for the Academicians, as it harbours standard and updated material pertaining to the concepts of Agricultural Marketing. I hope this volume will even serve the students, who are aspiring high in the field of Agricultural Economics and to be competent enough in various competitive examinations.

V. Nagi Reddy

ACHARYA N.G. RANGA AGRICULTURAL UNIVERSITY
ADMINISTRATIVE OFFICE, RAJENDRANAGAR, HYDERABAD-500030

Dr. Y. ESWARA PRASAD
Ph.D
Professor & University Head &
Hon. Director
Department of Agricultural Economics & CCS
College of Agriculture,
Rajendranagar – 500 030, Hyderabad
and
Member, Board of Management,
ANGRAU

Phone : +91-40-24015042
+91-40-24011561
Extn: 374
Mobile: 9441609517
Res: 040-23341406

Preface

Agricultural marketing system in our country has been criticized on several grounds for not providing the quality, variety and forms of commodities to local consumers have come to expect. Even the farmers complain frequently about the pricing mechanism of agricultural commodities and different forms of Government's interventions in the field of agricultural marketing. With these bottlenecks in the present agricultural marketing system in India, there is a need to revamp the existing system in tune with the challenges and opportunities offered by the global market. In critically assessing the basic marketing constraints and possible interventions to be made to understand, accept and meet the global challenges, a thorough knowledge is inevitable on the basic and fundamental concepts of Agricultural Marketing. In this context, this text book on 'Agricultural Marketing' deserves a special mention, as it harbours and deals with all the basic concepts in a practical and applicable approach for the improvement of present agricultural marketing system in the country. Further, to understand the dimensions and implications of the problems in present agricultural marketing system in India, a basic orientation is very useful. This text book also provides an opportunity to understand these basic concepts, so as to avoid many pitfalls previously encountered in the context of safeguarding the farming community from market volatilities. Considering these efforts made in this text book, I complement Dr. K. Nirmal Ravi Kumar in bringing out this important and timely publication. Further, it contains an up-dated and standard material on the various aspects of Agricultural Marketing duly covering the latest syllabi as prescribed by the several State Agricultural Universities in India. This text book also helps the Academicians to understand and explain the concepts in a more analytical manner.

(Y. ESWARA PRASAD)

ACHARYA N.G. RANGA AGRICULTURAL UNIVERSITY
AGRICULTURAL COLLEGE, MAHANANDI

M.C. FARM - 518 502, MAHANANDI, KURNOOL DISTRICT

Dr. K. NIRMAL RAVI KUMAR
Ph.D
Associate Professor & Head
Dept. of Agricultural Economics

Phone : +91-08179696983
Email: drknrk@gmail.com

Author's Note

I consider this as a bold attempt to present a comprehensive text material on the fundamental aspects of Agricultural Marketing. With a limited experience in the field of teaching (6 years) in the Acharya N.G. Ranga Agricultural University, I will not dare enough to make a tall claim regarding the originality of the material incorporated in this text book entitled, 'Agricultural Marketing', but I certainly admit that, the concepts are dealt in-detail in a lucid form, so as to enable the students and Academicians to learn, understand and conceptualize the same. The present volume can serve as a standardized text book for the graduates, post-graduates and doctoral students, who wish to pursue their career in the field of Agricultural Economics in different State Agricultural Universities.

This task of writing and publishing the text book on the advanced versions of 'Agricultural Marketing' has been one of the most challenging and satisfying experiences of my life. I am especially thankful for the friendship and unfailing support of Mr.M. Satish Rahul and Dr. K.V. Hari Prasad, who practically demonstrated the true meaning of friendship by standing with me during the pursuit of this work. Finally, last but by no means least, a grateful kiss goes to my wife, Meena, my son, Sri Ram and cheerful thanks to my mother Janaki for their love and support. Their devotion, patience and understanding were indispensable. I gratefully dedicate this piece of work to my late father, K. Rama Chandra Rao, who instilled in me at an early age the importance of education and the value of hard work. I only wish you could have been here to see me finish the journey I started with your efforts and initiatives. I know you would have been so proud.

K. Nirmal Ravi Kumar

Contents

Acronyms

AAY: Antyodaya Anna Yojana
ACC: Average Combined Cost
AEZ: Agricultural Economic Zones
AGMARK: Agricultural Produce (Grading and Marking) Act, 1937
AGMARKNET: Agricultural Marketing Information Network
AGPLM: Food grains to Adolescent Girls, Pregnant and Lactating Mothers
AIRCRC: All-India Rural Credit Review Committee
AIRCSC: All India Rural Credit Survey Committee
AMA: Agriculture Marketing Advisor
AMIS: Agricultural Market Intelligence System
AMS: Aggregate Measurement of Support
AoA: Agreement on Agriculture
APAGROS: Andhra Pradesh State Agro-Industries Development Corporation Limited

APC: Average Production Cost
APEDA: Agricultural and Processed Food Products Export Development Authority
APL: Above the Poverty Line
APMC: Agricultural Produce Market Committees
APSSDC: Andhra Pradesh State Seeds Development Corporation Limited
ARC: Average Revenue Curve
ASC: Average Selling Cost
ATC: Average Total Cost
ATF: Aviation Turbine Fuel
AVC: Average Variable Cost
BCR: Benefit-Cost Ratios
BICP: Bureau of Industrial Costs and Prices
BIS: Bureau of Indian Standards
BKS: Bharath Krishak Samaj
BOGOF: Buy-One-Get-One-Free
BoP: Balance of Payment
BoT: Balance of Trade
BOT: Build Operate and Transfer
BPL: Below Poverty line
CAC: Codex Alimentarius Commission
CACP: Commission for Agricultural Costs and Prices
CAN: Calcium Ammonium Nitrate
CAP: Cover and Plinth Storage
CBD: Conservation of Biological Diversity
CBOT: Chicago Board of Trade
CC: Combined Cost
CCI: Cotton Corporation of India Limited
CCS: Cost of Cultivation Scheme
CDB: Coconut Development Board
CEPC: Cashewnuts Export Promotion Council of India
CFMTTI: Central Farm Machinery Training and Testing Institute
CFQC&TI: Central Fertilizer Quality Control and Training Institute
CFSs: Co-operative Farming Societies
CFTRI: Central Food Technological Research Institute

CGIAR: Consultative Group on International Agricultural Research
CGPDTM: Office of the Controller General of Patents, Designs and Trademarks
CIAE: Central Institute of Agricultural Engineering
CIB: Central Insecticides Board
CIF: Cost, Insurance and Freight
CIL: Central Insecticides Laboratory
CIP: Central Issue Prices
CIPHET: Central Institute of Post Harvest Engineering and Technology
CMSs: Co-operative Marketing Societies
COP: Cost of Production
CORDET: Co-operative Rural Development Trust
COSAMB: National Council of State Agricultural Marketing Boards
CPP: Consumer's Purchase Price
CR: Concentration Ratios
CS Scheme: Comprehensive Scheme for studying the cost of cultivation of Principal Crops in India
CSB: Central Silk Board
CSC: Central Seed Committee
CSIR: Council of Scientific and Industrial Research
CSTL: Central Seed Testing Laboratory
CWC: Central Warehousing Corporation
DAP: Diammonium Phosphate
DCCB: District Central Co-operative Banks
DCP: Decentralized Procurement
DDA: Doha Declaration Agreement
DESMOA: Directorate of Economics and Statistics, Ministry of Agriculture
DGFT: Directorate General of Foreign Trade
DIP: Destructive Insects and Pests Act
DMI: Directorate of Marketing and Inspection
DMP: Degree of Market Power
DOF: Department of Fertilizers
DSB: Dispute Settlement Body
DSM: Dispute Settlement Mechanism
DST: Department of Science and Technology

E_c: Cross elasticity of demand
ECA: Essential Commodities Act
EEI: Economic Efficiency Index
EFP: Emergency Feeding Programme
EFTA: European Free Trade Association
EGS: Education Guarantee Scheme
EOAZ: Export Oriented Agri-Zones
EOUs: Export Oriented Units
EPCs: Export Promotion Councils
EPZ: Export Processing Zones
ER: Efficiency Ratio
EUREGAP: European System Related to Good Agricultural Practices
EWR: Electronic Warehouse Receipts
EXIM: Export-Import
FACT: Fertilizers and Chemicals Travancore
FAO: Food and Agriculture Organization
FAQ: Fair Average Quality
FCI: Food Corporation of India
FCO: Fertilizer Control Order
FCRA: Forward Contracts (Regulation) Act
FDI: Foreign Direct Investment
FLDs: Front Line Demonstrations
FMC: Forward Markets Commission
FNSP: Farmer's Net Selling Price
FOB: Free On Board
FPI: Food Processing Industry
FPS: Fair Price Shops
FPTC: Food Processing Training Centres
FRA: Food Regulation Authority
FSCR: Farmer's Share in Consumer's Rupee
FSSAI: Food Safety and Standards Authority of India
FTZ: Free Trade Zones
FYP: Five Year Plan
GAP: Good Agricultural Practices

GATT: General Agreement on Tariffs and Trade
GATS: General Agreement on Trade in Services
GBY: Gramin Bhandaran Yojana
GC: Gini Coefficient
GCMMF: Gujarat Co-operative Milk Marketing Federation
GDP: Gross Domestic Product
GFCL: Godavari Fertilizers and Chemicals Limited
GHP: Good Handling Practices
GI: Geographical Indications
GM: Genetically Modified
GMM: Gross Marketing Margin
GMP: Good Manufacturing Practices
GNFC: Gujarat Narmada Valley Fertilizers Limited
GSFC: Gujarat State Fertilizers and Chemicals Limited
GSP: Generalized System of Preference
HACCP: Hazard Analysis and Critical Control Points
HDPE: High-Density PolyEthylene
HHI: Herfindahl–Hirschman Index
HIMFED: The Himachal Pradesh State Co-operative Marketing and Consumer's Federation Limited.
HMS: Horizontal Marketing System
HOPCOMS: Horticultural Producers' Co-operative Marketing and Processing Society Limited
HPMC: Himachal Pradesh Horticultural Produce Marketing and Processing Corporation Limited
HRD: Human Resource Development
IASRI: Indian Agricultural Statistics Research Institute
ICAR: Indian Council of Agricultural Research
ICDS: Integrated Child Development Services
ICEX: Indian Commodity Exchange
ICRISAT: International Crop Research Institute for Semi-Arid Tropics
ICT: Information and Communication Technology
IEC: International Electrotechnical Commission
IFFCO: Indian Farmers Fertilizer Co-operative Limited
IFFDC: Indian Farm Forestry Development Co-operative Limited
IFS: Institute of Food Security

IGMRI: Indian Grain Storage Management and Research Institute
IGPB: Indian Grape Processing Board
IGSI: Indian Grain Storage Institute
IICPT: Indian Institute of Crop Processing Technology
IIP: Indian Institute of Packaging
IMF: International Monetary Fund
IMG: Inter-Ministerial Group
INM: Integrated Nutrient Management
IP: International Price
IPL: Indian Potash Limited
IPM: Integrated Pest Management
IPPC: International Plant Protection Convention
IPRs: Intellectual Property Rights
IPSM 15: International Phyto-Sanitary Measure No.15
IPSTA: India Pepper and Spices Trade Association
ISEPC: Indian Silk Export Promotion Council
ISI: Indian Standards Institute
ISO: International Organization for Standardization
ISOPOM: Integrated Scheme for Oil seeds, Pulses, Oil and Palm and Maize
ISTA: International Seed Testing Association
IT: Information Technology
ITC: Indian Tobacco Company
ITO: International Trade Organization
JCI: Jute Corporation of India Limited
KRIBHCO: Krishak Bharati Co-operative Limited
KVK: Krishi Vigyan Kendra
LDCs: Least Developed Countries
LDPE: Low-Density PolyEthylene
LDR: Law of Diminishing Returns
LNG: Liquefied Natural Gas
LPG: Liberalization, Privation and Globalization
LRATC: Long Run Average Total Cost
LRMC: Long Run Marginal Cost
LRS: Long Run Supply

LRTC: Long Run Total Cost
MANAGE: National Institute of Agricultural Extension Management
MAP: Mono-Ammonium Phosphate
MC: Marginal Cost
Mc: Marketing costs
MCX: Multi-Commodity Exchange of India Limited
MDM: Mid-Day Meal
MEI: Marketing Efficiency Index
MFN: Most Favoured Nation
MFP: Minor Forest Produce
MFPS: Mega Food Parks Scheme
MIP: Market Intervention Price
MIS: Market Intervention Scheme
MIU: Market Intelligence Units
MLs: Marketing Losses
MME: Modified Measure of Marketing Efficiency
MMs: Marketing Margins
MNCs: Multi-National Companies
MOFPI: Ministry of Food Processing Industries
MOP: Muriate of Potash
MPEDA: Marine Products Export Development Authority
MPS: Market Period Supply
MRP: Maximum Retail Price
MRC: Marginal Revenue Curve
MRTP: Monopolies and Restrictive Trade Practices
MSAMB: Maharashtra State Agriculture Marketing Board
MSC: Marginal Selling Cost
MSCS: Minimum Seed Certification Standards
MSP: Minimum Support Price
NABARD: National Bank for Agriculture and Rural Development
NAFED: National Agricultural Co-operative Marketing Federation Limited
NBA: National Biodiversity Authority
NBOT: National Board of Trade
NBHC: National Bulk Handling Corporation Limited

NBPGR: National Bureau of Plant Genetic Resources
NBS: Nutrient Based Subsidy
NCA: National Commission on Agriculture
NCAP: National Center for Agricultural Economics and Policy Research
NCDC: National Co-operative Development Corporation
NCDEX: National Commodity and Derivative Exchange Limited
NCDWB: National Co-operative Development and Warehousing Board
NCMSL: National Collateral Management Services Limited
NCF: National Commission on Farmers
NCT: National Capital Territory
NDDB: National Dairy Development Board
NERAMAC: North Eastern Regional Agricultural Marketing Corporation Limited
NFFWP: National Food For Work Programme
NFL: National Fertilizers Limited
NFPDB: National Fish Processing Development Board
NFSM: National Food Security Mission
NGOs: Non-Government Organizations
NHB: National Horticulture Board
NHM: National Horticulture Mission
NIAM: National Institute of Agricultural Marketing
NICNET: National Informatics Centre Network
NIFTEM: National Institute of Food Technology, Entrepreneurship and Management
NMCE: National Multi-Commodity Exchange of India Limited
NMFP: National Mission on Food Processing
NMM: Net Marketing Margin
NMPPB: National Meat and Poultry Processing Board
NNWR: Non-Negotiable Warehouse Receipt
NOAPS: National Old Age Pension Scheme
NOVOD: National Oil seeds and Vegetable Oils Development Board
NPPO: National Plant Protection Organization
NPSS: Non-Product Specific Support
NRFMTTI: Northern Region Farm Machinery Training and Testing Institute
NSAI: National Seed Association of India
NSB: National Seed Board

NSC: National Seed Corporation
NSP: National Seed Project
NSRTC: National Seed Research and Training Center
NTGF: National Tobacco Growers Federation Limited
NWR: Negotiable Warehouse Receipt
OECD: Organization for Economic Co-operation and Development
OGL: Open General License
OIE: International Office of Epizootics or International Animal Health Organization
OILFED: Oilseed Growers' Federation Limited
OMSS: Open Market Sale Scheme
OPEC: Organization of Petroleum Exporting Countries
OTC: Over The Counter
OTM: Oilseeds Technology Mission
PACS: Primary Agricultural Co-operative Credit Societies
PAMS: Present Agricultural Marketing System
PBRs: Plant Breeder's Rights
PCM: Price-Cost Margin
PDS: Public Distribution System
PE: Private Enterprise
PEG: Private Entrepreneurs Godown Scheme
PEST: Political, Economical, Social and Technological
PET: Poly Ethylene Terephalate
PFA: Prevention of Food Adulteration Act
PFEPC: Processed Food Export Promotion Council
PFS: Plants, Fruits and Seeds Order
PHM: Post-Harvest Management
PPA: Plant Protection Adviser
PPL: Paradeep Phosphate Limited
PPP: Public Private Partnership
PPV&FR: Protection of Plant Varieties and Farmers Rights
PR: Profitability Ratio
PS: Price Spread
PSB: Phosphate Solubilizing Bacteria
PSC: Phyto-Sanitary Certificate

PSE: Producer Support Estimate
PSS: Product Specific Support
PVR: Plant Variety Rights
QC: Quality Control
QCLs: Quality Control Laboratories
QRs: Quantitative Restrictions
R&D: Research and Development
RALs: Regional Agmark Laboratories
RAMS: Regulated Agricultural Marketing System
RBI: Reserve Bank of India
RCF: Rashtriya Chemicals and Fertilizers Limited
RFCL: Regional Fertilizer Control Laboratories
RIDF: Rural Infrastructure Development Fund
RPDS: Revamped Public Distribution System
RRB: Regional Rural Banks
RSP: Retailer's Selling price
RTC: Ready To Cook
RTE: Ready To Eat
S&D: Special and Differential Treatment
SAMBs: State Agricultural Marketing Boards
SAUs: State Agricultural Universities
SCPP: Market Structure-Conduct-Performance Paradigm
SEZ: Special Economic Zones
SFCI: State Farms Corporation of India
SGC: Save Grain Campaign
SGLs: State Grading Laboratories
SGRY: Sampoorn Gramin Rozgar Yojna
SHGs: Self-Help Groups
SMP: Statutory MSP
SPS: Sanitary and Phyto-Sanitary Standards
SRATC: Short Run Average Total Cost
SRAVC: Short Run Average Variable Cost
SRMC: Short Run Marginal Cost
SRS: Short Run Supply

SRTC: Short Run Total Cost
SRTVC: Short Run Total Variable Cost
SSCs: State Seed Corporations
SSM: Special Safeguard Mechanism
SSP: Single Super Phosphate
SSTLs: State Seed Testing Laboratories
STC: State Trading Corporation
STCL: Spices Trading Corporation Limited
SWCs: State Warehousing Corporations
SWOT: Strengths (S), Weaknesses (W), Opportunities (O) and Threats (T)
TAPE: Total Agricultural Products Exports
TBT: Technical Barriers to Trade
TC: Total Cost
TCOC: Total Cost of Cultivation
TEXPROCIL: Cotton Textile Export Promotion Council
TKK: Tata Kisan Kendras
TMC: Technology Mission on Cotton
TMc: Total Marketing Cost
TPDS: Targeted Public Distribution System
TPRM: Trade Policy Review Mechanism
TRC: Total Revenue Curve
TQM: Total Quality Management
TR: Total Revenue
TRIFED: The Tribal Co-operative Marketing Development Federation of India Limited
TRIMs: Trade Related Investment Measures
TRIPS: Trade Related Aspects of Intellectual Property Rights
TSP: Triple Super Phosphate
TTCL: Tea Trading Corporation of India Limited
TVAP: Total Value of Agricultural Production
UN: United Nations
UNDP: United Nations Development Programme
UNIDO: United Nations Industrial Development Organization
UPOV: International Union for Protection of New Plant Varieties
UTs: Union Territories

VER: Voluntary Export Restraints
VFC: Virginia Flu Cured
VGB: Village Grain Bank
VMS: Vertical Marketing System
WBNP: Wheat Based Nutrition Programme
WDRA: Warehousing Development and Regulatory Authority
WFP: World Food Programme
WHO: World Health Organization
WIPO: World Intellectual Property Organization
WRF: Warehousing Receipt Based Finance
WTA: World Trade Agreement
WTO: World Trade Organization

1

Definitions of 'Market'

The concept of a 'Market' is very confusing, as there are many usages of this term in economic theory, in business, in general and in particular. The first definition, in a study dating back to 1952, refers to marketing as a production-dependent activity. The definition states *'marketing is the means by which you dispose of the output of a farm, of factories, mines, quarries, forests, fisheries, hunting, oil, as well as everything imported'*. From the seller's viewpoint, *'marketing is the ability of the marketing system to transfer everything produced from the producer to the consumer'*. On the other hand, from the consumer's view point, *'marketing is simply the ability to transfer goods in which he is interested, in the form and the manner he desires, and at the lowest price to him'*. By making a synthesis of these two view points, it is clear that, *'marketing is in fact, a series of foregone decisions, based on a suitable market survey, as to what are the goods one should produce or import, and in what quantities'*. In fact, these definitions teach us the basic concept of marketing. Inside any marketing system, we find the producer on the one hand, and the consumer on the other hand, and in between them, we find the mechanism (market intermediaries), which causes the products and services to pass from the producer's side on to the consumer's side.

Another definition, which also dates back to the fifties, is *'production means the concentration of raw materials in one place, while marketing means that, this parcel of finished products is sent into farther and farther, far-flung regions'*. Another definition of the same period is, *'marketing, in general, may be defined as the whole process by which real marketable surplus, belonging to the producers, reaches the consumers.'* In the sixties, there is a shift in perspective of the marketing definition giving due emphasis on the market intermediaries that come in between producer and consumer *i.e.*, on the mechanism which enables goods and services to pass on from the producer to the consumer. This definition states that, *'the firm is the bridge of decision-making between the market and the sources from which it buys. The buyer's market constitutes the main source of income, whereas the selling market is the main absorber of the firm's outlays.'* We further find that, in the decade of the sixties, it was attempted to extend and refine the concept of marketing and the definition states that, *'market is an area for organizing and facilitating business activities for answering the basic economic questions of what to produce, how much to produce, how to produce and how to distribute production'*. This definition clearly indicates that, 'market' can be defined on the basis of:

a. A location (*e.g.* Nagpur market)

b. A product (*e.g.* the oranges market)

c. A time (*e.g.* July oranges market)

'A market may be defined as a place or a geographical area, where buyers and sellers meet and function, goods or services are offered for sale and transfers of ownership of title occur'. This definition implies that, buyers and sellers should involve in personal negotiations to facilitate exchange of goods or services, which ultimately results in transfer of ownership rights from sellers to buyers.

'A market is one of many varieties of systems, institutions, procedures, social relations and infrastructures whereby parties engage in exchange'. This definition infers that, the parties (buyers and sellers) may exchange goods for other goods (barter system) or sellers exchange goods or services for money from buyers. Through the exchange mechanism in terms of money, the prices of goods and services gets established.

'Market refers to arrangement, whereby buyers and sellers come in contact with each other directly or indirectly, to buy or sell goods.' This definition indicates that, face to face contact of buyers and sellers is not necessary for market. For example, in futures trading, the buyers and sellers exchange their transactions through internet. So internet, here forms an arrangement and such arrangement also is included in the market.

'A market also may be defined as an aggregate demand by potential buyers of a product or service'. According to this definition, the concept of a market also implies a demand for a product or service. In fact, the terms 'market' and '"demand' are often used interchangeably and they also may be used jointly as 'market demand'.

However, the above definitions of a 'Market' still may not be sufficiently explain the whole concept of marketing, as the process of marketing is dynamic with changing time. So, in this book, let us define 'Market' as *'people with needs to satisfy, the money to spend and the willingness to spend on the desired goods and services'*. So, this comprehensive definition includes three essential elements *viz.*, *consumer's needs, their purchasing power and their buying behaviour.*

1.1. Definitions of 'Marketing'

Emphasizing the purpose of marketing not simply an extension of the production process, Adam Smith quoted in his publication, *'The Wealth of Nations'* (1776) that, *'consumption is the sole end purpose of all production: and the interest of the producer ought to be attended to only so far as it may be necessary for promoting that of the consumer.'* 'Marketing' is a discipline that is often misunderstood. Some people view it is a tool that is designed 'to sell the things, as required by the potential customers', while others consider it as a 'promotion' activity. However, this misconception is fast changing with the changes in business dynamics. So, there exist today a wide variety of definitions about marketing and these definitions have also undergone considerable development especially in the second half of the 19th century and they are discussed here under.

According to Dixie, *'Marketing refers to the series of services involved in moving a product (or commodity) from the point of production to the point of consumption.'* This definition aptly describes the functions carried out by marketing parastatals with respect to food marketing in developing countries.

Marketing can be defined as *'all activities involved in the production and flow of goods and services from point of production to consumers.'*

Marketing can be defined as *'the commercial functions involved in transferring goods from producer to consumer'.*

'Marketing as a process aids in providing assistance in identifying, satisfying, and retaining customers'.

According to American Marketing Association, *'Marketing is the process of planning and executing the conception, pricing, promotion and distribution of ideas, goods, and services to create exchanges that satisfy individual (customer) and organizational objectives.'* Though this definition comprehensively covers different parameters of marketing, it does not provide a full picture.

Chartered Institute of Marketing defined 'Marketing' as the *'management process responsible for identifying, anticipating and satisfying customer requirements profitably'.*

'The all-embracing function that links the business with customer needs and wants in order to get the right product to the right place at the right time'.

'The achievement of corporate goals through meeting and exceeding customer needs better than the competition'.

'The management process that identifies, anticipates and supplies customer requirements efficiently and profitably'.

'Marketing may be defined as a set of human activities directed at facilitating and consummating exchanges'.

'Marketing is about meeting the needs and wants of customers'.

'Marketing is a business-wide function – it is not something that operates alone from other business activities'.

'Marketing is about understanding customers and finding ways to provide products or services which customers demand'

In the words of Cournot, a French Economist, *'A market is a set of conditions in which buyers and sellers meet each other for the purpose of exchange of goods and services for money'.*

According to Professor Converse, Huegye and Mitchell, *'marketing includes all the activities involved in the creation of place, time and possession utilities'.*

In the words of Pyle J.F., *'marketing is the phase of business activity through which human wants are satisfied by exchange of goods and services'.*

In the words of Prof. Cumdiff and Still, *'marketing is the business process by which products are matched with market and through which transfers of ownership effected'.*

In the words of Prof. Banerjee, *'marketing is the process of getting the right goods to the right consumes at the right place and time and at the right price'.*

In the words of Prof. Wheeler, *'marketing is concerned with all the resources and activities involved in the flow of goods and services from producer to consumer'.*

It is evident from the above definitions that, though the general tasks of marketing are somewhat straight forward, attaching an acceptable definition covering the whole concept is quite difficult. So, defining 'Marketing' is not so easy and hence, no one has yet been able to formulate a clear, concise definition that finds universal acceptance. Among all the above definitions, the centre point we can isolate is that, marketing involves the process of transferring goods and services from the producer to the ultimate consumer. This process is continuous because, the end consumer is in continuous need of goods and services offered by the producer. This continuity process gave more importance to the concept of marketing and it is highlighted from the following points:

- ✰ The consumers get an opportunity to know about various products that are available in the market.
- ✰ Provides employment to a large number of people and thus, sustains economic balance.
- ✰ Marketing fosters healthy competition among brands (fertilizers, pesticides, processed products etc) and this increases the level of competence among the firms, thereby, consumers get attractive products at affordable prices.
- ✰ Marketing helps other firms in getting more revenue. For example, 'advertising', a sub-function of selling influences television, newspapers, internet marketing, hoardings etc.
- ✰ In the absence of marketing, few firms dominate and enjoy monopoly status at the cost of consumers.
- ✰ Marketing promotes Research and Development (R&D) in the firms and thus make them to innovate both differentiated and quality products.

1.2. Evolution of Concept of Marketing

Marketing concept means the philosophy of an organization in relation to marketing of a product or service. With the passage of time, the concept of marketing also changed greatly in tune with the rapid changes in the society in terms of technology, tastes and preferences of the consumers etc. The various stages in the evolution of marketing concept are discussed here under.

- ✰ ***Self–sufficient Stage:*** In the past, agricultural production is only subsistence oriented *i.e.*, the farmers cultivate the crops as per their food requirements. So, there was no surplus produce for exchange. Therefore, the concept of marketing is absent in this stage.
- ✰ ***Exchange Oriented Stage:*** In this stage, the farmers enjoy surplus production beyond their actual production. The farmers exchange their surplus production with other goods from others. So, there is no involvement of money in the transaction of produce. This sort of exchange mechanism is called 'Barter system'. But, the demerit of this barter system of exchange is absence of double coincidence of wants. So, to overcome this, goods are brought to a common place so that, it facilitates their smooth exchange. Thus, the concept of 'market' came into existence in this stage.
- ✰ ***Production Oriented Stage:*** In this stage, the farmers cultivate the crops in accordance with the principle of comparative advantage *i.e.*, those crops which enjoy relative advantage in terms of yield or quality or price etc. Such commodities definitely enjoy trade advantage in the market. That is, if the product is of good quality and price in reasonable, there is no need of any special marketing efforts from the farmers. So, in this concept, it appears that, producers

gave more emphasis to production than consumption and production is the starting point of the business.

- ☆ ***Sales Oriented Stage:*** In this stage of marketing, mere production of commodities is not enough, but for their efficient marketing, the farmers and middlemen have to put several marketing efforts like grading, processing, advertizing, attractive packaging and labeling etc., so as to boost their sales. The main objective of these sales promotion techniques and scientific performing of marketing functions is to attract the consumers to purchase their products.
- ☆ ***Consumer Oriented Stage:*** In this stage, the marketing pattern of produce or products are in tune with the requirements of the consumers. That is, the approach is 'demand driven' and all the marketing strategies revolve around customer satisfaction. So, in this stage, consumer is the starting point, unlike production oriented stage.
- ☆ ***Social Oriented Stage:*** When compared to the above stages, this stage is much broader in its view, as it not only deals with the producers and consumers' needs, but the overall social welfare. That is, in the stage of marketing, besides commercial aspects due importance is given towards food quality and standards, pollution free environment and quality of human life. Every marketing agency and organization should adopt socially responsible marketing policies and plans in order to assure social welfare in addition to consumer welfare.

1.3. Why to Study Marketing as a Separate Discipline?

The following points highlights the necessity of studying the 'Marketing' subject as a separate discipline:

- ☆ The modern economies are customer-oriented and therefore, must be marketing-oriented.
- ☆ Of the four forms of utility *viz.*, time, place, possession and form utilities, first three are created by the marketing job. Though the creation of form utility, usually considered as a production activity, but it is directed by marketing. So, considering these different forms of utility, it is essential to study the subject of marketing. In the modern era of agri-business, another form of utility *i.e.*, Task utility is gaining more importance. This 'task utility' is provided, when someone performs a task for someone else. For instance, when a bank handles financial transactions for a farmers to offer pledge loan for efficient marketing of produce, it falls under task utility. But, just providing form and task utilities, it doesn't result in consumer satisfaction. The point is that, production and marketing are both important parts of a total business system aimed at providing consumers with need-satisfying goods and services. Thus, together, production and marketing supply five kinds of economic utilities *viz.*, form, task, time, place, and possession utility (Figure 1.1).
- ☆ Marketing functions like transportation, processing, storage, grading etc., add value to the commodities. The earlier Economists are of the opinion that, marketing concept is more theoretical and it should take place almost automatically without additional cost. If additional costs are incurred on the products, they considered them as a waste. However, the recent view is that, there are nine marketing functions *viz.*, Packaging and Packing, Transportation, Processing and Storage (Physical functions), Buying and Selling (exchange functions), Grading and Standardization, Market financing, Market Intelligence and Risk bearing (Facilitating functions) that are essential in boosting up the value of the products. For example, compared to raw mango, mango juice fetches more revenue to the seller, as here

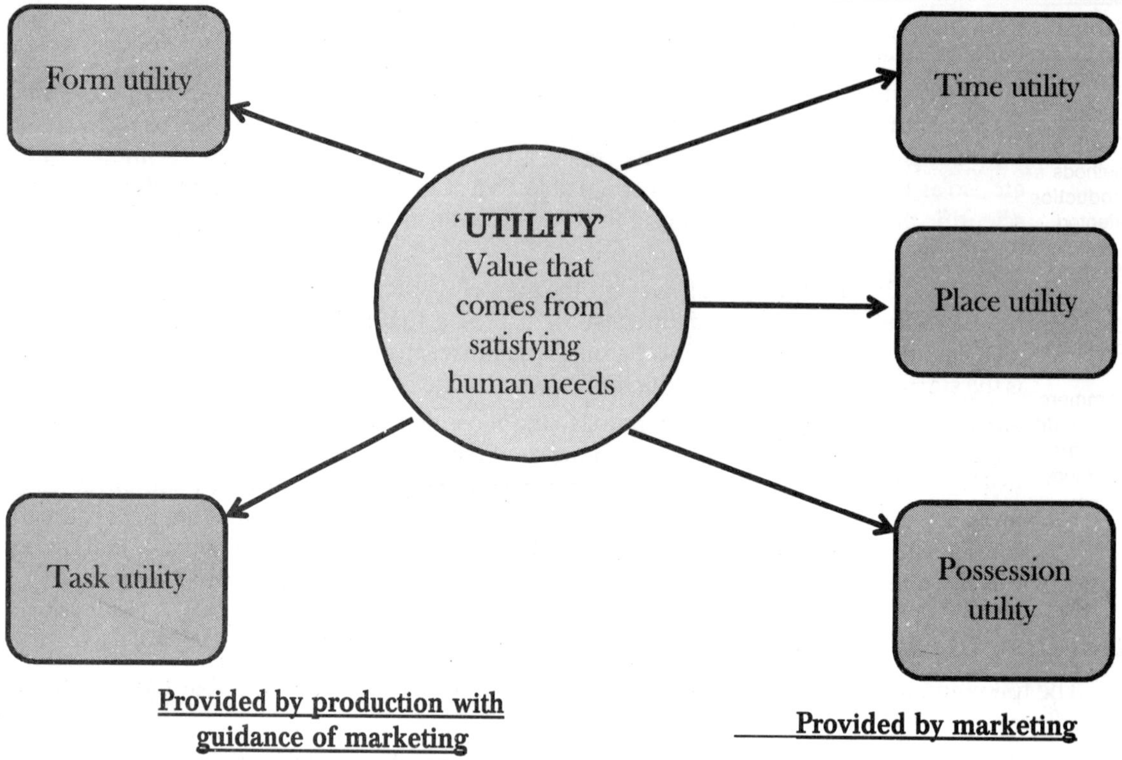

Figure 1.1: Types of utility and how they are provided.

processing function added value to the product. In case of commodities like rice and wheat, processing function is inevitable to convert the raw produce into palatable form. Similarly, a kilogram of vegetables fetches more price to the farmer in the town rather than in a local shandy. Here, the transportation function added value to the produce (place utility).

- ☆ Marketing guides the allocation of resources in different sectors and enterprises in the light of consumer demand. So, it guides the business firms in strengthening R&D to produce more innovative products. Thus, marketing servers as a stimulus, which encourages innovation.
- ☆ Marketing influences the distribution and size of income in the economy. If a business firm enjoys more profits, it will have a significant share in the National Income because, consumers are willing to pay more for the products or service offered by that firm and *vice versa*.
- ☆ Marketing is the most important 'multiplier' in economic development of the country. In a well-organized market environment, new investment opportunities, business expectations and level of employment are high and this leads to rapid economic growth. Thus, the prosperity of an economy is inherently linked to its marketing environment.
- ☆ Considering the extent of marketing orientation, a country can be classified as a Under developed or Developing or Developed category and the same is explained through Table 1.1.

Table 1.1: Relation between extent of market orientation and economic development of a country.

Nature of Production and marketing	*Market orientation*	*Function*	*Economic Development*
Traditional production methods are employed and production is subsistence oriented	No market orientation	Barter system of trade and State control	Under developed country
Mass production of commodities with little or increasing orientation towards market	Limited market orientation, but giving importance to open trade	Both Free and Regulated trade.	Developing country
Commercial production of commodities considering comparative advantage of commodities	Emphasis on export competitiveness	Free trade	Developed country

1.4. Importance of Marketing

Marketing is important to the business firm, consumer as well as the society. This is evident from the following points.

- ☆ Marketing helps business firm (say, food processing firm) to keep pace with the changing tastes, fashions, preferences of the consumers. This is essential because, the consumers' tastes and preferences are dynamic and hence, the business firm must keep on updating their products and services, so as to sustain in the business. Thus, marketing ensures the consumers in providing better products and services and improve their standard of living.
- ☆ Marketing helps in making products available at all places and throughout the year. For example, consumers in Andhra Pradesh are able to get basmati rice, Nagpur oranges, Kashmir apples, Assam Tea etc., and that too round the year due to proper warehousing, proper packaging etc. Thus, marketing creates form, place and time utilities.
- ☆ Various functions and sub-functions of marketing like packaging and packing, transportation, processing, storage, buying and selling (advertising, personal selling), grading and standardization etc., generate employment for a large number of people and accelerate growth of business.
- ☆ Marketing helps the business firm in increasing its sales volume, generating revenue and ensuring its success in the long run.
- ☆ All marketing activities are directed towards consumer satisfaction. In fact, marketing process starts with ascertaining consumer needs and produce goods that satisfy those needs most effectively. Not only that, the pricing and distribution functions of marketing are also planned accordingly.
- ☆ Through advertising and other sales promotional efforts, marketing aims at creating additional demand for their products. Satisfied consumers also help in creating new consumers. For example, if a consumer buy a Ready-to-Eat (RTE) and Ready-to-Cook (RTC) foods and feel satisfied, he will go for repetitive purchasing the same products and obviously guide others in purchasing the same.

- Marketing provide better quality products to the consumers, as the business firms keep on updating the products and services and upgrade their knowledge and technology to continuously provide better products. If they fail to do so, they will be phased out through competition.
- Efficient marketing of products create goodwill for the firm and industry as well. This also leads to developing loyalty towards the product and to the firm, and thereby, the new products of the same firm are readily accepted by the consumers.
- Marketing efforts generate profitable sales volumes for the business. Taking care of consumer needs and wants by providing the required goods and services at affordable prices, at convenient places and at right time ultimately lead to increased sales and profits.

As mentioned earlier, the marketing process is dynamic and it is fast moving in tune with the pace of economic growth. In the modern era of business, processes such as storage, transportation, processing and grading became more important. With this divergence in business processes, greater activity was moved away from production side towards the marketing side thereby, increasing the employment opportunities in the economy. Further, if the business firms produces the products having high income elasticity of demand, the pace of economic growth benefits such firms or sector. These firms also enjoy economies of large scale both in production and marketing process. For example, India is the leading producer of milk in the world and with the increased trade of this commodity, the livestock sector enjoys the establishment of large and scientific storage facilities, (which lowers per unit storage costs), better roads and communication facilities.

1.5. Types of Marketing

Marketing is of three types *viz.*, Non-agricultural Marketing, Agricultural Marketing and Agri-business Marketing.

- *Non-agricultural Marketing:* This type of marketing deals with goods and services pertaining to industrial sector like cars, computers, softwares etc.
- *Agricultural Marketing:* It deals with the series of functions involved in moving the agricultural produce from the farmer to the ultimate consumer. Thus, agricultural marketing is a series of inter-connected activities involving planning production, growing and harvesting, grading, packing, transport, storage, processing, distribution and sale. There is a fundamental difference in the marketing of industrial products (Non-Agricultural marketing) and agricultural products (Agricultural marketing). In case of industrial products, one manufacturer supplies the products to many wholesalers or dealers, as the quantity of production by each firm or factory is large enough to meet huge demands and later the same products are supplied to retailers and consumers. In case of agricultural products, since marketable surplus at each individual farmer is relatively low, many farmers supply the same type of products to one wholesaler, before the product can move to retailers and consumers. Further, the agricultural products are relatively more perishable compared to industrial products. These two fundamental differences necessitates the creation of special systems, institutions and infrastructure in the marketing of agricultural products that are not required in the marketing of industrial products.
- *Agri-business Marketing:* It involves sum total of all the operations or activities involved in the business of production and marketing of farm supplies and products. So, it is a blend of agricultural and non-agricultural marketing. Agri-businesses perform the functions of adding utility or value to the raw farm produce to enable consumers for having the preferred product, in the right place and at the right time.

2

Agricultural Marketing or Agri-Marketing

Now that, we know, what marketing is, and let us examine and learn, what agricultural marketing is. It is essential to note that, the basic concepts of marketing also holds good for agricultural marketing, but with slight differences between agricultural and industrial products (Table 2.1). In the past, traditional rural areas are distinguished by a subsistence economy. In such villages, the production unit is the family, which produces the food for its own consumption. The village economy was more or less self sufficient, thereby, the marketing of agricultural produce presented no difficulty, as the farmer sold his produce directly to the consumer on a cash or barter basis. Thus, agricultural marketing is in its simplest form regarding the buying and selling of agricultural produce. Surplus production is offered for sale, only after a particularly plentiful cultivation season. The family unit considered as a production unit, is quite small and such units operate separately. This situation makes it difficult to concentrate the produce for efficient marketing. In certain areas, the vast majority of the population is not at all used to thinking in terms of commerce and barter trading. However, in the modern agricultural marketing system, a series of functions and services are involved in moving the commodities from the producer to the final consumer. It includes not only the physical movement to the place, where the produce is wanted, but also putting it into the form and amount that is desired and having it ready at

Table 2.1: Differences between agricultural production and industrial production.

Sl.No.	Item	Agricultural Production	Industrial Production
1.	Way of life	Agricultural production is a way of life for farmers and it is only a means of livelihood for them	It is not a way of life, but it involves application of more of business principles to earn more profits
2.	Dependency on natural factors	Agricultural production is mainly based on seasonal and climatic conditions and hence, production is not guaranteed	Industrial production does not depend upon seasonal and climatic conditions and hence production is guaranteed.
3.	Production and consumption patterns	Production is seasonal, but consumption is regular	Production is not seasonal and products are purchased less frequently
4.	Bulkiness of produce	The products are bulky in nature and hence, creates transportation problems	The products are less bulky compared to agricultural products
5.	Perishability	Agricultural products are more perishable and hence, needs scientific storage	Industrial products are less perishable and more durable
6.	Joint products	Agricultural products are mainly joint products and they are produced through a single production programme. Hence, it is very difficult to workout cost and returns separately for each product. *e.g.* Paddy grain+straw, Cotton lint+seed etc.	Industrial products are not joint products and hence, it is convenient to work out cost and returns separately for each product
7.	Variations in standards	Agricultural product vary in their standards to a greater extent *e.g.* Regarding paddy, specifications like grain size, shape, colour etc., vary across different varieties, regions, countries etc.	Industrial products will have more or less same standards.
8.	Time lag	In agricultural production, the time lag is more between investment and production of output	Time lag is less between investment and production because, there is controlled production
9.	Price fluctuations	Price fluctuation are high in case of agricultural products	Price fluctuations are low in case of industrial products
10.	Production units or size of land holdings	Land holdings are small because of sub-division and more over, they are fragmented	Industrial production units are comparatively large and localized
11.	Plan of production	Agricultural planning is quite difficult for the farmers because, it mainly depends on seasonal and climatic conditions	Industrial planning is quite easier because of controlled production and non-dependency on seasonal and climatic conditions
12.	Labour	Less skilled and less organized	More skilled and well organized having trade unions
13.	Elasticity of demand for products	Less elastic	More elastic

Contd...

Table 2.1–*Contd...*

Sl.No.	*Item*	*Agricultural Production*	*Industrial Production*
14.	Law of diminishing returns	Operate at an early stage (*i.e.*, quickly) in agricultural production, because of excess dependency on natural factors	Operate at a later stage in industry because of controlled production
15.	Capital turn over	Capital turnover is slow	Capital turnover is quick
16.	Marketing	In agricultural marketing, more number of intermediaries exist between farmer and final consumer and hence, the Farmer's Share in Consumers Rupee (FSCR) is low and more over, the markets are not properly regulated	In case of industrial products, less number of intermediaries exists between manufacturer and final consumer and the market outlets are well organized and systematized.

the time it is wanted. This is clearly witnessed with the gaining popularity for the RTE and RTC foods. Another feature is that, many of the traditional peasants would be prepared to switch over to the cultivation of alternative crops, provided the prices are remunerative for them.

Further, there exist differences between agricultural marketing and urban marketing (Table 2.2). Agricultural marketing denotes marketing of agricultural produce of the rural areas to the urban consumers or industrial consumers, whereas urban marketing involves delivering manufactured or processed products or services to rural producers or consumers.

Table 2.2: Differences between agricultural marketing and urban marketing.

Agricultural Marketing	*Urban Marketing*
They are widely spread and scattered and hence, the farmers have to travel long distances to sell their produce at remunerative price	Mostly concentrated
They have poor amenities and facilities	They enjoy all facilities on scientific lines
These markets serve low density of population per sq km	These markets serve high density of population per sq km
Government intervention in these markets is essential (through announcing Minimum Support Prices (MSPs)) to safeguard the interests of farming community and to balance the interests of both farmers and consumers.	Less Government intervention in these markets
Acute seasonality in income receipts due to seasonality of production	Income received at regular time
Less convenient buying, less retail outlets per 1000 population	More convenient buying, more retail outlets per 1000 population

The performance of agriculture in India is important, as the sector not only contributes to overall growth of the economy, but also provides employment and food security to majority of the population in the country. The 11th Five Year Plan (FYP) also indicates that, agricultural development is an important component of inclusive growth approach. So, the economic performance of the Indian agricultural sector is fundamental to the well-being of hundreds of millions of people and pivotal to the security of their livelihoods and their food supply. Some 70 per cent of the population of approximately 1.1 billion lives in rural areas, 55 per cent of the population is employed in agriculture, a sector which contributes approximately 17 per cent to Gross Domestic Product (GDP). Until quite recently, the sector has been subjected to pervasive Government interventions at almost every level and almost completely insulated from global markets through numerous trade restrictions. It has been recognized that, in many respects, the performance of the sector has been quite poor and that Government policies have contributed to this poor performance. In particular, there is a general consensus that, agricultural markets in India suffer from serious inefficiencies and that the institutional and regulatory framework governing the internal agricultural markets constrain India's ability to adjust its agricultural markets with the evolving global environment (Chand, 2005). It has become widely accepted that, there is a need for urgent and major reforms and hence, structural reforms were introduced in India in a big way in 1991. In view of the crucial importance of this agricultural sector in the economic development of the country, marketing of agricultural products (Agricultural Marketing or Agri-Marketing) assumes considerable importance and in this context, it dealt in-detail in this book.

2.1. Definitions of Agricultural Marketing

The term agricultural marketing is composed of two words *viz.*, 'agriculture' and 'marketing'. 'Agriculture', in the broadest sense means activities aimed at the use of natural resources for human welfare, and 'Marketing' connotes a series of activities involved in moving the goods from the point of production to the point of consumption. The agricultural marketing system is a link between the farm and the non-farm sectors. There is hardly a single, universally accepted definition of agricultural marketing. This concept has been defined in a variety of ways. In simplest form, agricultural marketing is defined as *'the process of buying and selling of agricultural produce'*. This definition of agricultural marketing may be accepted in olden days, when the farming is more or less self-sufficient and subsistence oriented and the farmers used to transact their produce directly to the consumers with no difficulty on a cash or barter basis. But, in modern days of business, marketing of agricultural produce has to undergo a series of transfers or exchanges from one hand to another, before it finally reaches the ultimate consumer. So, it is often defined as *'the series of functions involved in moving an agricultural commodity from the point of production to the point of consumption'*. This definition takes into account the activities of many functionaries involved between farmers and consumers. Agricultural marketing, is otherwise, defined as *'the connecting link between farmers and consumers and this link involves two major activities viz., physical distribution and economic exchanges. The physical distribution is concerned with the physical handling, processing and transfer of raw and processed products, as they move from farmers to ultimate consumers. The economic exchange is concerned with the exchange and price-setting processes in the marketing system'*. The National Commission on Agriculture defined agricultural marketing *'as a process, which starts with a decision to produce a saleable farm commodity and it involves all aspects of market structure of system, both functional and institutional, based on technical and economic considerations and includes pre and post- harvest operations, assembling, grading, storage, transportation and distribution'*. The Indian Council of Agricultural Research (ICAR) defined agricultural marketing *'as involvement of three important functions viz., assembling (concentration), preparation for consumption (processing) and distribution'*.

Farmers, market intermediaries and consumers, all view agricultural marketing from different perspectives. So, agricultural marketing may be defined differently from the stand point view of different agencies involved in transacting agricultural produce.

- ☆ ***Farmers' definition:*** For farmers, agricultural marketing means the supply of requisite commodities for sale into the market. For example, farmers supply paddy to the nearby regulated market after harvest.
- ☆ ***Marketing intermediaries or middlemens' definition:*** To processors, wholesalers and retailers, agricultural marketing means the process of purchase and sale of agricultural produce with a profit-motive. For example, rice miller purchases paddy from farmers and sell rice to a wholesaler.
- ☆ ***Consumers' definition:*** For consumers, agricultural marketing means the purchase of requisite goods and services in the market. For example, consumers purchase rice in super market.

From the above definitions of agricultural marketing with reference to different marketing agencies, we can infer that, each agency has only a partial concept or view about the agricultural marketing process. Considering the above definitions, we can define agricultural marketing as '*the performance of all business activities involved in the flow of agricultural commodities and services from the point of initial agricultural production until they are at the hands of consumers*'. This definition has the following implications.

- ☆ ***Farm gate:*** It implies, the process of marketing begins at the farm gate, as no farmer produces the produce, unless it is accepted by the consumers. The farmers will produce such commodities that the consumers prefers rather than their personal interests.
- ☆ ***Interdependence:*** There is interdependence between farmers, market intermediaries and consumers, so as to ensure the process of marketing from the farm gate to the ultimate consumer.
- ☆ ***Decision making process:*** This process is involved at each and every stage of marketing. For example, a farmer will supply the produce into the market only, when he realize competitive price otherwise, he will stock the produce in the godowns. If the price of the commodity falls below the Reserve price, the farmer will not sell the commodity.
- ☆ ***Inclusion of farm supplies:*** It is essential to note that, agricultural marketing deals with both input marketing and output marketing. This agricultural input supply industry serves as the resource base for the agricultural production and consequent output marketing.
- ☆ ***Conflicts in the agricultural marketing system:*** The motives or self-interests of farmers and consumers are exactly opposite in the sense that, farmers prefer to have higher possible prices for the products, whereas consumers are interested in lower prices for the products. The market intermediaries like processors, wholesalers and retailer function with a profit motive in the process of marketing. So, this implies that, the agricultural marketing system reconciles these conflicting demands.

2.2. Objectives of Agricultural Marketing

The objectives of agricultural marketing vary from the stand point view of different stakeholders involved in the marketing process. The different stakeholders include, farmers, processors, wholesalers, retailers and consumers. Each stakeholder sees the agricultural marketing system from a perspective of his self-interest and their interests are sometimes in conflict. The most striking aspect is that, farmers' self-interest in the field of agricultural marketing is exactly opposite to other stakeholders' self-interests (Table 2.3). For the individual farmer, the objective is to get maximum price for the agricultural produce. For other stakeholders (processors, wholesalers and retailers), the objective is to purchase the agricultural produce at low price and transact the same to the succeeding agency at a high price with a view to make profit. From the consumers' point of view, the objective is to purchase the agricultural produce at affordable or reasonable price. However, from society point of view, the objective of agricultural marketing is to encourage efficient allocation of resources, so as to create wealth and promote economic growth in order to improve the general welfare of society.

Table 2.3: Conflict of self-interests among different stakeholders of agricultural marketing process.

Stakeholders	*Self-Interests*
Farmers	Maximum price and transacting large quantities of agricultural produce
Processors	Low purchase price, high quality of agricultural produce and reliable supplies from the farmers/ wholesalers
Wholesalers	Low purchase price, high quality of agricultural produce and reliable supplies from the farmers/ processors
Retailers	Low purchase price, high quality of agricultural produce and reliable supplies from the wholesalers/ processors
Consumers	Affordable/Reasonable (Preferably low) purchase price and High quality of agricultural produce

It is interesting that, the above listed stakeholders are interdependent on one another, as the survival of one stakeholder depends upon the survival of other. So, when the self-interest of each stakeholder fulfills in the market process, it leads to balanced and long term marketing. But, these self-interests of different stakeholders are protected only in the short run and may not in the long run considering dynamic changes in the field of agricultural marketing. For example, in Andhra Pradesh, there exist a number of middlemen between farmers and consumers regarding marketing of vegetables thereby, the farmers gets exploited. However, with the Government intervention in the form of establishing Rythu bazaars, farmers came in direct contact with the consumers in transacting vegetables and middlemen are completely eliminated from the marketing channel. So, in the long run, considering dynamic changes in the field of agricultural marketing, the self-interests of different stakeholders may be positively or negatively influenced.

2.3. Different Approaches to Study Agricultural Marketing

Agricultural marketing can be analyzed from the point of view of different approaches, as each approach involves a particular emphasis or focus. The different approaches are discussed here under.

1. Functional Approach
2. Commodity Approach
3. Institutional Approach
4. Behavioural Approach
5. Managerial Approach
6. The Structure-Conduct-Performance Paradigm (SCPP)

1. Functional Approach

This approach analyze the concepts of agricultural marketing from the point of view of different activities or functions involved in moving the agricultural produce from the farmer to the ultimate consumer. So, this approach involves the breaking down of entire marketing processes into different marketing activities or functions. A marketing function is defined as any activity that is involved in accomplishing the marketing process or in moving the agricultural produce from the farmer to the ultimate consumer. It is, otherwise, defined as *'an act or operation or service by which the farmer and the final consumer are linked together'*. The various activities or marketing functions can be conveniently categorized as Physical functions, Exchange functions and Facilitating functions (Table 2.4).

Table 2.4: Functional approach of agricultural marketing.

Type of Function	*Series of Activities Involved*
Physical functions	Packaging and Packing, Transportation, Processing and Storage
Exchange functions	Buying and Selling
Facilitating functions	Grading and Standardization, Market financing, Market Intelligence and Risk bearing.

a. Physical Functions

This includes activities like packaging and packing, transportation, processing and storage that ensure handling, movement and physical change of the actual commodity itself. They are involved in solving the problems of 'when', 'what' and 'where' in marketing. These functions relate to the physical handling of agricultural produce either in moving it from one place to another or in storing it over a period of time or changing its form or in convenient handling. Since, agricultural produce has to be moved from threshing floors to the consuming areas, as it is not consumed at the place of its production. We know, agricultural production is seasonal, but consumption is regular. So, it demands the essentiality of two important functions *viz.*, transportation and storage. Further, in case of raw produce like paddy, wheat etc., they are not fit for direct consumption. So, it again demands for 'processing' function that leads to change in form of the commodity and also add value to the produce. To fulfill these functions, the bulky produce is to be handled properly and the same is essential even at the time of supplying final product to the ultimate consumer. This emphasizes the importance of 'packaging and packing' function.

b. Exchange Functions

This includes two important activities *viz.*, buying and selling and they ensure equilibrium price determination for the produce under consideration and also in transfer of title to the produce. Thus, these functions have their primary objective of negotiating the price settlement between buyers and sellers. These exchange functions are considered to be the most important of all the functions of agricultural marketing. Buying and selling functions are complementary to each other and one cannot take place without the other. The selling function involves identifying and seeking out potential buyers, determining an equilibrium price, negotiating terms of sale etc. The ultimate aim for any seller is to meet its consumer's (buyer's) needs. Thus, a farmer can be said to be market oriented, when production is purposely planned to meet specific demands or market opportunities. For example, a contract farmer, who wishes to meet the needs of a fruit processor producing mango juice will only grow improved mango varieties.

c. Facilitating Functions

This includes important activities like grading and standardization, financing, market intelligence and risk bearing. These activities serve as catalyst for the smooth performance of the above physical and exchange functions. It is essential to note that, these activities are not directly involved in either physical handling of produce or exchange of title. But, in the modern era of agricultural marketing, these activities are of crucial significance and in the absence of these activities, it is difficult to imagine about smooth functioning of marketing system. These activities are popularly regarded as *'grease that makes the wheels of the marketing machine go round'* and help in the smooth discharge of the physical and exchange functions. Grading function helps in separating heterogenous lot of produce into different homogeneous lots based on scientific standards laid down. So, grading makes price determination easier for the produce and thereby, provides a fair return to the farmer and at the same time, it offers good quality produce to the consumer. We are aware that, considerable time lag is indispensable for the farmers for assembling the produce and their sale in the market. So, to prevent distress sales, market finance is to be provided to the farmers. To ensure effective marketing of produce at higher price, the farmer should be provided timely market information. Lastly, no business can be done without undertaking the inherent risk, which may be caused either due to a decline in price or deterioration of the produce itself by fire, flood etc. These risks have to be borne by someone in the marketing channel. Physical risks may be covered under insurance, while risks stemming from price fluctuations are handled through the hedging operation.

Thus, agricultural marketing functions are many and varied. The part played by each function varies widely as regards to the specific commodities and services. It may further be noted that, these functions are indispensable regardless of the institution or agency, which performs them or the commodity in connection with which they are performed. These functions are closely related to each other and cannot be isolated from one another.

Merits

The advantages of the functional approach in the study of agricultural marketing problems are:

- ✰ Inter-functional comparison of the marketing costs can be made.
- ✰ Inter-agency comparison of the cost of performing a marketing function can be made.
- ✰ Inter-commodity comparison of cost of performing the various functions can also be made.
- ✰ It is possible to avoid a marketing agency in a marketing channel for a commodity, but it is not possible to avoid the marketing functions.
- ✰ This approach also aids in analyzing the marketing costs at each and every stage of marketing process and thus, guides us to enhance operational efficiency in performing different marketing functions.

Demerits

The defects of this approach include:

- ✰ An undue emphasis on functions of marketing does not permit one to know, how these functions are applied to specific business operations.
- ✰ The marketing functions are so numerous that, it is difficult to eliminate the unnecessary from the necessary functions.

Kohls and Uhl suggested that, there are three characteristics of marketing that can be recognized from applying functional approach to marketing. They include:

- ✰ First, marketing not only adds costs, it adds value. The evaluation of marketing functions should be analyzed from the 'efficiency' point of view. That is, the ratio between the benefits offered and costs incurred in performing the marketing function must be considered. Higher the ratio, greater is the efficiency in executing a marketing function.

- ☆ The second important advantage of applying functional approach to marketing is that, though there are different agencies involved in performing marketing functions *viz.*, individuals (processors, wholesalers, graders etc), co-operatives, Government agencies, private firms etc., it is possible to eliminate one or more agencies, but it is not possible to eliminate performing marketing functions altogether.
- ☆ Third, the scale of performing or specializing in marketing functions differs across the agencies. For example, the big warehouses can be provided and handled by Government agencies and not by private traders. Similarly, distant transport of agricultural produce through ship or air is also in the hands of Government agencies.

2. Commodity Approach

In this approach, the agricultural produce is the pivot around which all institutional and functional aspects are studied in-detail. This approach is important in the sense that, the seasonality of production, marketing processes, marketing channels, number of market intermediaries involved, problems of marketing etc., vary across the commodities, thereby, commodity-wise process of marketing can be studied. However, this approach is more time consuming and moreover, some of the activities are repetitive across the commodities. For example, processing is a common marketing function for paddy, wheat, oilseeds etc. Storage is a common marketing function for rice, wheat, fruits, vegetables, tamarind, chillies etc. But, the method of storage may differ across the commodities. For example, paddy is stored in godowns, but vegetables are maintained in cold storage. So, considering these peculiarities across the commodities, it is difficult to follow this approach to analyze the marketing process. However, if the objective is to ascertain the marketing process commodity-wise, this approach is concrete and superior to other approaches. But, this approach is more time consuming and often results in excessive repetitions. Considering these limitations, Economists mainly use functional approach relative to the commodity approach in analyzing the marketing process.

It is essential to note that, agricultural produce differs widely compared to industrial products and these differences also highlight the importance of this approach. So, it calls for a completely different system of marketing for transacting agricultural produce compared to industrial products. The following are the important features of agricultural produce:

- ☆ Uncertainty in agricultural production due to dependency on natural and climatic factors
- ☆ Production of agricultural produce cannot be regulated and adjusted according to the existing or anticipated demand
- ☆ Heterogeneity of produce
- ☆ Perishability nature of produce
- ☆ Production is seasonal, but consumption is regular
- ☆ Processing is inevitable for some produce like sugarcane, cotton, jute, tobacco etc., before their final consumption
- ☆ Bulkiness of agricultural produce and this makes their storage and transportation difficult and costly.
- ☆ Demand is less elastic for agricultural produce

On account of the above features, agricultural produce can be classified into various categories based on the following criteria:

a. On the Basis of Cultivating Season

This criterion classifies the agricultural produce into the following two categories:

- ☆ ***Kharif crops***: These crops are cultivated with the onset of South-West monsoon. Sowing for these crops is done from June and harvesting is done from November. The various crops of *Kharif* season include paddy, jowar, jute, bajra, cotton etc.
- ☆ ***Rabi crops***: These crops are cultivated with the onset of North-East monsoon. Sowing for these crops is done from November and harvesting season is from February. The various *Rabi* crops include wheat, gram, peas, potatoes, barley etc.

b. On the Basis of Encashability

This criterion divides the crops into two categories as discussed below.

- ☆ ***Food crops:*** These crops include rice, wheat, barley, maize etc. For these crops, the prices are lower than cash or commercial crops, but these crops will give assured returns to the farmers, as their price volatility is low.
- ☆ ***Cash crops or Commercial crops:*** These crops include tea, coffee, rubber, oilseeds, tobacco, cotton, chillies etc. For these crops, the price volatility is high.

c. On the Basis of Perishability

According to this criterion, there are two categories of agricultural produce *viz.*, Perishable and Less-perishable produce.

- ☆ ***Perishable produce:*** Produce like fruits, leafy vegetables, tomatoes, cabbage etc., fall under this category, as they survive only for a few days.
- ☆ ***Less-perishable produce:*** Produce like cereals, pulses etc., fall under this category, as they survive for more number of days compared to perishable produce.

d. On the Basis of Processing

Some products require processing before they are finally consumed. For example, *sugarcane*, cotton, paddy, wheat etc. However, some produce like fruits, vegetables, milk etc., are directly fit for consumption and processing function is not required.

3. Institutional Approach

With the intervention of Government in the field of agricultural marketing in the form of establishing different institutions or organizations with the prime objective of safeguarding the interests of the farmers, this institutional approach gained more significance in the recent period. The agencies and institutions, which perform various marketing functions such as Food Corporation of India (FCI), National Agricultural Co-operative Marketing Federation (NAFED), Coffee Board, Tea Board, Coconut Development Board (CDB), Co-operative Marketing Societies (CMSs) etc., deserve special attention in taking up various marketing functions. Further, with the establishment of international organizations like World Trade Organization (WTO), Codex Alimentarius Commission (CAC), World Health Organization (WHO) etc., the scope of this approach is further widened. So, the marketing institutions are the wide variety of business organizations that have been developed to operate the marketing machinery and they vary widely regarding their size, organization pattern and ownership. This approach helps us to find answers to the problems of 'who does what' in the marketing process. This is where it differs from Functional approach. In Functional approach, attempt is made to answer 'what' in the question 'who does what', whereas institutional approach focuses its attention on the 'who'. In this approach, the human element receives primary emphasis and considers the nature and character of various middlemen and related agencies involved in the process of agricultural marketing. Thus, institutional approach implies a study of agencies and institutions, which perform various functions in the marketing process. These agencies vary widely in their size and ownership. They get their reward in the form of Net Marketing Margins (NMMs). The agencies and institution, which perform various marketing functions, are individuals, partnership, corporation, co-operatives, Government organizations etc. This approach focuses on the following aspects:

- ✰ whether the NMM of a market middleman or agency is commensurate with the services it rendered.
- ✰ whether Government intervention is needed to regulate the market practices, so as to eliminate unhealthy and unscrupulous practices of market intermediaries.
- ✰ how best to simplify the agricultural marketing system.
- ✰ helps in understanding the specialized role of different market middlemen in transacting huge quantities of agricultural produce. For example, we can imagine farmers can assume middlemen's activities like transportation, storage, grading etc. But, still the presence of marketing middlemen of different institutions can perform the same functions with much more efficiency than farmers. So, the recognition of specialized middlemen of different marketing organizations and the way in which they organize themselves to ensure smooth marketing process with greater efficiency furnishes the importance of this approach.
- ✰ takes into account the personal element in the marketing process and bring attitudes in the analysis of the agricultural marketing problems.

But, the serious demerit of this approach is that, it is purely descriptive thereby, it does not effectively explain the inter-relations of the different institutions under consideration.

4. Behavioural Approach

The earlier two approaches *viz.*, functional and institutional approaches helps to analyze the existing agricultural marketing system. But, the important aspect is that, both organization and functional aspects of marketing system are dynamic. Since the marketing system comprises of different firms or organizations, it is essential to understand their behaviour, so as to ascertain the overall functioning of marketing system. For example, in a market channel for paddy comprising of various agencies *viz.*, farmer, miller (processor), wholesaler and retailer, each one tries to formulate the decisions to solve their respective problems, so as to get profits in the marketing process. If these decisions of different agencies are analyzed, a greater' understanding can be made regarding the complexities involved in transacting paddy, thereby, guides us to know the market performance. Thus, this approach aims at studying the behaviour of firms, institutions and organizations, which exist in the marketing system for different commodities. An understanding of the behaviour of the individuals is essential, if changes in the behaviour and functioning of the system are to be predicted. The updating of this analysis considering the dynamic changes in the field of marketing further helps to formulate sound marketing policies.

5. Managerial Approach

This approach analyzes the concepts of agricultural marketing from management point of view. This approach deals with the management of 4Ps of marketing mix *viz.*, Product, Price, Place and

Promotion of a firm. The 4Ps of marketing mix were first expressed in 1960 by E J McCarthy and they are considered as 'Heart of marketing programmes'. The management of the 4Ps of marketing mix is influenced by the environment under which a firm operates in. Such environment can be influenced by Political, Economic, Socio-cultural and Technological (PEST) factors. It is essential to note that, among 4Ps, 'Price' is the only source of revenue to the business firm and thereby, fixing a remunerative price for the product is very important. Further, if the first 3Ps are in synchronizing with what consumers actually need, the 4th P (Promotion) makes the consumers aware about the availability of the product. According to Lauterborn, these 4Ps of marketing mix correspond to 4Cs of marketing from buyer's point of view (Table 2.5).

Table 2.5: 4Ps of marketing mix correspond to 4Cs of marketing from buyers' point of view.

4Ps of marketing mix	*4Cs of marketing*
Product	Customer satisfaction/needs and wants
Price	Customer's cost
Place	Customer's convenience
Promotion	Communication

Boom and Bitner further extended this 4Ps of marketing mix to 7Ps. The additional 3Ps include, Physical layout decisions, People (provision of customer service) and Processes. These additional 3Ps have been added because, today marketing is far more customer oriented than earlier. Among these 7Ps, first 4Ps are regarded as Product mix and last 3Ps are regarded as Service mix. These 7Ps of marketing mix comprise the modern marketing mix that is particularly relevant in service industry, but is also relevant to any form of business, where meeting the needs of customers is given priority. So, considering these 7Ps of marketing mix, the marketing manager of a firm analyzes the market, marketing prospects of product, develop market strategies both for the present and for the future.

6. SCPP

This is an analytical approach used to study, how the market structure and the behaviour of sellers of different commodities and their services (conduct) influences the performance of firms and consequently the welfare of the community as a whole. As mentioned earlier, the marketing environment is influenced by various PEST factors, factor prices, tastes and preferences of the consumers etc. These factors can be conveniently categorized into demand side and supply side conditions (Figure 4.1). These conditions influence the market structure, which comprises of relatively stable features like number of buyers and sellers, product differentiation, barriers to entry and exit, vertical integration, diversification, cost structures etc. Some examples of market structure include, the number of buyers and sellers of agricultural commodities, number of sellers of agricultural inputs like fertilizer and pesticides etc. This market structure determines the market conduct, which then sets the performance of the firms in the market. Market conduct refers to the behaviour of firms regarding, what they do with their production levels, nature of product, price setting behaviour and other aspects like advertising, R&D, setting prices for the product either independently or in collusion with other firms etc. The market conduct influences the market performance indicating that, how well the market fulfills certain social and private objectives like quality of agricultural output, profits of the firm both in the short run and long run, price stability, cost effective production, achieving allocative and production efficiencies etc. The following example clearly illustrates the relationship between S-C-P.

Market Structure

Assume there are only three farmers cultivating paddy in a village. This situation may arise due to one or more of the following reasons:

- ☆ Due to lack of adequate canal irrigation water in that village. So, only three farmers, who are having bore wells might have gone for paddy cultivation
- ☆ The sharp rise in cost of resources employed in paddy cultivation
- ☆ In the recent past, in that village, paddy crop is severely infected by rice blast disease.

Market Conduct

Considering the above market structure, the three farmers may behave in a number of different ways. They may collude and sell the paddy at a very high price or they may act independently or they may discriminate the prices depending upon the income status of the consumers (higher price to rich consumers and lower price to low-income consumers).

Market Performance

The expected effects (performance) from such a structure and conduct are that, the consumers have to pay higher prices due to less quantity of paddy supplied into the market. Had the number of sellers (farmers) are more, they would have paid lesser prices, as the supply of paddy into the market will be more. Further, it is difficult for the sellers to go for collusion, if their number is significantly high.

Thus, this approach of S-C-P helps to describe and analyze dynamics of agricultural marketing and provide useful indicators of evolving or future food security conditions. This concept of S-C-P is discussed in-detail in the Chapter 4.

So, the above discussed approaches outline the general framework of theoretical and practical thinking about agricultural marketing. Of all the approaches, considering their respective merits and demerits, 'Functional approach' is more important considering the following characteristics of marketing functions:

- ☆ The execution of marketing functions influences both marketing costs and value of products to the consumers. For example, processing function adds cost to the farmer, but it provides form utility to the product. Similarly, transportation and storage functions provide place and time utilities for the product. Even the exchange and facilitating functions also provide cost-effective services than farmers and consumers can perform them. So, in evaluating different types of marketing functions, the net gain from their costs and benefits should be given due consideration.
- ☆ It is very difficult to avoid performing marketing functions, but is quite easy to avoid marketing middlemen. For example, if the farmer is well-versed about packaging and packing of product, transportation of produce, grading and standardization, scientific storage of produce etc., he can minimize the marketing costs by eliminating some number of marketing middlemen. But, it is very difficult to eliminate these marketing functions in the marketing process. This is because, eliminating the marketing middleman involves the transfer of marketing function and costs to someone else.
- ☆ Marketing functions can be performed by anyone, anywhere in the marketing system. If the firm gains considerable control right from farm level to retail level, it can handle all the functions, thereby, it can eliminate the marketing middlemen and greatly reduce marketing costs. On the other hand, there are specialized firms, who can perform only one or two marketing functions with greater degree of efficiency. For example, fruit juice factories, oil mills, cotton ginning mill etc., specialize in processing function.

2.4. Different Views of Agricultural Marketing

There are two contrasting views regarding the concept of agricultural marketing. They include,

1. Parasitic View

This a pessimistic view, as some people consider several middlemen like commission agents, brokers, wholesalers, processors, retailers etc., depend on agricultural produce produced by the farmers for their business activities and reap profits at the cost of farmers and consumers. These intermediaries perform different activities or functions like processing, storage, transportation, grading etc., and they survive as 'parasites' on the farmers' produce that too taking away a part of the FSCR. But, this view is not altogether valid because, what farmers produce they do not consume and what they consume they do not produce. Say, a farmer producing cotton (kapas) has to necessarily depend upon the processing agency to produce cotton lint and with the income he derives from selling cotton lint, he has to spend on rice for his survival.

2. Utility View

Majority of the Economists are of this view that, the various marketing functions performed by the market intermediaries creates different forms of 'utility', as discussed here under.

- ☆ ***Form utility:*** The change in form of a commodity will bring change in utility to the consumer. For example, consumption of raw mango will give utility to the consumer. But, if the mango is processed into juice or jelly, its utility will change to the consumer. Similarly, paddy produced in its original form is not useful to the consumer directly, but if it is processed into rice, it will yield utility to the consumer. Forest wood when transformed into furniture, its utility will increase. The above examples imply that, 'processing function' add form utility to the commodities.
- ☆ ***Place utility:*** The change in place of availability of the commodity will add place utility to it. For example, farmers cultivate vegetables, fruits, food grains etc., in the villages and if they are made available in towns and cities, it adds utility to these commodities. So, 'transportation function' will add place utility to the commodities.
- ☆ ***Time utility:*** The change in time of availability of the commodity will add utility to the commodity. For example, farmers cultivate different crops like vegetables, fruits, food grains etc., in the villages and if they are made available to the consumers at the same time, due to increase in supply in the market, they fetch low prices. So, if the farmers store these commodities and made available to the consumers at times of scarcity, more utility can be realized. So, 'storage function' adds time utility to the commodities.
- ☆ ***Possession utility:*** We know, utility refers to want satisfying power of a commodity. So, anything, which satisfy human want is said to possess utility. For getting access to the commodities, they have to be exchanged in the market economy between sellers and buyers. So, the 'buying and selling function' leads to transfer of ownership of commodity from seller to buyer and this adds possession utility to the commodity. This increased usefulness created by marketing enable the consumer to own, use and consume the commodity. So, it is also called as Ownership utility. Thus, possession utility establishes legal ownership of a commodity. Since the possession utility is due to exchange function, only economic goods will have possession utility. For example, paddy in the hands of farmers has less utility. But, once it reaches the consumers after processing through exchange mechanism, it will add utility to the commodity and this increase in utility refers to possession utility.

☆ ***Task utility:*** This utility is provided when someone performs a task for someone else. For example, to offer pledge loan (market finance) to the farmer, a bank handles financial transactions.

It is to be noted that, both form utility and task utility are provided by production with guidance of marketing and other three types of utilities *viz.*, time utility, place utility and possession utility are provided by marketing (Figure 1.1)

2.5. Sub-systems of Agricultural Marketing

Agricultural marketing comprises of five sub-systems *viz.*, Production, Exchange, Distribution, Consumption and Regulatory (Figure 2.1). Among these five sub-systems, Consumption sub-system comprising 'consumers' are the focal points of all the other sub-systems. The three sub-systems *viz.*, production, exchange and distribution work in tune with the regulatory sub-system towards the end point *i.e.*, consumption sub-system. Each of the sub-system is independent of one another, but a change in any one of them impacts on the others, as well as upon the system as a whole.

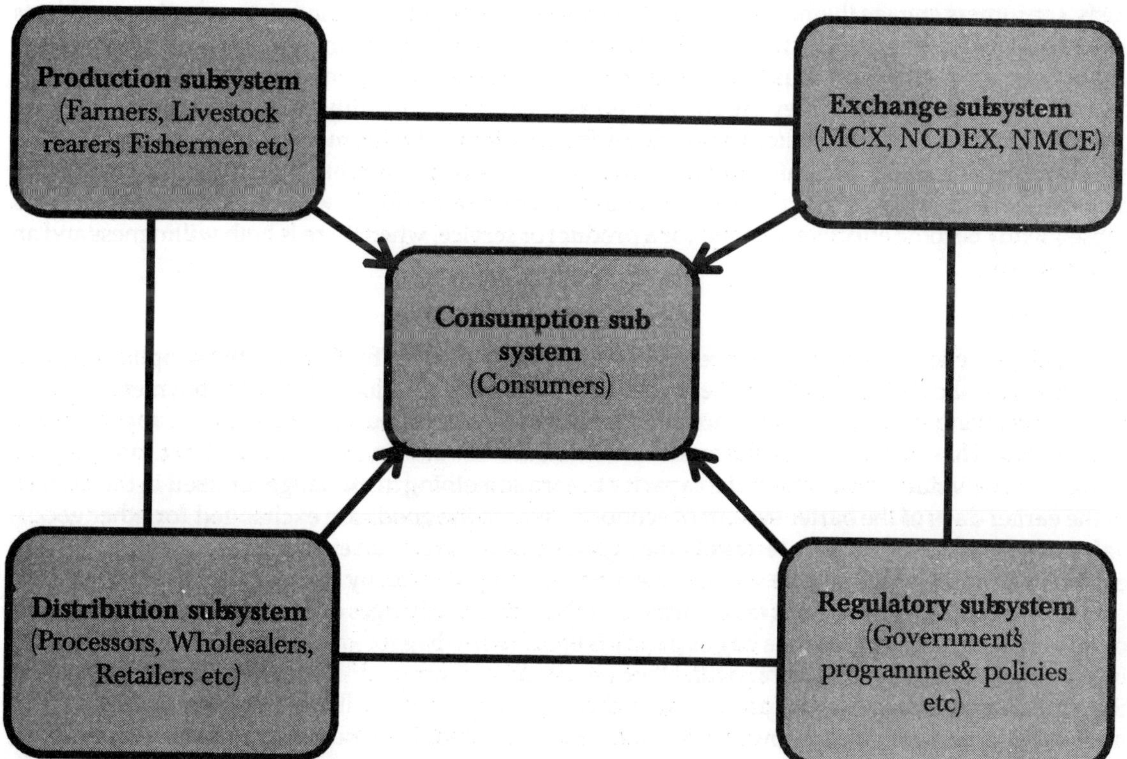

Figure 2.1: Agricultural Marketing sub-systems.

2.6. Key Concepts of Agricultural Marketing

There are four key concepts of agricultural marketing popularly regarded as heart of marketing process. They include,

1. Consumers

They constitute the critical element of the marketing process. Since, the goal of agricultural marketing system is to satisfy the ultimate consumer, he is regarded as the ruler of the marketing system. The tastes and preferences of the consumer direct the production and marketing activities and this act is referred as *'Doctrine of consumer sovereignty'*. That means, consumer is treated as 'king' in the economic system. If farmers fail to recognize the preferences of consumers, they may produce goods and services that nobody prefers or wants, thereby, they end up in losses and have to shut down their production activities. However, it is not always true that, business firms act according to the wishes of the consumers, as through various marketing functions and strategies like packaging, processing, grading, advertising, product design etc., the business firms may educate and influence the consumers in their buying decisions. So, considering these marketing strategies, the consumer loses a degree of sovereignty.

2. Wants

Consumers' wants are diversified and they are unlimited. To fulfill these wants on prioritization basis, consumers engage themselves in various economic activities. So, the existence of human wants is the basis for all economic activities. Human wants are not only unlimited, but they vary in their importance across the individuals in the society. That means, some wants are essential, in the sense that, they must be fulfilled. They are considered as necessaries and without fulfilling them, it is very difficult to imagine the human life. They include, food, clothing, shelter, medication at times of sickness etc. However, some wants (comforts and luxuries) are not so important, but their fulfillment makes the mankind to lead fuller life in the society. These wants also multiply with the civilization. Wants subsequently become effective demand for a product or service, when there is both willingness and an ability to pay.

3. Value

With reference to farmers, value refers to the payment received by them for the supplies of goods and services to the consumers. From the consumers' point of view, value refers to the payment made by them for enjoying the goods and services. The concept of 'value' is purely relative in its approach and not absolute. That means, we use the term 'value' in terms of value in exchange only. So, to say a good is said to have 'value', it must have the capacity to earn something in exchange for itself in the market. In the earlier days of the barter system of economy, where the goods are exchanged for other goods, value of a good is measured in terms of other goods that it earns in an exchange mechanism. Say, for example, if paddy is said to have value in the market, it implies that, by exchanging paddy the farmer gets something (say, jowar) from other farmers. If the farmer exchanges one bag of paddy for two bags of jowar, it indicates the value of a bag of paddy is equal to two bags of jowar. So, value of a commodity is expressed in terms of relationship with other commodity or commodities in an exchange mechanism. Say, in the above example, if one paddy bag fetches three bags of jowar, it implies, value of paddy in the market is increased or value of jowar in the market is decreased. That means, in valuation mechanism, if the value of one good is increased, the value of other good must necessarily be decreased and *vice versa*. So we can conclude that, there cannot be a general rice or fall in value of goods in the market. Similarly, in the monetary market, if a quintal of paddy fetches ₹1000/- in one situation and ₹800/- in another situation, it indicates, the value of paddy is decreased in the market in terms of money income or the purchase value of a 'Rupee' in the market is increased. A business firm can add value to its product by better specifying a product offer in accordance with customers' expectations, such as guarantee for the product, effective after-sales service, home delivery etc.

4. Exchange

The method of production and allocation of goods and services among the consumers differ across the economic systems. In capitalistic economies, automatic price mechanism directs the production and allocation of goods and services in the economy. The prices are determined by the forces of demand and supply. These market forces are the signals to direct the system. Government does not interfere in the fixation of prices of commodities. This automatic price mechanism directs consumption, production, exchange, distribution, investment, saving etc., without any central planning and control. In case of disequilibrium say, if demand exceeds the supply, new producers will enter the market and start producing that product and existing producers will increase their production, thereby, equality between demand and supply is attained. Similarly, if supply exceeds the demand, some firms which experience losses will leave the industry and again the equality of demand and supply forces is attained. Thus, shortages and surpluses in the economy are generally adjusted by the forces of demand and supply and the economy operates automatically through the price mechanism. Consumers are not dictated by any agency to purchase or not to purchase a particular type of good. They purchase the goods as per their will and choice and according to their purchasing power. However, in socialistic economies, all decisions regarding production and allocation of goods and services are undertaken by the Government or Central Planning Authority. Economic decisions about what, how, how much and for whom to produce are all made by the State through its Central Planning Authority and other administrative units. In modern days of business, the exchange of goods and services usually takes place between the firms and the consumers *via* money as medium of exchange.

2.7. 4As of Agricultural Marketing

They include,

- ☆ ***Availability:*** The first challenge of agricultural marketing is to ensure the availability of products and services to the potential consumers. For example, with the advent of Green Revolution, India could meet the food security needs of mounting population and actively competing in international trade. But, this does not mean the country is free of domestic problems like poverty, inequality of income distribution etc., and these are on declining trend during the period of independent India.
- ☆ ***Affordability:*** The second challenge is to ensure affordability of the products and services at the consumer's level. That means, the products offered by the firms must be well within the reach of the consumers' income levels. For example, the Government of India is supplying rice, wheat, sugar, edible oil, kerosene etc., through Fair Price Shops (FPS) to the Below Poverty Line (BPL) consumers at issue prices. Similarly, the Government supplies requisite fertilizers, pesticides, irrigation water, power supply etc., at subsidized prices to the farming community.
- ☆ ***Acceptability:*** It implies, the commodities and services offered by the firms must suit to the consumer's tastes and preferences, so that it is accepted by the majority population. For example, Aswini Homeo and Ayurvedic Products Limited firm (Hyderabad) supplies 'Diabetic Atta' suiting to the nutritious requirements of diabetic consumers.
- ☆ ***Awareness:*** The firms must engage in advertising about the products and services offered by them so that, the potential consumers are motivated and purchase the product. For example, Bayer company advertizes about its products (pesticides) to make the farmers aware and purchase them. Similarly, Brooke Bond Green Label firm advertizes about its coffee brand in televisions.

2.8. Features of Agricultural Marketing

They include,

- Agricultural marketing leads to optimization both in terms of inputs used and output produced. From inputs side, it aims at minimizing the losses during the process of performing different marketing functions. From output side, it aims at distributing the products to the ultimate consumers at affordable prices.
- Agricultural marketing aims at providing higher income to the farmers through offering higher prices to the produce. This further induces the farmers to generate higher marketable surplus.
- Agricultural marketing widens the markets, as the produce (say, paddy) produced by the farmers will move to distant areas. This infers that, the farmers get higher income through increased demand for their produce.
- Agricultural marketing offers requisite raw material to the agro-based industries and helps in producing processed products that enjoy comparative advantage in the international market. Thus, agricultural marketing helps in strengthening the establishment of agro-based industries.
- Agricultural marketing sends price signals of different commodities to the farmers. This enable them to plan their crop production programme in accordance with the principle of comparative advantage. Besides planning the cropping pattern, it also guides the farmers in adopting the production techniques that ensure cost-effectiveness. It also guides the processing firms to plan the production of processed products. Thus, agricultural marketing system can become a direct source of new technical knowledge and induce farmers to adopt up to date scientific methods of cultivation. So, it plays an important role in the economic development and stability of a country.
- Agricultural marketing offers large scale employment to the people through engaging them in various marketing functions.

Besides above, agricultural marketing also creates different types of utilities to the product *viz.*, form utility, place utility, time utility, possession utility and task utility (Figure 1.1). Through this, it ensures better living to the agencies involved in the marketing process of commodities. Thus, the features of agricultural marketing highlight its dynamic role and it has been described as the most important multiplier of agricultural development. Further, it is necessary to improve the agricultural marketing system to aid the process of agricultural development for two important reasons. Firstly, if the additional produce does not move to the market to bring additional revenue to the farmers, it may work as a disincentive to increase production by the farmers. Secondly, if the system does not support supplying the commodities at reasonable prices to the consumer at the time and place needed by them, increased production has no meaning and plays no role in the welfare of society. Thus, the importance of agricultural marketing lies in balancing the interests of both farmers and consumers.

2.9. Agricultural Marketing in India

The importance of marketing in agriculture is very well illustrated by saying that, *'a good farmer has one eye on the plough and the other on the market'*. This is true, when agriculture is subsistence oriented. But, today, agriculture is more market oriented and now, even Indian agriculture is becoming commercialized. In these days of commercial agriculture, it will be more fit to say, *'a good farmer has only his hands on the plough, but both the eyes on the market'*.

Economists and various studies conducted across the globe envisage India and China to rule the world in the 21st century. For over a century, the USA has been the largest economy in the world. But, major developments have taken place in the world economy and since then, the focus has been shifted from the USA and the rich countries of Europe to the two Asian giants - India and China. By 2025, the Indian economy is projected to be about 60 per cent the size of the USA economy. The transformation into a tri-polar economy will be completed by 2035, with the Indian economy only a little smaller than the USA economy, but larger than that of Western Europe. By 2035, India is likely to be a larger growth driver than the six largest countries in the EU, though its impact will be a little over half that of the USA. India, which is now the fourth largest economy in terms of purchasing power parity, will overtake Japan and become third major economic power within ten years.

As India prepares herself for becoming an economic superpower, it must expedite socio-economic reforms and take steps for overcoming institutional and infrastructure bottlenecks inherent in the system. India is the third largest economy in Asia after Japan and China, as measured in terms of its GDP and it is continuing to grow rapidly. The Indian economy has seen high growth rates of more than eight per cent since 2003. In 2005 and 2006, GDP grew at a rate of over nine per cent. Globally, India's growth is surpassed only by that of China. This is expected to continue with growth just under seven per cent by 2015. Agriculture occupies a prominent position in Indian policy-making not only because of its contribution to GDP, but also because of the large proportion of the population that is dependent on the sector for its livelihood. The economic performance of the agricultural sector of the Indian economy is fundamental to the well-being of hundreds of millions of people and pivotal to the security of their livelihoods and their food supply. Some 70 per cent of the population of approximately 1.1 billion lives in rural areas and 55 per cent of the population is employed in agriculture, a sector which contributes approximately 17 per cent to GDP. While looking at some of the agricultural products, India is the largest producer of tea, jute and jute like fibre. India is not only the largest producer, but also largest consumer of tea in the world. India accounts for around 14 per cent of the world trade in tea. Indian tea is exported in various forms such as bulk tea, packet tea, tea bags, instant tea etc., to more than 80 countries in the world. Among livestock, cattle and buffalo are found maximum in India. Indian total milk production is highest in the world. India has also the privilege of having the first rank in total irrigated land in area terms in the world. Among cereals production, India is placed third, having second largest production in wheat and rice and the largest production in pulses. India is the front ranking producer of mangoes (41 per cent of world's mango) and bananas (23 per cent of world's banana).

Indian agriculture employs 61 per cent of economically active population, which is nearly 15 per cent higher than world average (46 per cent). India shares 2.3 per cent of world land area, 11.8 per cent of world arable land and 21.3 per cent of world irrigated area. With these resources, India has to support 16.6 per cent of world population, who depended on agriculture. Food grains production in the country grew at an average 1.98 per cent during 2004-05 to 2008-09, which was higher than the average rate of population growth of 1.50 per cent during the same period. This implies, India made notable achievements in agriculture to meet the food security needs of the population, as the rural population constitutes 69 per cent of total population (2011 census), mainly dependent on agriculture and allied activities (Table 2.6). However, the per cent share of cultivators in terms of total rural population was declined from 23 per cent to 17 per cent during 1951 to 2001, unlike agricultural labourers. Similarly, the cumulative share of both cultivators and agricultural labourers population in terms of total rural population showed slight declining trend during the same period. This implies, much emphasis should be given for promoting agricultural and allied activities in rural areas, as they serve as major growth engine for overall development of the economy.

Table 2.6: Demographic details of total and rural populations in India

(Million)

Year	*Total Population*	*Average Annual Exponential Growth (per cent)*	*Rural Population*	*Cultivators*	*Agril. Labourers*	*Total*
1951	361.1	1.25	298.6	69.9(23)	27.3(9)	97.2 (32)
1971	548.2	2.22	439.0	78.2(18)	47.5(11)	125.7(29)
1991	846.4	2.14	628.9	110.7(18)	74.6(12)	185.3(30)
2001	1028.7	1.95	742.6	127.3(17)	106.8(14)	234.1(31)
2011	1210.2	1.64	833.1	NA	NA	NA

Note: NA: Not Available.

Figures in parentheses indicate per cent shares in rural population.

Source: www.agricoop.nic.in.

Indian agriculture has undergone significant changes and transformation since Independence. The underlying factors for these changes were different in different periods. During fifties and sixties, institutional reforms like land reforms and development of irrigation and other infrastructure played a major role in output growth. Technological breakthrough has been the prime mover during seventies and spread of technological changes to wider areas and crops has been the main factor during eighties. The decades of sixties and seventies also witnessed high growth in public investments in agriculture, which improved infrastructural base for growth of agricultural output in the country in the following decade. However, the decade of eighties showed decline in public investments in agriculture and sharp rise in level of input subsidies. This trend continued even during nineties (Chand, 2005). Thus, India's record of progress in agriculture over the first four decades since Independence has been quite impressive. The agriculture sector has been successful in keeping pace with rising demand for food. The contribution of increased land area under agricultural production has declined over time and increases in production in the two decades (late 1960s to 1980s) have been almost entirely due to increased productivity. Contribution of agricultural growth to overall progress has been widespread. Increased productivity has helped to feed the poor, enhanced farm income and provided opportunities for both direct and indirect employment. The success of India's agriculture is attributed to a series of steps that led to availability of farm technologies, which brought about dramatic increases in productivity in seventies and eighties often described as the Green Revolution era. The major sources of agricultural growth during this era were the spread of modern crop varieties, intensification of input use and investments leading to expansion in the irrigated area. Agricultural growth, measured as the annual rate of growth in net domestic product, increased from 1.9 per cent per annum before the Green Revolution period (1950-51 to 1966-67) to 2.3 per cent during the first phase of the Green Revolution (1966-67 to 1980-81). The growth rate accelerated further to 3.1 per cent during the second phase of the Green Revolution (1980-81 to 2000-01). At the same time, the share of agriculture in total GDP has fallen from about a little over half in the early fifties to around one-fourth in the late nineties.

The implementation period for the Agreement on Agriculture (AoA) coincided with the reform process, which has been under way since 1991 in response to the macro-economic crisis that occurred as a result of the depletion of foreign currency reserves and a high fiscal deficit. To address this crisis, key reforms were initiated, by way of ending the old industrial licensing regime, more liberal policies towards foreign direct investment, significant reduction in tariffs on industrial goods and currency devaluation. The focus of this reform process was initially on the manufacturing sector, but gradually these unilateral changes have been extended to the other sectors of the economy including agriculture. In agriculture, although there was no direct government intervention in production and investment decisions of farmers, government did, and still does, influence the legal and economic environment in which farmers and other economic agents operate. Most of these interventions in agriculture were conceived to deal with situations associated with famines and scarcities that existed before the Green Revolution period. In the case of the domestic market, the withdrawal of the restrictions on the movement of agricultural commodities is one of the major changes that have been brought about during the present reform process. As far as external trade is concerned, in the early nineties, a few steps to liberalize agricultural trade were initiated about a year and a half after the July 1991 reforms, and these have been followed by a number of prominent reforms thereafter. India, now, is playing a significant role in the international trade with the opening of economy in 1991. India exports a wide array of agricultural products like cereals, millets, groundnuts, soyabean, castor oil, soyabean cake, sugar, refined, horticultural produce like mangoes, grapes, apples, cashewnuts shelled, spices, coffee and tea, marine products, buffalo meat etc. Similarly, India imports major commodities like palm oil, soybean oil, cashewnut, cotton lint, pulses, peas, dry, silk, fatty acid oils etc., on priority basis. Due to the total removal of Quantitative Restrictions during April 2001, the pace of imports was increased. However, with the scientific advancement in R&D in agriculture and allied areas, the quantum jump in production leads to increased exportable surpluses. This is especially true for edible oils trade, as India gained net exporter status during early 21st century when compared to last decade of 20th century.

India successfully completed 16 years period as a member of WTO. During this period, the agriculture in the country was viewed more from the trading side, besides concentrating on the food security needs of the population. India actively represented in the multilateral negotiations during several rounds held by WTO in view of improving the trade from the country and raised several issues regarding the provisions that hamper the trade of both developing nations and LDCs in general and with reference to India in particular. For example, in the Sixth WTO Ministerial Conference held at Hong Kong, China during 13 to 18th, December, 2005, India actively represented the developing nations against the excess domestic subsidies and export subsidies offered by the developed nations for cotton, which depressed the prices in the international market and thereby, hampered the trade interests of developing nations (particularly India), to concentrate on flexibility issues in market access provision, special and differential treatment for developing countries etc. This multilateral trading experience has enabled the country to improve the agricultural trade balance significantly from 1995 onwards. Further, India benefited on several grounds during post-WTO regime *viz.*, viewing the agricultural sector more from commercial point of view rather than for subsistence purpose, promoting quality agriculture, access to and adoption of modern technologies, streamlining the trading practices both in domestic and external trades, strengthening both production and marketing infrastructure, increased emphasis on value addition of commodities etc. The World Trade Agreement (WTA) provisions like tariffication, Aggregate Measurement of Support (AMS), market access etc., were found congenial to India, if viewed from the imports side. This is because, the tariff rates imposed on the imports by India were far less than bound rates specified and still remains competitive in the domestic market. Regarding AMS, India could still exploit this provision, as the total AMS offered to the agricultural sector is still negative and below the *de minimis* level. India importing fertilizers, machinery (technology), pulses, edible oils etc., at significant level, as per the domestic demand, indirectly satisfying the market access provision. However, India could not fully exploit the opportunities and challenges offered by the WTA provisions from the exports side. This is clearly indicated by the declining share of agricultural exports in total exports of the country (Table 2.7). Moreover, agricultural sector is receiving less and less importance during successive FYPs and hence, the shares of both public and private sectors investments together in GDP of the country is showing declining trend. Hence, the desired positive impacts were not noticed in this sector even with the opening of trade environment. This is because, the changes in economic polices since early nineties in India were not successful in completely reforming protectionist policies. Relatively higher tariff rates on agricultural commodities remained one of the features of trade regimes. Regarding quality standards of agricultural commodities, India could not meet the requirements imposed by the developed countries. However, the institutional developments related to trade policy have paved the way to some liberalization of agricultural trade. The structural changes during eighties and nineties placed non-agricultural sector of the country in the driving seat of economic growth. Though the shares of agricultural exports in national exports have been declining, agriculture and agricultural trade still

Table 2.7: India's imports and exports of agricultural commodities vis-à-vis total national imports and exports

(Value in ₹ Crores)

Year	Agricultural Imports	Total National Imports	Percentage Agriculture Imports to Total National Imports	Agricultural Exports	Total National Exports	Percentage Agriculture Exports to Total National Exports
1990-91	1205.86	43170.82	2.79	6012.76	32527.28	18.49
1991-92	1478.27	47850.84	3.09	7838.04	44041.81	17.80
1992-93	2876.25	63374.52	4.54	9040.30	53688.26	16.84
1993-94	2327.33	73101.01	3.18	12586.55	69748.85	18.05
1994-95	5937.21	89970.70	6.60	13222.76	82673.40	15.99
1995-96	5890.10	122678.14	4.80	20397.74	106353.35	19.18
1996-97	6612.60	138919.88	4.76	24161.29	118817.32	20.33
1997-98	8784.19	154176.29	5.70	24832.45	130100.64	19.09
1998-99	14566.48	178331.69	8.17	25510.64	139751.77	18.25
1999-00	16066.73	215528.53	7.45	25313.66	159095.20	15.91
2000-01	12086.23	228306.64	5.29	28657.37	201356.45	14.23
2001-02	16256.61	245199.72	6.63	29728.61	209017.97	14.22
2002-03	17608.83	297205.87	5.92	34653.94	255137.28	13.58
2003-04	21972.68	359107.66	6.12	37266.52	293366.75	12.70
2004-05	22811.84	501064.54	4.55	41602.65	375339.53	11.08
2005-06	21499.22	660408.90	3.26	49216.96	456417.86	10.78
2006-07	29637.86	840506.31	3.53	62411.42	571779.28	10.92
2007-08	29906.24	1012311.70	2.95	79039.72	655863.52	12.05
2008-09	37183.03	1374435.55	2.71	85951.67	840755.06	10.22
2009-10 (P)	59367.62	1356468.65	4.38	89522.59	845125.21	10.59

Note: P: Provisional.

Source: Director General of Commercial Intelligence and Statistics, Ministry of Commerce, Kolkata.

play a very important role in the economic development country. This is because, India is the net agricultural exporter in the international market. As India possess lot of advantages in the international market regarding comparative advantage of majority of agricultural commodities, wider market for several commodities produced by the country, progressive nature of farming community in adoption of cost-effective modern technologies, capability of producing the commodities suiting to the importer's needs etc., the developed nations indirectly checked the India's potentiality through imposing more stringent measures on quality aspects of the commodities/products rather than through tariffication measures. It is very difficult for India to cope up with the standards imposed by the developed countries, as they are too stringent and as per the international standards and India needs altogether sophisticated technologies to meet their requirements. Moreover, the developed countries are providing higher export subsidies (Amber Box) to their commodities and showing the same through Green and Blue Boxes. This actually depressed the international prices of several commodities, thereby, the comparative advantage for India with reference to several commodities is on the decline. Hence, India rightly raised this issue in the recent Sixth Ministerial Conference of the WTO to reduce the export subsidies by the developed countries, as per the WTA provisions.

2.9.1. Commodity-Group Wise Performance of India in the Selected Country's Groups

India occupies a prominent place with respect to both area and production of cereals (Table 2.8), coarse grains (Table 2.9) and pulses (Table 2.10) among the South Asian countries, but ranks fifth (cereals) and sixth (coarse grains and pulses) positions respectively with respect to their productivities

Table 2.8: Average share(per cent) of cereals, total area, production and productivity at all-India level in different country's groups and world before and after WTO.

Sl.No.	*Item*	*Period*	*South Asia*	*Asia**	*Developing Nations*	*World*
1.	Area	1961-70	82(1)	–	25	14
		1971-80	82(1)	–	25	14
		1981-94	80(1)	–	24	14
		1995-05#	78(1)	31	23	15
		1961-05	80(1)	–	24	14
2.	Production	1961-70	76(1)	–	19	9
		1971-80	77(1)	–	18	9
		1981-94	77(1)	–	18	10
		1995-05	76(1)	23	19	11
		1961-05	76(1)	–	19	10
3.	Productivity	1961-70	–7(6)	–	–24	–37
	comparison	1971-80	–6(6)	–	–27	–39
	(per cent)	1981-94	–3(4)	–	–25	–32
		1995-05	–3(3)	–27	–17	–25
		1961-05	–5(5)	–	–23	–33

Note: *: Data not available from 1961-94; #: Period after WTO.

Figures in parentheses indicate rank among SAARC nations.

Raw Data Source: www.fao.org

Table 2.9: Average share(per cent) of coarse grains, total area, production and productivity at all-India level in different country's groups and world before and after WTO.

Sl.No.	Item	Period	South Asia	Asia*	Developing Nations	World
1.	Area	1961-70	94.19(1)	–	25.58	13.74
		1971-80	94.00(1)	–	24.26	12.61
		1981-94	92.58(1)	–	20.80	11.16
		1995-05#	90.52(1)	34.25	15.62	9.47
		1961-05	92.75(1)	–	21.37	11.64
2.	Production	1961-70	91.14(1)	–	15.01	5.04
		1971-80	91.41(1)	–	12.72	4.11
		1981-94	90.61(1)	–	10.41	3.77
		1995-05	87.88(1)	14.53	8.29	3.46
		1961-05	90.23(1)	–	11.43	4.05
3.	Productivity	1961-70	–3.35(7)	–	–71.04	–174.22
	comparison	1971-80	–2.84(6)	–	–91.36	–207.74
	(per cent)	1981-94	–2.18(5)	–	–101.04	–200.42
		1995-05	–3.05(5)	–137.12	–89.09	–175.13
		1961-05	–2.80(6)	–	–89.30	–190.04

Note: *: Data not available from 1961-94; #: Period after WTO.

Figures in parentheses indicate rank among SAARC nations.

Raw Data Source: www.fao.org

during the overall reference period, 1961-05. Both during pre and post-WTO regimes, India continued its top position with respect to both area and production of selected commodities, but in terms of productivity, not much progress was achieved during post-WTO regime. The productivity level of cereals, coarse grains and pulses in India were nearly five, three and one per cents respectively lower than average productivity of these commodities of South Asian countries during 1961-05. However, during post-WTO regime, there is slight improvement in productivity of cereals when compared to pre-WTO regime (sixties and seventies), unlike coarse grains and pulses, total. India continued its prominence even among the Asian countries with respect to both area and production of these commodities during the post-WTO regime. But, India's productivity levels of these commodities were 27, 137 and 19 per cent lower when compared to Asia's average productivity during post-WTO regime, indicating that, the country's production of these commodities is limited by productivity rather than area. Similar trends were observed with respect to shares of area and production and comparison of productivity of selected commodities of India with developing nations and world both during pre and post-WTO regimes.

Regarding oilseeds, total also, India occupies a predominant position with respect to both area and production of oilseeds among the South Asian countries, as indicated by the shares of 88 and 85 per cents respectively during the overall reference period, 1961-05 (Table 2.11). During pre and post-WTO regimes, India continued its predominant position both in terms of area and production and especially the shares improved slightly during post-WTO regime among South Asian countries. However, productivity of oilseeds posed worry some picture, as India occupies sixth position among

Table 2.10: Average share(per cent) of pulses, total area, production and productivity at all-India level in different country's groups and world before and after WTO.

Sl.No.	*Item*	*Period*	*South Asia*	*Asia**	*Developing Nations*	*World*
1.	Area	1961-70	91(1)	–	43	34
		1971-80	89(1)	–	43	36
		1981-94	89(1)	–	41	34
		1995-05#	90(1)	63	37	32
		1961-05	90(1)	–	41	34
2.	Production	1961-70	90(1)	–	36	26
		1971-80	87(1)	–	35	26
		1981-94	88(1)	–	35	24
		1995-05	89(1)	51	33	24
		1961-05	88(1)	–	35	25
3.	Productivity comparison (per cent)	1961-70	–1.11(5)	–	–17.61	–25.08
		1971-80	–2.19(7)	–	–19.25	–28.77
		1981-94	–0.77(6)	–	–14.48	–30.05
		1995-05	–1.04(6)	–18.57	–9.27	–25.04
		1961-05	–1.23(6)	–	–14.96	–27.44

Note: *: Data not available from 1961-94; #: Period after WTO.

Figures in parentheses indicate rank among SAARC nations.

Raw Data Source: www.fao.org

the South Asian countries lead by Maldives. The productivity of oilseeds in India is nearly three per cent lower than the average productivity of oilseeds in South Asian countries. Even though area and production of oilseeds in India occupy significant shares even at Asia, developing nations and world levels, the productivity shares exhibited disappointing picture. The average productivity of oilseeds in India was 58, 53 and 52 per cents lower than average productivities at Asia, developing nations and world during the post-WTO regime. Further, the productivity declined drastically during post-WTO regime when compared to pre-WTO regime. This signifies that, much progress was not made on productivity side of oilseeds, despite effective implementation of Oilseeds Technology Mission (OTM) in India since 1986.

Fruits production in India has brighter prospects, as the country leads other South Asian countries both in terms of area and production (Table 2.12). Similarly, in terms of productivity, India possesses advantage over the selected country's groups, as India's average productivity of fruits is always higher than all the selected country's groups both during pre and post-WTO regimes, except during seventies among South Asian countries. But among South Asian nations, India stood at fourth position during the overall reference period, 1961-05. During sixties, Bangladesh leads other South Asian countries and from seventies to mid eighties Sri Lanka dominated the productivity scenario of fruits. However, India occupied first position only during 1987 and finally stood at second position during post-WTO regime lead by Maldives in South Asia. This clearly implies that, boosting fruits production, strengthening cold storage and processing facilities are essential to meet the domestic needs and to capture the international market. Like fruits, for vegetables, India leads other South Asian countries

Table 2.11: Average share(per cent) of oilseeds, total area, production and productivity at all-India level in different country's groups and world before and after WTO.

Sl.No.	Item	Period	South Asia	Asia*	Developing Nations	World
1.	Area	1961-70	88.87(1)	–	29.37	20.23
		1971-80	87.80(1)	–	26.48	17.60
		1981-94	87.64(1)	–	25.45	17.06
		1995-05#	87.72(1)	37.95	23.53	15.97
		1961-05	87.97(1)	–	26.08	17.62
2.	Production	1961-70	84.21(1)	–	18.89	11.14
		1971-80	84.72(1)	–	16.42	9.42
		1981-94	85.84(1)	–	14.14	9.00
		1995-05	86.30(1)	16.03	11.18	7.69
		1961-05	85.34(1)	–	14.98	9.25
3.	Productivity	1961-70	–5.25(6)	–	–35.68	–44.97
	comparison	1971-80	–3.50(6)	–	–38.11	–46.64
	(per cent)	1981-94	–2.07(5)	–	–44.53	–47.38
		1995-05	–1.62(5)	–58.00	–52.83	–52.13
		1961-05	–2.99(6)	–	–43.17	–47.84

Note: *: Data not available from 1961-94; #: Period after WTO.

Figures in parentheses indicate rank among SAARC nations.

Raw Data Source: www.fao.org

both in terms of area and production (Table 2.13), but limited by productivity. Among South Asian countries, Maldives rank first in terms of productivity followed by Pakistan and India. In terms of selected country groups *viz.*, Asia, developing nations and world, the yield levels of vegetables in India is low, except compared to South Asia.

Considering the potentiality of India with regard to shares of area, production and productivity of major agricultural commodities in different country groups and world during post-WTO regime, it is essential to plan for marketing/trade reforms in relevance to the provisions prescribed under WTA. As agricultural marketing in India is undergoing a significant metamorphosis because of economic liberalization and globalization, it is imperative to plan and design market-oriented strategies for different agricultural commodities. In this context, studying 'Agricultural Marketing' as a separate discipline is essential, as it is considered as an area for the 'second generation' of green revolution problems and marketing of agricultural produce is considered as an integral part of agriculture. The clientele (farmers) being largely illiterate, an efficient and organized marketing system is necessary to enable them to realize a remunerative price for their produce and to reduce their exploitation by marketing middlemen. Further, the theoretical background on 'Agricultural Marketing' helps to understand and analyze the marketing prospects and retrospects both in domestic and international trade. The benefits of studying marketing as a subject are already discussed earlier under Sections 1.3 and 1.4 of Chapter 1.

Table 2.12: Average share(per cent) of fruits (excluding melons), total area, production and productivity at all-India level in different country's groups and world before and after WTO.

Sl.No.	*Item*	*Period*	*South Asia*	*Asia**	*Developing Nations*	*World*
1.	Area	1961-70	77.35(1)	–	11.92	6.38
		1971-80	83.54(1)	–	11.17	6.47
		1981-94	79.00(1)	–	10.13	6.70
		1995-05#	79.46(1)	16.73	10.07	7.58
		1961-05	79.75(1)	–	10.75	6.79
2.	Production	1961-70	81.03(1)	–	13.91	7.15
		1971-80	79.84(1)	–	12.26	6.70
		1981-94	80.49(1)	–	12.30	7.72
		1995-05	84.07(1)	22.10	12.66	9.30
		1961-05	81.34(1)	–	12.74	7.75
3.	Productivity	1961-70	4.46(4)	–	14.17	10.69
	comparison	1971-80	–4.65(5)	–	8.87	3.25
	(per cent)	1981-94	1.82(3)	–	17.50	12.76
		1995-05	5.47(2)	24.09	20.40	18.44
		1961-05	1.86(4)	–	15.55	11.57

Note: *: Data not available from 1961-94; #: Period after WTO.

Figures in parentheses indicate rank among SAARC nations.

Raw Data Source: www.fao.org

2.10. Growth of Agricultural Marketing in India

Economic development of a country provides the impulse towards more sophisticated and more efficient marketing systems. With increasing consciousness towards market-oriented agriculture from the farmers' side and increasing awareness about quality and differentiated (processed) agricultural products from the consumers' side, there is substantial growth in the pattern of agricultural marketing. These associated with trade reforms implemented by the Government of India further brought drastic changes in the field of agricultural marketing. At the time of Indian independence, farming is mainly subsistence oriented and less market-oriented (Figure 2.2). With the advent of Green revolution, the emphasis was shifted towards market from subsistence mode of farming. With the introduction of trade reforms since 1991, there is altogether shift in the pattern of agricultural marketing and now the agricultural markets are more export oriented considering the Principle of Comparative Advantage. Further, the pattern of marketing is more consumer-oriented *i.e.*, aiming at the needs and requirements at the grass root level consumers. For example, several firms are producing agro-based products like wheat atta, biscuits, sugar free tablets, sugar free sweets, Diabetic Protinules, Horlicks lite etc., suiting to the nutritious requirements of diabetic consumers. There is also rapid acceptance at the consumers' level with reference to RTE and RTC foods.

The growth of agricultural marketing in India can be conveniently categorized into four periods, as discussed here under.

Table 2.13: Average share (per cent) of vegetables (including melons), total area, production and productivity at All-India level in different country's groups and world before and after WTO.

Sl.No.	*Item*	*Period*	*South Asia*	*Asia**	*Developing Nations*	*World*
1.	Area	1961-70	89.15(1)	–	21.64	13.77
		1971-80	89.83(1)	–	24.65	16.46
		1981-94	88.21(1)	–	21.35	15.70
		1995-05#	87.05(1)	19.42	16.22	13.31
		1961-05	88.51(1)	–	20.89	14.86
2.	Production	1961-70	89.74(1)	–	17.55	9.34
		1971-80	89.88(1)	–	19.53	10.74
		1981-94	89.43(1)	–	16.52	10.64
		1995-05	89.08(1)	13.76	12.39	9.57
		1961-05	89.51(1)	–	16.41	10.12
3.	Productivity comparison (per cent)	1961-70	0.56(3)	–	–18.92	–32.03
		1971-80	0.08(3)	–	–20.78	–34.73
		1981-94	1.39(3)	–	–22.37	–31.87
		1995-05	2.34(3)	–28.96	–23.39	–27.82
		1961-05	1.15(3)	–	–21.50	–31.55

Note: *: Data not available from 1961-94; #: Period after WTO.

Figures in parentheses indicate rank among SAARC nations.

Raw Data Source: www.fao.org

1. Pre-Independence Era (Before 1947)

- ☆ Goods were scarce and buyers would accept virtually any goods that were produced.
- ☆ The central notion was that, products would sell themselves. Businesses focused on production, not marketing.
- ☆ Markets are unorganized, thereby farmers are at the mercy of traders.

2.Between 1947 to 1965

- ☆ More focus on boosting agricultural production, as achieving food security for the mounting population is the priority.
- ☆ Emphasis on regulating the market practices through establishing regulated markets with the main objective of safeguarding the interests of farming community in transacting their produce.
- ☆ State trading was introduced to check price volatility of commodities
- ☆ Establishment of CMSs to ensure the farmers realizes the benefits of large scale marketable surplus.

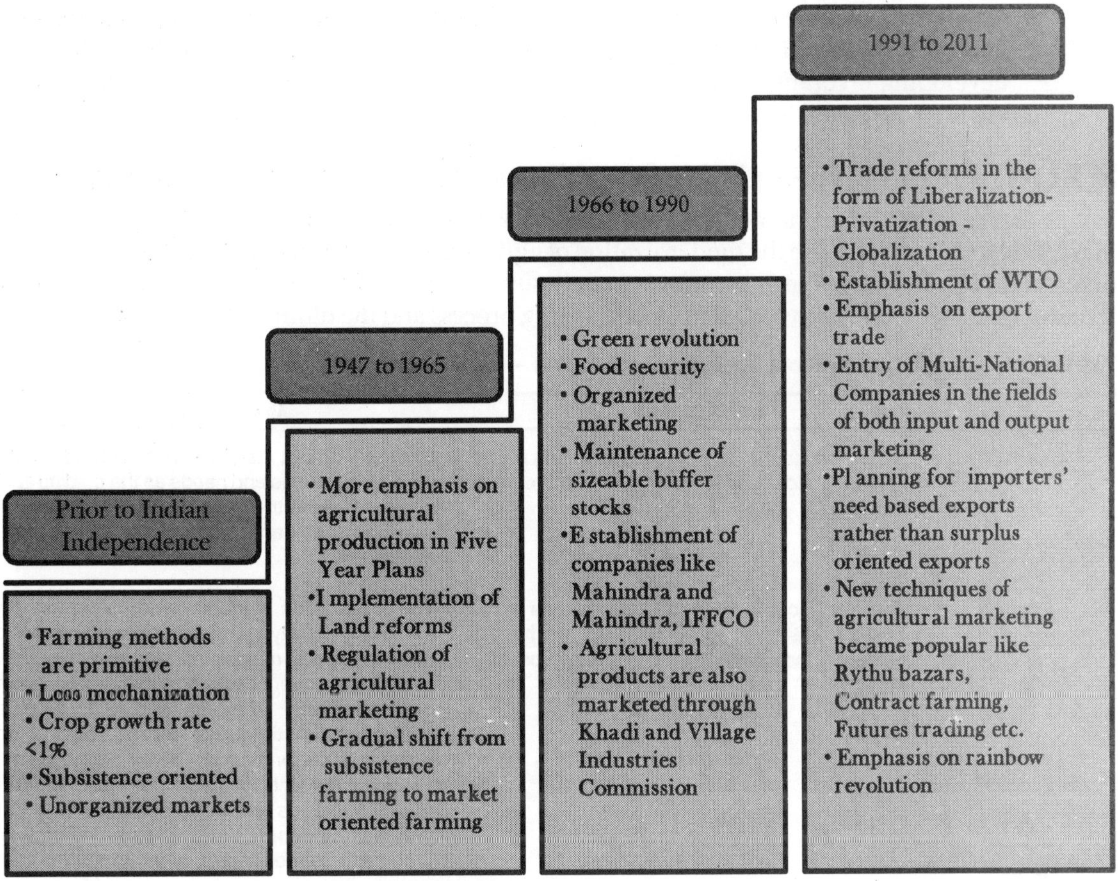

Figure 2.2: Growth of Agricultural Marketing in India.

3. Between 1966 to 1990

- ✰ Marketing of agricultural produce on commercial scale gained immense significance with the advent of green revolution
- ✰ Maintenance of sizeable buffer stocks

4.After 1991

- ✰ The country entered the Liberalization-Privatization-Globalization (LPG) phase and emphasis of production was now more on export orientation suiting to the importers' requirements
- ✰ Formation of WTO (01-01-1995) and this led to trading of commodities and processed products across the member countries on systematic lines with well established rules and regulations
- ✰ Quality standards of the commodities assumed significance
- ✰ Increased demand for RTE and RTC foods.

☆ An important outgrowth of this market orientation is customer relationship management *i.e.,* the process of identifying prospective buyers, understanding them intimately and developing favorable long-term perceptions about the firm and its products, so that the buyers will choose them in the market place.

2.11. Traditional vis-à-vis Modern Views of Agricultural Marketing

There are two different views of looking at the marketing of agricultural products *viz.,* Traditional view and Modern view. With the modernization of agri-business, the concept of marketing is a more demand driven and the same in the past is mainly supply driven. These two views mainly differ in terms of their focus, efforts involved in the marketing process and the ultimate goal (Table 2.14).

Table 2.14: Different views about agricultural marketing – Comparative analysis.

Item	*Traditional View*	*Modern View*
Approach	Supply driven *i.e.,* agricultural marketing means selling goods and services that have been produced.	Demand driven *i.e.,* the firm considers the consumers' wants and needs as the guiding spirit and focuses on the delivery of such goods and services that can satisfy those needs most effectively.
Focus	The main focus is on the product, *i.e.,* we have a product and it has to be sold. So, we have to persuade the consumers to buy the product.	The focus is on customer orientation *i.e.,* the marketing activity starts with an assessment of the consumers' needs and plan the production of goods and services that satisfy these needs most effectively. This also applies to all other marketing functions like pricing, packaging, transportation, processing, sales promotion etc. For example, the increasing demand for RTE and RTC foods among the consumers is a classical example for the release of such products into the market by various firms.
Means	All efforts of the marketing agencies are concentrated on selling the product. They adopt all means like personal selling and sales promotion to boost the sales.	All marketing activities like product planning, pricing, packaging, distribution and sales promotion are combined into one as coordinated marketing efforts. This is called Integrating marketing.
Goal	The ultimate goal of all marketing activity is to earn profit through maximization of sales.	The main aim of all efforts is to earn profits through maximization of consumer satisfaction.

3

Classification of Agricultural Markets

The word 'market' comes from the Latin word *'marcatus'*, which means merchandise or trade or a place, where business is conducted. A market is one of many varieties of systems, institutions, procedures, social relations and infrastructures where parties engage in exchange. To achieve an efficient system of buying and selling of agricultural commodities, there are different kinds of agricultural markets. These markets will function to ensure a reasonable amount of profits to the farmers engaged in transacting agricultural produce. In general, 'market' means a place, where buyers and sellers personally negotiate each other in transacting the produce. For a market to exist, certain conditions must be satisfied. These conditions should be both necessary and sufficient. They may also be termed as the components of a market. They include: the existence of a good for transaction, existence of buyers and sellers, trading operations between buyers and sellers and geographical demarcation of an area such as place, region, country or the whole world. The existence of a perfect competition or uniform price is not a necessary condition. But, in Economics, market may refer to a particular place or perhaps may not be. That means, from Economists point of view, a market can exist even without personal negotiations between buyers and sellers. Similarly, physical existence of goods is not necessary in the modern concept of marketing (Futures markets). These facts can be explained with the help of the following statements:

Market is not any particular place, where the goods are bought and sold, but the whole of any region, where the buyers and sellers are in close business interactions with each other, such that, the prices of the same good tend to equate easily and quickly. (Cournot)

'Market is not a particular place, but a commodity where buyers and sellers are in direct competition with each other'. (Chapmann)

'Market refers to arrangement, whereby buyers and sellers come in contact with each other directly or indirectly, to buy or sell goods.' This definition indicates that, face to face contact of buyers and sellers is not necessary for market. For example, in futures trading, the buyers and sellers exchange their transactions through internet. So internet, here forms an arrangement and such arrangement also is included in the concept of market.

So, considering the above statements, 'market' need not be a particular place, where buyers and sellers engage in transacting the produce through personal negotiations. Further, with the development of grades and standards for the commodities, the transactions between buyers and sellers can happen even without their personal interaction. Thus, in the modern era of agri-business, the term 'market' has a renewed meaning and implication. It is as diverse in meaning as it is in functionality. Thus, markets can be categorized based on various aspects like nature of product or service, place of transaction, process of transaction, number of commodities dealt with, nature of competition between the firms etc. This classification of agricultural markets based on various criteria guides us in the following aspects:

- ✰ Helps to conduct market research study for a product
- ✰ Guides the firms in market segmentation
- ✰ Helps to analyze about the target market of the product
- ✰ Facilitate the firms for market positioning of the products in the consumers' minds
- ✰ Helps to conduct market penetration studies for the product
- ✰ Guides the firms both in price and non-price competitions
- ✰ Guides the firms to produce the commodities suiting to the consumers' tastes and preferences

3.1. Classification of Agricultural Markets

The classification of agricultural markets based on various criteria ensures that, each business firm must customize its marketing efforts in response to its marketing environment and the exchange process. So, a business firm can plan the marketing strategies that tend to work towards achieving its objectives. Considering the dynamic changes in the economy, the business firm can also modify its marketing strategies. The various criteria based on which the agricultural markets are classified are here under.

I. On the basis of nature of markets
II. On the basis of position of seller
III. On the basis of area from which the buyers and sellers are drawn for transactions
IV. On the basis of competition
V. On the basis of location
VI. On the basis of time
VII. On the basis of number of commodities the markets dealt with
VIII. Based on nature of commodities
IX. On the basis of volume of transactions
X. On the basis of population the markets serve

XI. On the basis of State intervention
XII. On the basis of vision
XIII. On the basis of objectives of personnel involved in marketing process
XIV. On the basis of final use of product
XV. On the basis where the product is transacted
XVI. On the basis of number of intermediaries between farmer and final consumer
XVIII. On the basis of number of items available of different categories for transaction to the final consumers
XIX. Based on profit margin
XX. On the basis of market segmentation
XXI. On the basis of type of agency involved in marketing transactions
XXII. On the basis how consumers are approached
XXIII. On the basis of methodology of convincing the customers
XXIV. On the basis of promoting the products
XXV. On the basis of the macro and micro-economic views of the market economy
XXVI: Based on personal interaction between buyers and sellers

I. On the Basis of Nature of Markets

According to nature, markets can be categorized as follows:

1. Commodity Market

In this market, the transactions are with reference to goods or agricultural produce. We know, goods may be of consumer goods or producer goods depending upon the satisfaction they yield to the consumer. Consumer goods are the goods, which yield satisfaction directly to the individual or consumer. So, these goods help the individuals to satisfy their wants directly. Hence, these goods are also called as Goods of First order. These goods are cheaper compared to Producer goods. These goods can further sub-classified into Perishable goods (fruits, milk, vegetables etc) and Durable goods (tractor, implements, machinery etc). On the other hand, producer goods (also called Capital goods) yield satisfaction to the consumers indirectly, as they aid in producing consumer goods. Hence, these goods are also called as Goods of Second Order. These goods generally involve more investment compared to Consumer goods. The strengthening of these capital goods in the economy implies capital formation. For example, sugar is a consumer good. But, there must be some processing machine to process sugarcane into sugar and that machine is known as capital good. The commodity market can be further sub-divided as follows:

a. Produce Markets

These markets deal with raw agricultural produce such as fruits, vegetables, paddy, wheat, cotton (kapas) etc. The produce dealt in these markets is consumed directly or they need processing for consumption purposes or for producing further goods.

b. Processed Goods Markets

These markets deal with processed products produced from raw produce. Examples include, sugar, jaggery, jam, jellies etc.

c. Bullion Markets

Unlike above two markets, these markets deal with the trade of precious metals like gold, silver precious stone etc. These markets are highly price sensitive compared to above two markets.

2. Capital Markets

These markets deal with money, shares foreign exchange of currencies etc., and accordingly, they are further categorized as follows:

a. Money Markets

These markets deal with money transactions between lenders and borrowers. The financial middlemen act as intermediaries in bring the lender and the borrower on the same platform. So, lenders, borrowers and middlemen are the important constituents of money markets. The price for lending money is 'interest', which borrowers pay to lenders.

b. Share Markets

These markets are highly organized and deal with purchase and sale of stocks, shares and debentures. National Stock Exchange (NSE), Bombay Stock Exchange (BSE) etc., are the common share markets in India.

c. Foreign Exchange Market

These markets deal with trade in foreign currencies. They deal in purchase and sale of foreign exchange. These markets are highly useful from the international trade point of view. These markets also help in settlement of foreign debts. The area covered by these markets is the whole world.

II. On the Basis of Position of Seller

The markets can be classified into the following categories on the basis of position of seller:

1. Primary Market

These are periodical markets locally known as 'Haats'. They are generally held once or twice a week. Usually traders (local wholesalers) from local area or neighbouring villages participate in marketing transactions. They transport the produce to wholesale (Secondary) markets. A part of the produce is purchased by the small retailers, who further sell it to the non-farm rural population. The farmers also sell their produce directly to the consumers. These markets are generally held in the open area or at centrally situated localities. Besides agricultural produce, a number of other commodities like tobacco, oils, gur, fruits and vegetables, spices, cloth, hosiery products and ornaments of cheap metals are also sold in these markets. These markets usually serve an area of 10 kms radius, but it may be more, ranging from 10 to 50 kms, depending upon the availability of communication and transportation facilities, nature of commodities handled and the location of the market with respect to other markets. The basic function of these markets is to serve as assembling centers for the local produce, but they also function simultaneously as distributive centers for local consumption. In spite of being situated in rural areas, the prices ruling in these markets are influenced by those prevailing in the wholesale markets. They are the oldest trading institutions in existence. The commodities traded are generally of inferior quality and the volumes are low. These markets are largely unregulated in their marketing practices.

These markets are the first point of contact for the farmers with buyers for selling the produce. The small farmer, with limited marketable surplus, does not find it economical to go to secondary markets located at distant places. These markets are not equipped with basic facilities like platforms for sale or auction, electricity, drinking water, grading, storage etc. The participants generally sit in the open air, thus exposing themselves and their products to heat, dust, rain, cold etc. This, in turn, shortens the shelf life of their agri-produce and renders other food items unhygienic and unsafe for human consumption. Owing to these shortcomings, the potential buyers/traders do not operate in these markets and the producers do not get competitive prices for the sale of their produce.

2. Secondary Market

These markets, also known as 'Mandis'. These markets are regularly held and provide a permanent place for daily transactions. These markets are generally situated in the towns, districts and important trade centers. Usually, they are situated near railway stations for easy loading and unloading of produce. The farmers are provided with various facilities for scientific marketing of produce. In this market, the transactions of commodities are between farmers and wholesalers or processors and between wholesalers and retailers. The commodities from primary market reach the secondary market through wholesalers. Thus, primary markets cater to the local demand whereas, secondary market cater to the distant demand. Some traders in primary markets assemble produce from primary market, where farmers bring their produce and they transport the same to secondary market. The infrastructure available in these markets is generally poor (not on scientific lines) and suffers from the similar handicaps as do primary market.

3. Terminal Market

In this market, the transactions are between retailers and final consumers. The transactions are made in small quantities compared to primary and secondary markets. These markets are usually situated in metropolitan cities like Delhi, Mumbai, Chennai and Kolkata etc. In these markets, merchants are well-organized and use modern marketing methods.

III. On the Basis of Area from which the Buyers and Sellers are Drawn for Transactions

According to this criterion, the agricultural markets are categorized as:

1. Local or Village Markets

In these markets, the buyers and sellers involved in transacting the agricultural produce are drawn from local area or same village or nearby villages. These markets deal with same type of commodities over a period of time, as there is no much variation in the cropping pattern in and around the surrounding villages. In general, these markets deal with perishable commodities like milk, vegetables, paddy, pulses etc. In this market, the seller is able to learn a great deal about the consumers and make necessary changes quickly. Naturally, the total potential of the market is limited and there is also the possibility that a new competitor or any environmental risk factor can put a local seller out of business. These markets are small and cater for a limited number of near-by households. These markets are informal and emerge at cross-roads to facilitate the exchange of products among local farmers and local consumers using commonly money as a means of exchange. The access to the local market is easy, but the supply at the market is very limited and fluctuates according to the season. Thus, these local markets cannot be relied upon to guarantee food security during shortage of food in the local area. These markets are not connected to the national markets and little attempt is made to engage with the larger markets in the region. Even though the demand for better cold storage facilities for perishable products and working capital requirement for expanding the business at the local market is high, the investments are not viable considering higher running costs, no access to main grid electricity and frequent power cuts in the villages. Also the expansion of business with increased capital is unlikely to succeed, as the customer base at the local market is fixed and it becomes viable only, if there are other sources of cash income in the village.

2. Regional Markets

In these markets, the buyers and sellers are drawn from a larger area compared to local markets. Usually, the produce assembled at local markets is purchased by the local traders in large quantities and transport the same to regional markets. Thus, the volume of transactions of agricultural produce is more in regional markets compared to local markets. The number of buyers and sellers involved in transactions are considerable higher in regional markets compared to local markets. The produce transacted at regional markets is further transported to distant areas depending upon the extent of volume of business handled by the buyers and available market price information. The business is conducted on systematic lines compared to local markets with due emphasis on different marketing functions like efficient transportation, scientific storage, processing, grading and standardization, market information etc. These markets deal with commodities like rice, wheat, chillies, cotton (kapas) etc. When there is significant differences in demand for the commodities, the sellers in these regional markets has to make extensive adjustments in their marketing strategies. These markets are often the largest markets available to the consumers and what the regional markets do not supply, the consumers go without. The supply at the regional market is more reliable than at the local markets, even though temporary shortage of goods may arise during the year. The producers may come to these markets to sell their products, but most often the trade is run by professional traders, who collect the produce from the local farmers (either at the farm gate or at the market) and who come to the market every day of the week. The markets are well organized and regulated and the traders must often licensed to trade at the market. Traders in regional markets commonly collect goods from a large geographic area including other regional markets, local small and large-scale producers, as well as collection points in food surplus areas. The traders at the markets can be classified into large, medium and small-scale traders depending on the size of the business they handle.

3. National Markets

In these markets, the buyers and sellers are drawn from the entire nation for transacting the commodities. Thus, the buyers and sellers from all over the country take part in buying and selling activities. There will be stiff competition between the buyers and sellers in transacting the commodities and it ensures competitive prices for the commodities. The prices so settled at national level form the basis for assessing the export competitiveness of commodities in the international market. The marketing functions are so dynamic in the sense that, they are performed considering updated technology, so as to meet the quality consciousness of the consumers. The sellers in these markets enjoy a good distribution net work and extensively concentrate on 'marketing mix' strategies. This market offers tremendous profit potential to the sellers, but at the same time it exposes them to new aggressive competitors. The national market collects produce from all regions within a country. Even though the national market, in a broader sense, can refer to trade between the regional markets, the poor infrastructure between the regional markets is currently limiting such transactions. The national market is dominated by large-scale traders working with small-scale trade facilitators. The national market can be characterized by a large number of small-scale producers and local traders, a few large-scale traders (who are able to finance transport and marketing costs), a large number of small-scale retailers and final consumers. Due to the bottle neck of capital required to buy and transport large quantities of goods, the national market is more limited in access than the local and regional markets. At the same time, it is also larger in volume and provides producers with cash income and opportunities for traders to expand their business. Still, the link from the national market of food crops to the international market of processed food products needs to be strengthened to ensure the dynamic nature of the market, and possibilities for expansion in the future.

4. International Market

In this market, the buyers and sellers are drawn from the entire world. Thus, the buyers and sellers from the whole world meet and exchange their goods and services and they operate in more than one country. The commodities can be bought and sold from different places in the world. Compared to the above markets, this market is biggest in terms of number of commodities handled, number of firms involved in the transactions and from the area point of view. The export trade is dominated by large-scale traders as in the national market, with the distinction that, at the export market, the traders are usually foreign dealing with unconstrained supply of capital and considerable market power to affect prices. These traders collect large quantities of goods from established buying centers and transport them (shipped) out of the country. With the exposure to international market, several marketing functions like processing, grading and standardization, market intelligence etc., gained more significance. More emphasis is given for Sanitary and Phyto-Sanitary (SPS) Standards and Technical Barriers to Trade (TBT) regarding the trade in the international market. Besides quality standards of the products, socio-cultural differences also influence the marketing strategies. It is essential to note that, international marketing differs from global marketing in some very definite ways.

International marketing means that, marketing decisions are made in the individual countries suiting to the target customers with the staff, who is the well-versed about the target markets. For example, McDonald Company markets its products with a slogan, *'I'm lovin'* in USA, whereas, the same company uses the slogan, *'C'est tout que j'aime'* in France. Thus, McDonalds is a bit closer down to the people in each country. But, Global marketing views the whole world as one and creates products that will fit into any regional market place. For example, Nike's slogan *'Just Do It'* is same anywhere in the world *i.e.*, either in USA or France or India. Thus, Global Marketing is a lot different from International marketing. Global marketing is popularly defined as *'marketing on a world-wide scale reconciling or taking commercial advantage of global operational differences, similarities and opportunities in order to meet global objectives.'* So, basically, global marketing is selling a firm's product all over the world, as the company looks at the entire world as one market. There are no differences between a local market and a foreign market and it views everything in the same way in marketing the product. Global marketing is just like a general store marketing the same exact products all over the world and the products transacted in each store irrespective of the location of the market is same. Thus, Global marketing will not bring anything different or new to the store and the products transacted cater to the needs of different people irrespective of their religion or cultural group. To become a global company, a company will focus on the "Four P's of marketing" *viz.*, Product, Price, Place and Promotion. It is important that, it is not easy for a company to become a global company, as it has to pass through several steps like establishing a global team and to have a global marketing plan. So, it takes considerable time for a company to evolve from a local company to one that sells products all over the world suiting to the needs of the targeted customers all over the world irrespective of their religion and socio-cultural groups.

This is where, International marketing differs from Global marketing. International marketing is researching and communicating within each country the company is targeting. That is, in International marketing, a company that is based in one country decides to sell products in another. It sets up offices and headquarters in the other countries, but the headquarters in the specific country cater the business needs of the company. So, the product is still manufactured in the home country, sold by their people in other countries and the profits are taken back to that country. The products sold under International marketing are meant to suit to the requirements of the targeted customers across the countries. So, it is not like a general store, as discussed under Global marketing. The advantage of the International marketing is that, the marketing personnel are usually native to the country and so are familiar with traditions, customs and ideas that are best suited for the target market. They execute most of the business decisions, but they have to report to the main headquarters in the country, where the business is established. So, when a company is started in one location and decides to expand, it has to analyze the prospects of selling its products in different countries. Before that, it has to examine its products and how best it can portray the products suiting to the needs of different customers across the countries. So, considering the pros and cons for global marketing versus international marketing, the company has to decide, which would be better for the company and product.

IV. On the Basis of Competition

The detailed classification of markets based on market competition is depicted in Figure 3.1. It is clear that, the markets are broadly classified into two types *viz.*, Perfect markets and Imperfect markets. Under imperfect market structure, there are different forms of market competition *viz.*, Monopoly, Duopoly, Oligopoly and Monopolistic competition. The basic features of these different types of market structures are presented through Table 3.1.

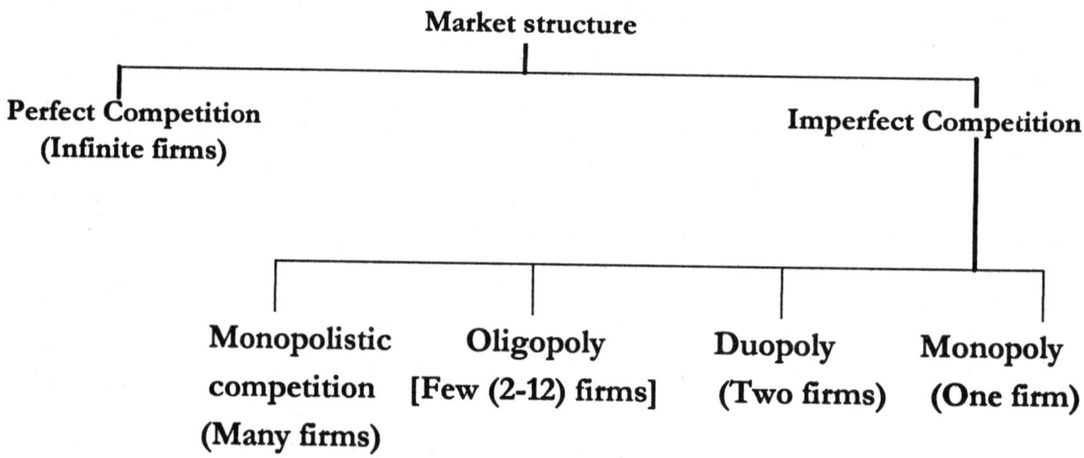

Figure 3.1: Different types of market structures based on competition.

The above classification of markets into different structures is with reference to sellers' side (firm's side). But, with reference to buyers' side, the following is the classification corresponding to sellers' view.

1. Perfect Competition

A market is said to be perfect, when all the potential buyers and sellers of the product have complete knowledge about the prices of the commodities of homogeneous nature at which transactions takes place. Under such circumstances, same price will prevail for the same commodity at the same time. No single buyer or seller has much effect on the going market price. A seller cannot charge more than the going price because, buyers can obtain as much they need at the going price. In a perfectly competitive market, marketing research, product development, pricing, advertising and sales promotion play little or no role. Thus, sellers in these markets do not spend much time on the marketing strategy. The following are the important features of perfect competition:

Table 3.1: Basic features of different types of market structures

Market Structure	*Number of Firms*	*Degree of Product Differentiation*	*Firms Degree of Control Over Price*	*Examples*
I. Perfect competition	Infinite number of firms	Homogeneous nature	No control over price	Agricultural commodities, Internet service
II. Imperfect competition				
a. Monopoly	One firm	Product with no close substitutes (homogeneous)	Considerable control over price	Microsoft is held to have monopolized market for computer operating systems, Public utilities like railways, electricity, irrigation etc.
b. Duopoly (A pure duopoly occurs very rarely in practical world and the more common is two dominate firms hold majority of the market share)				
◆ Perfect Duopoly	Two firms	Homogeneous	Some control over price	Visa and MasterCard control a large proportion of the electronic payment processing market, Hindustan Petroleum vs Bharat Petroleum
◆ Imperfect Duopoly	Two firms	Heterogeneous	Some control over price	Pepsi vs. Coca-Cola in soft drink market, Gillette vs. Wilkinson Sword in razor blade market, Kodak vs. Fujifilm in the color 135 film market
c. Oligopoly				
◆ Perfect Oligopoly	Few firms	Homogeneous	Some control over price	Organization of Petroleum Exporting Countries (OPEC), Steel industry in USA, Cement.
◆ Imperfect Oligopoly	Few firms	Heterogeneous	Some control over price	Automobile industry in USA, Computers, Electrical appliances, Television etc.
d. Monopolistic competition	Many firms (but less than perfect competition)	Heterogeneous	Some control over price	Retail outlets of fertilizers, pesticides, restaurants, clothing, shoes etc.

- ☆ Infinite number of buyers and sellers
- ☆ The products are homogeneous products and they are perfect substitutes. So, Cross elasticity of demand (E_C) is infinity.
- ☆ Free entry and exit of firms in the industry
- ☆ There is perfect knowledge about the product for buyers and sellers
- ☆ Firms are 'price takers' and each firm face a perfectly elastic demand curve for their product
- ☆ All buyers and sellers are completely aware of the price of product
- ☆ No individual firm or seller can influence the price of the product

Classification of markets

(Sellers' view)

- **Perfect competition**
- **Imperfect competition**
  - **Monopoly**
  - **Duopoly**
  - **Oligopoly**
  - **Monopolistic competition**

Classification of markets

(Buyers' view)

- **Perfect competition**
- **Imperfect competition**
  - **Monopsony** (*e.g.*, Cotton Corporation of India procuring cotton from farmers, Employment of nurses etc)
  - **Duopsony**(e.g., Two firms who dominate in selling the refined petrol also likely to dominate the input market for buying crude oil)
  - **Oligopsony**(e.g., Sugarcane market, Jute market, Few firms who dominate in manufacturing and selling the computers also likely to dominate the input market for buying computer components, such as silicon chips, hard disk drives, programmers etc.)
  - **Monopsonistic competition:**(e.g., Firms engaged in hiring skilled labour in agricultural sector)

- ☆ Same price will prevail for the same commodity at the same time
- ☆ No Government regulation on prices of the products
- ☆ Perfect mobility of factors of production
- ☆ Average Revenue Curve (ARC) or demand curve is perfectly elastic (parallel to X-axis), and Marginal Revenue Curve (MRC) coincides with ARC for a perfect competitive firm.
- ☆ Complete absence of transportation costs

As the firms under perfect competition are simply price takers, thereby, they have no market control and receive the same or going market price for all the output sold. So, at the same price, the firms have to transact all the output and this implies, the demand curve or ARC are perfectly elastic. Since, the firm has no control over the price, it has to sell the additional units of output at the same price, thereby, MRC is also constant and it is equal to ARC. So, for a perfect competitive firm, both ARC and MRC are equal and when ARC is constant, MRC must be constant. The same concept is shown through Table 3.2.

In the Table 3.2, it was shown that, the farmer transacted various quantities of paddy at the same price, as under perfect market competition, same price will prevail for the same commodity at the same time. From the quantity-price data, the Total Revenue (TR), Average Revenue (AR) and Marginal Revenue (MR) are computed. TR is computed in the third column as the product between quantity of paddy transacted and price received. AR is computed in the fourth column by dividing the TR with quantity of paddy transacted. The obvious point here is that, AR is equal to ₹ 1000 for each and every unit (quintal) of paddy sold. The reason is that, price is constant at ₹ 1000 for each and every unit (quintal) of paddy sold. So, AR is price or price is AR and these two are one and the same. MR is

Table 3.2: TR, AR and MR of a perfect competitive firm transacting paddy.

Output (Quintals)	*Price (₹/quintal)*	*TR (₹)*	*AR (₹/quintal)*	*MR (₹/quintal)*
0	1000	0	–	–
1	1000	1000	1000	1000
2	1000	2000	1000	1000
3	1000	3000	1000	1000
4	1000	4000	1000	1000
5	1000	5000	1000	1000
6	1000	6000	1000	1000
7	1000	7000	1000	1000
8	1000	8000	1000	1000
9	1000	9000	1000	1000
10	1000	10000	1000	1000

calculated by dividing the change in TR by the change in quantity of paddy sold. For example, when the farmer sells one quintal of paddy, he receives ₹ 1000 MR and if he sells 10th quintal of paddy, still he receives ₹ 1000 MR. So, the MR is equal to ₹ 1000 for every extra quintal of paddy sold. This results because, price is constant at ₹ 1000 for every quintal of paddy sold. So, for every extra quintal of paddy sold by the farmer, he receives an extra ₹ 1000, which is the going market price. So, MR is equal to price and AR for a perfectly competitive firm. The same results are shown through Figures 3.2 and 3.3, where the vertical axis measures the revenue and the horizontal axis measures the quantity of output (paddy) sold by the farmer. The ARC and MRC curves are perfectly elastic and hence, they lie parallel to X-axis. That means, if the farmer sells one quintal of paddy, he receives ₹ 1000 AR and ₹ 1000 MR. If the farmer sells 10 quintals of paddy, still he receives ₹ 1000 AR and ₹ 1000 MR (for 10th quintal), but the TR increases from ₹ 1000 to ₹ 10000. This implies that, TR continuously slopes upwards with increase in quantity of paddy sold by the farmer, but the AR and MR received by the farmer remains same irrespective of the quantity of paddy transacted by the farmer. So, MRC coincides with ARC. Another interesting aspect here is, under perfect competition, the AR and MR coincides with price of the output. So, AR is just another term for price and this ARC is actually the demand curve. Under the perfect market competition, the MRC of the farmer is also his demand curve, as MRC coincides with ARC. But in imperfect competition, ARC alone indicates the demand curve for the farmer, as MRC is negatively sloped and lies below the ARC (demand curve) and the same is discussed in the ensuing pages.

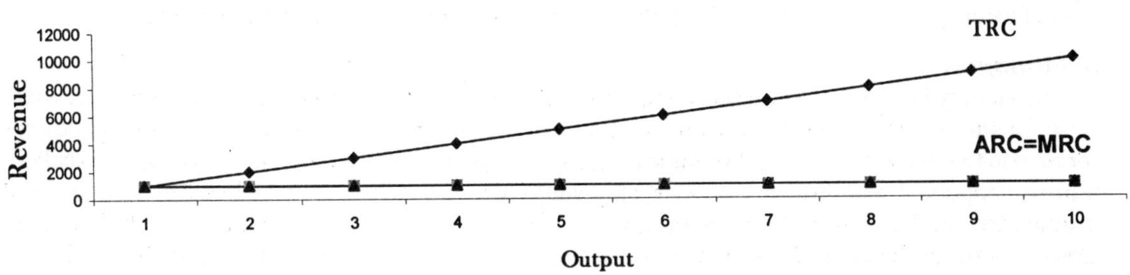

Figure 3.2: TR, AR and MR curves of a firm under perfect market competition.

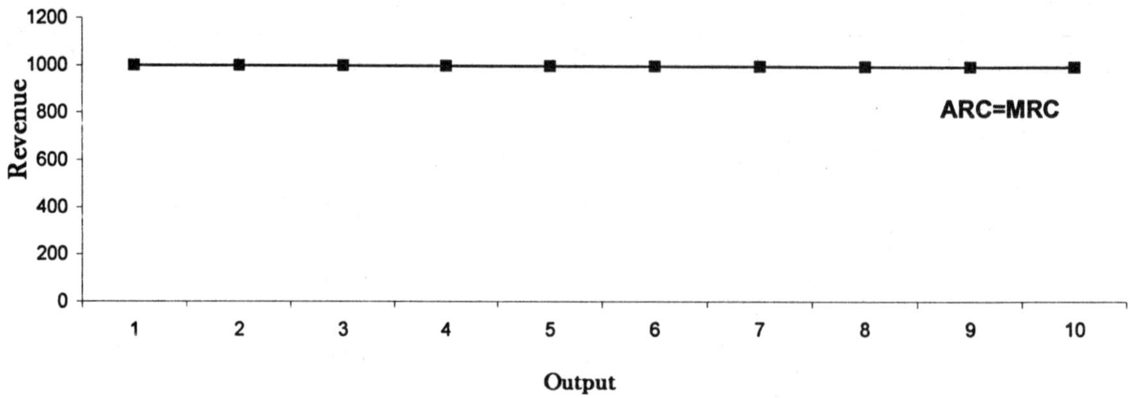

Figure 3.3: ARC and MRC of a firm under perfect market competition.

Pure Competition and Perfect Competition

Economists like Chamberlin and others often make a distinction between Pure competition and Perfect competition. The term 'Pure competition' (coined by Chamberlin) is used in a restricted sense. In order that competition be 'pure', it requires the fulfillment of first three features of perfect competition *viz.*, infinite number of buyers and sellers, homogeneous nature of the product and free entry and exit of firms. These conditions together mean that, no individual firm can exert any influence over the market price. But, the term perfect competition is a wider concept, in the sense, that it includes the features of pure competition and some additional features as mentioned earlier. This means, perfect competition requires that, there should be no imperfections in the market. Such imperfections arise due to imperfect knowledge or immobility of the factors of production. In fact, pure competition is a part and parcel of perfect competition. American Economists prefer to use the term pure competition, while the English Economists prefer the term perfect competition. However, both the terms are used interchangeably to analyze the features of perfect markets.

2. Imperfect Competition

As discussed earlier, perfect competition describes a market structure with extreme assumptions, thereby, it is highly unlikely to exist in the practical world. This is because, in reality, most markets are imperfectly competitive. So, any deviation from the conditions of perfect competition in a market leads to the existence of imperfect competition *i.e.*, less than infinite number of sellers, homogeneous or heterogeneous nature of the product, absence of price information, restrictions on free entry and exit of firms, existence of transportation costs etc. Depending upon the degree of imperfections, the different market forms include, monopoly, duopoly, oligopoly and monopolistic competitions.

a. Monopoly

Monopoly is a form of imperfect competition in which, a single firm produces the product with no close substitutes. That means, there are no other firms in the economy that make a similar product (or service) in the market, thereby, the single firm represents an industry and such monopoly is called as 'Pure monopoly'. Such 'Pure monopolies' are relatively rare, but they do exist. Local electric and water utilities are good examples for pure monopolies, as they produce a product for which there are no close substitutes. It is important that, a firm's degree of monopoly may vary among the markets. For example, an electric company enjoys monopoly status in the production of electricity meant for lighting purposes, but for heating a commodity or product, the firm has to face stiff competition from gas, coal and fuel oil supplying industries. So, the electric company may consequently be a 'pure monopoly' in the lighting market, but do not enjoy monopoly status in the heating market. Similarly, railway transport do not enjoy monopoly status, as there are other means of transport for the customers to move to distant places such as air lines, buses etc. To generalize, by law, a monopoly occurs, if a firm has a market share of more than 25 per cent.

The fundamental cause of monopoly is due to the existence of strong barriers for entry of new firms into the industry. The barriers may be technological or economical that raise the cost for the new firms to enter the industry above the cost of a firm enjoying monopoly status in the market. If the barriers to entry prevent others firms from entering the market, there is no competition and the monopolist remains the only seller for the product in the market. So, the monopolist can either decide the output or price for the product. Here lies the major difference between a monopoly and a perfect competition. A monopolist has the ability to influence the price of the product (price maker), whereas, the firm under perfect competition is simply a price taker rather than a price maker. This is because, under perfect competition, the competitive firm enjoys a very small share in the total market, thereby, the price of the product is determined by market conditions (industry) and this price is passed on to each individual firm. On the other hand, under monopoly competition, a single firm is the sole producer of the product in the market, thereby, it can often alter the price of the product by adjusting the quantity of supply of output into the market. The following are the important features of monopoly:

- ☆ A single firm or seller selling all output in a market *i.e.*, the product have only single control
- ☆ A unique product, thereby, the E_c for the product is zero.
- ☆ A monopoly often acquires and generally maintains single seller status due to restrictions on the entry of other firms into the market.
- ☆ The monopolist is either a price maker or output decider.
- ☆ The distinction between firm and industry will disappear in monopoly.
- ☆ Power to influence price is the essence of monopoly.
- ☆ There are absolutely no rivals, as monopoly firm represents the industry.
- ☆ A monopoly firm often possesses information not available to others. This specialized information comes in the form of legally-established patents, copyrights or trademarks etc.
- ☆ Because of the presence of only seller, the firm faces a downward-sloping demand curve (*i.e.*, industry demand curve) and this implies that, the monopolist sells more of the output at lower price.

Let us assume, a monopoly firm enjoys patent for marketing Bt cotton seed in the market and it offers various quantities of seed for sale at different prices and the same are shown through Table 3.3. From the quantity-price data, the TR, AR and MR are computed. TR is computed in the third column as the product between quantity of seed transacted and price received. AR is computed in the fourth column by dividing the TR with quantity of seed transacted. The obvious point here is that, AR decreases with the increase in quantity of seed transacted. Moreover, AR is also equal to the price of seed for each quantity sold. For example, when the firm sells one unit of seed, it receives ₹ 1000 as AR and it is equal to the price of the seed *i.e.*, ₹ 1000. Similarly, if the firm sells 10 units of seed, it receives ₹ 550/unit as AR and it is equal to the price of the seed *i.e.*, ₹ 550/unit. So, AR is price or price is AR and these two are one and the same.

Table 3.3: TR, AR and MR of a monopoly firm selling Bt cotton seed.

Quantity (Units)	*Price (₹/unit)*	*TR (₹)*	*AR (₹/unit)*	*MR (₹/unit)*
0	1050	0	–	–
1	1000	1000	1000	1000
2	950	1900	950	900
3	900	2700	900	800
4	850	3400	850	700
5	800	4000	800	600
6	750	4500	750	500
7	700	4900	700	400
8	650	5200	650	300
9	600	5400	600	200
10	550	5500	550	100
11	500	5500	500	0
12	450	5400	450	–100

MR is calculated by dividing the change in TR by the change in quantity of seed sold. For example, when the firm sells one unit of Bt cotton seed, it receives ₹ 1000 MR and if it sells 10th unit of seed, it receives ₹ 100 MR. So, the MR decreases with the increase in quantity of seed sold by the firm. Further, the MR is less than the price (AR) of seed for each quantity sold. This is because, any decline in an average, which is most certainly happening to AR, is associated with a marginal that is less than the average. That means, when AR is decreasing, MR must be less than AR. For example, the price at which three units of seed is sold is ₹ 900/unit, but the MR of the firm is only ₹ 800/unit. Similarly, the price at which ten units of seed is sold is ₹ 550/unit, but the MR of the firm is only ₹ 100/unit. And because, price is AR, MR is also less than AR.

The Table 3.3 further infers that, the rate of fall in MR is more compared to price or AR, and this gives more steepness to the slope of MRC compared to ARC (Figure 3.4 and 3.5). So, both AR and MR decreases with increase in quantity of seed sold by the firm, thereby, both ARC (demand curve) and MRC are negatively sloped. As the slope of MRC is steeper compared to ARC, it implies, ARC is more elastic compared to MRC. That means, if the price of seed falls in the market, MR falls more rapidly than AR, but the TR of the firm increases continuously and reaches maximum, when MR is equal to zero. Thus, ARC and MRC for the firm enjoying monopoly status look different when compared to perfect competition. For a monopoly firm, both ARC and MRC are negatively-sloped and the MRC lies below the ARC, which means that, MR is less than AR (and price) for any given quantity and larger quantities of output are sold only at lower prices, whereas under perfect market competition, MRC coincides with ARC and price and they are perfectly elastic.

b. Oligopoly

The word 'Oligopoly' is derived from the Greek word '*oligos*' which means few. When the whole market structure of few firms controls the major share of market sales, then the resulting structure is referred as Oligopoly. In this form of competition, the entire market is dominated by as few as two firms or as many as twelve firms. These few firms are interdependent on each other with respect to investment, output and pricing decisions. That means, each firm is aware of the actions of the rivalry firms.

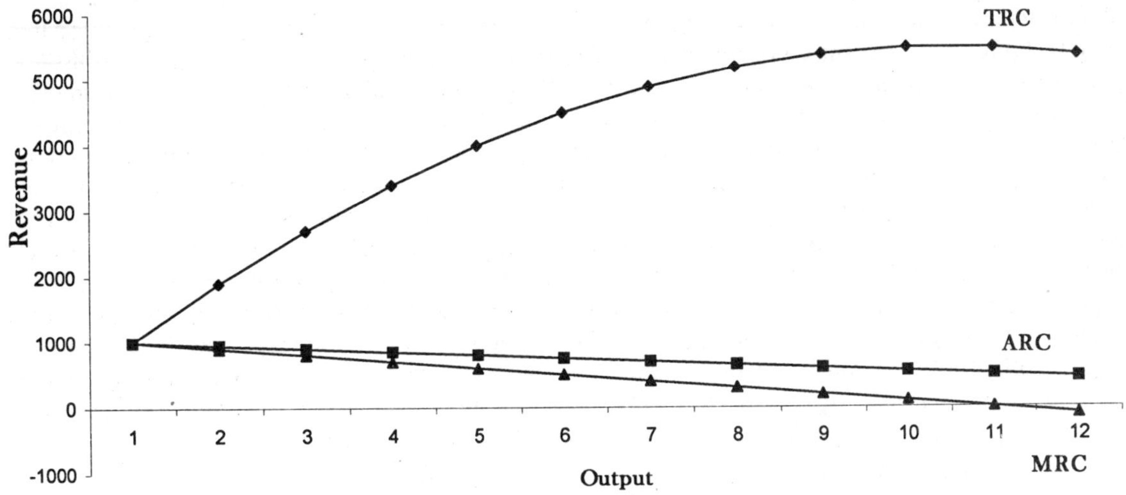

Figure 3.4: TR, AR and MR curves of a monopoly firm selling *Bt* cotton seed.

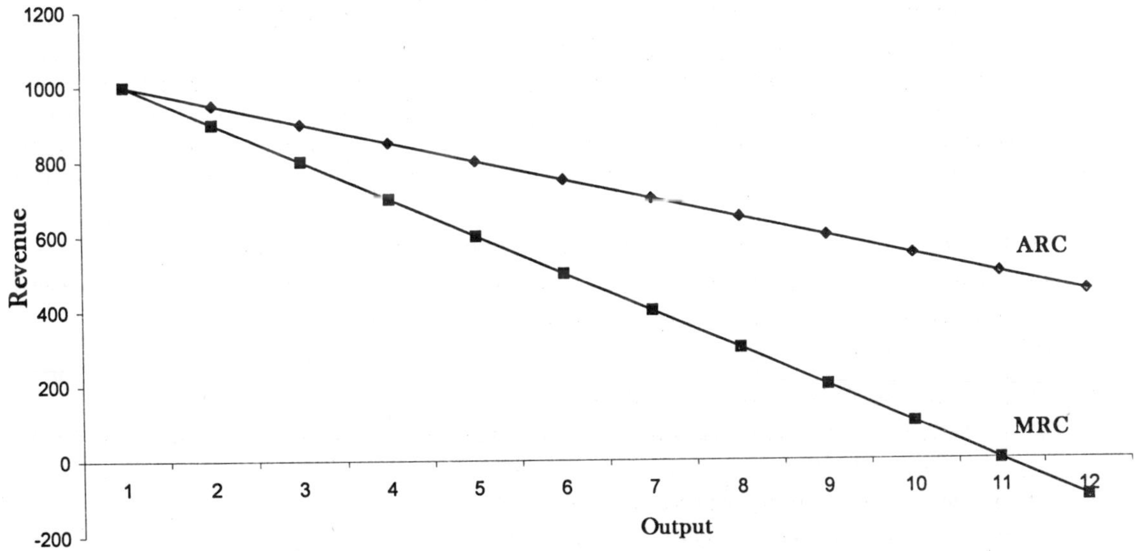

Figure 3.5: ARC and MRC of a monopoly firm selling *Bt* cotton seed.

Normally, an oligopoly market exists, when the top five firms in the market account for more than 60 per cent of the total market sales. With fewer than two firms, the market structure is Monopoly. The simplest form of oligopoly is Duopoly, where there are only two firms producing and selling the product. For example, Pepsi vs. Coca-Cola in soft drink market, Gillette vs. Wilkinson Sword in razor blade market, Kodak vs. Fujifilm in the colour 135 film market etc., are the good examples for Duopoly market. But, as the number of firms is many, then Oligopoly becomes Monopolistic competitions.

Oligopoly competition dominates the modern economic landscape, accounting for about half of all output produced in the economy. Oligopolistic industries deals with a wide range of products like breakfast cereals, cars, computers, television broadcasting, pharmaceuticals, petroleum etc. OPEC is an example of Oligopoly since few countries control the production of oil, the steel and the automobile industry in United States of America is another example. These industries tend to be a prime source of innovations that promote technological advances and economic growth. The oligopolies may be Perfect oligopolies and Imperfect oligopolies with respect to differentiation of products. In a 'perfect oligopoly' such as the oil industry, there is little or no differentiation in terms of the product sold. But, in an 'imperfect oligopoly' like the IT and automotive industry, there's lot of differentiation between the products. The pricing strategies followed in perfect oligopolies vary greatly compared to imperfect oligopolies. However, the differentiations across the products are often short-lived, as the market is attuned to replicating originality, thereby, imperfect oligopolies becomes more perfect in due course of time. So, oligopoly is a market structure characterized by a small number of large size firms that dominate the market, selling either identical or differentiated products, with significant barriers to entry into the industry. The following are the important features of oligopoly:

- Presence of few sellers in a market situation for a commodity
- Interdependency among the firms for decision-making
- The individual firm faces an indeterminate demand curve, as the price elasticity depends on the rival's reaction to changes in investment, output and price
- Firms sell branded products, which are close substitutes of each other
- High selling costs is the unique feature of this competition. So, the non-price competition is a consistent feature of the competitive strategies of oligopolistic firms. Examples of non-price competition includes *viz.*, free deliveries and installation, extended warranties for consumers and credit facilities, longer opening hours (supermarkets and petrol stations), branding of products and heavy spending on advertising and marketing, extensive after-sales service etc
- The existence of both the co-operative and competitive trends among the business firms
- Entry barriers for the other firms are high and the barriers can be due to patents, copyrights, Government rules and regulations, ownership of scare resources etc
- Lack of uniformity in size of firms
- Products may be homogeneous (standardized) or heterogeneous (differentiated).
- The sellers are the price makers and not price takers, since the few sellers mutually dominate the pricing decisions
- No unique pattern of price behaviour
- The firms tend to keep the prices at the same level or rigid.

Even in oligopoly competition, the ARC and MRC slopes downward from left to right and MRC lies below ARC (Figure 3.6). But, the interesting aspect is that, the ARC or demand curve will bear a kink such that, above kink 'K', the ARC is more elastic for price rise and below kink, the ARC is less elastic for price fall. Due to differences in elasticities along the ARC, the MRC will show discontinuity below the kink.

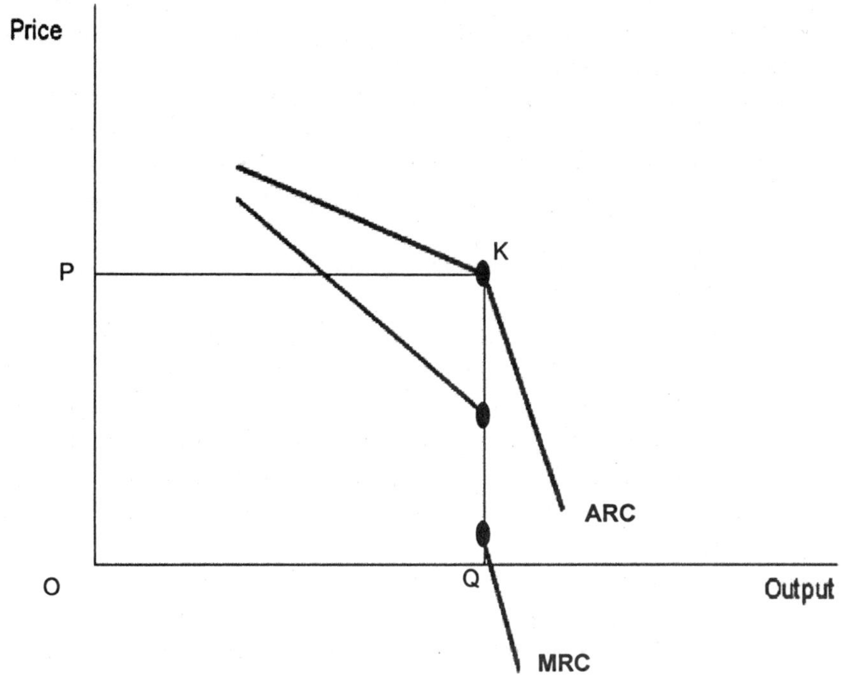

Figure 3.6: ARC and MRC of an oligopoly firm.

c. Monopolistic Competition

Monopolistic competition refers to a market competition, in which large number of firms sells differentiated products. Since the products produced by these firms are close, but not perfect substitutes to each other, there exists stiff competition among the firms. It is important that, though the number of firms are larger in monopolistic competition, but they are not as large as in perfect competition. This form of market competition is extremely common. Perhaps, half of the economy's total production comes from monopolistically competitive firms. Almost all retail operations in the economy are in this form of market competition. For example, retail outlets of fertilizers, pesticides, restaurants, clothing, shoes etc., fall under this category. Under this competition, starting a business is relatively easy, but staying in the business is quite difficult, as the firm requires an ability to convince the customers that the product is different and better than that of competitors. Considering the above features, it infers that, the market, which consists of large number of buyers and sellers who trade over a range of prices rather than a single market price is called monopolistic competition. A range of prices occurs because, sellers can differentiate their products offered to the buyers. This product differentiation is usually achieved in one of three ways *viz.*, physical differences (differences in size, shape, colour of the product), perceived differences (supplying the products in tins, bottles, packets etc) and support services (repairs and maintenance services, offering discounts etc). This product differentiation is the primary reason that each firm operating in a monopolistically competitive market is able to create a little monopoly all to itself.

Perfect competition and Monopoly are the two extreme forms of market competition, the former being extreme form of perfection and the latter being extreme form of imperfection. However, these two competitions are purely theoretical and not practically seen. Between these two extremes, there is a monopolistic competition, which is more common and practically seen in the economy. So, in effect, monopolistic competition is something of a hybrid between perfect competition and monopoly and this combines the features of both perfect competition and monopoly. Monopolistic competition is the best approximation of perfect competition that the real world offers. Edward Hastings Chamberlin is regarded as the 'Founding Father' for studying the concepts of monopolistic competition in his famous work, '*Theory of Monopolistic Competition* (1933)'. The following are the important features of monopolistic competition:

- There are large number of sellers producing differentiated products and no firm has total control over the market price.
- Product differentiation is the essence of monopolistic competition *i.e.*, every producer tries to keep his product dissimilar than his rival's product in order to maintain his separate identity.
- There is more freedom of entry and exit of firms into the market. This leads to stiff competition and helps to maintain normal profits in the market for a longer span of time.
- Due to product differentiation, every firm has to incur some additional expenditure in the form of selling cost. This cost includes sales promotion expenses, advertisement expenses, salaries of marketing staff etc.
- The firms produce very close substitutes, thereby, E_C is very high.
- Large numbers of firms are different in their size. Each firm has its own production and marketing policies. So, no firm is influenced by other firm and each firm is independent in its output-pricing policies.
- There is two dimensional competition in this market *viz.*, price competition (firms compete with each other on the basis of price) and non-price competition (firms compete on the basis of brand, product quality, advertisement etc).
- The firm faces a downward sloping demand curve *i.e.*, elastic demand curve. It means, the firm can sell more output at lower price and *vice versa*. The demand curve in monopolistic competition is more elastic compared to monopoly.
- Each firm produces similar, but not identical goods. So, each firm is an industry itself. Since products are not homogeneous, there is no industry, as in the case of perfect competition. So, Chamberlin introduced the term 'group' under monopolistic competition. A group is a cluster of firms producing very closely related, but differentiated products. That means, the collection of firms that produce same type of products with high E_C will constitute a 'group' or a 'product group'.
- The firm under monopolistic competition is a price maker.

So, the features of a monopolistically competition are almost the same as in perfect competition, but the difference is, in monopolistic competition, the products are heterogeneous and thereby, involves a great deal of non-price competition. Further, the individual firm's demand curve is downward sloping in monopolistic competition, in contrast to perfect competition, which is perfectly elastic.

Let us assume, a retail firm operating under monopolistic competition selling fertilizer in the market and it offers various quantities of fertilizer for sale at different prices and the same are shown through Table 3.4. From the quantity-price data, the TR, AR and MR are computed. TR is computed in the third column as the product between quantity of fertilizer transacted and price received. AR is computed in the fourth column by dividing the TR with quantity of fertilizer transacted. The obvious point here is that, AR decreases with the increase in quantity of fertilizer transacted. Moreover, AR is also equal to the price of fertilizer for each quantity sold. For example, when the firm sells one unit of fertilizer, it receives ₹ 1040 as AR and it is equal to the price of the fertilizer *i.e.,* ₹ 1040. Similarly, if the firm sells 10 units of fertilizer, it receives ₹ 950/unit as AR and it is equal to the price of the fertilizer *i.e.,* ₹ 950/unit. So, AR is price or price is AR and these two are one and the same.

Table 3.4: TR, AR and MR of a firm selling fertilizer under monopolistic competition.

Quantity (Units)	*Price (₹/unit)*	*TR (₹)*	*AR (₹/unit)*	*MR (₹/unit)*
0	1050	0	–	–
1	1040	1040	1040	1040
2	1030	2060	1030	1020
3	1020	3060	1020	1000
4	1010	4040	1010	980
5	1000	5000	1000	960
6	990	5940	990	940
7	980	6860	980	920
8	970	7760	970	900
9	960	8640	960	880
10	950	9500	950	860

MR is calculated by dividing the change in TR by the change in quantity of fertilizer sold. For example, when the firm sells one unit of fertilizer, it receives ₹ 1040 MR and if it sells 10th unit of fertilizer, it receives ₹ 860 MR. So, the MR decreases with the increase in quantity of fertilizer sold by the firm. Further, the MR is less than the price (AR) of fertilizer for each quantity sold. This is because, any decline in an average, which is most certainly happening to AR, is associated with a marginal that is less than the average. That means, when AR is decreasing, MR must be less than AR. For example, the price at which three units of fertilizer is sold is ₹ 1020/unit, but the MR of the firm is only ₹ 1000/unit. Similarly, the price at which ten units of fertilizer is sold is ₹ 950/unit, but the MR of the firm is only ₹ 860/unit. And because, price is AR, MR is also less than AR.

The Table 3.4 further infers that, the rate of fall in MR is more compared to price or AR, and this gives more steepness to the slope of MRC compared to ARC (Figure 3.7 and 3.8). So, both AR and MR decreases with increase in quantity of fertilizer sold by the firm, thereby, both ARC (demand curve) and MRC are negatively sloped. As the slope of MRC is steeper compared to ARC, it implies, ARC is more elastic compared to MRC. That means, if the price of fertilizer falls in the market, MR falls more rapidly than AR, but the TR of the firm increases continuously and reaches maximum, when MR is equal to zero. Thus, ARC and MRC for the firm under monopolistic competition look different when compared to perfect competition and monopoly. For a firm operating under monopolistic competition, both ARC and MRC are negatively-sloped and the MRC lies below the ARC, which means that, MR is less than AR (and price) for any given quantity and larger quantities of output are sold only at lower prices. The same is the case under monopoly, but the AR and MR curves are more steeper compared to monopolistic competition (Figure 4.10). As the firm itself represent the industry under monopoly, the steepness of the slopes fetch the firm higher price for the product, if the firm wishes to maximize the

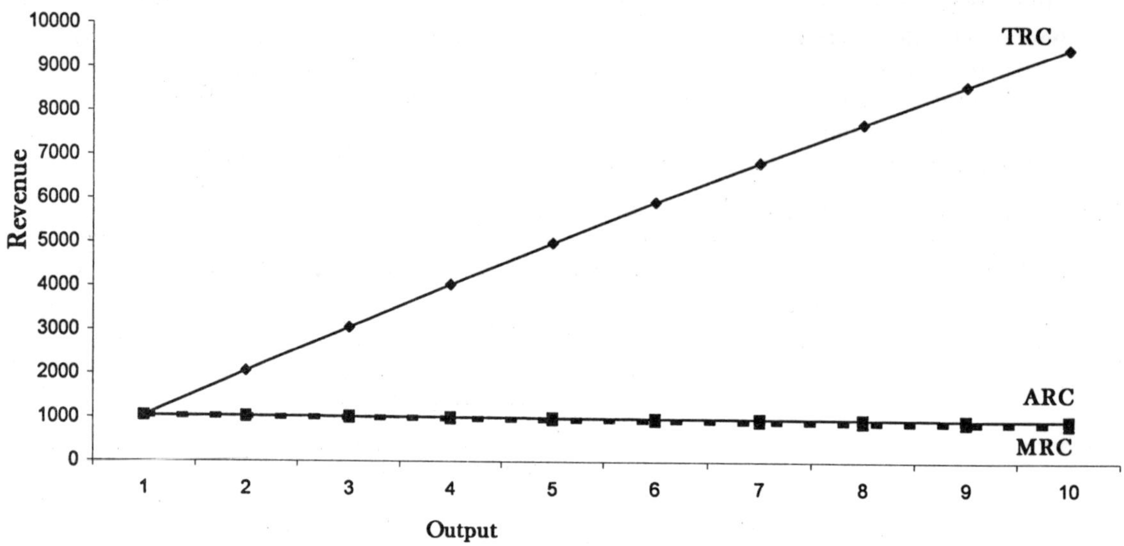

Figure 3.7: TR, AR and MR of a firm selling fertilizer under monopolistic competition.

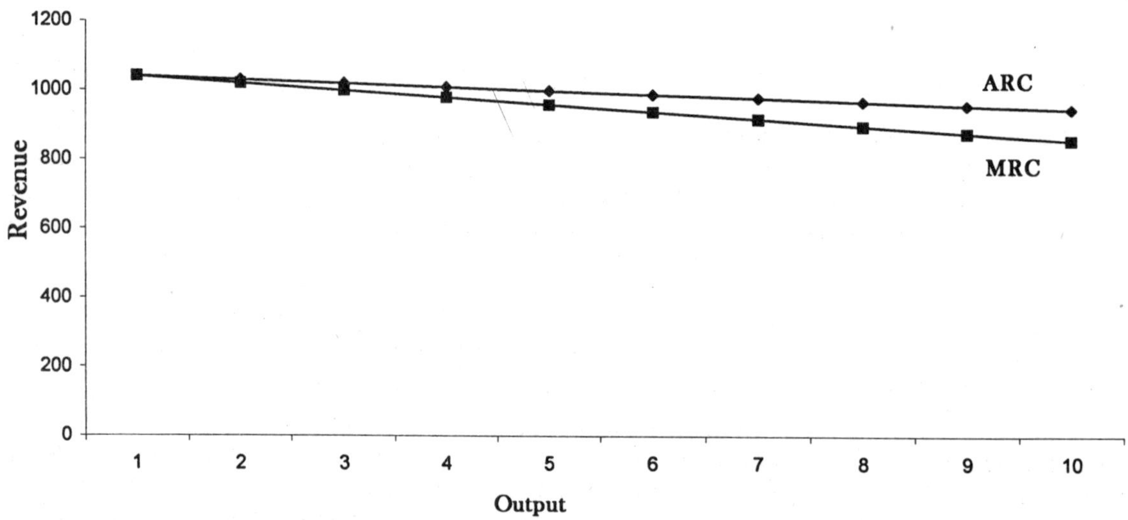

Figure 3.8: ARC and MRC of a firm selling fertilizer under monopolistic competition.

profits through increasing the price of the product. However, under perfect market competition, MRC coincides with ARC and price and they are perfectly elastic.

Competitive Structure vis-à-vis Competitive Behaviour

A perfect competitive market structure is characterized by the presence of infinite number of firms producing and selling a homogeneous commodity. But, each individual firm is unable to influence the price of the commodity in the market, as they are simply price takers. The price of the commodity is determined by the demand and supply forces at the industry level and the same is passed on to the individual firms at which, each individual firm has to transact any amount of output. So, each firm does not need to compete with other firms to obtain a market for its produce. This is clearly the opposite of the meaning of what is commonly understood by the competitive behaviour. For example, in case of imperfect competitive market structure, say duopoly, where two giant firms like Bayer and Syngenta compete with each other in a variety of ways to achieve a higher level of sales or profits. Thus, intense competitive behaviour exists between these two giant firms to gain significant share in the market. Conversely, we do not find individual farmers competing among themselves to sell a larger amount of produce. This is because, both Bayer and Syngenta firms possess the market power to influence the price of pesticides, while an individual farmer does not. Thus, competitive market structure and competitive behaviour are, in general, inversely related. That is, the more competitive the market structure, the less competitive is the behaviour among the firms, as in case of perfect competition. On the other hand, the less competitive the market structure, the more competitive is the behaviour of firms towards each other for gaining significant market share. However, pure monopoly is the most visible exception.

V. On the Basis of Location

Markets on the basis of location they serve can be categorized as follows:

1. Village Markets

Marketing of agricultural produce starts from this market. These markets serve the transactions of one or two villages. Usually, the produce produced in these villages will be brought and transacted in these markets. These markets generally deal with perishable produce like leafy vegetables, fish, milk, flowers etc. These markets are not held on continuous basis, but they are more common at times of harvesting season of agricultural produce. The buyers and sellers are in close private negotiations with each other in market transactions. In these markets, the transactions were made in small quantities, as per the requirements of direct consumers. Itinerant traders may also join the market and purchase the produce at low prices at times of peak harvesting season and transact the same in the neighbouring villages.

2. Primary Wholesale Markets

These markets are held weekly or bi-weekly in mandal/taluq headquarters or nearby towns. They are popularly called as 'Shandi' or 'Haat'. The produce from different villages assembles at these markets. In these markets, the important market intermediary is village trader or primary wholesaler and they directly purchase the fresh produce from farmers and transport the same to distant areas or to processors.

3. Secondary Wholesale Markets

These markets are located at district headquarters. They are popularly klnown as 'Mandi' or 'Gunj'. In these markets, the important market intermediary is secondary wholesaler and they purchase the produce from village trader or primary wholesaler or from processors. Thus, a secondary wholesale market is one, in which, the original buyers of the product resell the product to a third party *i.e.*, secondary wholesalers. The major distinction between the primary wholesaler and the secondary wholesaler is that, primary wholesalers are those wholesalers in the initial phase of distribution who typically purchase fresh produce from the farmers directly. In this primary wholesaling sector, processing activity is involved in transforming the fresh produce into palatable form (if the commodity demands processing function). Thus, primary wholesaling sector also includes processing function.

But, secondary wholesalers are those wholesalers, who purchase the produce (or processed products) from other wholesalers or processors and make final distribution to retailers. Thus, this functionary (secondary wholesaler) is often viewed more as a distribution agency. These secondary wholesalers handle large quantities of products compared to primary wholesalers and they are more specialized in distribution activity.

4. Terminal Markets

Terminal Market is a central site, often in a metropolitan area or cities or big towns, that serves as an assembly and trading place for agricultural commodities in large-scale. Here, produce is either finally disposed off to the consumers or processors or assembled for export. Usually, terminal market is the final consumer market. In this market, the produce or processed product is distributed to final consumers through retail outlets like retail markets, supermarkets, restaurants and hotels. These terminal markets are widely scattered in urban areas thereby, the consumers can have easy access for the products. It is essential to note that, these markets are not specialized, but deal with various kinds of food products like fruits, vegetables and other perishables. In these markets, the prices of products are generally 40-50 per cent higher than in local markets, as these markets perform specialized marketing functions like packaging and packing, scientific storage, grading and standardization etc. These markets may also arrange futures trading for the commodities they dealt with. In India, the Government has taken the initiative to modernize the terminal markets in providing modern infrastructural facilities for electronic auction, cold chain and logistics and operate through primary collection centres conveniently located in producing areas to allow easy access to farmers. The terminal markets are envisaged to operate on a 'hub-and-spoke' format, wherein the terminal market (the hub) would be linked to a number of primary collection centres (the spokes).

5. Seaboard Markets

These markets are located near the sea ports for the purpose of import and export of agricultural commodities or products. These markets are often specialized in dealing with marine products. With the advent of trade liberalization, these markets assume crucial significance in the free trade of commodities that enjoy more comparative advantage in the international market.

VI. On the Basis of Time

On the basis of 'time', markets are classified into the following three types:

1. Very Short Period Market

Very short period is the time period in which supply of commodity remains constant. That means, demand is the sole factor (since supply of commodity is fixed) in determining the pice of the commodity. The price so determined in a very short period market refers to Market price. This market deals with perishable commodities like fish, milk, leafy vegetables, eggs and slightly durable commodities like rice, wheat etc., which can be stocked for a limited period.

a. Very Short Period Market Dealing with a Perishable Commodity

In case of a perishable commodity, supply into the market is fixed and cannot be changed at all. Since supply is fixed, demand plays a decisive role in determining the price of the commodity and it is explained through Panel A of the Figure 3.9. The Panel A reveals that,

- ☆ On X-axis, quantity demanded and supplied of milk and on Y-axis, price of the milk are taken.

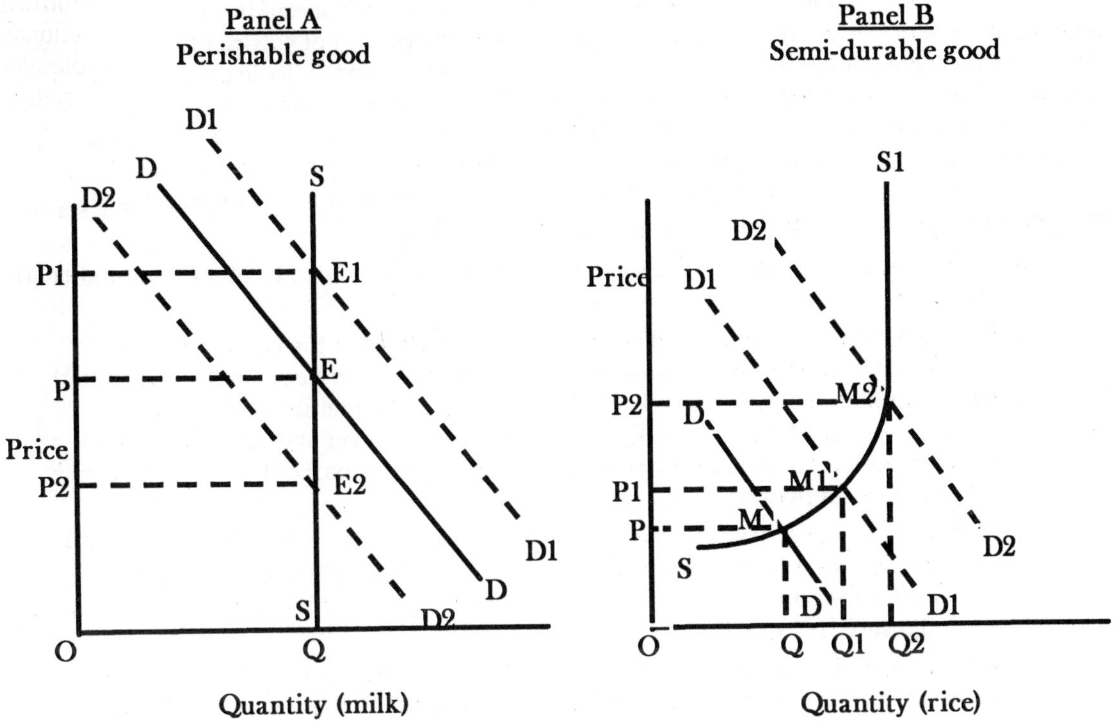

Figure 3.9: Market period (Very short period) price determination under perfect market competition.

- ☆ SS is the supply curve of commodity (milk) and DD is initial demand curve of the commodity (milk). Since, the supply of milk cannot be increased in the market in a very short period, the supply remains fixed and thereby, the supply curve SS remains parallel to Y-axis.
- ☆ These two curves, DD and SS intersect at point E, which indicates that, OQ quantity of milk is supplied and demanded at OP equilibrium price.
- ☆ If the demand for milk is increased, D_1D_1 is the new demand curve (shifts upwards) and it intersects the SS supply curve at E_1, thereby, the equilibrium price is increased to OP_1 (*i.e.*, same quantity purchased at higher price).
- ☆ If the demand for milk is decreased, D_2D_2 is the new demand curve (shifts downward) and it intersects the SS supply curve at E_2, thereby, the equilibrium price is decreased to OP_2 (*i.e.*, same quantity purchased at lower price).
- ☆ Thus, demand plays a decisive role in determining the price of milk in a very short period market.

b. Very Short Period Market Dealing with a Semi-durable Commodity

In this case, the commodity is not perishable, but semi-durable and hence, it can be stocked for a certain period. So, supply of the commodity can be slightly varied in the market depending upon the demand situations, thereby, supply has little influence on the price of the commodity. This market deals with the commodities like rice, wheat, pulses etc. Since, the commodity is semi-durable, the sellers can stock or withhold the commodity, if the market price is not sufficiently high. However, demand still plays a dominant role in influencing the price of the commodity in the market. There is no question of distress sale of the commodity, as the seller can stock the commodity at times of fall in prices. The seller can fix a minimum price below which he will not sell the commodity and that minimum price is called Reserve price or Floor price. If the seller anticipates increase in demand for the commodity in the future, he fixes higher reserve price for the commodity and *vice versa*. This is explained through Panel B of the Figure 3.9. It is clear from the Panel B that,

- ☆ On X-axis, quantity demanded and supplied and on Y-axis, price of the commodity (rice) are taken.
- ☆ SS_1 is the supply curve and DD is initial demand curve of the commodity (rice). The SS_1 supply curve initially slopes upwards indicating that, upto a certain quantity or output of the commodity (OQ_2), the supply can be increased in the market in response to rise in price of the commodity. But, beyond OQ_2 quantity of rice, what ever may be the increase in demand for the commodity, the supply of rice into the market remains fixed, as it cannot be increased in the very short period.
- ☆ Initially, at OP level of equilibrium price, OQ is the quantity demanded and supplied. This OP is the reserve price below which the farmer is not willing to sell the commodity. So, if the demand for the commodity is decreased below DD, price will not fall much and it will be around reserve price only.
- ☆ If the demand for the commodity is increased, D_1D_1 is the new demand curve (shifts towards right). Farmer will increase the supply from OQ to OQ_1 and the new equilibrium price is OP_1. So, the price increases from OP to OP_1.
- ☆ If the demand is still further increased, the new demand curve is D_2D_2 (shifts towards right). The farmer further increase the supply of the commodity to OQ_2 and the new equilibrium price is OP_2. Beyond OQ_2 supply, the farmer cannot increase the supply, as it is relatively fixed now. Beyond this point *i.e.*, at OQ_2 quantity of supply, if demand increases, price increases greatly, as supply has no influence on price. So, the supply curve (SS_1) has three phases:
  - ❖ *Minimum price phase:* SM phase of the curve
  - ❖ *Maximum price phase:* S_1M_2 phase of the curve
  - ❖ *Change in price due to changes in demand and supply of commodity phase*: MM_2 phase of the curve.

So, we can infer that, the farmer sells OQ quantity of rice at minimum price OP and withhold the stock QQ_2 in anticipation of higher prices in the future. At higher price level, OP_2 the farmer disposes all the stocked portion of the commodity (QQ_2).

2. Short Period Market

This market is held for few months period in which supply can be changed in accordance with demand. This is possible by changing the variable factors only in the production programme. In the short period, it is not possible to change the fixed factors, the scale of production and organization. Therefore, supply of output of durable good (cloth) can be increased or decreased to match the increase or decrease in demand by changing the variable factors alone.

In short period, price is determined by the forces of demand and supply. The Short Run Supply (SRS) curve slopes upward from left to right like the ordinary supply curve. It establishes short-run equilibrium price when it is intersected by the demand curve. The short-run price is also known as the short-run normal price. Its determination is shown in Figure 3.10.

- In Panel A, at industry level, DD is the original demand curve and MPS is the Market Period Supply (MPS) curve, which lies parallel to Y-axis, as discussed earlier (Panel A of the Figure 3.10). Both the curves intersect at P and establishes OP=PQ equilibrium price and OQ equilibrium quantity.
- Suppose the demand (for cloth) rises, as indicated by the D_1D_1 curve. The immediate effect will be for the market price to increase from PQ to P'Q. This is because, since supply is fixed in the market period, it is not possible to increase the supply beyond OQ.
- But, in short period, it is possible to increase the supply with the help of existing plants and machinery by employing extra labour, raw material and other variable factors. Thus, by increasing the variable factors, the supply can be increased along the SRS curve. This SRS curve will intersect the new demand curve D_1D_1 at P_1. In this way, the short-run normal price $OP_1=P_1Q_1$ is determined at which OQ_1 quantity of commodity is bought and sold. This short-run price ($OP_1=P_1Q_1$) is higher than the original market price (OP=PQ), but it is lower than the new market period price OP'=P'Q, which is prevailed immediately after the increase in demand.
- Suppose, the demand for cloth falls in the market. The demand curve falls from DD to D_2D_2. The market period price immediately falls from PQ to P″Q. So, all the firms in the industry

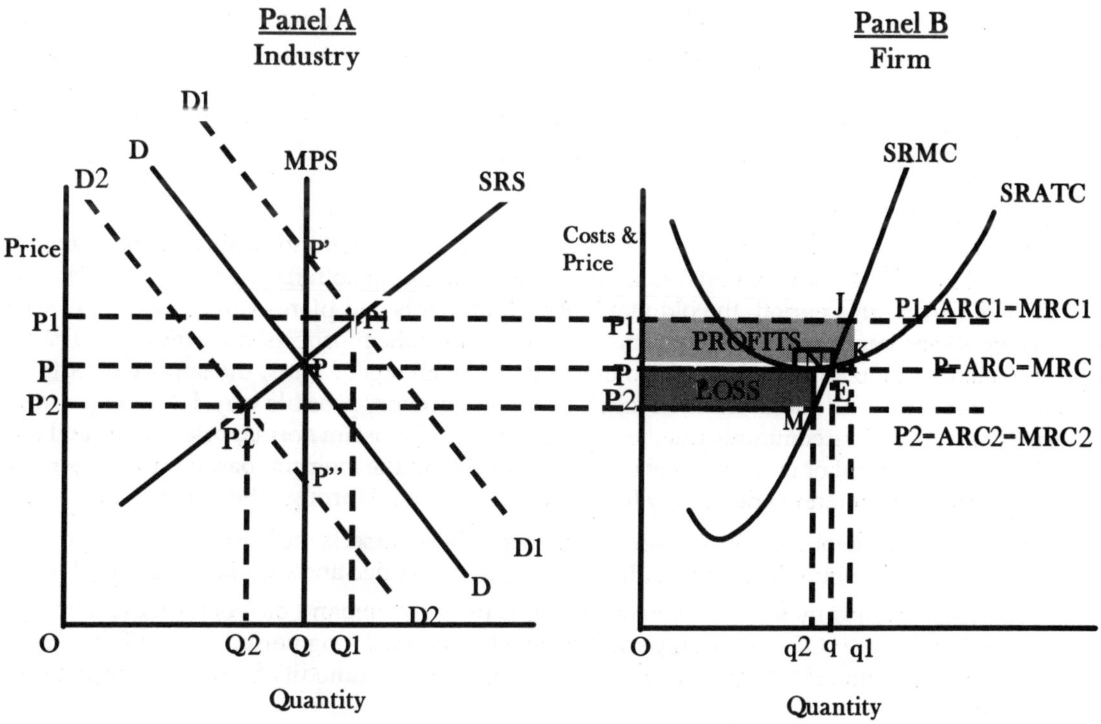

Figure 3.10: Short period price determination for a durable good under perfect market competition.

will reduce the output by reducing the amount of variable factors like labour, raw materials etc., in the production programme and also reduce supply by stocking the goods, thereby, the new supply is decreased along SRS. As a result, the new equilibrium is established at P_2, where the SRS and D_2D_2 intersect. Now, OQ_2 quantity is bought and sold at $OP_2=P_2Q_2$ price. This short-run price $OP_2=P_2Q_2$ is less than the original market price OP=PQ, but higher than the new market price OP″=P″Q established after fall in demand.

- ☆ Thus, supply is more important than demand in the short period because, supply can be increased or decreased to match the rise or fall in demand by changing the variable factors in the production programme.
- ☆ As explained earlier, the same prices QP=OP, $Q_1P_1=OP_1$ and $Q_2P_2=OP_2$ determined in the industry due to changes in the supply of the commodity in the short run, will be passed on to the individual firms and as shown in the Panel B, they are horizontal lines parallel to X-axis (as individual firms are price takers) and these price lines coincides with respective ARC and MRC of the individual firms under perfect market competition.
- ☆ With the original price OP=QP, we will compare SRATC and SRMC curves. This SRMC curve will cut P=ARC=MRC curve from below and hence, the firm reaches the equilibrium at 'E' and the corresponding point on X-axis represent the Oq quantity of output supplied and demanded in the market at which the firm earns normal profits, as price=SRATC.
- ☆ At $OP_1=P_1Q_1$ price, for the firm, $OP_1=ARC_1=MRC_1$. This MRC_1 is intersected by SRMC at J, which implies that, the firm is again at equilibrium at point J and the corresponding point on X-axis represent Oq_1 level of output supplied and demanded in the market at OP_1 price, where the firm earns super-normal profits, P_1JKL because, price is above SRATC by JK. So, rise in price above SRATC enables the firm to earn super-normal profits.
- ☆ At OP_2 price, for the firm, $OP_2=ARC_2=MRC_2$. This SRMC intersects the MRC_2 curve at M, which implies that, the firm is in equilibrium at point M and the corresponding point on X-axis represent Oq_2 level of output supplied and demanded in the market at OP_2 price, where the firm incurs losses, PP_2NM because, price OP_2 is less than SRATC by NM.

3. Long Period Market

This market is held for many years, as it deals with durable goods and supply can be changed or fully adjusted to changes in demand. This is possible by changing both fixed and variable factors of production. During this period, the old machines, equipments and plants can be replaced by new ones. There will be free entry and exit of firms depending upon the prospects in the industry. The scale of production, organization and management can also be changed. Thus, the supply of output of durable goods can be fully adjusted to changes in demand in every possible way, in the long run. The price that prevails for the commodity in the long run is called Long run normal price or Normal price. It represents the normal or natural value of the commodity in the long run based on the interaction between permanent demand and supply forces. This is explained through the Figure 3.11.

- ☆ In Panel A, at industry level, the demand curve DD intersects the MPS curve at P, where OQ quantity of commodity (Tractor, Implements) is demanded and supplied at OP level of price.
- ☆ When there is increase in demand, D_1D_1 is the new demand curve, thereby, the market period price is OP′, the short period price is OP_1 and the long period price is OP_2. Hence, a rise in demand in the market, increases the price of the commodity from OP to OP_2 in the long run.

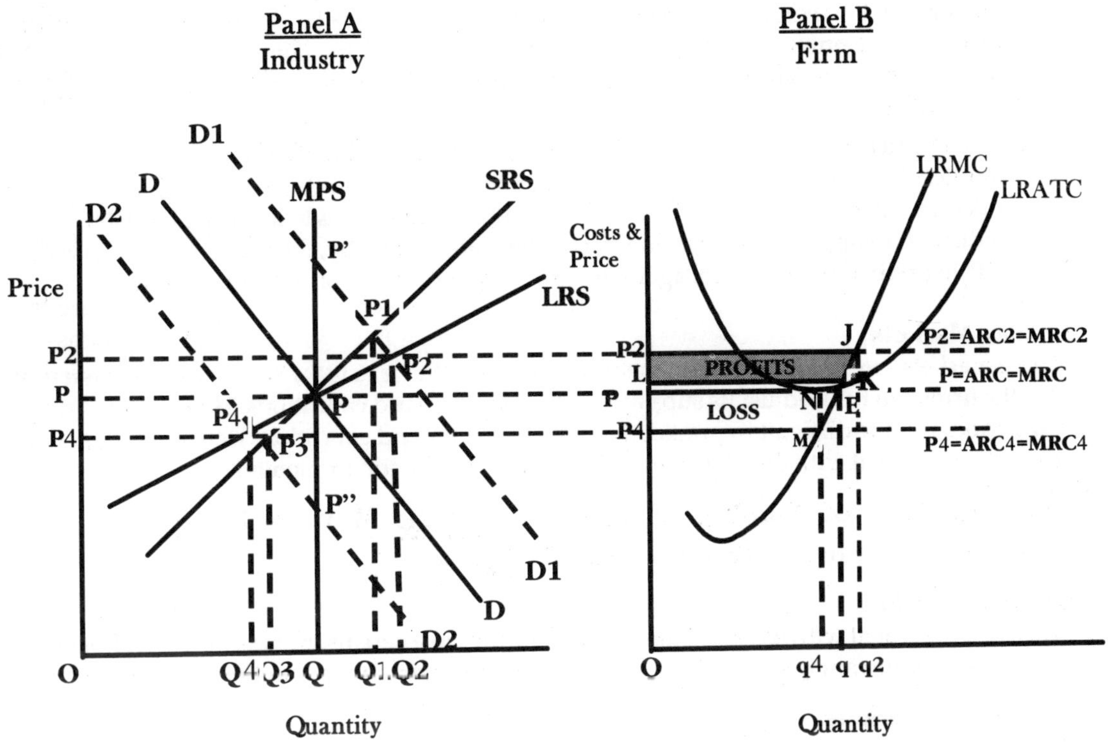

Figure 3.11: Long run price determination for a durable good under perfect market competition.

- If the demand for the commodity falls, D_2D_2 is the new demand curve and hence, the market period price falls to OP", the short period price is OP_3 and the long period price is OP_4. Hence, a fall in demand in the market, leads to fall in price of the commodity from OP to OP_4 in the long run.
- The same prices OP, OP_2 and OP_4 determined in the industry due to changes in the supply of the commodity in the long run will be passed on to the individual firms and as shown in the Panel B, they are horizontal lines parallel to X-axis (as individual firms are price takers) and these price lines coincides with their respective ARCs and MRCs of the individual firms under perfect market competition.
- With the original price OP, we will compare LRATC and LRMC curves. This LRMC curve will cut P=ARC=MRC curve from below and hence, the firm reaches the equilibrium at 'E', where the firm sells Oq quantity of output at price OP and the firm earns normal profits, as price=LRATC.
- At OP_2 price, for the firm, OP_2=ARC_2=MRC_2. This LRMC intersects the MRC_2 at J, which implies that, the firm is again at equilibrium at point J, thereby, it sells Oq_2 output at OP_2 price and earns extra (super-normal) profits, P_2JKL. But, under perfect market competition, the firms enjoy only normal profits in the long run instead of super-normal profits. This is because, if the existing firms enjoy extra (super-normal), it will attract new firms to enter the industry, thereby, the demand for factors increases and hence, their prices will increase in

the market. Further, the existing market of the industry will be shared by more number of firms. So, the entry of new firms will increase supply of the commodity into the market, thereby, the price falls from OP_2 to original price OP. So, again the firms earn normal profits in the long run.

☆ At OP_4 price, for the firm, $OP_4=ARC_4=MRC_4$. The LRMC intersects the MRC_4 at M, which implies that, the firm is again at equilibrium at point M, thereby, it sells OQ_4 output at OP_4 price and incurs losses PP_4MN. These losses make some firms to leave the industry. So, this reduces the supply of the commodity into the market, thereby, the price gradually rises from OP_4 to original price OP. So, again the firms earn normal profits in the long run.

4. Secular Markets

These markets are also called Permanent markets and they exists for even more than ten years, thereby, the firms can fully adjust its supply to changes in demand conditions. However, during this lengthy period, is quite difficult to predict the changes in demand conditions due to various factors. These markets also deal with durable goods like machinery, tractor, implements etc.

VII. On the Basis of Number of Commodities the Markets Dealt with

Based on this criterion, the markets are classified as follows:

1. General Markets

These markets deal with a wide range of commodities like cereals, pulses, oil seeds, fiber crops, jaggery etc. So, a wide range of marketing functions and intermediaries function in these markets.

2. Specialized Markets

These markets specialize in dealing with one or two specific commodities like rice or wheat or jaggery or fruits or vegetables etc. Since the marketing transactions are specialized with reference to one or two specific commodities, the marketing middlemen are highly professionalized in performing the marketing functions.

VIII. Based on Nature of Commodities

This criterion classifies the markets into the following categories.

1. Commodity Markets

Commodity markets are the markets, where raw or primary produce are exchanged. So, this market deals with commodities like rice, wheat, maize, cotton, chillies etc. Existence of a vibrant, active, liquid, and transparent commodity market is normally considered as a sign of development of an economy. It is, therefore, important to have active commodity markets functioning in a country. Commodity trading markets involves different types of transactions like Spot trading, Forward trading and Futures trading.

a. Spot Trading

In this method of transaction, delivery of the product either takes place immediately or with a minimum lag between the trade and delivery due to technical constraints. The spot market is also called the 'Cash market' or 'Physical market' because, prices are settled in cash on the spot at current market prices (as opposed to forward prices) and the products are transferred from sellers to buyers immediately. So, a market place that facilitates cash transaction is referred to as the cash market and the transaction price is usually referred to as the cash price. Buyers and sellers meet face to face and deals are struck. These are traditional markets. Example of a cash market is a *mandi*, where food grains are sold in bulk. Farmers would bring their products to this market and merchants/traders would immediately purchase the products and they settle the deal in cash and take delivery immediately. Cash markets thus, call for immediate delivery of commodities after cash payment. Contracts bought and sold on these markets are immediately effective. This method of trading generally involves visual inspection of the commodity or a sample of the commodity by the buyers. These types of transactions are also common in wholesale and retail markets. However, with the advent of SPS and TBT provisions in WTO, the buyers and sellers are exchanging the goods even without visual inspection.

b. Forward Contracts or Trading

A forward contract is an agreement between two parties (buyer and seller) to exchange at some fixed future date a given quantity of a commodity for a price defined today. The fixed price today is known as the Forward price.

c. Futures Contracts

A futures contract has the same general features as a forward contract, but the commodity is transacted through a futures exchange. In essence, a futures contract is a standardized forward contract in which the buyer and the seller accept the terms in regards to product, grade, quantity and location and are only free to negotiate the price.

Thus, in forwards and futures markets, the agreements are normally made to receive the commodities at a later date in future for a pre-determined consideration based on agreed upon terms and conditions. Forward and Futures markets reduce the risks by allowing the trader to decide a price today for goods to be delivered on a particular future date. These markets allow delivery at some time in the future, unlike cash markets that call for immediate delivery. These advance sales help both buyers and sellers with long-term planning. Forward contracts laid the groundwork for futures contracts. The main difference between these two contracts is the way in which they are negotiated. For forward contracts, terms like quantity, quality, delivery date and price are discussed in person between the buyer and the seller. Each contract is thus, unique and not standardized, since it takes into account the needs of a particular seller and a particular buyer only. On the other hand, in futures contracts, all terms (quantity, quality, and delivery date) are standardized in an exchange. The transaction price is discovered through the interaction of supply and demand in a centralized marketplace or exchange. Forward contracts help in arranging long-term transactions between buyers and sellers, but could not deal with the financial risk that occurred with unforeseen price changes resulting from crop failures, inadequate storage, bottlenecks in transportation, factors beyond human control (floods, natural calamities, etc.), or other economic factors that may result in unexpected changes and hence, counterparty default risks for parties are involved. This led to the development of futures market. As mentioned above, since futures are standardized contracts that are traded through an exchange, they can be used to minimize price risk by means of hedging techniques. Since the exchange standardizes the quality and quantity parameters and offers complete transparency by using risk management techniques (such as margining system), the counterparty default risk has been greatly minimized.

2. Capital Markets

Capital markets are like any other markets, but differ in terms of the products traded and their organization. These markets deal with the trading of securities, stocks, bonds and shares.

- ☆ Securities are financial instruments or legal documents signifying either an ownership position in a company (*i.e.* shares) or a creditor relationship with a company or government (*i.e.* stocks and bonds).

- ☆ Shares represent part-ownership in a business concern. So, shareholders have a vote in running the business and if the business makes profit, they are entitled to a share of it. However, the dividend which shareholders receive is dependent on the business's profitability and management decisions such as business shares, bonds issued by governments in private companies, units in Collective Investment Schemes, debentures, commercial paper and notes.
- ☆ Stocks and bonds are long-term fixed interest bearing securities issued by Government and companies. When one invests in stocks and bonds, one gets interest income, which is paid periodically until the loan matures or is called back by the issuer. The holder of stocks and bonds gets interest, even if the issuer does not make a profit.

Capital market can be further classified into primary and secondary capital markets. The primary capital market is a market for new shares, where as in the secondary capital market, the existing securities are traded.

3. Money Market

Money market facilitates dealings in short-term financial instruments such as inter-corporate deposits, certificate of deposits, treasury bonds, commercial papers, commercial bills etc. There exists several differences between capital market and money market. The main difference between them is made on the basis of maturity period of instruments and claims. Short-term instruments maturing within a period of one year are traded in money market, whereas the capital market deals with longer maturity financial assets and claims. Further, money market is a wholesale market and the participants in money market are large institutional investors, commercial banks, mutual funds and corporate bodies. However, in case of capital market, even a small individual investor can deal by sale/purchase of shares, securities, stocks and bonds. Unlike capital market, the money market cannot be sub-classified. The total volume of trade/day in money market is many fold that of the volume per day taking place in capital market.

4. Service Market

This market facilitates the beneficiaries in having requisite information or in enjoying services to carry out the marketing process. For example, traders can avail useful marketing information on prices of different agricultural commodities across different agricultural markets through AGMARKNET, MCX etc. Similarly, researchers in the field of agricultural marketing can have access to useful information through <indiastat.com>.

IX. On the Basis of Volume of Transactions

According to the volume of transactions, markets can be of two types as discussed below.

1. Wholesale Markets

In these markets, the commodities are bought and sold in large quantities. The important market intermediaries in these markets are primary wholesalers, processors and secondary wholesalers. The primary wholesalers purchase the produce in large quantities from farmers in primary wholesale market and transact the same to processors (if the produce demands processing to convert into palatable form) or secondary wholesalers. These secondary wholesalers sell the products in terminal markets to retailers for final transfer to consumers. These wholesale markets are mostly located in big cities and they have been in existence for a longer period of time and thus, wholesalers have good trading relationships with retailers. Retailers confine their business permanently to dealing with selected wholesalers.

2. Retail Markets

In these markets, the commodities are bought and sold in small quantities compared to wholesale markets. The secondary wholesalers or processors sell insignificant quantities of products to retailers and they in turn transact the same to final consumers. So, retailers are closest to consumers in a marketing channels. They often serve as personal representatives to consumers.

The wholesalers serve as the vital link between the farmers and the final consumers. It is the presence of wholesalers and processors, that let farmers breathe a sigh of relief, as they can sell their entire quantum of harvested produce and get back into production activities in the ensuing season. So, one just cannot imagine the plight of the farmers, if there had been no wholesalers, as then farmers would have had to wait for the next production season until they sold their produce directly to customers. Further, for some agricultural produce, processing is inevitable and the prior indebted farmers are no where in a position to spend on processing function. So, the presence of wholesalers is essential from the point of view of farmers, as they purchase the bulk of produce at a remunerative price negotiated between them. From wholesalers, the produce may reach the processors or sometimes, the processor himself acts as a wholesaler in purchasing the produce from the farmers directly. From the wholesalers or processors, the produce (products) reach to retailers, who sell the same at a greater margin to the final consumers in requisite quantities. Though the chain of Farmer → Wholesaler → Processor → Retailer → Consumer looks similar across majority of the agricultural commodities, there are many differences between wholesale and retail activities and the same is presented through Table 3.5.

Table 3.5: Wholesalers vis-à-vis Retailers in purchase and sale of agricultural produce.

Wholesalers	*Retailers*
A wholesaler purchases the produce in large quantities from the farmers for transacting the same to retailers	A retailer purchases the produce from a wholesaler in comparatively less quantities for end consumers.
A wholesaler buys in bulk thereby, he get the produce of mixed qualities and varieties.	A retailer buys according to his needs and requirements depending upon his stock at his retail store and the produce he purchases will be more or less of same quality and variety.
Wholesaler mainly aims at purchasing the produce at low price from the farmers and though he aims at profit making, his profit margin is small compared to retailer.	A retailer always has an eye on both the selling price as well as the profit margin he wishes to gain from the final consumers. But, the retailer is more worrisome about his margin compared to wholesaler, as if the margin is too high, the consumer's purchase price will be much higher and then, he finds it difficult to sell the produce to the final consumers.
The wholesaler can stock the produce anywhere thereby, the storage costs can be lowered. This is because, for wholesalers, the potential buyers are retailers and they are interested more in their profit margin and not the place from where they are purchasing.	For retailers, the market is in front of their shop outlets, thereby, they have to incur lots of expenses in maintaining the retail space, so as to attract the final consumers.
The wholesalers have to pay cash immediately to the farmers after purchasing the produce, as the farmers have to clear their prior debts.	The retailers do not buy the products from wholesaler strictly on cash basis and they take a period of one to two months for clearing the invoice depending on the nature of commodities transacted and revenue they earn.
The profit margin of a wholesaler per unit quantity of produce transacted may be too small compared to a retailer. But, the wholesaler makes more money compared to a retailer, as wholesaler transacts large quantities of produce.	The profit margin of a retailer per unit quantity of produce transacted is higher compared to a wholesaler.

X. On the Basis of Population the Markets Serve

We know, the population in the economy differs in terms of their standard of living. Due to differences in standard of living, the consumption requirements also differ across the consumers and this leads to existence of different types of markets as discussed here under.

1. Urban Markets

In general, urban consumers have more awareness about the latest products or different product types, their standards of living are higher and more comfort demanding compared to rural consumers. Thus, market competition is intense in urban areas compared to rural areas thereby, brand awareness, brand recognition, variety, choices and options are higher in urban market. This market mainly deals with processed products like jams, jellies, fruit juices, RTE, RTC foods etc., as the urban consumers are rich enough to puechase them even at higher prices.

2. Rural Markets

Compared to urban markets, pricing of commodities or products exert more influence on the demand in rural markets, as the income levels of rural consumers are relatively lower. This market mainly deals with fresh produce like fruits, vegetables, paddy, wheat, chillies, cotton etc.

So, it is clear that, what works with one demographic may not work with another. This calls for the existence of different types of markets dealing with different categories of consumers and commodities or products. With the advent of trade liberalization and most notably the advent of the internet, the differences across these two markets are clearly demarcated. So, this calls for executing different types of marketing strategies for the farmers and traders wishing to sell to both the markets.

XI. On the Basis of State Intervention

This criterion classifies the markets into the following two categories:

1. Regulated Markets

In these markets, the marketing transactions are carried out as per the rules and regulations framed by the Government or a Statutory organization (Agricultural Produce Market Committees (APMCs)). Thus, the Government intervenes in the market mechanism by regulating the market prices and marketing costs, so as to eliminate the unhealthy practices followed by the traders to exploit the farming community. The form of protection to the farmers is not only in terms of regulating the market prices, but also in terms of providing a remunerative price to them. Even at times of distress sales of produce, the Government agencies like FCI, Cotton Corporation of India (CCI) etc., intervene the market and purchase the commodities at the announced MSPs. It is essential to note that, the market (equilibrium) prices are allowed to determine based on the two market forces *viz.*, demand and supply and the prices are not regulated by the Government. All the commodities are notified in the regulated markets and these markets ensure prompt payment of sales proceeds to farmers immediately after the transactions are over.

2. Unregulated Markets

On contrary, in unregulated markets, there is no Government intervention in the form of regulating market practices and marketing costs. So, the farming community are exploited to a greater extent, as they are relatively less educated about the modern marketing techniques as compared to the marketing middlemen. Though prices are settled based on demand and supply forces, the traders often cheat the farmers on the ground of stating that, the produce is of inferior in quality. At times of bumper harvests, the farmers are greatly exploited and they often resort to distress sales (Sale price<Cost of production

of commodity). Further, the farmers have no guarantee of getting the sales proceeds immediately from traders after transacting their produce. The traders further exploit the consumers by creating artificial scarcities of commodities through charging indiscriminately higher prices. Thus, both the farming community and final consumers gets exploited due to non-regulation of marketing practices and marketing costs.

XII. On the Basis of Vision

It is essential to note that, this criterion does not refer to physical entities based on which the markets are classified, but it is the situation on how the marketing or trade is conducted. According to this criterion, the markets are classified as follows:

1. Open Market

A market is said to be open, if it is free from all the trade restrictions like imposition of non-tariff barriers (quotas, SPS, TBT), tariffs etc., thereby, it ensures free flow of goods and services across the countries. Thus, in open market situation, there is no restrain on the part of the farmers and traders in carrying out the marketing transactions. With the advent of WTO, 159 member nations as on 2nd March, 2013 have open markets and are freely exchanging their goods and services with low or minimal restrictions. For example, Indian market is an open market, as it imports several commodities like pulses, edible oils etc., from the member nations of WTO.

2. Closed Market

It is also called as Protected market, in the sense that, there is strict imposition of trade restrictions regarding the movement of goods and services across the countries. For example, at times of dumping the produce, the importing countries impose higher tariffs or fix quotas, so as to safeguard the interests of domestic farmers. However, in WTO, the provisions of tariffication and market access enable the markets to be open rather than closed. Thus, in protected markets, it is not possible for traders to execute trading actions freely. This form of protection may also be given to deliberately keep some players out of the market or may be a case, where entry criteria are high or difficult. Sometimes, these closed markets are also called as Black markets, as when demand is more than supply for a commodity (the traders may also create artificial scarcity of commodity), it leads to greater rise in its prices.

However, it is hard to classify a market as an open market or a closed market, as Economists have their own interpretations according to which they judge the openness or lack of it in a market. So, they have opted for a new term which is 'free competition'. Further, for some commodities, the market is kept closed and for some, the market is open considering the comparative advantage the commodities enjoy across the countries. For example, India imports rye, barley, pulses etc., without imposing any import tariff *i.e.*, under free trade. But, for commodities like wheat, maize, tea, milk powder etc., it imposes import tariff to restrict free trade.

XIII. On the Basis of Objectives of Personnel Involved in Marketing Process

This criterion classifies the markets into the following categories:

1. Consumer Markets

These markets deal with the consumers, who purchase the produce either from the farmer directly (vegetables and fruits) or from middlemen (rice, pulses etc) for their personal consumption. The consumers are not interested in reselling the product or setting themselves up as a processor. So, their main objective is to derive maximum satisfaction from the commodities or services purchased in the marketing process. However, the tastes and preferences, capability of consumers to afford various products, marketing avenues available at the consumer level etc., decide the growth and size of business firms.

2. Farmers' Markets

These markets consist of firms or organizations, who conduct business considering farmers as their clientele. For example, seed production firms, fertilizer companies, pesticides companies, firms selling irrigation equipments, machinery etc., deals with the farmers. So, the farmers purchases the requisite inputs and services in these farmers markets. Unlike consumers markets, the farmers markets function with a profit-motive and the pattern of transactions are entirely different.

3. Institutional Markets

This is another important market sector, which includes various types of institutions or corporations or organizations that involve in the process of agricultural marketing. For example, FCI, CCI, APMCs etc., involve in the process of marketing to safeguard the interests of the farming community. Unlike Farmers' markets, these markets function with a non-profit or welfare motive. The income generated by these institutions will be further utilized for the improvement of marketing of agricultural produce.

4. Reseller Markets

These markets include the marketing agencies that exist between farmers and final consumers in transacting the agricultural produce. They purchase raw produce from the farmers and finally transact the product in final (palatable) form to the ultimate consumers. These agencies execute the marketing functions efficiently and they function with a profit motive.

XIV. On the Basis of Final Use of Product

This criterion classifies the markets into input markets and output markets and they are discussed here under.

1. Input Markets

These markets deal with the sale of inputs employed in the crop production programme. When compared to output markets, input markets are more organized with a well distribution net work. The target customers for these markets are farmers. Examples include fertilizers market, pesticides market, seed market etc.

2. Output Market

These markets deal with the sale of output (agricultural produce or processed products) in the marketing process. When compared to input markets, output markets are less organized thereby, market imperfections are more common. The target customers for these markets are market intermediaries (wholesalers, processors and retailers) and final consumers. Examples include markets dealing with agricultural commodities like rice, wheat, barley, pulses, oilseeds, fruits, vegetables etc.

XV. On the Basis 'where the Product is Transacted

Traditional agriculture is subsistence oriented. But, there is gradual shift from subsistence farming to market-oriented farming. Further, with the advent of trade liberalization, the focus is shifted from domestic marketing to international marketing. So, considering this criterion, the markets are classified into the following categories:

1. Domestic Market

This market is restricted to the political boundaries of a country. So, the farmers and traders transacting the products within nation's boundaries only has to consider domestic competition. Even if they face competition from foreign firms, they have to focus on the competition that exists in the domestic market.

2. International Market

In this market, the produce arrives from different countries. That is, for a particular country, if the trade is open, the farmers and exporters export the produce into foreign countries considering the comparative advantage of the commodities in the international market. Similarly, if a country is shortage of a particular commodity, it will import the same from the foreign countries. However, compared to domestic market, international market involve more trade regulations and difficulties like restricted movement of factors of production, differences in currencies, markets are totally separated etc.

XVI. On the Basis of Number of Intermediaries between Farmer and Final Consumer

This criterion classifies the markets into the following categories:

1. Mass Market

In this market, there exists more number of market intermediaries between the farmer and the final consumer. So, these markets are characterized by wide separation of farmer and final consumer and they have indirect communication through market intermediaries. For example, a cotton farmer (called as mass marketer) has very little direct contact with the final consumer wearing clothes and he distribute the produce through various intermediaries like wholesalers, processors (Ginning millers) and retailers to final consumers. The communication between farmer and final consumer is impersonal through advertisements in televisions, movie slides, news papers etc.

2. Direct Markets

In these markets, the farmers directly negotiate with the final consumers in transacting the agricultural produce. So, there exist no marketing middlemen between the farmers and final consumers. Thus, direct marketing establishes a personal relationship between the farmers and final consumers, as they are in personal contact with each other. Presently, this type of marketing is experiencing tremendous growth, as it yields 100 per cent FSCR. For example, in Rythu bazaars established in Andhra Pradesh (popularly called as Apni Mandi in Punjab and Haryana, Uzhavar Santhaigal in Tamil Nadu, Krushak Bazaars in Odisha, Hadapsar Vegetable Market in Pune and Raithara Santhegalu in Karnataka), the farmers are in direct contact with the final consumers in transacting vegetables.

3. Internet Market

This is a modified and more advanced type of direct market, that is evolved with the advent of internet technology. With the growing importance towards grades and standards for agricultural commodities, this internet market gained significant attention in the recent period and besides domestic transactions, international exchanges of commodities are also taking place. This is especially gained significance with the advent of WTO as trading platform between 153 member countries. The major advantage of this internet market is that, it ensures quick transactions between the farmers and final consumers even they are separated by distant places, thus avoiding all possible delays involved in the marketing process.

XVIII. On the Basis of Number of Items of Different Product Categories Available for Transaction to the Final Consumers

This criterion guides to divide the markets into the following two categories:

1. Super Market

The scenario of marketing was changed with the advent of super markets. Earlier, for purchasing the required total stock of commodities by the consumer, he has to move to different shops and it is more time consuming in having them. Say, the consumers have to purchase fresh foods like rice, pulses, wheat atta, edible oils etc., in one shop and other items like cherries, bread, jams, jellies, ice creams, processed food products etc., from other shops. The produce like fish, vegetables, flowers etc., was hard to find in the same market. Further, they have to wait at these shops till all the items are weighed and packed. So, there is no much of a choice for the consumers with regard to choosing and comparing about different brands of the products. But, the scenario changed with the arrival of first the supermarkets. In super markets, all the items with reference to particular category are available at a particular shop and they are ready made packed with all the details (firms logo, quantity, date of packing, date of expiry, maximum retail price including taxes etc) and technical ingredients quoted on the covers of the packs. So, the shopping has turned from drudgery and a boring experience into something that is pleasurable and relaxing for the consumers. For example, all the items with reference to food category say, rice, wheat atta, pulses, edible oils, ice creams, biscuits, cherries, bread, jams, jellies, processed food products, RTC, RTE foods etc., are available at one shop.

2. Hyper Market

The scenario of marketing of products was further changed with the advent of hyper markets in the recent period. In hyper markets, the consumers get a wide range of products of different categories under a single roof. For example, in hyper market, the consumer can have all food items like rice, wheat aata, pulses, edible oils, cherries, bread, jams, jellies, ice creams, processed food products, fruits, flowers, vegetables, fish, milk etc., along with a wide range of products of electronics, clothes, furniture, toys, video games, audio and video CDs, restaurants, magazines, internet cafes, beauty parlours etc., thereby, enable them to fulfill all their requirements under one roof. So, a hyper market is bigger in size and stores a more number of commodities of different categories than a super market. It looks more like a warehouse than a store. Further, the prices at a hyper market are cheaper than at super market. But, compared to hyper markets, super market provides more warm services and has a personal touch, which are absent in a hyper market.

XIX. Based on Profit Margin

According to this criterion, markets are classified as Profit markets and Non-profit markets.

1. Profit Markets

These are the markets, in which the primary objective of marketing transactions is to gain profits by the marketing agencies. That means, the success of marketing is measured in terms of extent of profits realized in the marketing transactions. Examples include, super markets, hyper markets, local (village) markets, speculators etc., function with a profit motive.

2. Non-Profit Marketing

On the contrary, in these markets, the main objective of marketing is welfare motive. That means, the success of marketing is measured in terms of extent of welfare realized by the concerned beneficiaries. For example, the Government has established regulated markets with the main objective of regulating the market practices and checking the irregularities in the agricultural marketing system. This ensures welfare on the part of the farmers in transacting their produce in the regulated markets through checking the unhealthy practices of marketing middlemen and by realizing a remunerative price. What ever the income earned by the regulated markets in terms of market fee, license fees from commission agents etc., are further spent for creating and strengthening amenities and facilities in the market yards for the welfare of the farming community. Similarly, CMSs functioning in India also work on non-profit basis towards the welfare of the farmers-members.

XX. On the Basis of Market Segmentation

We know that, with the advent of WTO, it became relatively easier for the business firms to freely trade the commodities across the countries. However, for trading the commodities, the business firms must analyze the key similarities and differences among the customers across the countries. This results in the determination of two types of markets *viz.*, Global market and Multi-domestic market.

1. Global Market

In this market, the business firms will not segment the customers across the countries, thereby, they offer a homogeneous product in the market. This type of marketing has distinct advantages like, it facilitates centralized management and ensures coordination among different business functions such as human resources, finance and product development. For example, Nike Shoe Company market shoes under a slogan, *'Just Do It'* and is same anywhere in the world *i.e.*, either in USA or France or India. The features of this market are already discussed under Global marketing in the earlier pages.

2. Multi-Domestic Market

In this market, the business firms will segment the customers across the countries, thereby, they offer a heterogenous products in the market suiting to different customers' tastes and preferences. This type of market calls for decentralized marketing, location specific decision-making because, management must be able to respond to local segmented customers. For example, 'Ashwin' atta selling firm in India offers normal atta and diabetic atta to non-diabetic and diabetic consumers respectively. But, this type of marketing involves more costs, as for each category of customers, separate advertising, packaging, promotional literature etc., are required.

XXI. On the Basis of Type of Agency Involved in Marketing Transactions

According to this criterion, the markets can be categorized as follows:

1. Primary Market

A primary market is a formal market place that brings together original sellers (farmers) and original buyers (say, wholesaler) of products. For example, a wholesaler purchasing paddy from the farmer directly in a village constitutes a primary market.

2. Secondary Market

Secondary market is a market place in which the original buyers (wholesalers) of the product resell the produce to a third party. For example, a rice miller purchasing paddy from the wholesaler constitutes a secondary market. Similarly, a retailer purchase paddy from rice miller also constitutes a secondary market.

It is essential to note that, the distinction between a primary and secondary market is not the same as the difference between wholesale and retail markets, as both wholesale and retail markets can have primary and secondary markets within them.

XXII. On the Basis How Consumers are Approached

According to this approach, the markets are classified as Push markets and Pull markets and they are discussed here under.

1. Push Markets

In these markets, the firms approach the consumers by pushing its products on to the consumers. The firms goes for heavy advertising of their products and try to stimulate and motivate the consumers towards purchasing the products. The other common sales tactics like sales persons negotiating with consumers directly, opening small retail outlets in the villages, displaying the products at public centers etc., fall under push marketing. The common example for push marketing is sales displays at nearby grocery store. We also frequently see the advertisements about saffola edible oil, Aashirvad atta, Kissan jams, fruit juices etc., in televisions and this make the consumers to stimulate and purchase the product.

2. Pull Markets

In these markets, the entire approach as discussed above is in the opposite direction. That means, the business firms try to get the consumers to their shops and hence, the term 'pull', where the marketers are attempting to pull consumers in. The only aspect the business firms will do is to make the consumers to know about the place of location of the shops so that, they are pulled towards the shops and purchase the product. So, from a business perspective, pull marketing attempts to create brand loyalty and keep consumers coming back, whereas push marketing is more concerned with short-term sales. Initially, the business firms go for advertizing about the product and it is only an Informative advertising, but not competitive advertising. Once the consumers gets awareness about the products and convinced about their meritorious aspects (from fellow consumers), they themselves moves towards the shops of these firms to purchase the products. For example, firms supplying drip and sprinkler irrigations equipment will adopt pull marketing strategy.

XXIII. On the Basis of Methodology of Convincing the Customers

It is essential to note that, marketing is not as simple as putting out a bunch of advertisements before the customers to convince them for purchasing the products. It requires different approaches to achieve maximum effectiveness and to attract the most customers. Considering the methodology followed in convincing the customers towards purchasing the products, the markets can be classified as Traditional markets and Experiential markets and the process of marketing in these two markets are effective and can be used in conjunction with each other.

1.Traditional Markets

In these markets, the firms convince the customers by mass advertising about their products. Frequents advertising through audio-visual aids, engaging sales persons for personal solicitation about the features of the product, displaying the products at the public centers etc., fall under this traditional marketing. For example, advertising about various agro-based products like wheat atta, jams, fruit juices, basmati rice etc., through television falls under traditional marketing. No doubt, these promotional measures are effective in conveying the firms' mission to the public, but attaining a personal touch is difficult.

2. Experiential Markets

Unlike traditional markets, experiential markets focus on the experiences and the emotions of the customers. A good example of experiential marketing is, say a fertilizer selling firm takes up demonstrations in the farmers' fields and convince the farmers in getting higher yields through the application of recommended dosage of fertilizer. The farmers with this sort of firsthand experience remember the fertilizer brand and hopefully, they will give a good recommendation to their neighbouring farmers. So, instead of wasting huge amount of money resources through advertising, experiential marketing is a more viable option.

XXIV. On the Basis of Promoting the Products

We know business firms pursue different forms of marketing to make the products to reach the potential consumers effectively. The traditional marketing has been around for a long time and is still used today, but with the access of internet facility, the pattern of marketing of products was greatly revamped. Considering the different ways of promoting the products, there are different types of markets and they are discussed here under.

1. Internet Market

This market exists on the online and hence, it is also called as Online market. This market encompasses a variety of marketing forms like video advertisements, search engine marketing and e-mail marketing. For example, in futures trading, ITC's e-choupal marketing is carried through internet. The process of marketing is rather quick through internet and with the advent of grades and standards, SPS and TBT, this internet marketing is gaining rapid significance.

2. Offline Market

It is the opposite of Online market and it includes all forms of marketing that are not done on the internet. For example, the advertisements we see through news paper and audio-visual aids for different agricultural products like tea, coffee, basmati rice, edible oils etc., fall under offline marketing. To ensure good reputation for the products and about the firms among the most of the customers, the firms in today's marketing world are using both online and offline marketing techniques as complements to each other.

3. Outbound Market

This is nothing but a traditional market, which is commonly followed by the business firms to reach majority of the consumers. This type of marketing includes all forms of marketing efforts that are taken by the firm to introduce a product or service to the customer, who isn't looking for that product or service. For example, advertisements through local news papers, audio-visual aids, websites etc., for organic food products fall under outbound market.

4. Inbound Market

It is opposite to outbound market, in the sense that, the products are found by the customers through searching websites, rather than the business firms approaching the customers. For example, the buyers of the products like rice, wheat, cashew, turmeric etc., search through search engines like AGMARKNET, NCDEX, MCX etc. Instead of using paid advertisements, inbound marketing is the search engine optimization part of web marketing.

5. Direct Market

In this market, the farmers directly negotiates with the consumers in transacting the agricultural commodities like vegetables, fruits, flowers etc., without the involvement of marketing middlemen. For example, with the establishment of Rythu bazaars in Andhra Pradesh in 1999, the marketing channel is direct between farmers and the consumer in transacting vegetables, fruits etc. With the advent of internet facility and the gaining popularity of grades and standards, SPS and TBT, direct marketing through email is gaining rapid significance.

6. Niche Market

In this type of market, the product or service is not readily supplied to the entire market. That means, initially, for a portion of the total market, the product is supplied and its performance is assessed among the customers. For example, we commonly see several pesticide firms or seed firms, will test their products at selected locations in the farmers' fields. If the performance is found fruitful and if the product is proved meritorious among the farmers, the firms will extend the supply to the entire market. This targeted marketing is generally successful because, the firm has identified a need at the farmers' level. Moreover, this niche marketing is also beneficial for the firm to plan production or output considering the actual demand in the market economy.

XXV. On the Basis of Macro and Micro-Economic Views of the Market Economy

This classification of marketing into macro marketing and micromarketing is a fairly recent one. These divisions are proposed considering the influences of marketing, 'what is being studied,' whether it is being viewed from the perspective of society or the firm,' and 'who receives the consequences of the activity'.

1. Macro-marketing

It refers to the performance of all business activities involved in the forward flow of goods and services from the producer to the consumer. This view of marketing is concerned with 'The WHO and WHAT of Marketing'. Thus, macro marketing deals with the influences or consequences of marketing on large aggregates like society, a nation or even at global level. For example, in Macromarketing, we deal with different types of marketing systems of different nations, the consequences of marketing on society as a whole, impact of WTA on different nations etc.

2. Micro-marketing

It refers to the performance of business activities with reference to an individual firm or small aggregate firms involved in the forward flow of goods and services from the producer to the consumer and accomplish the objective of the firm. This view of marketing is concerned with 'The HOW and WHY of Marketing'. Thus, in Micro-marketing, the consequences of marketing are studied with reference to particular firm or product or service. For example, if we study the market penetration of Nokia cell phone, Diabetic wheat atta, Fair and Lovely fairness cream in their respective market segments, it implies, Micro-marketing.

XXVI: Based on Personal Interaction between Buyers and Sellers

This criterion classifies the markets as follows:

1. Physical Markets

Physical market is a set up, where buyers can physically meet the sellers and purchase the desired produce from them at an equilibrium price. For example, in regulated markets, farmers and wholesalers interact with each other for transacting the produce.

2. Non-Physical Markets/Virtual Markets

In these markets, buyers purchase goods and services through internet. In such a market, the buyers and sellers do not meet or interact physically, instead the transaction is done through internet. For example, Futures trading, Rediff shopping, eBay etc., fall under this category.

3. Knowledge Market

Knowledge market is a set up, which deals in the exchange of information and knowledge about agricultural based products. For example, through www.indiastat.com, we can access through useful database about agricultural products.

This discussion about different types of markets based on several criteria is only for the purpose of convenience. Whatever may be the criterion, the marketing process includes different agencies to perform the functions efficiently. But, as discussed earlier under Table 2.3, the self-interests of farmers, traders and consumers are exactly opposite in the marketing process. So, the marketing process must be executed on efficient scale considering the prospects of different agencies, as even the presence of marketing middlemen is essential in moving the produce from the farmers to the final consumers, especially for the produce that demand processing function. Other marketing functions like transportation, storage, grading and standardization, packaging and packing, market information, risk management etc., are equally essential in the marketing process. So, these aspects are discussed in-detail in the ensuing Chapter 5.

4

Market Structure-Conduct-Performance Paradigm (SCPP)

Every industry or market has its own set of characteristics (structure) that determine its competitiveness (conduct) and performance in business. The SCPP is an approach used to analyze this relation among market structure, market conduct and market performance. So, it serves as a principal approach to analyze an industry or more specifically, the market power-profitability relationship in an industry. Before a marketing system is said to be good, its structure, conduct and performance must be critically be examined. This SCPP has been widely used in agricultural marketing studies and it postulates that, under given basic exogenous conditions, the market structure determines the market conduct of the participants (buyers and sellers), which in turn influences the performance of the market. Working backward, we find that, market performance is determined by market conduct, which in return depends on market structure. Among these three components of SCPP, market structure is the relatively permanent strategic element that influences and are influenced by the conduct and performances of firms in the industry in which it operates. Thus, this approach aims to study the market performance through analyzing the relationships that exist between structural characteristics of a market and the competitive behaviour of market participants. So, this paradigm is based on the proposition that, structure, conduct and performances are linked in a deterministic way *i.e.*, an industry's basic conditions determine its structure, which in turn determine conduct and conduct determines performances (Stead: 1996, p. 6). Research suggests that, both the environment and the firm's characteristics play a role in determining the firm's specific level of profitability, which proves that, there is likely a reciprocal relationship between the environment and the firm's strategy, thereby affecting the firm's performance (Hitt: 2003, p. 19). Thus, the industry environment which comprises structure, conduct, and performance has a direct impact on the performances of a firm (Jain, p. 75).

The SCPP approach was initially started in the 1930s, when a group of Economists attempted to analyze the relationships among a firm's environment, its behavior and performance. It was derived from the neo-classical analysis of markets and it was the brainchild of the Harvard school of thought. It was developed by Edward Mason and Joe Bain in the 1940s and 1950s and was popularized during 1940-60 in studying the correlations between industry structure and performance. Later, this approach has been applied to a diverse range of problems ranging from helping businesses become more profitable to solving the market crisis. This approach of market analysis has been further developed into a new body of theory called the *"New Industrial Economics"* that include the dynamic nature of the interaction of the elements (Stead: 1996, p. 6).The original objective of this work was to describe conditions under which perfect-competition dynamics would not develop in an industry and thereby, it guides to design and implement the strategies that enhance competition in the industry. We know, a market is considered perfect competitive, if there are infinite number of buyers and sellers in the market, no single firm in the market can influence the price of the commodity nor engage in unethical marketing practices, there is free entry and exit of firms in the industry, there is perfect mobility of factors of production, the product is homogeneous so that, customers are indifferent between supplies offered by alternative channels etc. So, this approach postulates that, as market structure deviates from the paradigm of perfect competition as characterized above, the extent of competitiveness of the market will decrease and consequently a decline in market efficiency will take place (Scarborough and Kydd 1992; Scott 1995). This effort led to the development of theoretical framework popularly called as SCPP.

However, Economists have diverse opinions about the importance of SCPP. Some opine that, studying the SCPP is important because, sellers' concentration adversely influences the market's social performance. Theoretical Economists emphasize that, the monopoly gains in the market are due to severe concentration of firms. However, Industrial Organization Economists opine that, market concentration leads to allocative inefficiency. Some Economists opine that, both market structure and market conduct equally influence the market performance. However, others argue that, market conduct is largely determined by market structure, and hence, market performance depends heavily on market structure, and that leads them to pay little attention on market conduct.

4.1. Components that make up the SCPP

Basically, SCPP implies a relationship between the three components *viz.*, market structure, conduct and performance in a way that, the performance of an industry is influenced by the behaviour (conduct) of the firms in the industry, which on the other hand, is determined by their power (structure). A simple diagram can be used to illustrate the inter-connections between the key variables (Figure 4.1). Thus, the SCPP shows specific causal relationships between market structure, conduct and performance. Further, two more components *viz.*, basic market conditions and Government policies are also included in the paradigm. In Economics perspective, market structure is first determined by the two basic market conditions *viz.*, demand conditions and supply conditions. Demand conditions include price elasticity of demand, availability of substitutes, market growth, type of good, cyclical

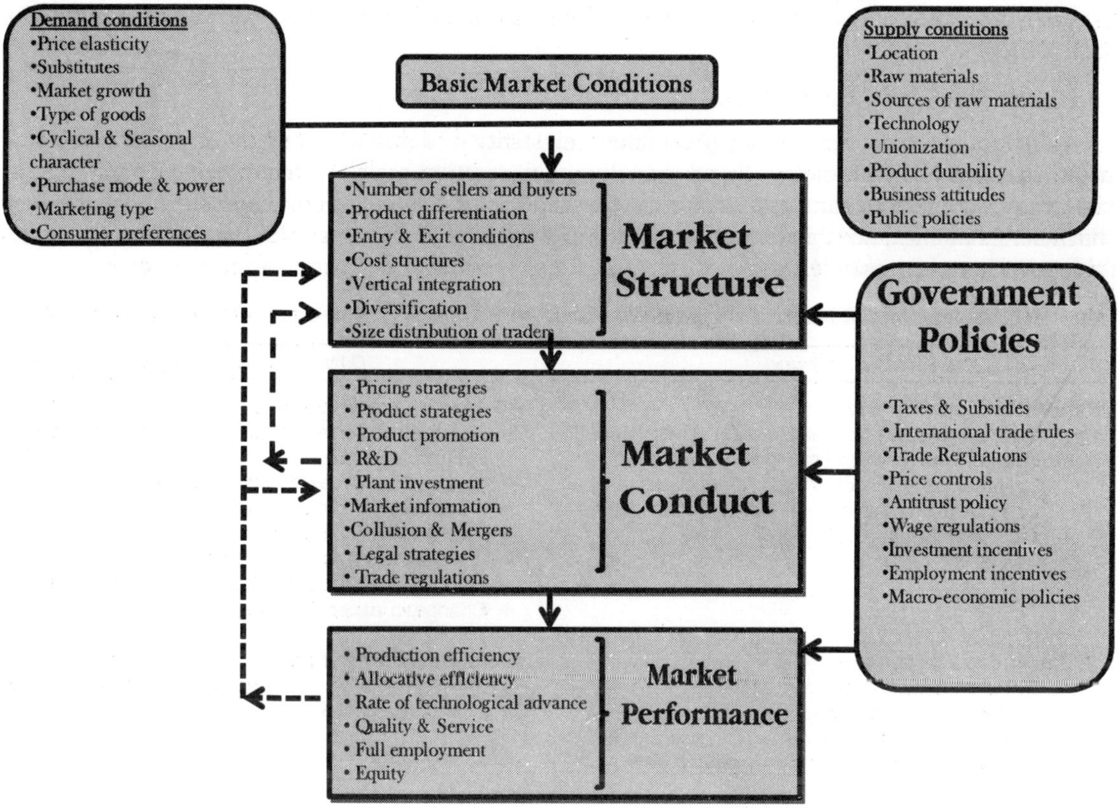

Figure 4.1: The SCPP model.

and seasonal character etc. Supply conditions include the location and availability of essential raw materials, nature of production technology, degree of work force unionization, durability of product etc. However, this SCPP suffers from a criticism that, it assumes the causation to be unidirectional, as indicated by the thick lines *i.e.*, S→C→P and not in the reverse direction. However, some Economists are of the opinion that, the causality between structure and conduct can run the other way round *i.e.*, firm's conduct (*e.g.* predatory behaviour or restrictions on entry of other firms into the industry) can shape the market structure within which the firm operates in. This implies that, market structure is endogenously determined. Some Economists also opined that, performance can also affect conduct. For example, production efficiency of a firm can influence its pricing strategies. Thus, the causality between SCP can also move in the reverse direction. In the Figure, the thick (bold) lines shows the primary relationships, where basic market conditions determines market structure, market structure determines conduct and conduct determines performance of firms in the industry. In addition to these, the Government's policies have direct influence on these basic three components *i.e.*, market structure, conduct and performance. Additionally, the broken line in the Figure points out the possibility of feedback or secondary influences between the components of the paradigm in a reverse way *i.e.*, in a direction from 'Performance' towards 'Structure'. That means, market performance can have feedback effects into market conduct and market structure. Thus, SCPP is a road map for identifying the factors that determine the competitiveness of a market, analyzing the behaviour of firms and assessing the success of an industry in producing benefits for consumers. Let us focus on each component of the SCPP in-detail.

1. Basic Market Conditions

It include, consumer demand, production, elasticity of demand, technology, substitutes, raw materials, seasonality, rate of growth, product durability, location, scale of economies etc. To summarize, basic market conditions can be studied from the standpoint view of demand and supply conditions. The factors that influence the demand for and supply of a product in an industry are shown through Table 4.1. These factors determine position and slope of both demand and supply curves.

Table 4.1: Determinants of position and slope of supply and demand curves of a product in an industry.

Determinants of Demand	*Determinants of Supply*
1. Position of demand curve ❖ Change in number of buyers (market growth) ❖ Change in buyers' tastes and preferences ❖ Change in income (income elasticity of demand) ❖ Change in the prices of related goods, substitutes and complements (EC)	**1. Position of supply curve** ❖ Change in number of firms (more or less competition in the industry) ❖ Change in prices of raw materials and labour wages employed in production programme ❖ Change in technology ❖ Changes in taxes and subsidies ❖ Change in prices of other related goods ❖ Change in expectations
2. Slope of demand curve ❖ Buyer price sensitivity (price elasticity of demand) — Degree of substitutability — Proportion of buyer income — Type of product (normal necessity or luxury) versus inferior good — Buyer time to purchase product	**2. Slope of supply curve** ❖ Time period taken by the firm to respond to a change in the price of the product

The interaction between demand and supply forces determines the equilibrium output and equilibrium price of the product in a given market structure. In a perfectly competitive market (industry), equilibrium price and output in a market (industry) are determined, where quantity supplied equals quantity demanded and the same price is passed on to the individual firms, as they are simply price takers. However, in other market structures (Imperfect competition), determination of output and price is a bit more complex, as the individual firms are price makers and in case of oligopoly market structure, the firms may even go for tacit collusion in fixing the price of the product (Kumar, 2012).

Analysis of External Environment

In addition to the above two market forces within the industry (*i.e.*, demand and supply), it is essential to monitor the external forces that may impact the industry on an ongoing basis and this constitutes the external environment to the industry. In general, the firm's manager may have a very good understanding about the internal factors (demand and supply forces), but he may be blindsided by external environment. In the words of David Needle, *'the external environment of a business organization comprises of all factors that exist outside the business enterprise, but that interact with it (the enterprise)'*. This external environment is rather dynamic, complex, consisting of many factors, which makes it difficult to be analyzed and predicted. But, in spite of this, it cannot be ignored in any meaningful analysis of business activity. Hitt divides this external environment of a firm into three major parts: general, industry and competitor environments (Figure 4.2).

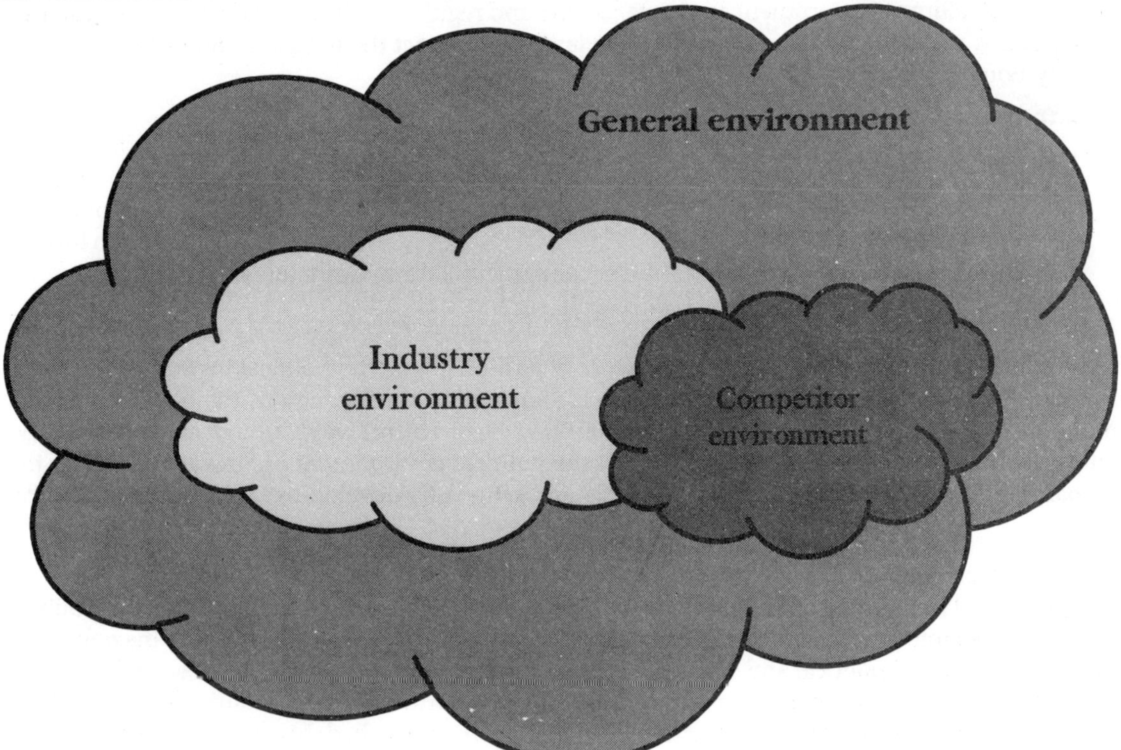

Source: Hitt *et al.*

Figuro 4.2: The External Environment.

a. General Environment

The general environment is composed of dimensions in the broader society that influence the industry and the firms within it. According to Hitt, they are popularly named as PEST* factors. PEST is an acronym for the Political, Economic, Socio-cultural and Technological factors and these PEST factors together with the internal factors (*i.e.*, demand and supply forces) influence the overall environment of an industry or a business. The possible influences of these PEST factors are discussed here under.

i. Political Factors

The first element of a PEST analysis is a study of political factors. When examining political factors, it is essential to look at any political changes that could affect the business, as these changes influence organizations in many ways. The political factors can create advantages and opportunities for an industry. Conversely, they can even place some restrictions on the industry. For a multinational business, these factors become much more complicated. So, the industry must continuously monitor

---

* - PEST analysis (also sometimes called a STEP, PESTLE or STEEP analysis) looks at the external business environment.

- the second **E** in STEEP indicate ecological/environmental factors and the **L** in PESTLE indicate legislative factors. The analysis examines the impact of each of these factors (and their interplay with each other) on the business. The results can then be used to take advantage of opportunities and to make contingency plans for threats.

the political stability of Government, understand rules and regulations imposed by the Governmental, EXIM policies and be knowledgeable about the laws that impact the industry and each firm in the industry. Some important political factors include:

- ☆ Legislation such as the Minimum Wage Laws
- ☆ Market regulations
- ☆ Trade agreements, tariffs or restrictions
- ☆ Tax levies and tax breaks
- ☆ Type of Government regime (Mixed economy, Capitalism, Communism, Socialism)
- ☆ Trends towards globalization of the markets

It is essential to note that, in the modern era of agri-business, the political institutions (mainly Government, which is the major institution in a country and can be seen as the biggest business enterprise at national or local level), may affect the business in a direct way, through above mentioned factors. Hitt also mentioned another feature of the political environment *i.e.*, the active role of the companies to influence the Governmental policies and actions, by developing their corporate political strategy.

ii. Economic Factors

The second element of a PEST analysis involves a study of economic factors. Often the political factors spill over into economic factors. For example, decisions on tax are usually taken by the politicians based on a mixture of political and economic factors. Interest rates, in many countries are decided by a central bank, but political factors may still be important. Other economic factors include unemployment rates, exchange rates, inflation levels, income growth, debt and saving levels (which impact available money), business expectations etc. So, all businesses are affected by national and global economic factors. The current state of Indian stock market is a typical example of the volatility of economic factors. For example, if an economy is undergoing recession, it will have high unemployment, low spending power and low business expectations. Conversely a 'booming' or growing economy will have low unemployment, high spending power and good business expectations.

iii. Socio-Cultural Factors

The third aspect of PEST focuses its attention on socio-cultural factors - the elements that build society. These factors influence people's choices and include the beliefs, values and attitudes of society. So, understanding changes in this area are crucial for the industry, as such changes can impact the purchasing behaviour of people. The following areas should be looked into by the industry:

- ☆ Consumer attitudes to firm's product
- ☆ Environmental issues (especially, if the product involves hazardous or potentially damaging production processes)
- ☆ Role of women in society
- ☆ Demographic changes, as changes in the structure of a population will affect the supply and demand of goods and services within an economy. For example, falling birth rates will result in decreased demand and greater competition as the number of consumers fall. Conversely, an increase in the global population and world food shortage predictions calls for greater investment in food production. With gradual changes in socio-cultural factors,

many of the developing countries and Least Developed Countries (LDCs) are now reconsidering their rejection of genetically modified foods.

- Growth rate of population, age distribution, changes in tastes and buying patterns etc.

These influences and developments are strategically important for the firms, as on one hand, people are a key organizational resource for them, and on the other hand, people are the consumers of their production. Therefore, the number and quality of people are essential for the business firms. Further, the firm manager must effectively monitor these trends, so as to make sure that, the product will continue to meet the needs and demands of the consumers. New attributes may emerge and the importance of existing attributes may change as a result of changes in society. Similarly, as the business enters new markets or countries, expectations may be very different based on the social structure and societal expectations of the specific culture or cultures. So, the industry must be vigilant on such changes and must be able to produce products and services that aim to complement and benefit people's lifestyle and behaviour. Otherwise, it will lose market share and demand for their product or service.

iv. Technological Factors

Advances in technology can have a major impact on business success. If the firms or industry could not maintain the pace of technological development, they have to go out of the business. Further, the technological changes influence socio-cultural attitudes. For example, the way people spend their leisure has changed dramatically over the last 25 years. As well as advances in the industry, the business must keep on watching the technological revolutions like internet, nanotechnology, mobile phones, computer software etc. Besides these technologies, the businessmen must be vigilant on latest production technologies that ensure economies of large scale, production of quality output etc., as they provide a competitive edge in the market.

b. Industry Environment

Compared to general environment, the industry environment will have a direct bearing on the firm's competitive behaviour and its profitability in the business. The framework that is commonly employed to analyze the industry environment is the Five Forces Model, developed by Michael Porter. This analysis guide the business firm to formulate effective business strategies, so that it develop opportunities to excel in its industry environment and also protects itself against possible competitive threats in the industry. Porter's Five Forces include bargaining power of buyers, bargaining power of suppliers, barriers to entry, threat of substitutes and intensity of rivalry among existing firms. The way these five elements interact was illustrated through Figure 4.3 and it shows that, the intensity of rivalry among the firms in the industry is the central force and is directly affected by the other four forces of the model. It is important to note that, all these five forces should be considered together, so as to formulate effective business decisions for the firm.

i. Intensity of Rivalry among Existing Firms

This is the central force and the rivalry between the firms in the industry is in terms of both price and non-price competitions. Non price competition may take place on the basis of product differentiation, branding, advertising about the products, product quality, innovations etc.

ii. Bargaining Power of Suppliers

It refers to the power or ability of the firms to fix higher prices for the products that ensures higher profits to them. There exists inverse relation between number of firms and bargaining power. For example, a monopolist enjoys highest market power compared to the firms under perfect market

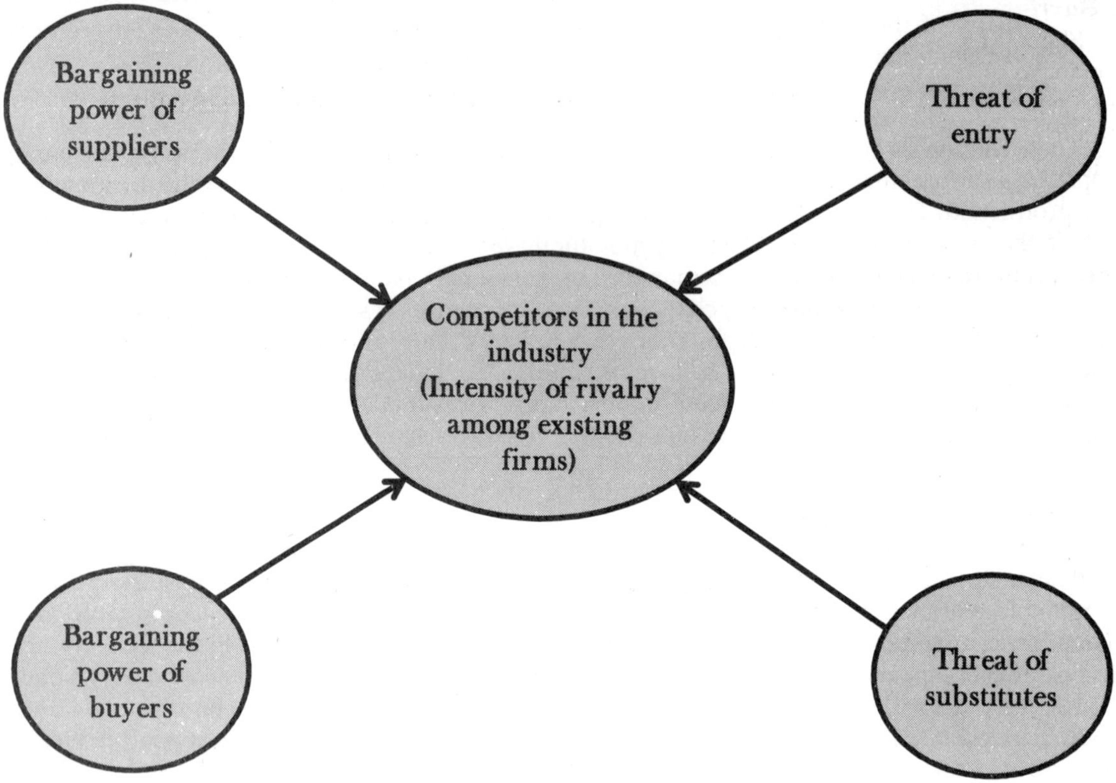

Figure 4.3: Porter's Five Forces of industry environment.

competition. Product differentiation, quality of product, availability of substitutes for the product, advertising etc., also influences the bargaining power of supplier in the market.

iii. Bargaining Power of Buyers

This force is viewed as a mirror image of the supplier's power. If the bargaining power of the buyer is high, it can become a threat for the industry. For example, if the number of firms are large and selling a homogeneous product (*i.e.*, perfect competition) and if one of the firms quote a high price for the product considering its average total cost of production, then there is greater chance on the part of buyers to switch over to another firm producing and selling the same product.

iv. Threat of Substitutes

The presence of substitutes for the products in the industry increases the rivalry between the firms. In general, new products (substitutes) produced by the firms are cost-effective in the context of producing them with modern technologies that ensure economies of large scale and further, the learning experience of the firms also leads to decline in costs over time. So, availability of substitutes among the products guides the business firms to keep a close watch on the updating of technologies that result in cost reduction per unit of product.

v. Barriers to Entry

Shepherd defined barriers to entry as *'anything that decreases the likelihood, scope or speed of entry'*. If an industry is earning super normal profits, it is attractive for the new firms to enter the industry. But, their entry will squeeze out the super normal profits and hence, the existing firms will impose certain barriers against the entry of new firms. They include capital requirements, economies of scale, differentiation of the products through branding, the degree of government protectionism etc. Thus, there exists an inverse relation between the entry barriers and competition in the industry *i.e.*, the lesser the barriers for entry into the industry, the more is the competition among the firms.

c. Competitor Environment

It refers about the rivalry firms, with which a firm competes in the industry. Porter's Generic Strategies approach (Figure 4.4) is commonly employed to analyze the competitor environment. As shown in the matrix, it has two dimensions *viz.*, Competitive advantage and Competitive scope. The competitive advantage dimension offers two strategies for the firm to gain profits in the business *viz.*, lower cost (cost-effectiveness in production and marketing) and product differentiation. Similarly, competitive scope also has two components relating to market focus *viz.*, broad target and narrow target. If the market focus is broad, then the firm enjoying economies of large scale gains through cost advantage (Cost leadership). Otherwise, it has to opt for product differentiation strategy to maximize the profits. On the other hand, if there is narrow market focus, the strategy is the Focus strategy. It is characterized by narrow target combined with both cost advantage and differentiation. Since, the market focus is narrow, firms pursuing a focus strategy have lower output supplies and therefore, less

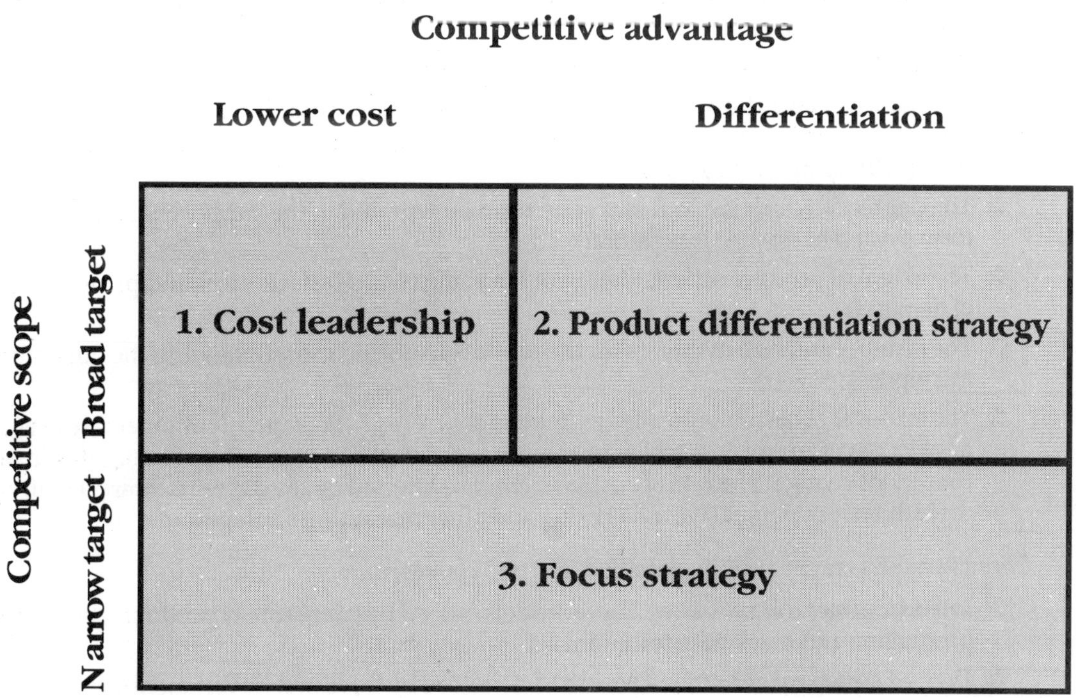

Source **: Grant (2005)**

Figure 4.4: Porter's Generic Strategies.

bargaining power with their input suppliers. According to Porter, the features of these three strategies *i.e.,* Cost leadership, Product differentiation strategy and Focus strategies cannot be combined, as the last strategy represents only narrow market focus. In his words, *'a firm stuck in the middle is almost guaranteed low profitability'*. This signifies the importance of competitor environment in influencing the firm's strategies in the business.

2. Structure

Market structure refers to the organizational characteristics of a market, which seem to influence strategically the nature of competition and pricing within the market. It consists of the relatively stable features of the market that influence the rivalry among the sellers operating in a market. The market structure is important not only because, it has implications on the conduct and performance, but also because, it impacts the strategic possibilities which face the organization, its ability to act strategically and the likely effects of such strategic behaviour. For example, a firm with a market share of 75 per cent will enjoy a different situation than a firm that has only five per cent share of the same market. The first will have a market power and the second will experience a rigorous competitive pressure. Market structure often differs across industries, because of variations in basic market conditions including the underlying technological base, the legal environment, demand conditions, and economies of scale. All of these basic conditions tend to affect the number and size distribution of firms in an industry. According to Mason, market structure is a multi-dimensional concept that is specified and measured by the following variables:

- ☆ The number of firms and the extent of overseas competition (*e.g.* from within the single market or in global market)
- ☆ The market share of the largest firms (measured by the concentration ratio)
- ☆ The nature of costs both in the short run and long run (including the potential of firms to exploit economies of scale and also the presence of sunk costs, which affects market contestability in the long run)
- ☆ The degree to which the industry is vertically integrated in the supply chain (for example, forward and backward integration).
- ☆ The extent of product differentiation or branding (which affects cross and price elasticities of demand)
- ☆ The number and size of buyers for the product in the industry (including the possibility of monopsony power)
- ☆ The turnover of customers (called as 'market churn') – *i.e.,* how many customers are prepared to switch over from one supplier to another supplier in a given period of time, when market conditions change. The rate of customer churn is affected by the degree of brand loyalty and the influence of persuasive advertising and other marketing strategies.

These variables can be further classified into two classes, namely

- ☆ ***Intrinsic structural variables:*** These are determined by the nature of products and available production and marketing technologies.
- ☆ ***Derived structural variables:*** These are determined by firms and Government such as barriers of entry, seller and buyer concentration and product differentiation.

The variables of the market structure are discussed in-detail under the following sub-heads:

A. The Number of Firms or Sellers

The number of firms is the most obvious structural measure of market power in the market. In case of perfect competition, there is infinite number of sellers, while monopoly has only one firm. In between these two extremes, there are oligopoly and monopolistic competitions, which have few and large number of firms respectively in the industry (Table 3.1 and Figure 3.1).

B. The Market Share of the Largest Firms (Market Concentration)

We know, the least competitive market structure is monopoly, as there is only one firm in the industry and the most competitive market structure is perfect competition, as it includes infinite number of firms, each with a very small share in the total market sales. Between these two extreme market structures, there are (imperfect) market structures that are more competitive than monopoly, but less competitive than perfect competition. So, it would be useful to have some measures of how competitive an industry is and they include counting the number of firms, Concentration Ratios (CRs), Herfindahl–Hirschman Index (HHI) and Lorenz Curve and Gini Coefficient (GC). Among these, the important measures like CRs and HHI falls under Absolute concentration measures and the last one (Lorenz Curve and GC) fall under Relative concentration measure. The CRs and HHI are the two most commonly used concentration indices used in empirical SCP studies. Lorenz curve and GC are used to measure the degree of inequality in firms size distribution. Debates on the choice of which concentration measures to be used have revolved around correlation between the different measures and the sensitivity of these measures to changes in the number of firms and market shares and these measures are discussed in-detail here under.

a. Counting the Number of Firms

When we say an industry is 'more concentrated', it implies, it is less competitive and *vice versa*. So, a monopoly is the most concentrated industry and a perfectly competitive industry is the least concentrated. One simple, but not a very good measure of how concentrated an industry is would be simply to count the number of firms in an industry. A monopoly has only one firm, a duopoly has two firms, an oligopoly has few firms, a monopolistic competition has large number of firms and a perfectly competitive industry will have infinite number of firms. But, this measure is not valid, as explained through the following example. Suppose, a fruit processing industry comprises of five firms. Among these, one big firm accounts for 96 per cent of total processed fruit products marketed in the country and each of the other four firms account for only one per cent of the sales. On the other hand, the fertilizer industry in that country has ten firms, each of which accounts for ten per cent of the total fertilizers sales in the country. So, we cannot say, the fertilizer industry is more concentrated, based on the number of firms. Though fruit processing firms are few in number compared to fertilizer firms, the fruit processing industry is more concentrated compared to fertilizer industry, as a single fruit processing firm accounts for greater share (96 per cent) of total market sales of fruit products. Thus, fruit processing industry is considered to be less competitive than fertilizer industry. So, simply counting the number of firms in an industry is not a correct measure of the concentration in an industry. A slightly better idea is to look at the shares of the market, which the largest firms have. An example of this sort of measure is the CRs.

b. CRs

Market concentration also refers to the sellers' concentration. It is, otherwise, referred as proportion of sales made by the largest firms in the industry. So, it is expressed as the fraction of industry's total sales accounted by one or more largest firms in the industry. The CR expresses the total sales of the 'N' number firms as percentage of industry's aggregate sales (Stead, 1996). In the words of Jacobson, it refers to the *'cumulative market share of the leading 'N' firms in the industry'* (Jacobson, 1996). A high market concentration indicates that, a few traders handle a large proportion of the quantities of the commodities transacted in the market. In case of monopoly competition, the market share is 100 per cent, as the single firm represents the industry. The market share gradually decreases from oligopoly competition to perfect competition. Thus, as the competition approaches towards perfection, the market share of the dominant firms goes down. Thus, in case of perfect competition, each firm will have an insignificant market share in the industry output. We know, in between monopoly and perfect competition, there is oligopoly competition and this concept of market share is especially important in case of oligopoly competition. This is because, in oligopoly, there are few firms and there is tendency of collusion among these firms due to their mutual interdependence in terms of output and pricing decisions. Thus, market share is important in measuring the degree of concentration of business with respect to dominant firms and their market power in the oligopoly industry. It is essential to note two important aspects in the context of market (sellers') concentration that, there exists inverse relation between the level of market concentration and market competition and there exists positive association between level of market concentration and profitability of the firms in the industry. So, a higher market concentration implies, non-competitive behaviour of the firms and thus, inefficiency. Kohl and Uhl (2002) concluded that, *'there exists higher level of market concentrations in food processing, wholesaling and retailing than in farming'*. Many studies indicated that, there exists positive association between market concentration and Gross Marketing Margins (GMMs).

CR is also called as Market concentration index, as it is the most widely used device for determining the seller concentration in an industry *i.e.*, it indicates, how much market share is concentrated in the hands of different number of firms. It is to be noted that, this CR can be used to measure different parameters like market sales or assets or employment or value added etc., and it is also used to classify how competitive an industry is. If an industry is characterized by low concentration, it will have a large number of firms with small market shares. On the contrary, if the industry is characterized by high concentration, it will have a small number of firms with relatively high market shares. Industries with high concentrations are more likely to have significant market power, *i.e.*, the ability to set price and *vice versa*. To compute the CR, the firms are ranked in order of size 'usually measured in terms of sale', starting from the largest in the industry at the top and going down to the smallest firm at the bottom. The CR for the 'N' number of largest firms in an industry is calculated simply by adding up the market shares (S) of these 'N' number of firms. This can be represented as $CR_N = \Sigma S_N$. Concentration ratios are usually given for the largest four firms (CR4 represented as $CR4 = S_1 + S_2 + S_3 + S_4$), largest eight firms (CR8) and sometimes, the largest 20 firms (CR20). If market share data of each firm is not readily available, the researcher must calculate market shares atleast for the largest firms in an industry. An individual firm's market share is calculated as Individual firm sales/Total industry sales. The arbitrary selection of number of largest firms is contingent on the data used and it can make a fundamental difference to empirical findings (Stead, 1996). But, the two most common CRs used are CR4 and CR8. The four-firm concentration ratio (CR4) is the proportion of total sales shared by the four largest firms in the industry and the eight-firm concentration ratio (CR8) is proportion of total sales shared by the eight largest firms in the industry. These CRs are commonly used to indicate the degree to which an industry is oligopolistic and the extent of market control of the largest firms in the industry. In general, the industries that are highly concentrated in one advanced economy tend to be highly concentrated in another. The CRs can be calculated in one of the following two ways:

- ☆ The first is to sum total sales of the top four (CR4) or eight (CR8) firms in the industry, then divide by the total and express the same in percentage.
- ☆ Alternatively, the market shares of the top four (CR4) or eight (CR8) firms can be calculated individually and then summed.

i. CR4

This ratio measures the total market share held by the top four largest firms in the industry. This ratio is designed to measure industry concentration, as indicated through the Table 4.2. Thus, CR4 ranges between 0 to 100. When the CR4 is zero, it implies, there is perfect competition among the firms or atleast monopolistic competition that is largely competitive. It also indicates that, the number of firms in the industry or group is very large such that, the share of the four largest firms is zero per cent. At the other extreme, if the CR4 is 100 per cent, it implies, there is an extremely concentrated oligopoly (because, the number of firms are more than one). If the calculation is based on a one-firm concentration ratio, then a 100 per cent value means monopoly competition. Between these two extremes *i.e.,* CR4 ratio is between zero to 100, it implies there is low or medium or high concentration. The Table 4.2 is only a rule of thumb, as there is no consensus among Economists on using the CR4. The same concept is explained through Figure 4.5 and it shows the inverse relationship between CR and market competition. That is, as we move from left (perfect competition) to right (monopoly) in Figure 4.5, market concentration increases. So, the market power (the ability to fix price) of firms in the industry also increases. Typically, profit margins increase, as market power increases. Another important reason to classify industries into different market structures based on CRs is that, it indicates whether the largest firms in an industry explicitly take other firms' actions into account in deciding policies regarding output, pricing, non-price competition etc.

Table 4.2: Classifying Industries with the CR4.

Level of Concentration	*CR4 (per cent)*	*Nature of Market Structure*
Low	0-50	Perfect competition to slightly concentrated Oligopoly/Oligopsony
Medium	50-80	Slightly concentrated to highly concentrated Oligopoly/Oligopsony
High	80-100	Highly concentrated Oligopoly/Oligopsony to Monopoly
Zero concentration	Zero	Perfect competition
Maximum concentration	100	Monopoly

Low Concentration

If the CR4 is in between 0 to 50 per cent, it implies the industry is with low concentration. The lowest percentage (towards zero) implies monopolistic competition and the highest percentage (towards 50) implies, emergence of an oligopoly form of competition.

Medium Concentration

If the CR4 is in between 50 to 80 per cent, it implies the industry is with medium concentration. This further implies, the industry is very much oligopoly. From these market share numbers, it is easy to see that, Firm A is the dominant firm (Table 4.3) in the pesticide market (Medium concentration). While Firm A does not totally dominate the market, the other firms like B and C undoubtedly look to Firm A to compete in the industry. So, they plan for advertising strategies, new product innovations and even resort to price competition to acquire a portion of Firm A's market share.

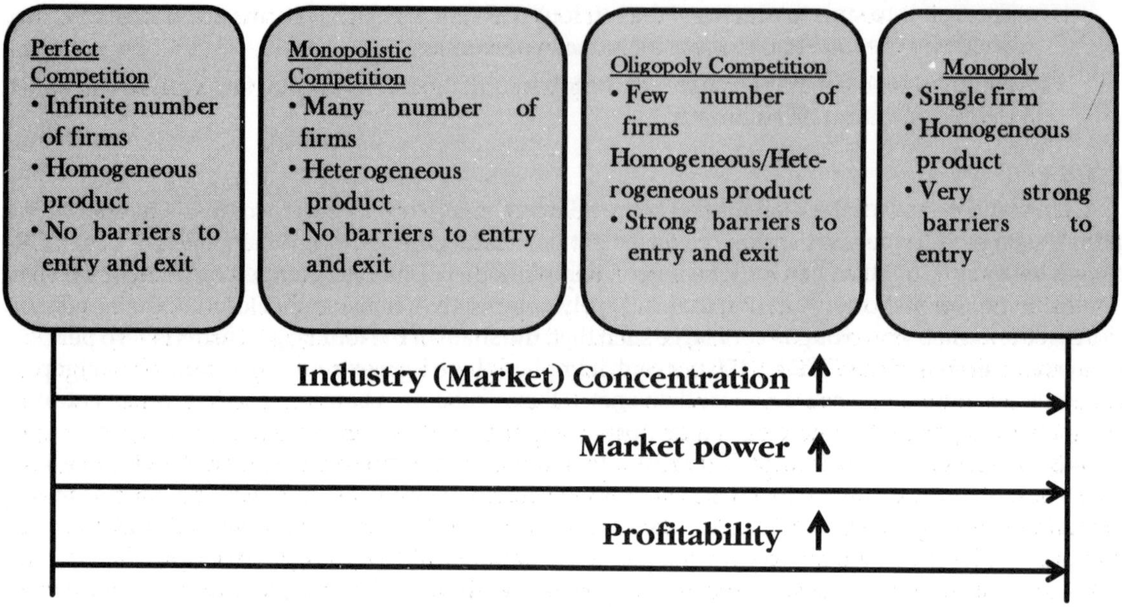

Figure 4.5: Market structure, Concentration and Profitability.

High Concentration

If the CR4 is in between 80 to 100 per cent, it implies the industry is with high concentration. This high concentration, calls for regulating the industry by the Government, as the top four firms are aiming at monopoly gains against the interest of the consumers.

Table 4.3: CR4 and market concentration in a pesticide market.

Firms	*Low Concentration*		*Medium Concentration*		*High Concentration*	
	Sales (₹ Lakhs)	*Market Shares (per cent)*	*Sales (₹ Lakhs)*	*Market Shares (per cent)*	*Sales (₹ Lakhs)*	*Market Shares (per cent)*
A	397	12.30	287	19.20	500	24.83
B	345	10.69	243	16.25	497	24.68
C	318	9.85	215	14.38	432	21.45
D	298	9.23	116	7.76	328	16.29
E	287	8.89	109	7.29	54	2.68
F	286	8.86	101	6.76	52	2.58
G	265	8.21	95	6.35	43	2.14
H	262	8.12	87	5.82	40	1.99
I	261	8.09	86	5.75	33	1.64
J	259	8.02	79	5.28	21	1.04
Others	250	7.74	77	5.15	14	0.70
Total	3228	100.00	1495	100.00	2014	100.00
CR4		42.07		57.59		87.24

It is essential to note that, as shown in the Table 4.3, CR4 is 57.59 and it suggests a medium level of concentration and a modest degree of competition. This ratio 57.59, however, can be achieved in a number of ways. If each of the top four firms has an identical ₹ 215.24 lakhs worth of sales, then the CR4 is 57.59 per cent. Alternatively, the concentration ratio is 51 per cent for Firm A, if it had ₹ 762.45 lakhs worth of sales and remaining three firms (B, C and D) each account for only ₹ 32.89 lakhs worth of sales. Though CR4 is the same in both cases, the degree of competition is likely to differ. The oligopolistic industry is likely to be more competitive, if four firms have nearly equal sales than if one firm has significantly more sales than the others.

ii. CR8

The CR8 is also calculated in a similar manner, as discussed under CR4, but here, the shares of market sales of top four firms listed above (A,B, C and D) plus the next four firms (E, F, G and H) are also taken into account to compute the total market share. Thus, the proportion of total sales of product in an industry transacted by the eight largest firms in an industry is taken into consideration to compute CR8. The CR8 also indicate the degree of market concentration and the extent of market control held by the eight largest firms in the industry. The classification of markets as low, medium and high concentrations on the basis of market shares held by top eight firms (CR8) was same as discussed under CR4 with reference to Table 4.2. As discussed under CR4 (Table 4.3), the same conclusions can be drawn for CR8 with reference to Table 4.4, but considering the shares of top eight firms in the industry. However, CR4 measure is more meritorious over CR8 in assessing the market concentration because, if top four firms had significant market share (high market concentration), definitely CR8 will be further high. Further, when CR4 itself is significantly high, it sends warning signals to the Government to plan for regulation measures, so as to check the monopoly practices resorted by top four largest firms. Further, as discussed earlier, markret concentration and competition are inversely related to each other.

Table 4.4: CR8 and market concentration in a pesticide market.

Firms	*Low Concentration*		*Medium Concentration*		*High Concentration*	
	Sales (₹ Lakhs)	*Market Shares (per cent)*	*Sales (₹ Lakhs)*	*Market Shares (per cent)*	*Sales (₹ Lakhs)*	*Market Shares (per cent)*
A	397	7.58	220	15.59	500	16.22
B	345	6.59	217	15.38	497	16.12
C	318	6.07	215	15.24	432	14.01
D	298	5.69	98	6.95	328	10.64
E	287	5.48	90	6.38	299	9.70
F	286	5.46	87	6.17	287	9.31
G	265	5.06	80	5.67	198	6.42
H	262	5.00	79	5.60	178	5.77
I	261	4.98	76	5.39	154	5.00
J	259	4.95	71	5.03	112	3.63
Others	2259	43.14	178	12.62	98	3.18
Total	5237	100.00	1411	100.00	3083	100.00
CR8		46.94		76.97		88.19

Demerits of CRs

The CRs, however, suffers from certain drawbacks. They include,

- ☆ Though they provide useful indication about the nature and degree of competition in an industry, it does not provide information about competitiveness of the industry.
- ☆ There is no obvious justification for focusing on the market share of the top four firms rather than the top three or six. Thus, the choice of the CR4 as the appropriate measure of concentration is somewhat arbitrary.
- ☆ More importantly, the CR4 provides only limited information about actual market structure. To take an extreme example, a market, where the top four firms each had a market share of 22.5 per cent would have a CR4 of 90 per cent. But, in another market, one firm had a market share of 60 per cent and remaining three firms have with 10 per cent shares each, thereby, CR4 is again 90 per cent. But, in the former case, the market structure is more competitive, whereas in the latter, there is monopoly. Thus, though the CRs remains same, the markets may have very different competitive outcomes.
- ☆ Further, CR is only one objective factor in classifying the market structures. There are many more factors, both objective and subjective, that a researcher must take into account before classifying the market structures.

c. Herfindahl Index or Herfindahl–Hirschman Index (HHI)

Competition arises when two or more firms offer their products as substitutes, to buyers in the same market. Competition can be researched from various angles. First, it is important to establish the *incidence of competition i.e.,* is there competition? Second, one needs to ascertain the *intensity of competition*. Third, the *basis of competition* must be also considered. Coming to the *incidence of competition* (*i.e.* is there competition?), we know that, if a market involves collusion of firms (anti-competitive behaviour), competition is difficult to enforce. Furthermore, when firms are of similar sizes, competition increases because, none of them can dictate to the market (Sharon, 1999). The HHI is a concentration measure that can be used as a tool for assessing the incidence of competition. This index is named after Economists, Orris C. Herfindahl and Albert O. Hirschman. Not only in analyzing the market concentration, this index is widely applied in competition law, antitrust and also technology management. It is computed as the sum of the squares of the market shares of the 50 largest firms (or summed over all the firms, if there are fewer than 50) within the industry, where the market shares are expressed as fractions. It is given by,

$$HHI = \Sigma(\text{Market share}_N)^2$$

$$H = \sum_{i=1}^{N} S_i^2$$

where,

'S_I' is the market share of firm (in percentage) and N is the number of firms. For example, if two firms A and B each have 50 per cent market share, the HHI equals 5000. So, the HHI can range up to 100^2, or 10,000. That is, if there is only one firm, it will have 100 per cent market share, thereby, HHI is 10000. The HHI can also be expressed as

$$HHI = (\text{Share 1})^2 + (\text{Share 2})^2 + (\text{Share 3})^2 +. + (\text{Share N})^2$$

Thus, HHI is calculated by summing the squares of the individual market shares of all the firms in an industry. In the above equation, the terms in parentheses (Share 1, Share 2 etc.) are the market shares for each firm in total industry sales. The last term (Share N) indicates the market share of the last (N^{th}) firm. For an industry with twenty firms, this formula includes twenty share terms and N is equal to twenty. The formula further denotes that, the market share of each firm is squared. This measure is certainly meritorious over CR because, unlike the CR4, the HHI is influenced both by the number of firms in the market and differences in their relative size.

Table 4.5: HHI and market concentration.

Level of Concentration	*CR4 (per cent)*	*HHI*	*Nature of Market Structure*
Low	0-50	0 to 1000	Perfect competition to Slightly concentrated Oligopoly/ Oligopsony
Medium	50-80	1000 to 1800	Slightly concentrated to highly concentrated Oligopoly/ Oligopsony
High	80-100	1800 to 10000	Highly concentrated Oligopoly/Oligopsony to Monopoly
No concentration	0	0	Perfect competition
Total (Maximum) concentration	100	10000	Monopoly

As shown through Table 4.5, HHI ranges from 0 to 10000. If the HHI is Zero, it implies perfect competition and if it is 10000, it indicates monopoly. So, higher values of HHI mean greater concentration, less competition and more market control held by individual firms. On the contrary, the lower values of HHI indicate lesser market concentration, more competition and less market control by individual firms. Between these two extremes, the HHI can fall into low, medium and high market concentrations.

Low Concentration

A HHI ranging between 0 to 1,000 is commonly interpreted as an industry with low concentration. The extreme zero HHI implies perfect competition, low values HHI implies monopolistic competition and as the value approaches 1000, it implies oligopoly is emerging. Thus, a small index indicates a competitive industry with no dominant players. In general, the industries with CRs ranging between 0 per cent to 50 per cent will have HHI values between 0 and 1,000.

Medium Concentration

A HHI ranging between 1000 to 1800 is commonly interpreted as an industry with medium concentration. This range of HHI implies the range of competition from slightly concentrated oligopoly/ oligopsony to highly concentrated oligopoly/oligopsony. In general, the industries with CRs ranging between 50 per cent to 80 per cent will have HHI values between 1000 to 1800.

High Concentration

A HHI ranging between 1800 to 10000 is commonly interpreted as an industry with higher concentration. This range of HHI implies the range of competition from highly concentrated oligopoly/ oligopsony to monopoly. In general, the industries with CRs ranging between 80 per cent to 100 per cent will have HHI values between 1800 to 10000.

No Concentration

At one extreme, zero HHI implies perfect competition. That means, the number of firms is so large that, the sum of the square of the market shares is zero.

Total (Maximum) Concentration

At the other extreme, HHI of 10000 implies monopoly. This is achieved, if there is only one firm with a market share of 100 per cent. Note that, this is slightly different from CR4 and CR8, which can achieve a 100 per cent value even for oligopolistic industries comprising only four and eight firms respectively in their respective industries.

Example 1

Suppose, there are five firms in fruit processing industry and their market shares are as follows: Firm A has 30 per cent market share, Firm B has 30 per cent market share, Firm C has 20 per cent market share, Firm D has 15 per cent market share and Firm E has five per cent market share. The HHI is given by, $(30)^2 + (30)^2 + (20)^2 + (15)^2 + (5)^2 = 2450$. Since, HHI is slightly above 1800, it implies the market structure is highly concentrated oligopoly.

Example 2

Assume there are twenty firms in a pesticide industry and their sales in the industry are shown through Table 4.6. The market shares for each firm are computed in the third column of the table. Firm A's market sales worth of ₹ 485 lakhs per year gives it a market share of 20.59 per cent. It is followed by Firm B with market sales worth of ₹ 401 lakhs per year, has a 17.02 per cent market share. The twentieth and least popular Firm T has ₹ 6 lakhs worth of market sales per year and a market share of only 0.25 per cent.

Table 4.6: Computation of HHI for a pesticide industry with 20 firms.

Rank	*Firms*	*Sales (₹ Lakhs)*	*Market Share (per cent)*	*Market Share²*
1	A	485	20.59	423.77
2	B	401	17.02	289.69
3	C	378	16.04	257.41
4	D	275	11.67	136.24
5	E	187	7.94	63.00
6	F	87	3.69	13.64
7	G	75	3.18	10.13
8	H	71	3.01	9.08
9	I	66	2.80	7.85
10	J	61	2.59	6.70
11	K	54	2.29	5.25
12	L	44	1.87	3.49
13	M	39	1.66	2.74
14	N	31	1.32	1.73
15	O	27	1.15	1.31
16	P	23	0.98	0.95
17	Q	18	0.76	0.58
18	R	17	0.72	0.52
19	S	11	0.47	0.22
20	T	6	0.25	0.06
		2356	100.00	HHI = 1234.39

The HHI index is calculated by squaring each market share and then summing all twenty. The square of Firm A's 20.59 per cent market share is 423.77. The square of Firm B's 17.02 per cent market share is 289.69. And the square of the 0.25 per cent market share of the twentieth Firm T is only 0.06. Summing all twenty squared market shares generates an HHI of 1234.39. This places the pesticide industry in the medium concentration range, as the top few firms have some market control, but not a whole lot.

It is essential to note that, if all firms have an equal share in the market, the reciprocal of the HHI multiplied by 10000 shows the number of firms in the industry. For example, if there are two firms and each have 50 per cent share in the market, then HHI is 5000. So, the number of firms in the industry is given by [(1/HHI)*10000] = 2. Similarly, if HHI is 2500, then the number of firms in the industry is [(1/2500)*10000] = 4. On the other hand, when firms have unequal shares, the reciprocal of the HHI multiplied by 10000 indicates the 'equivalent' number of firms in the industry. For example, if market shares of four firms A, B, C and D are 30, 25, 22 and 33 per cent respectively, then HHI is 3098. So, the equivalent number of firms is given by, [(1/3098)*10000] = 3.23. Similarly, a HHI of 2455 is equivalent to a market in which there are 4.1 equally sized firms, while a HHI of 4170 equates with a market in which there are 2.4 equally sized firms. Thus, comparing the number of equally sized firms corresponding to the value of the HHI with the actual number of firms in the market provides further information about the structure of a particular market.

It is also essential to note that, the value of the HHI decreases, as the number of firms in a market rises. Similarly, the value of the HHI will be greater, the larger the degree of inequality in firms size. To explain the same, consider the following two cases.

Case 1

There are four firms in a food processing industry each with 22.5 per cent market share.

Case 2

There are four firms in a food processing industry, where one firm has 60 per cent and three others have a 10 per cent market share each.

Assume that, in both the cases, the remaining 10 per cent market share is divided between two firms each with a 5 per cent. On computation, the HHI in the first case will be 2075 and 3950 in the second case. So, the greater the inequality in firm size in the industry, the higher the HHI.

From the above methodology, two key features of HHI can be outlined. First, the market shares are squared before summing and information about every firm in the industry is used in computing the HHI. This is where, it is meritorious over CRs. As discussed earlier, in case of CRs, only information (market shares) of the top four or eight firms are considered and not of all the firms. Further, the squaring of the market shares ensure the following advantages in computing the HHI:

- ☆ First, it gives relatively more importance to firms with greater market shares than the firms with low market shares. We know that, the firms with greater market shares, have greater market control. For example, the top two firms in a market have a combined 50 per cent market share. The HHI computed is significantly different, if the individual market shares of the top two firms are 30 per cent and 20 per cent than if they are 40 per cent and 10 per cent. The first combination contributes 1300 points to the HHI and the second contributes 1700 points. But, the CR remains same in both the cases.
- ☆ Second, even though the computation of HHI includes every firm in the market, the inclusion of small size firms have a minimal impact on the calculation and can be largely ignored. For

example, firms with less than one per cent market share add less than one point to the HHI that falls in the range of 1,000 to 5,000 points.

Merits of HHI

The following are the merits of HHI over CRs:

- HHI uses the information of each firm in the industry, unlike CRs.
- The HHI gives proportionately greater weight to the market shares of the larger firms. This recognizes the larger firms' relative importance in competitive interactions. Let us examine an interesting case. Assume the following two cases in which, the six largest firms produce 90 per cent of the total market share.
  - ***Case 1***: All six firms sell 15 per cent each, and
  - ***Case 2:*** One firm sells 80 per cent, while the five others sell two per cent each.

  Further, assume that, the remaining 10 per cent of sales is divided among 10 equally sized firms. So, the six-firm CR (CR6) would equal to 90 per cent for both case 1 and case 2. But, Case 1 implies there is perfect competition whereas, the Case 2 approaches monopoly. The HHI computed for these two cases makes the lack of competition in the second case (*i.e.*, monopoly) is strikingly clear:
  - ***Case 1:*** $HHI = 6 * 15^2 + 10 * 1^2 = 1350 + 10 = 1360$
  - ***Case 2:*** $HHI = 1 * 80^2 + 5 * 2^2 + 10 * 1^2 = 6400 + 20 + 10 = 6430$

  This behavior rests in the fact that, the market shares are squared prior to being summed, giving additional weight to firms with larger size.
- If there are twenty firms in a pesticide industry (as shown through Table 4.7) and if the sales of S and T firms are raised over Firm G but less than Firm F, then there is no much improvement in CR4, as it considers only the market shares of top four firms *i.e.*, A, B, C and D. But, with these changes in market sales of S and T firms, there will be a definite significant change in HHI, as the increased shares of S and T firms are also considered and they are further squared. Further, if the market shares of the first and second firms in an industry change, then the HHI also changes. This is not true for CRs. For example, in the Tables 4.6 and 4.7, the sales of Firm A increases from ₹ 485 lakhs to ₹ 600 lakhs and the sales of Firm B falls from ₹ 401 lakhs to ₹ 286 lakhs. This causes an increase in market share to 25.47 from 20.59 for Firm A and fall in market share from 17.02 to 12.14 for Firm B. These changes in market shares increase HHI from 1234.39 to 1316.85. So, there is significant change in HHI due to changes in market shares of only two firms. However, this is not the case with reference to CR4 and CR8. In both the cases (Tables 4.6 and 4.7), the CR4 and CR8 remains same around 65 and 83 per cent respectively. But, the increase in sales of Firm A at the expense of Firm B, shows Firm A gained more market control. This change in market control is reflected by the HHI, but not either by CR4 and CR8.
- Unlike the CR4, the HHI is influenced both by the number of firms in the market and differences in their relative size.
- The HHI reflects both the distribution of the market shares of the top four firms and the composition of the market outside the top four firms.

Table 4.7: Computation of HHI for a pesticide industry with 20 firms with changes in market shares of Firm A and Firm B.

Rank	Firms	Sales (₹ Lakhs)	Market Share (per cent)	Market Share^2
1	A	600	25.47	648.56
2	B	286	12.14	147.36
3	C	378	16.04	257.41
4	D	275	11.67	136.24
5	E	187	7.94	63.00
6	F	87	3.69	13.64
7	G	75	3.18	10.13
8	H	71	3.01	9.08
9	I	66	2.80	7.85
10	J	61	2.59	6.70
11	K	54	2.29	5.25
12	L	44	1.87	3.49
13	M	39	1.66	2.74
14	N	31	1.32	1.73
15	O	27	1.15	1.31
16	P	23	0.08	0.95
17	Q	18	0.76	0.58
18	R	17	0.72	0.52
19	S	11	0.47	0.22
20	T	6	0.25	0.06
		2356	100	HHI=1316.85

Demerits of HHI Measure

HHI measure suffers from the following limitations:

- The value of HHI does not carry a meaning. For example, if CR4 is 75 per cent, it implies, the top four firms in the industry account for 75 per cent of total industry sales. But, a HHI value of 5610 does not mean anything for interpretation, except for inferring the level of concentration in the market structure.
- There is no special reason, why the market shares of each firm is doubled. Further, the squaring of market shares of firms, especially firms having larger shares will raise the HHI to a significant level.
- The computation of HHI demands more data or information compared to CRs. In case of CR4, a researcher requires the market shares of top four firms or total industry sales and market sales of top four firms. But, in case of HHI, the researcher must need the information pertaining to market sales of each and every firm and total industry sales. So, if the industry is monopolistic, it is labour some for the researcher to compute HHI. So, in defense of the HHI, summing the squares of the market shares for only the top five firms can provide a relatively close approximation of the actual HHI. In the given example (Table 4.6), the HHI

for the top five firms is 1170.12, which is close to 1234.39 when the shares of all the firms are considered. The bottom firms can be excluded, as their market shares are relatively insignificant compared to the top five firms and thereby, they are of diminishing importance in measuring the HHI. Thus, the firms with small market shares do not greatly affect the value of HHI because, larger firms have a much greater weightage in the calculation of the HHI. In other words, if one knows the market shares of all of the large firms, then it is possible to make a reasonably accurate estimate of the HHI. Thus, when looking at a merger between two firms A and B, one can estimate the change in the HHI on the basis of the market shares of just the two merging firms. Pre merger their contribution to the HHI is $A^2 + B^2$, post merger their contribution is $(A + B)^2$. The difference between these two HHI values is 2AB, *i.e.* twice the product of their market shares.

☆ The HHI is, however, an absolute and not a relative measure and is therefore, not suited for investigating the degree of disparity or irregularity in the shares of total firms that produce for the market. Such disparities can be analyzed using Lorenz curve and associated GC. They are based on the shares of the market that accrue to different groups of firms starting with the smallest and working up to the biggest.

d. Lorenz Curve and GC

These two statistical indicators measure the relative concentration in terms of analyzing the degree of inequality in firm size distribution. In order to calculate the GC, it is important to first understand the Lorenz curve, which is a graphical representation of measuring the extent of inequality in the distribution of volume of business. The Lorenz curve indicates the extent of inequality in which market intermediaries (commission agents, traders etc) are counted cumulatively in percentage along the horizontal axis, while market share is counted again cumulatively up the vertical axis (Figure 4.6). The percentage share of market sales are cumulated from the smallest-sized firm to the largest and the percentages of the number of firms, cumulated again from the smallest-sized firm to the largest. The Lorenz curve drawn may be compared with a diagonal line which indicates equal size distribution. The more unequal the distribution of sales firms, the greater the divergence between the Lorenz curve and the diagonal line. The area between the diagonal line and the Lorenz curve is called Area of concentration or Area of inequality. The GC summarizes the degree of inequality, since this statistic is the ratio of the area of concentration to the total areas under the diagonal (Figure 4.6).

The Lorenz curve, named after Statistician, Max Otto Lorenz thus, shows the portion of quantum of arrivals handled by different sizes of a market intermediary (say, commission agents) of the total quantum of arrivals transacted in the market and is expressed considering the cumulative percentage of a market intermediary on the horizontal axis and cumulative percentage of quantity of market arrivals handled by that intermediary on the vertical axis. If the market arrivals of produce were distributed equally across the different size-groups of a selected intermediary, the Lorenz curve would be a straight line at a 45-degree angle. But, in general, this is not the case in any market thereby, the Lorenz curve is bowed or curved, falling below the 45-degree line and creating what is called the inequality gap. The more is the gap between 45 degrees line and Lorenz curve, the more is the inequality in the distribution of business for the selected commodity thereby, more is the market concentration with respect to handling that commodity by the intermediary of different size groups. Once a Lorenz curve is constructed, calculating the GC is pretty straightforward.

The GC was invented by the Italian Statistician, Corado Gini in his paper, '*Variabilità e mutabilità*' ('Variability and Mutability', 1912) and it measures the degree of inequality in the distribution of business across the selected market intermediaries. As the Lorenz curve approaches the diagonal line,

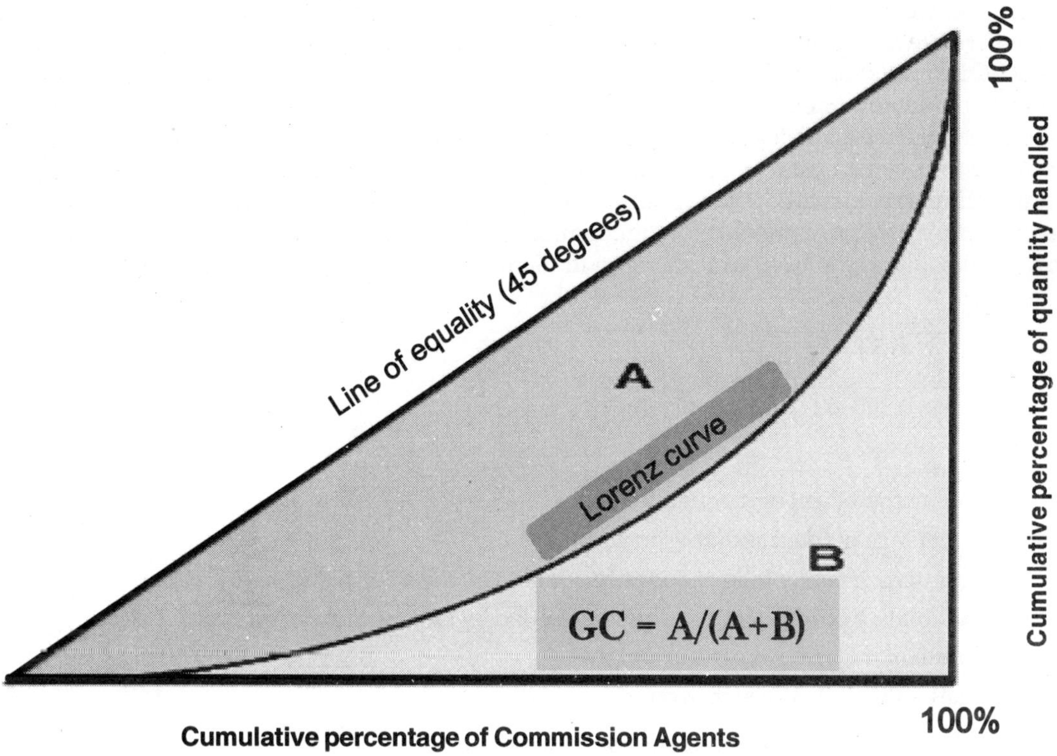

Figure 4.6: Lorenz curve for commission agents handling paddy in a regulated market.

area of concentration shrinks and the GC approaches zero. Conversely, as the inequality expands, the area of concentration, the GC approaches a value of one. Thus, GC ranges between zero to one or between zero to 100, if expressed in percentage.

As shown in the Figure 4.6, GC is measured as a ratio: the numerator is the area between the Lorenz curve of the distribution and the uniform distribution line (diagonal line) and the denominator is the entire area under the uniform distribution line or diagnol line. (The GC is equal to half of the relative mean difference). It is often used to measure the extent of inequality in the volume of business handled by the market intermediaries like commission agents, wholesalers etc in handling agricultural commodities. Here, zero corresponds to perfect business equality (*i.e.*, everyone *i.e.*, commission agents, wholesalers etc., handle the same quantity of market arrivals) and one corresponds to perfect business inequality (*i.e.*, one intermediary *i.e.*, commission agent, wholesaler etc., handle all the market arrivals into the market, while others handles zero quantum of market arrivals). The GC can be calculated by fitting the ratio of the areas on the Lorenz curve diagram (Figure 4.6). It is given by,

$$GC = \text{Area of inequality} / \text{Area below the diagonol line}$$

If the area between the line of perfect equality (diagnol line) and Lorenz curve is A, and the area under the Lorenz curve is B, then the GC is A/(A+B). Since A+B = 0.5 then, GC = 2A = 1-2B. As shown in the Figure 4.6, if A = 0, it means the Lorenz Curve is actually the Line of Equality (diagonal line). In this case, the GC is 0 and it means there is "perfect" distribution of arrivals across the market intermediaries (everyone handles the same quantity). If A is a very large area (making B very small), then the GC is large (almost 1) and it means, there is very uneven distribution of handling of commodities across the market intermediaries. So, the higher the GC, the higher the level of concentration and consequently, higher inefficiency in the market structure and *vice versa*. Markets with a high GC are more likely to become unstable, since there is a large mass of market intermediaries handling less quantities of market arrivals of commodities against few intermediaries handling large quantities. Thus, no competitive price prevails for the commodities, as there is higher market concentration. The following formula is employed in computing the Gini ratio or Lorenz coefficient of inequality:

$$L = 1 - \frac{\sum_{i=1}^{n} P_1(Z_l - Z_{l-1})}{10000}$$

where,

L = Lorenz coefficient of inequality

P_I = Percentage of intermediaries in i[th] class

Z_I = Cumulative percentage of quantity handled by the intermediaries upto ith class

Z_{I-1} = Cumulative percentage of quantity handled by the intermediaries upto i-1 class

n = Number of firms/size groups/classes

i = takes value 1, 2, 3,. n size groups

When, L=0, there is a perfect equality in the distribution

When, L=1, there is perfect inequality in the distribution

Examples

The concepts of Lorenz curve and GC can be better understood from the following hypothetical examples:

Table 4.8: Size distribution of commission agents handling paddy in a regulated market (2010-11).

Class (Quantity in Quintals)	*Per cent to Total Commission Agents (P_I)*	*Cumulative Per cent of Total Commission Agents*	*Per cent of Total Quantity Handled*	*Cumulative per cent of Total Quantity Handled (Z_I)*	*(Z_I+Z_{I-1})*	*PI (Z_I+Z_{I-1})*
<150	18.67	18.67	2.10	2.10	2.10	39.21
150-300	12.00	30.67	3.89	5.99	8.09	97.08
300-450	14.67	45.34	6.50	12.49	18.48	271.10
450-600	8.00	53.34	8.46	20.95	33.44	267.52
600-750	9.33	62.67	11.06	32.01	52.96	494.12
750-900	10.67	73.34	14.39	46.40	78.41	836.63
900-1050	12.00	85.34	16.14	62.54	108.94	1307.28
1050-1200	9.33	94.67	18.06	80.60	143.14	1335.50
>1200	5.33	100.00	19.40	100.00	180.60	962.60
			Gini ratio = 0.44			

Table 4.9: Size distribution of wholesalers handling paddy in a regulated market (2010-11).

Class (Quantity in Quintals)	*Per cent to Total Qholesalers (P_i)*	*Cumulative Per cent of Total Qholesalers*	*Per cent of Total Quantity Handled*	*Cumulative per cent of Total Quantity Handled (Z_i)*	*(Z_i+Z_{i-1})*	*PI (Z_i+Z_{i-1})*
<300	30.00	30.00	7.01	7.01	7.01	210.30
300-600	31.43	61.43	14.15	21.16	28.17	885.38
600-900	24.29	85.72	21.29	42.45	63.61	1545.09
900-1200	5.71	91.43	26.00	68.45	110.90	633.24
>1200	8.57	100.00	31.55	100.00	168.45	1443.62
			Gini ratio = 0.53			

It could be seen from the Table 4.8 that, only 5.33 per cent of commission agents were in the group of >1200 quintals and their share of quantity handled was 19.40 per cent of the total sales, whereas 18.67 per cent of commission agents were in the group of <150 quintals and their share of quantity handled was only 2.10 per cent of the total quantity handled. The Gini ratio was found to be 0.44. This situation revealed that, there exists higher degree of competitiveness for paddy in the selected regulated market, as the market concentration is low (Figure 4.7). On the other hand, Table 4.9 shows that, 8.57 per cent of wholesalers were in the group of >1200 quintals and their share in total quantity handled was 31.55 per cent, whereas 30 per cent of wholesalers were in the group of <300 quintals and their

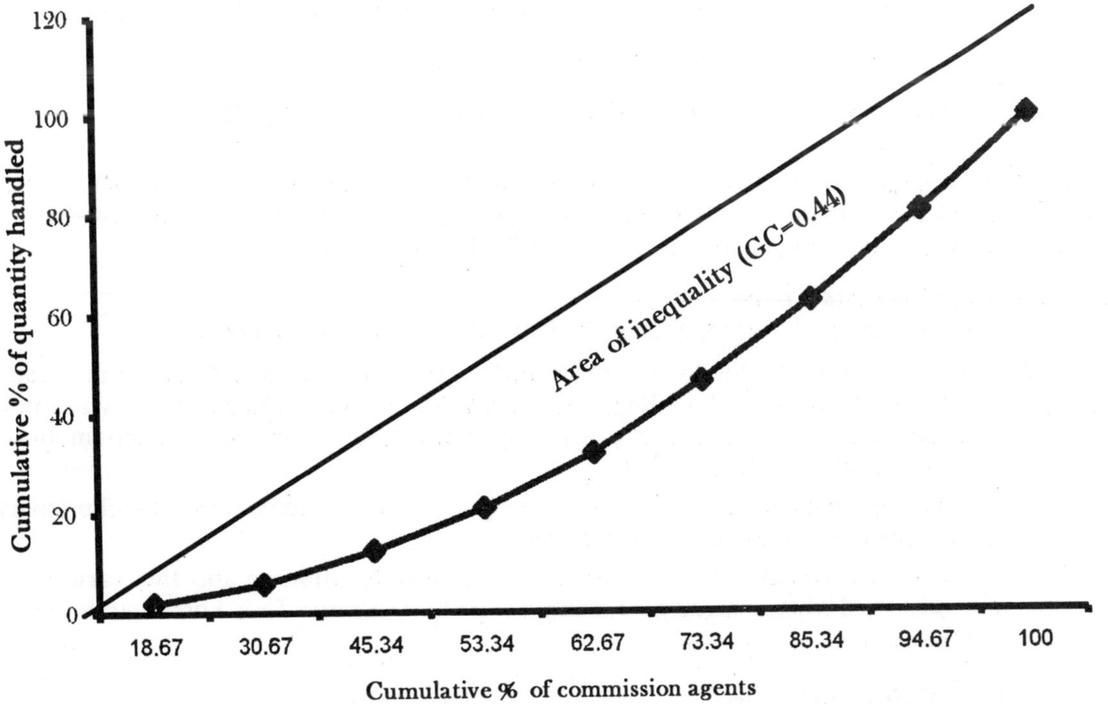

Figure 4.7: Lorenz curve for commission agents handling paddy in a regulated market (2010-11).

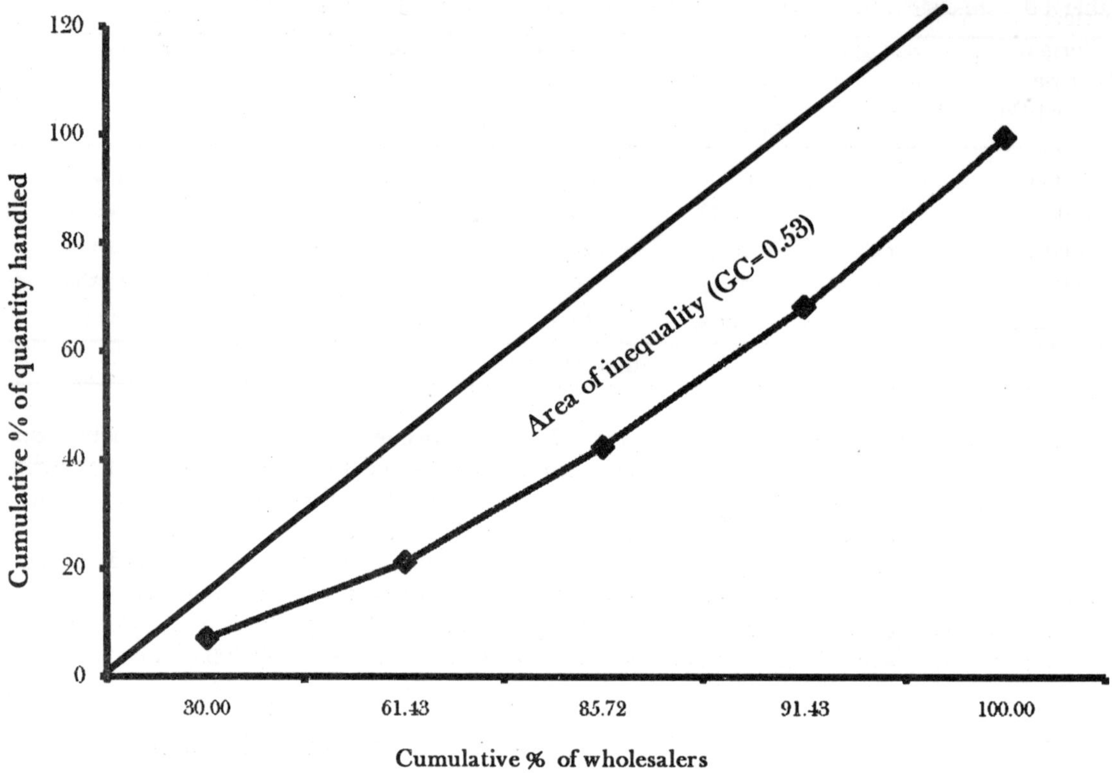

Figure 4.8: Lorenz curve for wholesalers handling paddy in a regulated market (2010-11).

share in total quantity handled was only 7.01 per cent. The Gini ratio was found to be 0.53. This situation revealed that, there exists lesser degree of competitiveness for paddy in the market, as the market concentration was high with respect to large wholesalers (Figure 4.8).

Limitations of Lorenz Curve

This relative measure of concentration suffers from the following limitations:

- The amount of inequality may be misleading, if the market intermediaries (commission agents, wholesalers etc) handling large quantities of commodities function with more efficiency than other intermediaries handling relatively less quantities, then the amount of inequality could be understated.
- When comparing two Lorenz curves, it is not possible to determine which distribution has more inequality, if the two curves intersect.
- A market intermediary's performance varies over his lifetime and this variation is not considered when analyzing inequality using a Lorenz curve. Thus, Lorenz curve ignores life cycle effects.

C. Product Differentiation

This is one important strategy that ensures profitability to the firms. Product differentiation refers to distinguishing a product from other products of other competing firms. It makes competition less perfect by reducing the degree of substitutability between products. This serves as one of the important criteria for classifying the market structures. The products produced by different market structures may be homogeneous (Perfect competition and Perfect oligopoly), differentiated (Monopolistic competition and Imperfect oligopoly) and unique (Monopoly).

A product is considered homogeneous (in perfect competition), if every firm in the industry sells exactly the same product. Most agricultural products are undifferentiated and thus they are operating in a close to perfect competition. (Food processors, however, try to differentiate their products in order to increase their competitiveness in the market (Kohl and Uhl, 2002). The products are perfectly same in terms of size, shape, taste, colour, ingredients, quality etc. So, the products are perfect substitutes, thereby, E_c is infinite. Since the product is homogeneous, no individual firm has any control over its price in the market. In perfect competition, the price of the product is determined at the industry level and the same is passed on to the individual firms (Figure 4.9). This ensures each firm being passive 'price taker' and the existence of a uniform price in the market. So, non-price competition will not influence the buyers to purchase the product and the buyers are simply attached to the sellers only on the basis of prevailing uniform market price. Due to product homegenity, all the buyers and sellers have readily available information about prices and products from competing suppliers. So, no buyer will pay a price higher than market price and no seller will accept a price less than the one prevailing in the market. The seller must know the price at which the product is offered for sale by the other sellers in the market. Similarly, buyers must know the price being charged by the different sellers. So, this

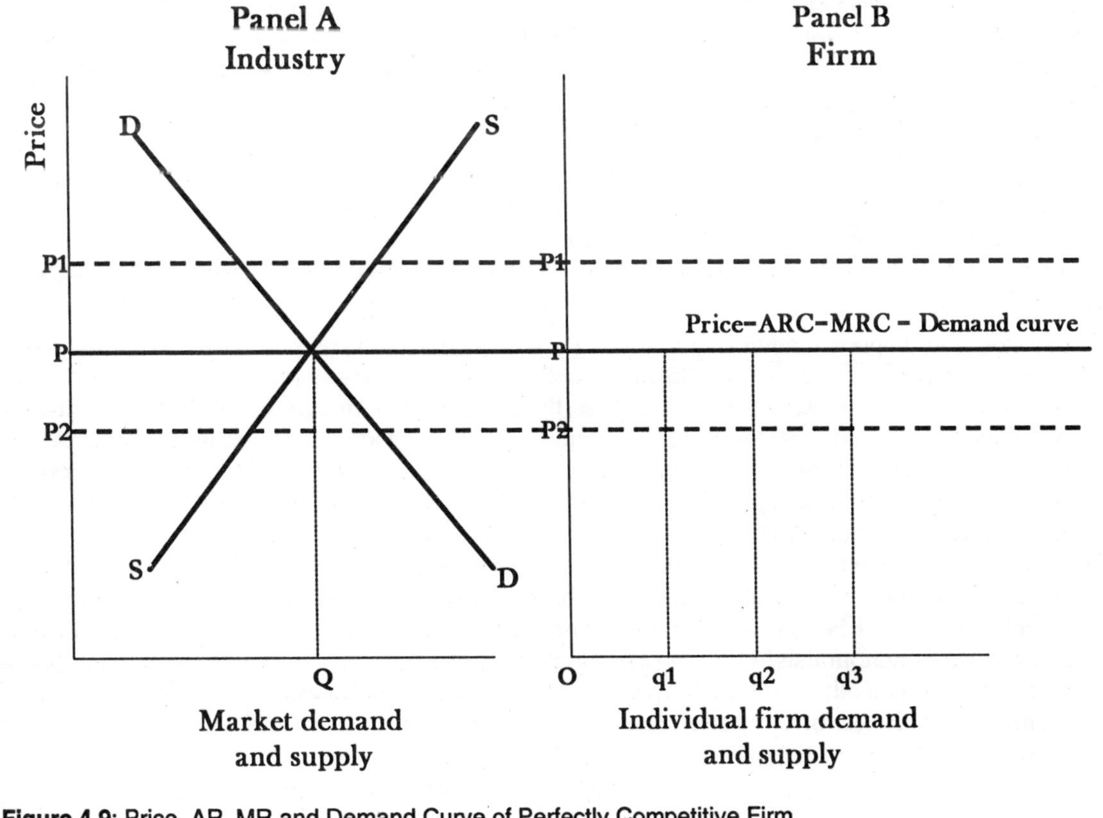

Figure 4.9: Price, AR, MR and Demand Curve of Perfectly Competitive Firm.

feature facilitates to have a single price for the product at the same time and no buyer will purchase the product from other sellers who quote a higher price for the product. So, this perfect knowledge about the output, prices etc., among the sellers and buyers ensures prevalence of uniform price for the product in the market. Further, in perfect competition, there is complete absence of Government regulation regarding the supply, prices and pricing mechanism of the products. There is also no collusion among sellers or buyers. The Government does not interfere in economic activities of the firms and price mechanism is influenced only by the two market forces *viz.*, demand and supply. Since, an individual firm cannot influence the price of the commodity, it has to take the market price as given, constant and uniform in nature. As shown in Panel B of the Figure 4.9, the firm sells OQ_1, OQ_2 and OQ_3 quantities of paddy at the same price OP. The firm cannot charge a higher or lower price than the OP. If it attempts to charge a some what higher price like OP_1, the firm will be able to sell nothing because, the buyers prefer the same commodity from other firms at price OP. On the other hand, if the firm charges a somewhat lower price OP_2, the firm will unnecessarily suffer losses. Thus, under perfect competition, an individual firm can sell as much as it wants at the given market price, thus each individual firm in a perfectly competitive industry faces a horizontal demand curve *i.e.*, the ARC for a perfect competitive firm both in the short run and long run is perfectly elastic *i.e.*, it lies parallel to X-axis. Since, ARC is constant, MRC must be constant and coincides with ARC (Figures 3.2 and 3.3). This is so because, an individual firm under perfect competition cannot influence the price of the product and hence, additional units are also sold at the same price as in the case of previous units. So, under perfect market competition, Price=ARC=MRC=Demand curve (Panel B of the Figure 4.9). However, at industry level, the demand curve slopes as usual from left to right (Panel A of the Figure 4.9). Since all the firms deal with homogeneous and standardized products with respect to quality, taste, shape, colour etc., there is no scope for product differentiation. More over, since buyers have perfect knowledge about the features of products and their prices in the market, there is no need for the firms to incur any expenditure on advertisement and other selling costs.

Since, the firms in an industry selling homogeneous products, they do not brand their products. Thus, an industry consisting of infinite number of firms with insignificant market shares and no branded products is likely to be a perfectly competitive market structure. On the other hand, an industry with a few firms with large market shares and no branded products is likely to be a perfect (pure) oligopoly.

In case of monopolistic competition and imperfect oligopoly, the product is differentiated so that, the firm can increase price without losing all of its sales (which means the individual firm faces a downward sloping demand curve). This implies, there is some differentiating feature of a product that some segment of consumers is willing to pay more for. This product differentiation is the primary reason that each firm operating in a monopolistically competitive market and imperfect oligopoly is able to create a little monopoly all to itself. In these market structures, each firm sells differentiated product, which are close substitutes to one another. The product of one firm is different from that of the other firm in terms of physical differences, quality differences, imaginary differences, purchase-benefit differences, features etc. So, each firm has an absolute monopoly in the production and sale of that differentiated product. So, the firms under monopolistic competition and imperfect oligopoly are also called as Quasi-monopolists. However, these monopoly positions are always threatened by the competitive actions of other firms in the industry. Especially in monopolistically competitive industries, there are large numbers of competing firms, as there are low-cost entry and exit into and out of the industry.

The main aim of product differentiation across the firms under monopolistic competition is to capture a large number of customers by the firm towards its products. Since, the products are close substitutes across the firms, the E_C is high, unlike monopoly. This product differentiation further helps the firms to enjoy the trade protection in terms of copy rights, trademarks, industrial designs etc., and it can be brought by the firms in various ways as given below.

- By using different types and quality of raw material and technology in producing the product.
- Differences in features, attractiveness, durability etc (*Physical product differentiation).* For example, electronic goods can easily be physically differentiated.
- By offering various kinds of benefits to customers like free service, home delivery of product, acceptance of returned goods, discounts on price, distinctive packaging etc (*Marketing differentiation).* For example, breakfast cereals can easily be differentiated through packaging.
- By effective advertisement and propaganda about the product.
- Special significance to products through enjoying copy rights, trademarks, industrial designs etc.
- The firms can bring differentiation through the enhancing the skills of their employees, imparting trainings etc (*Human capital differentiation).*
- The firms can bring differentiation in the mode of distribution of their products to the customers (*Differentiation through distribution). For example,* distribution via mail order or through internet shopping such as Amazon.com.

A product is differentiated only, if the consumer perceives and values the differentiating feature of the product. The degree of differentiation, and thus the individual firm's market power, depends upon the degree of substitutability for the product. Many close substitutes implies little differentiation and little market power and *vice versa*. Generally, firms in an industry selling differentiated products brand their products. An industry consisting of many firms with small market shares and branded products is likely very close to a monopolistically competitive market structure. An industry with few firms and large market shares and branded products is likely an imperfect (impure) oligopoly. In these competitions, compared to pricing, product differentiation is the main essence to stay in the industry. That means, with product differentiation only the firm can survive in the industry.

Demand curve or ARC of firms under monopolistic competition slope downward from left to right indicating that, the firms can sell more output by reducing price. It is relatively elastic in monopolistic competition because, each firm faces competition from a large number of firms producing the products, which are very close substitutes. However, ARC is not perfectly elastic (as in perfect competition) because, the output of each firm is slightly different from that of other firms and monopolistically competitive goods are close substitutes, but not perfect substitutes. It is not highly inelastic (as in case of monopoly) because, there is product differentiation in monopolistic competition. So, the ARC in monopolistic competition is rather more elastic (flat slope) *i.e.,* less elastic compared to perfect competition and more elastic compared to monopoly market situation (Figure 4.10). So, when a firm raises the price, the buyers may switch over to other firms product and when the firm lowers or decreases the price, the demand for the firm's product will increase, as other firms' products may become more costly to the customers. So, the demand curve of a firm in monopolistic competition appears to be more elastic with a flat slope, mainly due to product differentiation. The extent of flatness of slope of ARC depends upon the following aspects:

- ☆ ***E_C for the product:*** If the product has more E_C, the demand curve is more flat and *vice versa.*
- ☆ ***Number of firms selling the same product and their output share in market demand:*** If this is more, then again the E_C is more, and hence, the ARC of a firm for the product is more elastic.

As shown in the Panel A of the Figure 4.10, in monopolistic competition, ARC slopes downward from left to right and hence, MRC also slopes downward from left to right with a smooth and flat slope and lies below it. The ARC or demand curve is more elastic in monopolistic competition compared to monopoly market situation (Panel B) mainly due to product differentiation. Thus, monopolistic competition resembles monopoly in the aspect that, firm exercises some control over price, but there is more control in case of monopoly and less control in case of monopolistic competition. This product differentiation also leads to price differentiation across the firms.

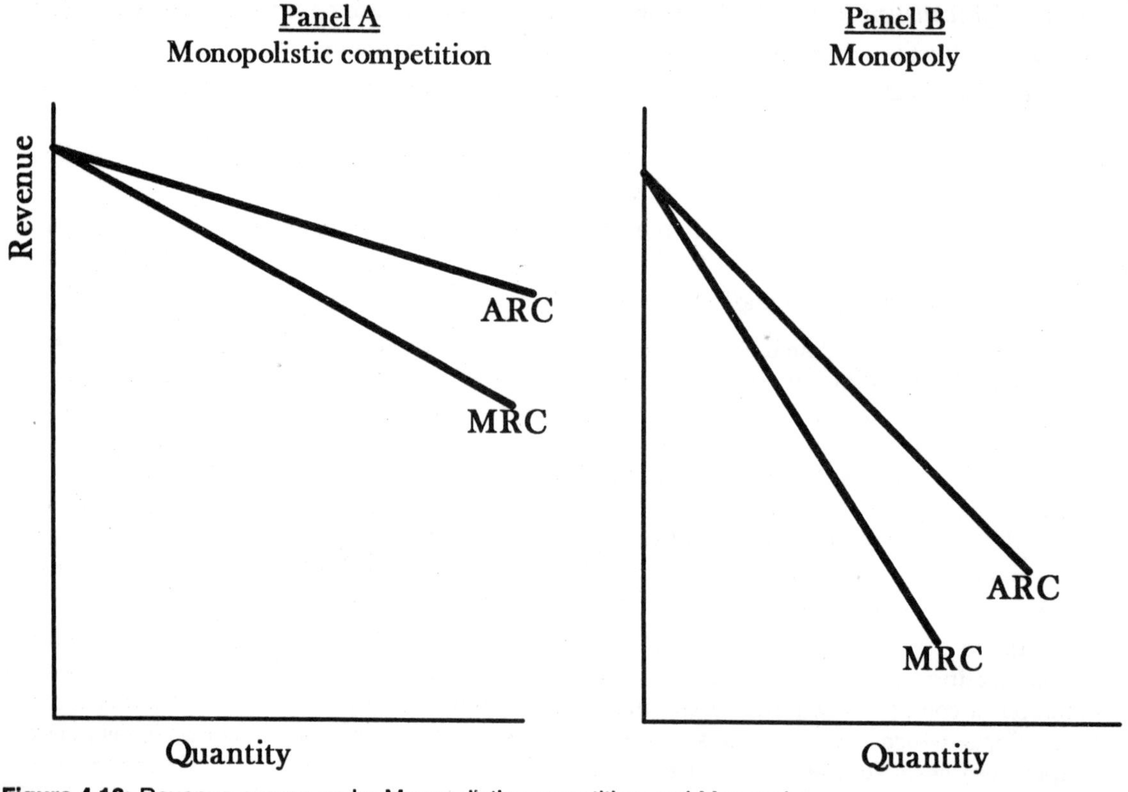

Figure 4.10: Revenue curves under Monopolistic competition and Monopoly.

In case of monopoly, the product is unique with no close substitutes (at least in the short run). As discussed above, a monopoly firm faces the downward sloping market demand curve which allows the monopolist to have market power. Since monopoly firm itself represents the industry, the firm demand curve itself is the industry demand curve. This single-seller status gives monopoly firm an extensive market control. He is either price maker or quantity adjuster in the market and not a price taker, as in perfect market competition. But, the monopolist cannot control demand for the product, as there are many buyers. Since, there are no close substitutes for the product produced by the monopoly firm, the E_C is zero. The buyers have no alternatives or choices for having the product. Either they have to purchase the product or go without it. The absence of substitutes makes the demand for the good relatively inelastic enabling monopolist to extract positive profits. If there are close substitutes, the monopoly power is lost because, the buyers can choose the substitute commodity produced. So, to enjoy the monopoly status, the product produced by the monopolist must be unique. The output-price policy of a monopoly firm does not influence firms of other industries nor it is affected by others, as the commodity produced by a monopolist will have no close substitutes. In the words of Prof. Boulding, *'a pure monopolist, therefore, is a firm producing a product which has no effective substitutes among the products of any other firm'*.

Thus, the above discussion infers that, product differentiation strategy to increase the profits is seen under two types of market structures *viz.*, monopolistic competition and imperfect oligopoly. If consumers consider products or brands of various firms to be imperfect substitute, a firm may raise its price above that of its rivals without losing its customers. In other words, product differentiation is said to exist, when products sold within the same market are not regarded by buyers as perfect substitutes. So, product differentiation assists to improve performances of firms, if appropriate marketing strategy is well implemented to bring about successful positioning in the minds of target markets. Further, it insulates sellers from the discipline of price competition. It can either shift a firm's demand curve outward, enabling it to sell a larger quantity at a give price or tilt its demand curve to a steeper slope, and therefore, lower its elasticity, so that, it enables the business firm to raise price of the product without losing many customers (Greer, 1992). This product differentiation may be of three categories as given below.

- ***Horizontal differentiation:*** When products are different according to features that can't be ordered in an objective way, it is said to be horizontal differentiation. That means, at the same price, some consumers would prefer the product, while others would prefer a different substitute. For example, considering milk products like ice cream, it is offered in different colours and tastes like chocolate, mango, vanilla, butter scotch etc., and we cannot say, chocolate flavour is better than mango and mango flavour is better than vanilla etc.
- ***Vertical differentiation:*** This occurs in a market, where the products can be ordered according to their objective quality. That means, we can rank the products from the highest to the lowest in terms of quality. So, by this differentiation, we can easily say that, one good is 'better' than other. For example, Plasma television is superior to its earlier versions. Similarly, MS Office 2007 is superior in its features compared to MS Office 2003 version.
- ***Mixed differentiation:*** Some products are differentiated both horizontally and vertically. For instance, apparel and shoes have a rich combination of shapes, colours, materials and appropriateness to social events. In such markets, the differences in colours or shapes are horizontal differentiation, while the quality of the materials is usually perceived as vertical differentiation.

D. Barriers to Entry

They refer to the stringent conditions of entry of new firms into the industry and these conditions are considered important in the assessment of competition for two reasons. First, the number of firms in an industry is influenced by cost of entry and consequently influences the level of concentration. Secondly, the conditions of entry determine the extent of potential competition. The barriers to entry also refer to a set of economic forces that create a disadvantage to new firms or competitors attempting to enter the industry.

In general, when the firms in the industry earn super normal profits in the short run (say, in perfect and monopolistic competitions), new firms gets attracted to these super normal profits and enter the industry. With every new entrant, the rivalry increases and the market shares of the already existing firms are jeopardized. There is a relation between the entry barriers and competition *i.e.*, 'the easier it is for a firm to enter a market, the greater the degree of competition is likely to be'. This feature of free entry of firms into the industry ensures stiff competition among the firms thereby, excess profits are eroded and only normal profits prevail in the long run. The ease of entry of firms is also facilitated by the absence of sunk costs on the part of the firms. However, in oligopoly and monopoly competitions, the barriers to entry are very high, as the existing firms in the industry aim at super normal profits even in the long run. In case of oligopoly, if the few firms predict in advance the threat of entry of new firms into the industry, then the existing firms merge and formulate a joint policy regarding the investment, output and pricing of the products. This joint action of the few big firms discourages the entry of new firms into the industry. There are several other sources of barriers to entry in oligopoly competition like, economies of scale, patents, access to expensive technology, existing firms gained control over essential inputs, heavy capital investments due to huge plant costs, substantial advertising budgets, traditional brand loyalty etc., and these effectively check the entry of new firms into the industry. Further, the mutual interdependence of small number of firms in an oligopolistic industry, make each firm to have a strict watch on the price charged by the rival firms in the industry. The firms generally avoid price wars and try to create conditions of mutual interdependence and to avoid entry of new firms into the industry.

In case of monopoly, the barriers to entry are more stronger compared to oligopoly, as the existing single firm will have full control on the market. Some of the key barriers that ensure monopoly status to a firm include,

- ✰ ***Natural resources:*** Monopoly arises due to naturally available resources in a particular area. So, the use and control of natural resources that are critical to the production of a final product ensures monopoly status to the firm. Examples include, South Africa has the monopoly of diamonds, Canada enjoys monopoly status with reference to nickel element, Middle East on the oil resources etc.
- ✰ ***Technological superiority***: Sometimes, monopoly arises due to technical reasons and such technicalities in the production of a product are not enjoyed by other firms. For example, the firm producing Coco Cola soft drink enjoys monopoly status for the formula involved in the product.
- ✰ ***Legal protection:*** The firms can even claim for legal protection in the form of Patents, Copy Rights, Trade Secrets etc., for the products, so as to enjoy the monopoly gains in the market. Through having legal protection, monopolist can gain exclusive control over the production and selling of certain products. For some firms, they enjoy legal protection over the materials or resources necessary to produce a product. For example, Monsanto Company enjoys patent protection with respect to *Bt* Cotton seed.
- ✰ ***Natural barriers to entry:*** For some products, the production processes involve huge amount of capital requirements, R&D costs, substantial sunk costs, which the small and medium firms could not afford to invest. In such cases, the large firm that is capable of producing that product enjoys monopoly status in the industry, as it is difficult for small and medium firms to enter the industry and expand. Examples include, public utilities like railway, electricity etc., and these are undertaken by the State. Economists, generally, refer to these monopolies as Natural monopolies.

- ☆ ***Deliberate actions:*** Some firms through their deliberate actions eliminate market competition. For example, the firms may collude, so as to enjoy monopoly gains in the market.
- ☆ ***Location advantages*:** Due to specific geographic locations or agro-climatic conditions or special skills of the people in a particular location, the product may have uniqueness in the market. For example, Gadwal sarees from Mahboobnagar district, Nirmal toys from Adilabad district in Andhra Pradesh enjoy trade protection for their uniqueness in the market. A popcorn shop in a cinema theatre is another example.
- ☆ ***Quality advantage:*** A firm may produce the product with superior quality may be due to quality resources, superior technology, skills of the labour etc., thereby, they compete out all rivalry firms from the market and enjoy monopoly status.
- ☆ ***Economies of scale:*** Monopoly firm enjoy economies of large scale and this leads to decline in cost of production of the product. It contributes to decline in price of the product. The monopoly firm gains advantage over the new entrants in fixing the price of the product below their Average Variable Cost (AVC) incurred in producing the same product. This infers that, a monopoly firm generally goes for destroyer pricing, so as to eliminate the rivalry firms entering the industry.
- ☆ ***Control over raw materials:*** If a firm gains control over the purchase and use of raw materials in producing a product, it eliminates other firms in employing the same raw material, thereby, the firm enjoys monopoly status in producing the product.
- ☆ ***Brand loyalty:*** This is one important source that provides monopoly power to the business firms. If the customers become loyal for a particular brand of the product of a firm, it makes difficult for other existing firms to survive in the competition, thereby, they leave the industry. Similarly, if new firms enter the industry, they also find it difficult to compete with the well established firm. For example, with reference to computer software, Microsoft is presently enjoying brand loyalty from the customers.
- ☆ ***Network externalities:*** We know, the use of a product by a person influences the value of that product to other people. This influence or effect is called network effect. There is a direct relationship between the proportion of people using a product and the demand for that product. That means, when more number of people are using a product, there is a greater probability of any individual starting to use the product. This leads to both spill in and spill over effects of the use of the product across the customers. This leads to the development or acquisition of market power. A classical example is the market dominance of the Microsoft operating system in personal computers.
- ☆ ***No substitute goods:*** The absence of substitute goods makes the demand for the good relatively inelastic, thereby, enable the monopolist to extract positive profits.
- ☆ ***Cost advantages:*** If the monopolist enjoys the cheaper way of producing the product, it can push its competitors out of the market.
- ☆ ***Limit pricing:*** Sometimes the monopolist sets the price of the product so low such that, it becomes unattractive for the competitors to enter the market. Thus, limit pricing refers to the situation in which a monopoly firm charges a price lower than the monopoly price (and earns less than the monopoly profit) to keep potential competitors out of the market. This strategy is especially effective when the existing firms have a cost advantage over potential entrants.

It is essential to note that, not only for other firms, the monopolist himself had to face barriers in the production of a product. If the product or service produced by the monopolist falls under public utilities, then the Government may impose certain restrictions on the industry in the interests of the public. Examples include, electricity services, natural gas distribution companies, railway service etc.

This discussion infers that, the barriers to entry for new firms are strict in oligopoly and monopoly competitions and infact, because of these barriers to trade, these structures can be sustained over time. This concept of barriers to trade applies only in the long run and not in the short run. By definition, the short run excludes entry. It is to be noted that, it is very easy to classify the market structures considering the number of sellers and product characteristics. But, the importance of studying market structure is to understand its implications for analyzing the competitive strategies of firms to maximize their long-run profits. So, analyzing the long run changes in the market structure is essential to study the profitability of firms in the industry. In this context, studying the barriers to entry for new firms into the industry assumes more significance. Though firms under perfect and monopolistic competitions gain super normal profits in the short run, they gain only normal profits in the long run, due to absence of strict barriers to entry. However, due to strict barriers to entry into the industry in case of oligopoly and monopoly competitions, the firms enjoy super normal profits even in the long run and these market structures are relatively stable over time compared to perfect and monopolistic competitions.

E. Barriers to Exit

They refer to a set of economics forces that influence the firm's decision of exiting the market and such forces make it cheaper for the existing firm to stay in the market than to exit the market. These exit conditions are important because, it influences the original entry decision of a firm. If firms anticipate that the cost of future exit, perhaps due to unfavorable business conditions, is likely to be high, then they may not enter the industry at all. This is likely to diminish the threat of potential competition. Although sunk costs could be barriers to entry, especially when the sunk costs are too large, they could even be a huge barrier to exit as well, because large investments on fixed plant and equipments commits the firm to stay in the industry. These barriers to exit increase the intensity of competition in an industry because, existing firms have little choice, but to stay and fight when market conditions have deteriorated. The loss of business reputation and consumer goodwill could be a barrier to exit especially, if the firm is planning on re-entering the market later or when the firm exits a specific market, but still operating in other markets. In such a situation, the decision to leave the market can seriously hurt the reputation of the firm among current consumers in other markets and affect the goodwill among previous customers, as when the firm withdraws its product, it will affect the availability of its replacement parts for the customers.

F. Number of Buyers in a Market

It refers to buyers concentration. This is equally important as sellers concentration, especially in markets with a few buyers. The term 'buyer concentration' was first used by Michael E. Porter in 1979 in his 'Five Forces Analysis'. According to Porter's analysis, in markets with high buyer concentration, the firms earn lower level of profits than in markets with low buyer concentration. As discussed earlier, the number of firms or sellers in an industry is important because, a small number of sellers with relatively large market shares may have more market power. However, this market power may be offset, if the number of buyers are small compared to the number of sellers. A small number of buyers may be able to demand lower prices. Identifying buyers (or segments of buyers) for an industry is probably best accomplished by identifying buyers for individual firms in the industry.

G. Growth Rate of Market Demand

The growth of market demand for the product(s) influences the market structure. That means, the market structure in industries with a relatively static demand or low growth rate of demand for products is different from the market structure in industries with an accelerated demand growth. This is because, if there is an accelerated growth in demand for the product(s), the firms in the industry expand their production to their fullest capacity and enjoy super normal profits. But, the existing firms face the threat of entry of new firms into the industry thereby, the existing firms tend to oblige collusive agreements and keep barriers against the entry of new firms into the industry, so as to earn super normal profits even in the long run. Such market structures tend to be stable over time. So, growth in market demand for the product(s), barriers to entry, collusive agreements existing market size etc., are all interrelated. Any change in one tends to bring about changes in another. By realizing this relation among the different elements of market structure, it becomes easier to understand, why market structures change over time.

It is essential to note that, many industries may not clearly fit the criteria for any single market structure. That means, the industry under consideration may have the criteria relevant to different market structures. We know, a market structure is named considering the level of competition. But, the competition can be viewed from two sides *viz.*, one from the firm level and another from the product level. In the modern days of business environment, there are some industries with small number of firms, but producing several brands of the products that are relatively close substitutes. So, at firms level of competition, the market structure appears to be oligopoly, but at the product (brands) level of competition, it appears to be a monopolistic competition. This type of market structures are referred as Hybrid market structures. The good examples for this type of market structure include cell phone industry, biscuit manufacturing industry etc. CR4 at the firm level is higher compared to at brand level thereby, indicating increase in competition at brand level compared to firms' level.

Implications of Market Structure for Pricing Strategy

The basic implications of each market structure for a firm's competitive strategy is depicted through Figure 4.11.

Perfect Competition

In perfect competition, the individual firm is a price taker and quantity adjuster. The price is determined at the industry level based on demand and supply forces and the same is passed on to the individual firms, thereby, individual firm faces an horizontal demand curve or perfectly elastic demand curve (Figure 4.9). So, a perfect competitive firm has to sell as much as it wants at the given price and it has no power to fix the price for the commodity. Since, the perfect competitive firms deal with homogeneous commodities, there are no different brands for the same and hence, same price will prevail for the same commodity at the same time. Further, improvements in operations that lower costs are ineffective in the long run because, other firms and new firms that freely enter the industry can quickly copy these cost-effective innovations, thereby, no firm enjoys super normal profits in the long run. In some cases, firms that join together in perfectly competitive industries can form associations to improve market demand conditions. These associations can conduct R&D and promotions to increase demand for the industry product, which can shift the demand curve to the right and/or make demand less elastic.

Monopoly

As there is no distinction between firm and industry under monopoly, the firm's demand represents the industry or market demand. In monopoly, there is a single firm producing the product and there is

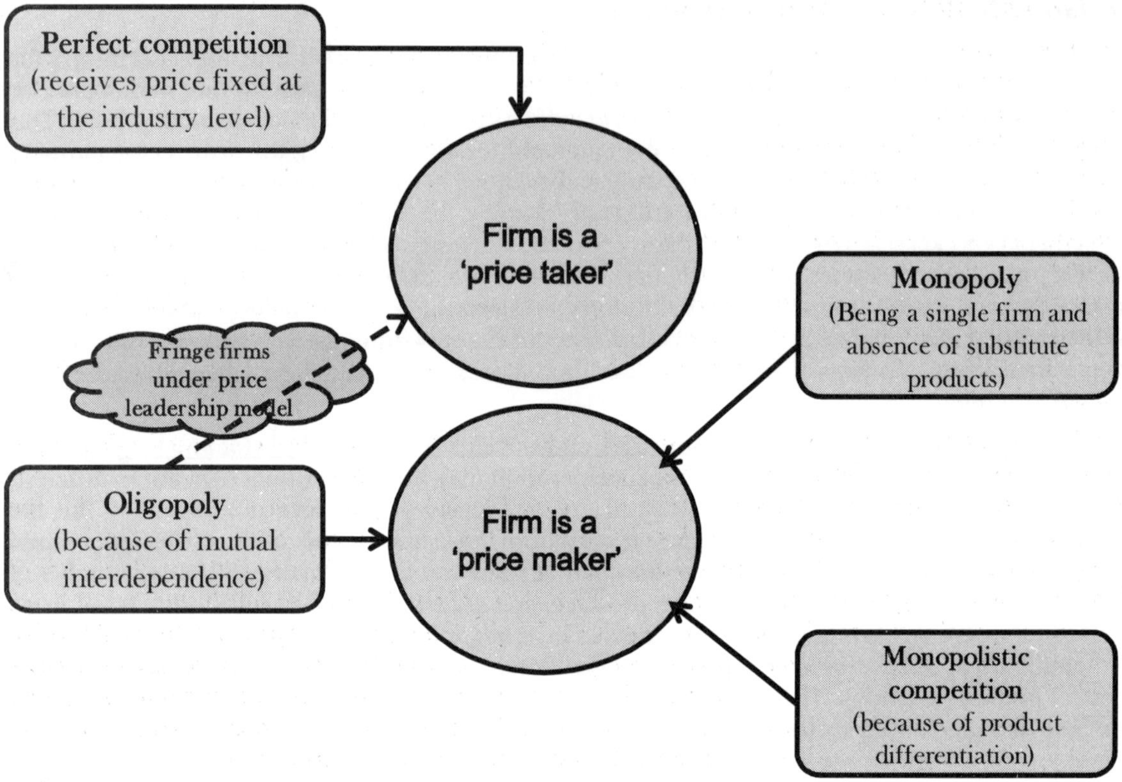

Figure 4.11: Price fixation strategies by firms across different market structures.

large number of buyers for the product. We know, the individual's demand curve for a product slopes downward from left to right. So, the aggregate demand curve of all the buyers for the product also slopes downward from left to right, as the demand curve for the monopolist is the summation of the demand curves of all the buyers for the product. Thus, the industry or firm's demand curve or ARC is downward sloping from left to right indicating that, monopolist sells large quantities of a commodity or output at lower prices or price reductions. Since, the monopolist is the sole producer of the product, he enjoys the market power. So, power to influence the price is the essence of monopoly. Hence, monopolist is a price maker and buyers are price takers. As discussed earlier, to support his pricing strategies and to enjoy super normal profits even in the long run, the monopolist create and sustain different barriers to entry. The monopolist can even go for advertising (Informative advertising) the product, as the consumers need to be aware of the product and its benefits in order for them to consider a purchase. Sometimes, exclusive distribution supports the perception of the uniqueness of the product. The goal of marketing is still to increase demand (shift the demand curve to the right) and to reduce the elasticity of demand *i.e.,* rotate the demand curve clockwise (as in monopolistic competition). To maximize the profits, the monopolist also resort to price discrimination strategy.

Oligopoly

An oligopolist faces a downward sloping demand curve, but its price elasticity depends upon the rivals' reaction to changes in its investment, output and price. So, the demand curve of a firm under oligopoly is indeterminate because, any strategy formulated by the rival firms, may affect or change the demand curve. It is more elastic than under simple monopoly and not perfectly elastic as under perfect market competition. This implies that, no single firm in oligopoly market situation can exactly determine the demand for its product by formulating its own output and price policies. This is because, the firm's output and pricing policies will definitely influence the fortune of other firms and hence, they counter move the strategies to survive in the market. That means, when one firm decides and implements a course of action, there will be a quick and equal reaction from the rival firms. Hence, by just changing the output and pricing policies, no firm can determine exactly their influence on demand or sales of the product. For example, if a firm reduces the price of a product, other firms will also reduce the price for the product. Hence, demand for the product may not increase as expected by the business firm, which reduces the price of product initially in the market. If the firm increases the price of the product, other firms will not increase the price, thereby, the firm which increases the price will suffer from huge losses. Hence, the demand curve or ARC is indeterminate, which is the characteristic feature of oligopoly. This is mainly due to interdependence among the firms.

Considering the indeterminate demand curve, the only way the firms can enhance the sales of the product is by resorting to non-price competition when compared to price competition. So, in order to increase the market share, the oligopolistic firms engage in aggressive and extensive sales promotion efforts by means of advertisement and by changing the design and improving the quality of their products, as these aims at stimulating and motivating the customers to purchase the products. One main objective of making higher selling costs or advertising costs is to make the demand curve more decisive and also favourable. Under monopolistic competition also, advertisement plays a considerable role because, the product is highly differentiated and it is necessary on the part of the firms to educate the customers about the features of the product. But, in oligopoly, advertisement still plays a bigger role, even when compared to monopolistic competition, mainly to make the output and price policies effective and to make the demand curve decisive or deterministic, as there is mutual interdependence among the firms regarding their investment, output and pricing policies. So, in oligopoly, non-price competition becomes a life and death matter for the survival of the firms.

Understanding the interdependency and indeterminate demand curve behaviour of the firms, they generally resort for a mutual co-operation or collusion to their common or best advantage such that, all the business firms will prosper. So, they tend to collude in terms of output and price policies to grab the maximum from the market and all will mutually benefit. But, at times of rivalry, they compete each other for gaining maximum profit share and try to capture other firms' market by increasing the selling costs and by formulating suitable output and price policies. Such collusion or group behaviour is not seen in perfect competition, monopoly and monopolistic competition. Since, there exists mutual interdependence among the firms and the price-output policies of one firm influences the fortunes of other firms, there will not be any unique pattern of price behaviour for each firm in the oligopoly market situation. Sometimes, all the firms join as a group and each firm wishes to co-operate with its rivals on price-output policies. They may even recognize one firm as a leader of their group and under that leader's directions, other firms will plan their price-output policies for their mutualism. In such case, individual seller's demand curve is a part of industry demand curve. In case of conflicting situation, each firm plan its own price-output policies and try to gain maximum share from the market. Taking into consideration the possible behavioural pattern of firms, it is not possible to predict any unique pattern of pricing behaviour in the oligopoly market situation.

Monopolistic Competition

As discussed earlier, Demand curve or ARC of firms under monopolistic competition is rather more elastic (flat slope) *i.e.*, less elastic compared to perfect competition and more elastic compared to monopoly market situation (Figure 4.10). So, a firm under monopolistic competition is a price maker with some degree of control over price. So, unlike perfect competition, a monopolistically competitive firm has the ability to raise or lower the price a little, but not at a significant level. As in monopoly, the price received by a firm under monopolistically competition is greater than its MR, as ARC is above MRC. In monopolistic competition, due to product differentiation, if one firm lowers the price, there is increase in demand for the product offered by that firm, thereby, it attracts the customers of other firms. Contrary wise, if a firm increases the price of the product, the customers will shift to other firms' close substitute product. So, the firm enjoys some control over the price of the product in monopolistic competition. unlike in perfect competition, where the individual firm has absolutely no control on the price of the product, because of homogeneous nature of the product. So, in monopolistic competition, a firm has to choose a proper output–price policy that maximizes the profits. However, in monopoly, the firm enjoys more control over the price compared to monopolistic competition, as there are absolutely no substitute goods in monopoly situation.

Under this market structure, competition among the firms takes place in two different ways *viz.*, price competition and non-price competition. In price competition, firms compete with each other on the basis of price of the product. Each firm has its own output-price policy, thereby, it fixes the price of the product, so as to take advantage of higher sales. In non-price competition, to increase the sales, the firms compete by adopting product differentiation or through increasing selling costs. Selling costs are the costs incurred by the firm on advertising and other forms of sale promotion. This non-price competition has become an integral part of marketing of products and through this competition, the firm tries to make a favourable shift in demand for the product and tries to capture the market. So, any action a firm takes to increase demand for its product other than cutting its price refers to non-price competition. The important types of non-price competition include, advertising the product, better service, product guarantees, free home delivery, more attractive packaging etc. Due to the presence of more number of firms producing the same product, but with differentiation, the non–price competition is significant among the firms. Among different types of non-price competition, advertisement expenditure is the prominent one. The costs incurred towards advertising and sales promotion by the firm will influence both the costs and returns of the business. Hence, the firm will adjust its outlay towards selling costs, besides planning output–price policy. This adjustment of funds towards selling costs is one of the characteristic features of monopolistic competition, unlike perfect and monopoly competitions, where there are no selling costs. However, compared to oligopoly, non–price competition is lower in case of monopolistic competition, as there is absence of mutual interdependency of firms in monopolistic competition.

3. Conduct

The market conduct refers to the pattern of commercial behaviour that firms follow in adjusting to the markets in which they sell or buy the commodities. Thus, conduct means what firms do to compete with each other. That is, it infers, how a firm struggles to attain due market share. In reverse words, conduct is a reaction pattern and adjustments by firms in an industry to achieve their goals and survive in the competition. We know that, in an industry, each firm will have different characteristics of conduct or behaviour. So, it is essential to perceive the conduct of firms considering several elements like price determination behaviour, decisions on product designs and features, investments on R&D, innovation, advertising, sales promotion strategies, terms of sales (cash or credit terms), financial policy, collusion (explicit or tacit) etc.

Relation between Market Structure and Conduct

The relationship between market structure and conduct is two-way. That is, conduct is influenced by market structure (For example, firms under oligopoly and monopolistic competition plan for heavy investments on non-price competition, unlike perfect competition and monopoly) and also influence market structure (For example, few firms under oligopoly competition can place strict barriers to entry for new firms by different ways like the existing firms gained control over essential and specialized inputs, high capital requirement due to huge plant costs, exclusive patent rights on the raw materials used in the production programme or products produced, less unused capacity, which makes the industry less attractive, economies of scale enjoyed by only few large firms, heavy advertising expenditure, existing firms join as a group and go for destroyer pricing etc). Further, conduct is also related to market performance. For example, the firms that enjoy higher profits (super normal profits) can afford higher advertising costs and in order to sustain these profits, they prevent the entry of new firms into the industry through showing advertising investments as endogenous sunk costs.

Market conduct of firms can be explained in terms of their marketing strategies that influence their performances in order to succeed in the desired objectives and to gain competitive advantages. While pursuing a particular line of conduct, firms either take defensive or offence measures against their rivals depending on the market structure and its competition. In general, pricing and promotional strategies are the major tools of line of conduct the firms choose, among other things, that influence their performances in an industry.

A. Price

Pricing is an important marketing tool that influences profits of the business. Among different forms of market structure, except in perfect competition, in other competitions, firms enjoy certain degree of market power in fixing the prices of the products. In perfect competition, the price of the commodity is determined at the industry level and the same is passed on to the individual firms. So, in perfect competition, the individual firms are simply price takers and quantity adjusters and not price makers. However, in imperfect competition, this is not the case. In extreme form of imperfect competition *i.e.*, monopoly, the single firm represent the industry and it enjoys high degree of market power in fixing the price of the product. Further, the firm has complete control over the supply of the product. But, the firm cannot fix the price and decide output simultaneously. That means, the firm can either fix the price and leave the output to be determined by the buyers' demand at that price. Or, it can fix the output to be produced and leave the price to be determined by the buyers' demand for that output. So, monopoly firm is a price maker or output decider. Thus, whatever price a monopolist fixes or whatever output he decides to produce, it is influenced by buyers' demand in the market. But, in general, the monopolist decides the output and leaves the price to be decided in the market. Although a monopoly's market power is high, it is not absolute. This is because, a monopolist faces a negatively sloped demand curve (ARC) and it is not a perfectly inelastic curve (Figure 4.10). Consequently, any price increase will result in the loss of some customers. One of the essential features of monopoly competition is that, the firm can charge different prices for the same product from different customers and this type of monopoly is called Price discriminating monopoly. The main objective of price discrimination is that, it is more profitable than charging a single price. It occurs, when a firm charges different prices from different customers for an identical product or service, for reasons not associated with costs. It is, otherwise, defined as 'the sale of technically similar products at prices, which are not proportional to Marginal Cost (MC)'. That means, the price variation of the product across the customers is not related to the cost of the product produced. In the words of Joan Robinson, *'the act of selling the same article, produced under single control at different prices to different customers is known as price discrimination'.*

According to J.S. Bain, *'Price discrimination refers strictly to the practice by a seller of charging of different prices from different customers for the same good'*. Price discrimination is also called as Discriminating monopoly or Yield management and the monopolist practicing price discrimination is called Discriminating monopolist. The classification of customers into different groups for whom the product is offered at different prices refers to Market segmentation. This price discrimination is a strategy that is employed by almost every industry that has some power to determine prices. Theoretically, it is an important feature of monopoly markets, but in practice, it occurs with oligopolies such as, the airlines and even in fully competitive retail or industrial markets. This price discrimination is possible, when the monopolist sells the product in different markets in such a way that, it is not possible to transfer the product from cheaper market to the dearer market. It is, however, not possible under perfect competition, even if the two markets could be kept separate. This is because, it is based on the assumption that, same price will prevail for the same commodity at the same time. Moreover, the firm's demand curve is perfectly elastic, and hence, every seller would try to sell in the market as much as he can at the given price. Considering different elasticities of demand for the product in different markets, the monopolist may adopt price discrimination. He may even fix different prices for the same product among different categories of customers in the same market segment. This price discrimination ensures the monopolist

- to convert consumer surplus to producer surplus.
- to increase the total revenue and profits
- to off-load excess capacity
- to take market share away from rival firms.

In oligopoly competition, firms have strong interdependence between them. That means, each firm must take into account the likely reactions of other firms in the market, while making investment, output and pricing decisions. So, considering this characteristic feature of mutual interdependence, the oligopoly competition can lead to a wide range of outcomes. Sometimes, the firms collude each other and restrict the output into the market, with a view to enjoy monopoly profits in the business. In some cases, there will not be any formal agreement, but there may be an acknowledged market leader which informally sets prices to which other firms will respond, known as price leadership. Sometimes, the competition between the few firms is so severe that leads to drastic fall in prices and increased production, thereby, approaching the features of perfect market competition. In some cases, considering mutual interdependence between the firms, they tend to keep the price of the product at the same level. This is because, if price is increased by one seller, other firms will not increase the price and thereby, the firm which increased the price will lose. If one seller reduces the price, other firms also reduce the price and thereby, the firm which reduced the price initially will not get the expected profits. Hence, the firms tend to keep the prices at the same level and this is called as price rigidity. Considering these diversified possible outcomes from the oligopoly market structure, there is no single, unified theory to explain its price-output mechanism. So, when compared to other forms of market competition, oligopoly is the least understood market structure.

A firm under monopolistic competition is a price maker with some degree of control over price. So, unlike perfect competition, a monopolistically competitive firm has the ability to raise or lower the price a little, but not at a significant level. As in monopoly, the price received by a firm under monopolistically competition is greater than its marginal revenue, as ARC is above MRC. In monopolistic competition, due to product differentiation, if one firm lowers the price, there is increase in demand for the product offered by that firm, thereby, it attracts the customers of other firms. Contrary wise, if a firm increases the price of the product, the customers will shift to other firms' close substitute product. So, the firm enjoys some control over the price of the product in monopolistic competition. unlike in perfect competition, where the individual firm has absolutely no control on the price of the product, because of homogeneous nature of the product. So, in monopolistic competition, a firm has to choose a proper output–price policy that maximizes the profits. However, in monopoly, the firm enjoys more control over the price compared to monopolistic competition, as there are absolutely no substitute goods in monopoly situation.

B. Advertisement

It is otherwise termed as non-price competition. This non-price competition makes the firm's ARC more inelastic (or) less elastic, as it increases the cost of production. This feature of non price competition vary across the market structures. It will not arise under perfect competition, as all the firms deal with homogeneous product and a single firm has to sell any quantity of the product at the ruling market price. Even if competition arises between the firms to sell the product, it leads to reduction in price until the new equilibrium price is reached. If a firm under perfect competition goes for advertising the product, the buyers will get stimulated and purchase the product from any of the sellers, as the product is homogeneous, thereby, the firm which advertises about the product may lose in the form of advertising costs. However, if all the firms in an industry commonly advertise about the product, they all may get mutually benefitted through increased sales of the product. Such advertising under perfect competition is called as Promotional advertising or Informative advertising and not Competitive advertising. For example, a Dairy Co-operative, Cashew Board, Coffee Board, Tea Board etc., advertises about their products and this advertising is called Promotional advertising or Informative advertising. Thus, under perfect competition, if there is any role for advertising, it would be to get consumers aware of the product rather than to persuade them to change their preferences. Within this context, the role which is expected to be played by advertising is an informative one, but not persuasive, and it is just for those firms, which are newcomers to the market.

Under monopoly, as a single firm itself represents the industry and the product produced by that firm has no close substitutes, the concept of non-price competition will not arise. But, the firm will undertake some advertisement, so as to make the customers aware about the product and its features. Thus, the advertisement costs constitute only insignificant proportion of total costs incurred by the firm on the product. So, the advertising even under monopoly also refers to Promotional advertising or Informative advertising and not Competitive advertising. For example, Monsanto firm advertises about *Bt* cotton seed and this refers to Promotional or Informative advertising. We know, pure monopoly and perfect competition are opposite in the sense that, the firm in a perfectly competitive market has so many competitors, whereas the firm under pure monopoly market structure has no competitors at all. So, there is complete absence of competition for a firm under pure monopoly. It should be acknowledged that, the firm operating under the pure monopoly market structure is not completely insulated from the actions taken by other industries in the economy. It is assumed that, all commodities are rivals for the consumer's favour. For example, if a firm is a monopolist in the supply of steel in a certain market, it would still face considerable competition from other firms producing aluminum or plastics. So, the product of the monopoly firm also faces a competition from other products produced by other industries. So, to influence the consumer tastes, a monopolist often spends considerable expenditures on advertising in order to influence the demand curve and shift it to the right.

In oligopoly market structure, there exists small number of firms and a great deal of interdependence among them. So, it is assumed that, each firm designs its policies with the consideration of their effect on its competitors. Competition in the oligopolistic market structure is non-price oriented. The firm tends to use advertising and differentiation in many product characteristics rather than price, as a competitive weapon because, it seems to view price cutting as a dangerous tactic. The firm as a less risky way to keep customers away from competitors uses advertising and product differentiation strategies. Firms also use advertising to differentiate their products from those of their competitors. In this way customers may be oriented to become brand loyal and so to stick to it, even though the products of all firms in the industry are much the same. The example, which can be effectively cited here, is the cigarette case, where various brands of cigarettes are quite similar, although not identical. The cigarettes industry spends a large amount on advertising to build a particular brand image in the consumers' minds. Within this context, advertising can promote a psychological differentiation. Similarly, advertising can perceptually differentiate life insurance services delivered by different insurance companies. So, considering the mutual interdependence among the firms regarding their investment, output and pricing policies, they rely heavily on the advertising as a main weapon (even when compared to monopolistic competition) to make the demand curve decisive or deterministic. However, the advertising expenditures of some oligopolistic industries have been expanded beyond the levels that are socially optimal. Sometimes, advertising expenditures only have the effect of increasing the costs of the entire industry, since one firm's advertising causes other firms to increase their advertising.

As in oligopoly, even under monopolistic competition, the non-price competition or selling costs assume crucial role in boosting the sales of the product. Chamberlin opined that, firms under monopolistic competition often resort for non-price competition only and price-competition will not arise much. But, compared to differentiated oligopoly, the non-price competition is low in monopolistic competition because, the concept of mutual interdependence of firms will not arise under monopolistic competition and each firm will not closely watch the output-price policies of rival firms, as they are relatively larger in number. But, still the non-price competition plays a crucial role in determining the price-output policies of the firms operating under monopolistic competition.

So, both under monopolistic competition and differentiated oligopoly competition, the advertising about the product serves as an important competitive weapon that increases the sales of the product of one firm at the cost of the rivalry firms. Hence, the advertising under monopolistic competition and differentiated oligopoly competition refers to Competitive advertising rather than simply promotional advertising. For example, various pesticide firms, fertilizer firms etc., advertises about their products and such advertising refers to competitive advertising.

It is essential to note that, of these two types of advertising *viz.*, Promotional advertising and Competitive advertising, the former is socially advantageous, as it educates the customers about the technical aspects of the product. But, there are both positive and negative views regarding competitive advertising. The positive views supporting competitive advertising include,

- ☆ it is informative to the buyers about the products and their features. This information allows customers to make better choices about what to buy and thus, enhances the ability of markets to allocate resources efficiently.
- ☆ it increases sales of the product
- ☆ it contributes to economic growth of the firms
- ☆ it increases competition between the firms and lowers the prices
- ☆ it provides consumers with the valuable information on product availability, quality and price that they need to make efficient choices in the market place.

- it fosters competition. Because advertising allows customers to be more fully informed about all the firms in the market, customers can more easily take advantage of price differences. Thus, each firm has less market power. In addition, advertising allows new firms to enter more easily because, it gives entrants a means to attract customers from existing firms.
- the willingness of the firm to spend a large amount of money on advertising can itself be a signal to consumers about the quality of the product being offered.

However, there are arguments against competitive advertising *viz.*, advertising costs raises the price of the product, the firms may use their influence to bias the media, affect the entry of new firms into the industry, it is not a productive activity, advertising often tries to convince customers that products are more different than they truly are, by fostering brand loyalty advertising makes buyers less concerned with price differences among similar goods thereby, each firm charges a larger markup over MC, much advertising is psychological rather than informational, it manipulates customer's tastes etc.

Measures of Market Conduct

The most commonly used measure of conduct is Advertising intensity. It is measured in terms of advertisement expenditure to sales ratio. According to the Dorfman and Steiner (1954) and Needham (1978), advertising to sales ratio or advertising intensity can be used as a proxy to market conduct. The advertising to sales ratio is related to the firms conduct and has relationships with the barriers to entry and market power. We studied earlier that, advertising influences the structural and performance characteristics of an industry and it also influences the prices of products.

$$\text{Advertising} - \text{Sales ratio} = A/S$$

where,

A = Advertising expenditure

S= Total sales

Firms with low advertising to sales ratios will have little market power and low price cost margin. *Ceteris paribus*, oligopolies have larger advertising-sales ratios compared to monopoly and competitive firms. However, a monopolist's advertising-sales ratio is greater than firms in a perfect competition.

4. Performance

Performance is the result of work, which is influenced by the industry's structure and conduct and it is usually indicated by how big is the market share or how big the profit is gained by a firm in an industry. That means, given that structure and conducts reflect how the game is played in the market, performance reflects how well it is played. So, market performance is the assessment of how well the process of marketing is carried out and how successfully the business firms' aims are accomplished. It is concerned with technological progressiveness, growth orientation of agricultural firms, efficiency of resource use and product improvement and maximum market services at the least possible cost. In other words, market performance is the appraisal of the extent to which the interactions of buyers and sellers in a market stimulate result that is consistent with social purposes. It is quite easy to distinguish between market conduct and market performance. Market conduct refers to the price and other market policies pursued by the firms in executing their activities in the business. Market performance refers to the end results of these policies *i.e.*, the relationship of selling price to costs, the size of output, the efficiency of production, adoption of modern technology etc. The performance of a firm in an industry or industry is measured by profitability. We know, profit is the difference between total revenue and total costs. Total revenue of the firm is influenced by the output and the price it fetches in the market. Total cost is influenced by the level of output and the costs incurred on the raw material or services employed in the production programme. Thus, profits earned by the firm are influenced by both prices and costs. Further, profits are determined by other factors like ability to of the firm to adjust to changes in market demand, R&D, availability of capital etc.

In order to compete, firms must make their products more attractive in the eyes of buyers. For this end, improved quality and / or reduced prices are the most direct and effective competitive instruments that sellers can use. These actions immediately yield a better efficiency performance. It is clear now that, the mechanism that drives firms to offer what consumers desire – more value for money – is the *sine qua non* for market competition to enhance economic performance. It follows that, different market structures (*i.e.* perfect competition, monopolistic competition, oligopoly, duopoly, and monopoly) cause individual sellers to show varying degrees of aggression to furnish what buyers demand, which dictates the state of efficiency (performance) of the market. This account is at the core of the SCPP therefore, any proposition about market performance has to be consistent with this premise in order to be workable.

In perfect competition, in the long run, the firm earns only normal profit. The super normal profits enjoyed by the firms in the short run will be eroded in the long run, as new firms enter the industry. When compared to monopoly, under perfect competition, output is higher and price is lower. In monopoly, the firm can earn supernormal profits even in the long run, as there are no competitors for the firm producing the same product and there are strong barriers for the entry of firms into the industry. In oligopoly market structure, since barriers to entry for new firms are very strong, the existing firms can earn super-normal profits even in the long run. The firms generally collude to maximize the profits either by fixing higher price for the product or sharing the market or fixing the output across the firms. In monopolistic competition, the firm enjoys super normal profits in the short run. In long run, the firm is in equilibrium, where its ARC is tangent to the downward sloping long run average total cost curve, thereby, the firms enjoy normal profits only. Firms may, however, retain some profits by using more non-price action.

Measures of Performance

These measures are primarily used to monitor the outcomes resulting from competition among firms within an industry, market and the entire economy. Thus, they indicate, how a firm has executed its activities and guides in its future course of actions. The performance of firms is directly influenced by the market structure and its conduct. Most common measure employed is the computation of Accounting profits and Economic profits realized by the business firms (Figure 4.12). Further, total revenue earned by the firm is the simple and key performance indicator at firm level. Average revenue earned by the firm is an important performance measure at industry level. However, a number of measures of performance will be used and they are discussed here under.

i. Lerner's Index

A firm is said to have market power, if it is able to increase the price of its product without losing all demand for it. If a firm is not able to do so, it has no market power, as is the case under perfect competition. In perfect competition, firms are price takers and can either incur losses, make normal profits or make economic profits in the short run, but in the long run, they can only make normal profits. Also, in the long run, perfectly competitive firms are both production and allocation efficient. But, we know, imperfect competition (firms enjoy market power) is the rule and perfectly competition

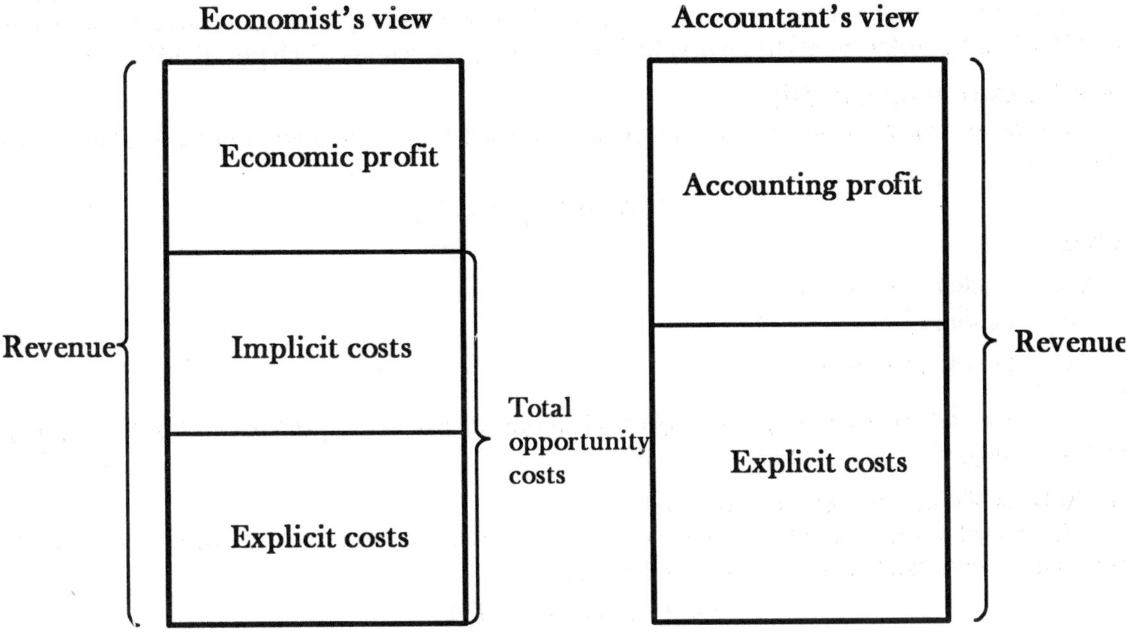

Figure 4.12: Economists vs. Accountants view of costs of the firm.

is seldom found in reality. This market power of firms under imperfect market competition serves as one of the important measures of market performance. This is simply because, the presence of market power impedes attainment of the market efficiency. A market is said to perform poorly, if undue market power is experienced. To detect market power, one commonly used indicator is the price-cost markup or the Lerner's index. This measurement expresses how far the market price (P) of a good exceeds the MC of production. So, Lerner's index to measure the Degree of Market Power (DMP) is computed by formulating the relationship between P and MC and it is given by,

$$DMP = [(P - MC)/P]$$

The index value ranges between (and including) zero and one. So, when the Lerner's index is > 0, a firm is said to have market power. The higher the value of the index, the more market power a firm has. The size of a firm's market power is *inter alia* determined by the elasticity of the demand curve it faces. The more elastic the demand curve, the more sensitive consumers are to price changes, which implies that the producers only have limited market power. There is an inverse relationship between the elasticity of a demand curve and the extent of a firm's market power. The elasticity of a demand curve is in turn influenced by the factors like, the number of close-substitute products that are available, whether the product is a necessity or a luxury item, the number of firms in the industry and the way in which those firms in the relevant market interact with one another. In general, there exists inverse relation between number of firms in a industry (or market competition) and market power exercized by each firm in the industry. So, as the extent of imperfection increases, market power enjoyed by the firm also increases. However, having a relatively large number of firms in a market offers no guarantee that competition levels will be high. A market may have of a relatively large number of small firms, but may be dominated by a few firms with market power, which lowers the level of competition. This is the case of oligopoly.

However, it is not always possible to derive the Lerner index empirically, as it is quite difficult to obtain MC data. Furthermore, firms may have numerous products and each priced differently.

ii. Price-Cost Margin (PCM)

A measure of performance that is conceptually closest to the Lerner index is the PCM. The formula for PCM is:

$$PCM = [(VA_J - W_J)/O_J],$$

where,

VA_J = Added value of industry j

W_J = Labour cost j

O_J = Output of industry j

iii. A number of accounting measures of performance like (Profit/Revenue), (Profit/Capital), (Profit/Equity) etc., can be employed.

iv. Rate of Return on Assets after Tax

The rate of return on assets after tax measures the overall ability of the firm to utilize the assets in which it has invested to earn a profit. It is given by,

$$\text{Return on assets} = (P\text{-}T\text{+}I)/A$$

where,

P = Net profits

T = Tax on profits

I = Interest payment to debt holders

A = Total assets

It is essential to note that, interest payments must be added to the numerator because, debt holders are paid interests, whereas profits are paid to shareholders, Total assets include the values of both equity and debt capital.

v. Rate of Return on Share Holders Equity after Tax

Rate of return on share holders equity after tax is used to measure profitability. It is given by,

$$\text{Rate of return on shareholders equity} = (P\text{-}T)/E$$

where,

P= Net profit

T= Tax on profits

E= Shareholders equity.

This measure is better because, it corresponds with what individual investors are trying to maximize. In addition, competitive industries with the same risk will have the same rate of return on equity in the long run.

vi. Sales Ratio or Return on Sales after Tax

This ratio is derived from the income statement and it measures the overall profitability of a business. It is given by,

Return on sales after tax = (P-T)/S

where,

P= Net profits

T= Tax on profits

S= Total sales

The advantage of this ratio is that, it measures the allocation efficiency more directly than Rate of return on assets after tax and Rate of return on shareholders equity after tax ratios.

vii. Market Margins and Marketing Efficiency

The high market margin is a reflection of imperfectly competitive market condition, which is detrimental to the final consumers.

- ☆ ***Conventional approach:*** This approach measures the Marketing Efficiency Index (MEI) by employing the following formula:

  MEI = (O/I) *Equation 1*

  where, 'O' is the value added for the commodity in the marketing channel *i.e.,* difference between CPP – FNSP (*i.e.,* Farmer's Net selling Price), 'I' is the Marketing Costs (Mc) incurred by all the agencies
- ☆ ***Shepherd's formula:*** MEI = (V/I) – 1 *Equation 2*

  Where, 'V' = Value of the commodity sold at the consumer level or CPP, 'I' = Mc incurred by all the agencies.
- ☆ ***Acharya's method:*** Acharya suggested Modified Measure of Marketing Efficiency (MME) and this is given below.

  MME = [FNSP/(Mc + MM)]

  Where MME is the Modified Measure of Marketing Efficiency, FNSP is the Farmer's Net Selling price, Mc and MM are the Mc and NMMs of all the intermediaries respectively.
- ☆ ***Acharya's Modified MEI:*** Considering Marketing Losses (MLs) during the transactions, the following is the formula:

  MEI = FNSP/(NMMs + Mc +MLs)
- ☆ $NMM_I = PS_I - Mc_I$ or $GMM_I - Mc_I$ or $SP_I - (PP_I + Mc_I)$
- ☆ $GMM_I = NMM_I + Mc_I$

  where, 'NMM_I' stands for NMM of 'i'th intermediary, 'PS_I' indicates the price spread of that specific intermediary, GMM_I refers to GMM of 'i'th intermediary, 'Mc_I represents the Mc incurred by ith intermediary, SPI refers to Selling price of 'i'th intermediary, PPI refers to Purchase price of 'i'th intermediary.
- ☆ FSCR= (FNSP/CPP) x 100

Marketing efficiency can be multiplied by 100 to convert it into percentage. The marketing efficiency is similar to rate of return on investment and if it is greater than bank lending rate, it implies that, the marketing of products is profitable.

viii. Gross Margin (GM) (or Contribution margin)

Data on MC is rarely available. In practice, researchers often assume that, MC can be approximated by Average Variable Cost (AVC). Gross margin is calculated as GM = [(Price - AVC)/Price]. If an industry or firm sells one product at a single price, then this measure of gross margin is equivalent to GM = [(Revenue - Cost of Goods Sold)/Revenue].

ix. Extent of Social Welfare

Performance can also be measured by the social welfare that the firm or industry yields. In perfect market competition, price=MC, thereby, the firm produces most efficient output and consumer's welfare is safe guarded. But, under monopoly competition, the resources are misallocated thereby, leads to loss of social welfare. We know, a monopolist is a price maker, thereby, he tries to limit the output and increase the price of the product, so as to maximize the profits. When a monopoly firm is under equilibrium, price>MC and this implies that, the monopolist gains at the expense of consumers. Thus, as the imperfection increases in the market competition (say, due to more concentration), social welfare gets affected.

However, there is no consensus on which is the best measure of performance. The choice of measure obviously depends on data availability and the desired aggregation level of analysis *i.e.*, industry, firm or plant.

Problems in Performance Evaluation of a Firm or an Industry

Though several measures, as discussed above, can be employed for studying the performance of a firm/industry, there involves certain constraints in employing these measures and they are discussed here under.

- ✰ Inadequate data availability on certain measures of conduct
- ✰ Significant variation in incomes across the firms in industries of perfect competition, monopolistic completion and oligopoly and these variations may be due to product differentiation or price differentiation or technology or economies of scale or market power or market concentration or advertising etc., and it is difficult to quote uniform reasons across the firms.
- ✰ The conduct or behaviour of rivalry firms also varies across the industries and across the market structures.

Considering the above detailed analysis of SCPP, it is possible to summarize the elements of structure, conduct and performance across different forms of market competitions and it is shown through the following Table 4.10.

Relationship between Market Structure, Conduct and Performance

The discussion revealed that, the relationship between market structure, conduct and performance is complex and has a lot of interdependencies. Edward S. Masson opined that, *'if you want to observe events at the market, where the price is high, you have to evaluate its performance first. The performance itself reflects the structure of the industry'*. So, it can be concluded that, if the researcher want to identify, whether the market's performance is good or bad, it is important to study the market structure that affect the market conduct first. Market structure influences the market conduct, and on the other hand, market conduct can have feedback effects on the market structure. These two inter-related components will exert influence on market performance. The relationship between market structure and performance can be identified quantitatively by using econometrics analysis with regression analysis.

Table 4.10: Structure – Conduct - Performance associated with different market forms.

Behaviour Attributes	*Perfect Competition*	*Monopolistic Competition*	*Oligopoly*	*Monopoly*
Market Structure				
No. of Firms	Infinite	Numerous (not as large as in perfect competition)	Few	One
Entry conditions	Easy	Easy	Moderate	Blocked
Product differentiation	Undifferentiated	Some differentiation	Variable	Unique product
CR4	Zero	0 to 80	80 to 100	100
HHI	Zero	0 to 1800	1800 to10000	10000
Market Conduct				
Recognition of interdependence	Unrecognized	Unrecognized	Recognized	None
Market Performance				
Technical efficiency	High	Moderate	Variable	Variable
Earnings	Normal	Normal	Above Normal	Above Normal

5. Government Policies

After identifying the market structure, conduct, performance and their inter-relationships, the next step is to identify and evaluate the Government policies in order to stabilize and sustain the market structure under consideration. The possible policies of the Government were already shown through Figure 4.1 and their influences should be studied from time to time.

5

Agricultural Marketing Functions

5.1. Producer's Surplus

Agricultural marketing involves all types of business activities that ensure the flow of agricultural produce from farmers to ultimate consumers. It is a known fact that, agricultural marketing begins at the level of individual farmers and they are widely dispersed. The agricultural produce is characterized by seasonality of supply, bulkiness, perishability, joint products etc. For some kinds of produce, processing is indispensable to convert the same into palatable form. Considering the persishability of produce, transportation and storage functions also assume greater significance. Thus, agricultural marketing system is comprised of several elements *viz.*, farmers, agricultural produce, middlemen involved in performing different marketing functions, consumers, physical location and personal interactions between the sellers and buyers. However, in the modern era of agricultural marketing (say, e-marketing), even in the absence of the last two elements, marketing process is ensured. It is essential to note that, the concepts of agricultural marketing deals with that portion of agricultural produce actually transacted in the market. This indicates that, whatever the quantity of produce brought by the farmers into the market is dealt under the concepts of marketing. That means, the field of agricultural marketing dealt with the volume of produce actually transacted in the market economy, rather than total quantum of agricultural production. To simplify, say for example, a farmer produced 100 quintals of paddy in a village. To meet domestic (family) needs, the farmer retained 30 quintals of paddy and the remaining amount (70 quintals) was transacted to meet his financial obligations. So, the process of agricultural marketing deals with only 70 quintals of produce. So, in analyzing the marketing process, the proportion of quantities of commodities transacted of their total production is important and it is named as Producer's surplus. Thus, producers' surplus refers to the quantity of produce offered by the farming population to the non-farm population of the country. Thus, the process of agricultural marketing mainly deals with the producers' surplus and not with the entire production. This analysis about the concept of producers' surplus is important considering the following points:

- ✩ The Government will formulate sound price policies to encourage the farmers in continuing the agri-business.
- ✩ It helps the Government to minimize price fluctuations of agricultural commodities by transporting them from surplus areas to deficit areas.
- ✩ It enables the Government to plan for procurement of produce for its re-distribution through the Public Distribution System (PDS) to the people lying BPL.
- ✩ Advanced estimates of producers' surplus of different commodities help the Government to plan for exports and imports of agricultural commodities.
- ✩ Guides the Government for the establishment of requisite infrastructure like cold storage, transportation, processing etc.
- ✩ Guides the traders to plan their trading strategies.

Considering the above meritorious aspects of analyzing the producers' surplus and the involvement of various marketing functions in dealing with this surplus, it is essential to analyze the same before discussing various marketing functions in-detail. This producers' surplus is of two types *viz.*, Marketable surplus and Marketed surplus.

1. Marketable Surplus

The term 'marketable' implies that, the agricultural produce that it is fit to sell in the market. 'Surplus' is the amount of produce that is left over with the farmers after meeting their retentions like seed requirements, wages paid in kind to agricultural labour, domestic consumption requirements, payment to the landlord as rent, feed requirements to livestock, social and religious contributions etc. Thus, marketable surplus deals with the surplus produce left over after meeting the total retentions or requirements of farmers. It is, otherwise, defined as the quantity of the produce, which can be made available to the non-farm population of the country. Marketable surplus is a theoretical concept of surplus. This is expressed as:

$$MS = P - C$$

where,

MS = Marketable Surplus
P = Total Production of the given crop in the year of reference
C = Total requirements of a farm family in the same year.

The various requirements include:

- ✩ Consumption by the farm family
- ✩ Consumption by permanent labour engaged on the farm

- Consumption by the temporary labour occasionally employed on the farm
- Quantity retained for seed
- Quantity retained as feed for farm animals
- Quantity retained for barter,
- Payments in kind to permanent labour, to temporary labour, for hiring machinery and equipment, for customary payments, to land owners as rent, to land owners as share of produce, for re-payment of loan, land revenue, irrigation charges etc.
- Physical losses during threshing and winnowing, transport from threshing floor to storage and in storage at producer's level.

Consumption by the Farm Family

The term 'consumption by the farm family' of the farmer-households has two distinct connotations in so far, as its impact on marketable surplus is concerned. In one case, it refers to the quantity actually retained for consumption by the family irrespective of the actual total requirements for the purpose. Another connotation refers to the quantity that ought to be retained by the farm family for its consumption or the quantity required for consumption. In case of marginal and small farmers, the quantity actually retained is usually less than the quantity actually required for consumption owing to the compulsions or constraints of the size of holding and production. They are, therefore, required to buy back quantities by which they fall short of their consumption requirements. They may make up the deficit with the help of borrowings, wages or gifts etc. In any case, they buy back some quantities from the total stocks, which move out of the farm. The term family consumption, in case of such farmers therefore, denotes the quantity that *ought to be retained* by a farm family for its consumption requirements for the whole year. The use of the term 'Surplus' would thus be justified only, if the quantity actually required for consumption is taken into account rather than the quantity actually retained for consumption, for arriving at the quantity of marketable surplus actually available for non-farm consumption. In case, the quantity actually retained for consumption (and not the quantity actually required for consumption) is taken into account, the quantity of marketable surplus gets unduly inflated because, the quantity sold will not include the buy backs by the producers. Hence, instead of quantity retained for family consumption, the quantity required for consumption is taken into consideration for calculation of marketable surplus. The quantity required for family consumption has been calculated by adding the 'Quantity retained for family consumption + Quantity purchased for family consumption + Total receipts in kind for family consumption'. The marketable surplus will thus be according to the formula, MS = P – C, where, P stands for production, and C includes all the items mentioned earlier except that, 'quantity required for consumption' has been treated to include the 'quantity required for family consumption' as explained above. This quantity is actually available for non-farm consumption and is, therefore, true Marketable Surplus.

The Directorate of Marketing and Inspection (DMI), Ministry of Agriculture has been conducting surveys on marketable surplus and post-harvest losses of food grains. The surveys provide information on marketable surplus ratios as well as on a variety of other important items like farm retention for family consumption, for seed, feed and wastage etc. The data collected were analyzed and published through Indian Agricultural Statistics Research Institute (IASRI) reports. This information is used in the National Accounts Statistics, and Ministry of Commerce and Industry in fixing the weights for certain agricultural commodities while compiling the All-India Index Number of Wholesale Prices in addition to its uses in planning and procurement operations and market development programmes. The concept of marketable surplus is used in a variety of senses, as discussed here under.

i. Marketable Surplus of the Rural Sector

This is the surplus of production left over after meeting the estimated consumption requirements of households within the rural sector and that is finally available for urban consumption.

ii. Marketable Surplus of the Agricultural Sector

We know, rural sector comprises of households engaged in agriculture as well as non-agriculture, and the latter depend upon the surpluses of the agricultural sector in the same manner as the urban sector. This marketable surplus of the agricultural sector is the surplus of production left over after meeting the estimated consumption requirements of households of the agricultural sector and thereby, it is larger than the surplus of the rural sector. But, the market dependence is also larger.

iii. Marketable Surplus of the Farm Sector

We know, the agricultural sector comprises of farmers and agricultural labourers. It is the former that control the marketable surplus. The agricultural labourers may receive their payments in kind or cash from the farm-entrepreneurs. This marketable surplus of the farm sector is the surplus of production left over after meeting the estimated consumption requirements of farmers of the farm sector and thereby, it is larger than the surplus of the agricultural sector. This further increases the market dependence even when compared to agricultural sector. It is common in the villages that, agricultural labourers receive wages both in terms of cash and kind and in some cases only in terms of kind component. But, if there is shift from kind wages to cash wages, it further increases the marketable surplus of farm sector and thereby, the farm sector and agricultural sector become more distinct.Thus, the concept 'marketable surplus' is studied from the following two view points:

- ☆ Surpluses of agricultural produce of the agricultural sector for use by the non-agricultural population and
- ☆ Surpluses of agricultural produce of the farm sector for use by the non-producing population.

The non-producing population includes besides the non-agricultural population, the labour population within agriculture. The non-agricultural sector comprises the non-agricultural population of both the urban and the rural areas. This marketable surplus is measured to assess whether a farmer has the real capacity to sell above his total retentions or consumption needs, irrespective of whether he actually sells or not. However, in fully market oriented production systems, the importance of marketable surplus as a concept becomes less relevant because, it is assumed that, the farmer produces only those commodities that have a market and are profitable, irrespective of whether it is a food grain or something else. Further, in the modern day context of agri-business, the farmer will produce what he does not consume and consume what he does not produce. So, the principle of comparative advantage guides the farmer in practicing the enterprises, rather than his total domestic requirements.

5.2. Factors Influencing Marketable Surplus

Marketable surplus differs from region to region and within the same region, from crop to crop. It also varies from farm to farm. On a particular farm, the quantity of marketable surplus depends on the following factors.

- ☆ ***Size of holding:*** These exits a positive relationship between size of holding and marketable surplus. That is, if size of holding increases, the production will increase and hence, marketable surplus will also increase.
- ☆ ***Production:*** As discussed above, the higher the production, the higher the marketable surplus and *vice versa* (Table 5.1).

Table 5.1: Production and marketed surplus of agricultural commodities at All-India level.

(m. tonnes)

Commodity	*Marketed Surplus Ratio (per cent)*	*Production*		*Marketed Surplus*	
		1999-00	*2006-07*	*1999-00*	*2006-07*
Rice	43.00	87.50	103.50	37.63	44.50
Wheat	51.50	68.70	84.30	35.38	43.41
Coarse Cereals	43.10	39.20	34.40	16.89	14.83
Total Cereals	–	195.40	222.20	89.90	102.74
Pulses	72.40	13.50	21.00	9.77	15.20
Total Food grains	–	208.90	243.20	99.67	117.94
Oilseeds	79.60	21.60	33.80	17.19	26.90
Groundnut	68.30	5.90	11.50	4.03	7.85
Mustard and Rapeseed	84.30	6.10	8.90	5.14	7.50
Other oilseeds	86.30	9.60	13.40	8.28	11.56
Sugarcane	92.90	315.10	352.80	292.72	327.75
Cotton	100.00	2.10	3.20	2.10	3.20
Vegetables	83.00	85.00	110.70	70.55	91.88
Fruits	97.00	49.50	70.50	48.02	68.38

- ***Family size:*** If the size of family is large, the consumption requirements will be more thereby, marketable surplus will be low.
- ***Farm requirements:*** If the farm requirements are more, the marketable surplus will be low.
- ***Nature of commodity:*** For non-food crops like mesta, jute, rubber etc., the domestic requirements are totally negligible and hence, the entire production can be considered as marketable surplus. Even in case of food crops, some crops need processing before final consumption like sugarcane, oil seeds etc., and for these crops, the marketable surplus is relatively more when compared to staple food crops like wheat, maize, paddy etc. As discussed earlier, the marketable surplus varies across the crops and following are the interesting points:
  - Marketable surplus for food grains, particularly in a deficit state, are low and such crops may not be considered as commercial crops in that area.
  - All fruit crops are definitely commercial crops because, their marketable surpluses are considerably high. Similarly, vegetables are also regarded as commercial crops.
  - In the past say, about 30 years back, dairy activity was just carried out as subsidiary to crop production to meet the family requirements of milk and there is no surplus. But, after the development of new high milk yielding cow and buffalo breeds, improvement in feeding and management practices of milk animals, strengthening of marketing facilities through establishment of Government Milk Schemes and Milk Producer's Co-operatives, the milk production has increased very rapidly and it has now become an important commercial activity.

- Some oil seed crops like groundnut, sunflower, safflower, soyabean, castor and other crops like red chillies, turmeric, tobacco, cotton and sugarcane are also recognized as commercial crops, as the marketable surpluses for them are almost 100 per cent.

☆ ***Consumption habits of the people:*** The consumption habits of the individual often determine the extent of marketable surplus. For example, in Punjab, rice forms a relatively small proportion of total cereals consumed by farm families and hence, the output of rice in Punjab will have more marketable surplus when compared to in Southern states, where rice accounts for major share in total cereals consumption.

☆ ***Cash requirements:*** If the cash requirement of the farmers is not fixed, the marketable surplus will increase in response to increase in price of commodities, as farmers, in general, are price conscious.

☆ ***Mode of production:*** Under traditional methods of production, farmer mainly employs owned resources like retained seed, owned labour and further, the production is subsistence oriented thereby, marketable surplus is low or insignificant. However, in the modern era of agri-business, capitalist system of farming is practiced and the farmer employs high yielding variety seeds, fertilizers, pesticides and hire machines and this increases production and hence, marketable surplus.

☆ ***Price of the commodity:*** It is essential to note that, price of the commodity under consideration may show positive or negative relationship with marketable surplus as discussed below.

- ***Negative relationship:*** According to P.N. Mathur and M. Ezekiel, price of the commodity shows inverse relationship with the marketable surplus. That means, as price of the commodity is increased in the market, less quantity of produce is offered for sale by the farmer or the marketable surplus is low and *vice versa*. This is based on the assumption that, farmers will have inelastic cash requirements. For example, assume the farmer's cash requirement is ₹ 10000. If the price of the commodity is ₹ 500/quintal, he will offer 20 quintals for sale in the market. On the other hand, if the price of the commodity falls to ₹ 400/quintal, he will offer 25 quintals for sale in the market, to meet the same cash requirement of ₹ 10000. A further assumption was that, farmers save in kind and not in cash. Given these assumptions and given a short-run situation in which the output was assumed to be fixed, it was easy to draw the conclusion that, regulating prices at a lower level will induce a higher quantum of marketable surplus. Khatkhate (1962), pushed this analysis a step further by assuming that, not only the marketable surplus responded negatively to prices, but even output response to prices was negligible. This is because, in an economy characterized by subsistence type of farming, prices do not exercise any influence on production, since the farmers are already producing to the maximum capacity and thereby, marketable surplus responds negatively to price and this is termed as 'Stinted consumption paradox'. It is essential to note that, though big farmers respond normally to price changes, the influence of the (majority of) small farmers on production and marketable surplus was assumed to have an adverse effect on the total. Further, a high price to the farmers was not only not helpful in inducing more marketable surplus, but was considered to have adverse effects on the overall level of savings in the economy. That is, whatever helpful effects the price rise might have on savings in the agricultural sector, it was likely to be more than offset by opposite changes in savings in the non-agricultural sector and as a result, the aggregate saving-income ratio might well decline. For example, with a rise in agricultural prices relative to non-agricultural prices, there will be a shift of income in favour of the agricultural sector, where average propensity to save is smaller than in the urban sector. Furthermore, the price rise of agricultural goods tends to raise the wage-profit ratio in the industrial sector and lowers down the saving ratio. Thus, lower agricultural prices help to reduce the wage-profit ratio in the industrial sector and at the same time induce more marketable surplus in the agricultural sector, of course, at the cost of welfare of the farmer.

However, Dandekar and Raj Krishna criticized the Mathur-Ezekiel proposition and opined that, it is not applicable to the small farmers because, in general, they have a net deficit and they purchase from the market. So, prices affect them not as producers, but as consumers. It does not apply to the large farmers because, their income elasticity of demand for home produce is likely to be zero. They further opined that, the relative prices between agricultural and non-agricultural sectors are less relevant to the farm production and well-being of farmers.

- ***Positive relationship***: According to V.M. Dandekar and Raj Krishna, marketable surplus varies directly with the price. That means, if the price of the commodity is increased, the farmers will increase the marketable surplus *i.e.*, they will offer more quantity for sale in the market. This is because, the farmers are more price conscious. That means, if the prices are higher, they will sell more of the produce, so as to get more revenue. In the present day context of agri-business, this relationship is more valid compared to the concept of inelastic cash requirements. In most cases, high prices may act as incentives to more production and may induce more marketable surplus. But, this raises the cost of production of agricultural commodities (due to increase in demand for resources and resource services), thereby, stimulate inflation and reduce the rate of growth. Thus, high prices of agricultural produce shift income to those, who have more marketable surplus, from urban to rural, from industrial to agricultural entrepreneurs, from urban and salaried workers to big landlords. On the contrary, low prices have opposite effects. The adverse or favourable effects on distribution of incomes depend upon the extent of distribution of marketable surplus among different strata of farmers, relative income position of farmers *vis-a-vis* the non-farmers and relative income position of non-farmer, rural *vis-a-vis* non-farmer, urban.

- ***Large-scale imports:*** Import surge at concessional rates act as a substitute for domestic products and this limits the quantum of production offered for sale in the market.
- ***Nature of trade:*** If there is Government intervention in the field of marketing of agricultural produce through regulating marketing practices, it ensure fair trade on the part of farmers and thereby, they offer more quantum of produce for sale. On the contrary, in the absence of regulation, private traders dominate the marketing scenario and they exploit the farmers. So, in the absence of regulation, farmers will offer less quantity for sale and retain more till prices shoot up in the market.

Significance of Analyzing Marketable Surplus of Agricultural Commodities

It is a known fact that, farmers can never know exactly, how much production their efforts will yield for a particular season until it is over. For this reason, a farmer may not know, what the marketable surplus will be until he actually harvests the crops. This uncertainty of the size of a crops' marketable surplus lends a certain level of uncertainty to the farmer's income expectations. But, from a livelihood and food security perspective, for small and marginal farmers, who may make distress sale, marketable surplus studies may indicate the welfare implications of their sales. However, the relevance of this concept declines along with increased commercialization of agriculture.

2. Marketed Surplus

It is the quantity of produce that was actually marketed by the farmer irrespective of his requirements like seed, payment of wages in kind, payment of rent to landlord in the form of kind, family consumption requirements, feed requirements to livestock, social and religious contributions etc. Thus, it refers to the actual quantity of produce marketed at a given point of time by the farmers. The marketed surplus may or may not be equivalent to marketable surplus depending upon changes in seed requirements, changes in payment of wages in kind, changes in payment of rent to landlord in the form of kind, changes in family consumption requirements, changes in feed requirements to livestock, changes in social and religious contributions etc. Marketed surplus of the farm sector is sometimes used to refer to actual sales of produce plus all disposals in kind. Thus, marketed surplus may include entire marketable surplus of the current year (if entire marketable surplus of current year is marketed) plus previous (old) stock disposed and all other disposals in kind. So, marketed surplus can be written as,

Marketed surplus = Marketable surplus of the current year *i.e.*, if entire marketable surplus of currentyear is marketed + Past stock sold out + Other disposals in terms of kind – Losses, if any.

From the above formula, we can infer that, if marketed surplus of a farmer is positive, marketable surplus may or may not be positive. Sales under condition of negative marketable surplus (consumption requirements are more than production) may have welfare implications, if the sales are of a distress nature and arise due to compelling cash needs that can't be generated otherwise. The marketed surplus may be more, less or equal to the marketable surplus and these cases are discussed here under.

- ***Marketed surplus is more than the marketable surplus:*** This is the case, when the farmer retains a smaller quantity of produce than his actual family and farm requirements. This situation holds well under the following conditions:
  - Holds good for small and marginal farmers, as they are in immediate requirements of liquid cash may be due to their prior debts and pressing family needs. This situation of selling more than marketable surplus is termed as Distress or Forced sale. Such farmers generally buy the produce from the market in a later period to meet their family and other requirements.
  - The farmers lack retention capacity
  - When price of produce is very high in the market, the farmers might have reduced their domestic requirements.
- ***Marketed surplus is less than the marketable surplus:*** This is the case, when the farmer sells only a part of the marketable surplus or when the farmer retains some of the surplus produce. This situation holds well under the following conditions:
  - Holds good for large farmers, as they enjoy more retention capacity and thereby, can stock the produce in the godowns, till prices shoot up in the market.
  - Farmer may substitute one crop (paddy) for another crop (jowar) either for family consumption purpose or other farm requirements because of the variation in prices. However, with the fall in the price of the crop (paddy) in the market relative to a competing crop (jowar), farmer may consume more of the first and less of the second crop.
  - Large scale domestic requirements of farmer.

- ☆ ***Marketed surplus may be equal to the marketable surplus:*** This is the case, when the farmer neither retains more nor less than his requirements. This holds true for perishable commodities and agricultural raw materials like mesta, cotton, jute etc., as for these commodities, the domestic retentions are nil.

5.3. Issues Under Consideration to Estimate Marketed Surplus

The following are the some of the important issues that need careful consideration for proper estimation of marketed surplus:

- ☆ ***Net output:*** Total output refers to the actual quantity of production realized by the farmer in his production programme. Net output is the difference between total output and 'seed, feed, waste'. from gross output.
- ☆ ***Gross marketed surplus:*** In general, we consider only sold amount is equivalent to marketed amount. But, the question arises, whether outgoings in terms of kind (payment of wages in kind, retainment for seed purpose, payment of rent to landlord in kind etc.), except own consumption, could also be treated as marketed. This is because, in reality, anything that leaves the farm over own consumption (whether in the form of sale or in-kind payment) could be treated as sales, as in-kind outgoings could be sold and the payments could be made in cash.
- ☆ ***Net marketed surplus:*** In general, we get net marketed amount, by deducting purchases made if any, from total sold amount. But, it is essential to note that, besides the purchases and sales, non-purchase (collecting produce from tenant farmer as rent) and non-sale (giving produce for social and religious ceremonies) transfers may also arise in the marketing process. If these non-purchases and non-sales transfers and transactions are in significant amount, the question arises then, whether they have to be included in computing net marketed surplus. Further, these non-purchases and non-sales transfers and transactions are not in equal amounts so that, they do not cancel each other, either for an individual farm or even if we calculate them at the aggregate level. So, from the accounting point of view, for an individual farm, net marketed amount should be derived by taking into account all incomings and outgoings rather than only purchases and sales.
- ☆ ***Gross marketed surplus ratio:*** It is computed by studying the ratio between gross marketed amount to gross output or net output and expressed as a percentage. But, the question arises here, whether to consider gross output or net output in the denominator. In reality, the sale of produce by the farmer is not influenced gross or net output, but by the overall supply *i.e.,* own production plus incomings. So, the appropriate denominator for estimation of gross marketed surplus ratio is total available supply rather than just output.
- ☆ ***Net marketed surplus ratio:*** It is computed by studying the ratio between net marketed amount to gross output or net output and expressed as a percentage. In line with the gross marketed surplus ratio, here also the appropriate denominator should be total available supply rather than just output.

Significance of Analyzing Marketed Surplus of Agricultural Commodities

The analysis of marketed surplus of different types of agricultural produce is of interest to researchers and policy makers because, market participation is essential for farmers for raising farm income and welfare. Further, adequate regular supply of agricultural commodities into the market is essential to keep food prices at affordable levels for consumers. As the economy develops, number and proportion of population engaged in agriculture usually decline though the size of the overall population increase. Consequently, the farmers need to produce not only enough for themselves, but also for an expanding non-producing consumers in both rural and urban areas so that, dependence on import or aid can be avoided or at least minimized. On the other hand, if surplus above national food need is produced, then there will be scope for export of agricultural produce to raise both producers' and National Income. Further, understanding marketed surplus and marketing behaviour of farmers helps design technology, policy and institutions to facilitate the process of commercialization of agriculture.

The above discussion infers that, in general, production, consumption, sales and purchases have been considered as elements in defining marketed and marketable surplus. In addition to sales and purchases, transactions and transfers may also take place in the form of payment of rent, labour wages and gifts in terms of kind. Moreover, agricultural sector is characterized by seasonality of production and regular consumption and this feature coupled with significant inventory changes between two seasons or years may not allow the complete sale of output within the season or year. Thus, the total volume of output or produce available on a farm over time depends on the volume of production due to new harvest, purchases made due to other reasons, harvested produce of the past seasons or years, cash requirements of the farmer etc. In fully commercial production systems or systems in which non-sale transactions and transfers and inventory changes is zero or negligible, sales as a percentage of net output is a good measure of marketed surplus. However, where non-sale transactions and transfers and inventory changes involve a significant proportion of output, accurate estimation of marketed surplus will require proper treatment of non-sale inter-farm transactions and transfers.

5.4. Process of Agricultural Marketing

As discussed in the earlier pages, the two terms marketable surplus and marketed surplus are different in the sense that, first refers to the saleable amount of the agricultural produce in a given period of time and the second to the actual amount sold at a particular time of sale. In the past, farming is mainly subsistence oriented and the farmers used to produce agricultural produce primarily for home consumption with an insignificant amount of marketable surplus. This preference for subsistence farming and for home produced food emanated from their efforts to avoid market risks, especially risks of possible high prices in the off season, when they needed to buy food. So, agricultural marketing is sometimes thought of as simply the process of buying and selling. However, with the recent advances in the fields of production and marketing, there is a marked shift from subsistence type of farming to market-oriented farming. Thus, a fully market oriented farmer will produce anything that is profitable, and buy requisite goods and services from the market. So, in the modern era of business, the marketing activities are much more extensive than those dealt under subsistence type of farming. For a modern marketing system to be operative and effective, it should involve various kinds of activities of economic nature and they create different types of utilities to the consumers like form, place, time and possession utilities. These economic activities in the marketing process of agricultural commodities are carried out by various market intermediaries, but this does not reduce or lessen the importance of such functions in any way. So, in modern marketing system, the agricultural produce has to undergo a series of transfers or exchanges from one hand to another before it finally reaches the ultimate consumer and this process is achieved through three important steps namely Assembling or Concentration, Equalization and Distribution or Dispersion.

1. Assembling or Concentration

Assembling refers to collection of produce or marketable surplus of individual farmers at a common place. It also refers to physical collection of a large quantity of produce of the same type, which may have been produced in small lots in different places. This concentration to a certain extent embraces various other functions like transportation, storage, weighment, processing etc. Assembling also refers to bringing the buyer, seller and commodity at a particular place to ensure marketing process. Thus, under this process, the agricultural produce from different farmers is collected at a central point to facilitate further action upon them. In marketing, the assembling helps in:

- ✩ Reducing the handling costs,
- ✩ Regulating supply,
- ✩ Widening the market,
- ✩ Reduces transportation costs,
- ✩ Facilitates scientific storage of produce in large quantities, and
- ✩ Making grading possible.

Assembling is of two types:

- ✩ ***Primary Assembling:*** In primary assembling, the produce is assembled either in the village or primary markets. The important functionaries in these markets include village merchants, Itinerant traders (they are the merchants who don't have any shop of their own in the village, but they will purchase the produce in one village market, transport the produce through bullock carts and sell the produce in other village market). The following are the possible reasons for assembling the produce in village or primary markets:
    - ◆ The farmers were in immediate requirement of liquid cash
    - ◆ Prior indebtedness of farmers
    - ◆ Transportation costs to transport the produce to distant markets may be too high
    - ◆ High perishability of the produce
    - ◆ Due to lack of sufficient transportation facilities
    - ◆ The farmers may not aware about the prices of the produce in the distant (major) markets.
- ✩ ***Secondary assembling:*** In this assembling, the produce will be pooled in wholesale markets. So, in these markets, the concentration or assembling is higher than in primary assembling. Nearly, 70-75 per cent of marketable surplus will be assembled in wholesale markets. The important market functionary in secondary assembling is 'commission agent'.

2. Equalization

According to Clark, '*equalization consists of the adjustment of the supply to the demand on the basis of the time and quantity, so as to ensure an equilibrium price for the commodity under consideration*'. This sort of adjustment can be done through storage and transportation respectively in market centers. We know, in general, the demand for agricultural commodities is regular, but their production and supply are seasonal. So, to ensure an equilibrium price, the commodities should be stored at times of peak marketing season and they should be timely released in accordance with the demand conditions in the market. Similarly, by transporting the commodities from surplus areas to deficit areas, the equilibrium can be maintained between demand and supply forces. For example, by transporting wheat from northern states to southern states, the equilibrium price can be achieved in both the regions.

3. Dispersion or Distribution

It refers to distribution of commodities to the consumers by the farmers through various marketing channels. The commodities, which do not require processing like leafy vegetables, milk, fish etc., can be directly sold to the consumer by the farmers. But, the commodities, which need processing like paddy, cotton, sugarcane etc., they should be distributed through the processors and not to the consumers directly. This dispersion also involves the operations of wholesaling and retailing at various points. By a series of indispensable adjustments and equalizing functions, it is the task of distribution system to match the available supplies with the existing demand.

5.5. Classification of Agricultural Marketing Functions

The marketing function is an act or operation or service in order to link the original producer (farmer) and the ultimate consumer. The process of marketing could be split down to certain major economic activities that are inherent in the marketing process *viz.*, transportation, processing, buying and selling etc. These functions become specialized through continuous division of labour. No doubt, these functions are independent in their own way, but when marketing process is considered as a total system, each function becomes essential and assumes continuity. It is essential to note that, the marketing process of a commodity can avoid a market intermediary, but not the market functions. So, we can define the term 'marketing function' as the function, which cannot be avoided in the marketing process. We know, marketing of agricultural produce is regarded as unique and deserve special attention due to the perishability and bulkiness of the products involved. The fact that, most of these products are basic foodstuffs, whose price and distribution are considered essential from the point of views of farmers, consumers and Government. This calls for efficient execution of certain activities that ensure the movement of agricultural produce from the farmers to the ultimate consumers. These activities that ensure the movement of agricultural produce from one agency to another are called agricultural marketing functions. Since, there are several marketing functions in the process of agricultural marketing, it is essential to classify them into different categories (Table 5.2). This classification is not unique, as even now, there is no consensus regarding the functional areas of agricultural marketing and the given classification is often regarded as widely accepted classification. Considering the different views of agricultural marketing experts, the marketing functions may range

Table 5.2: Functional approach of agricultural marketing.

Type of Functions	*Series of Activities Involved*
I. Physical functions	A. Packaging and Packing B. Transportation C. Processing and D. Storage
II. Exchange (Trading) functions	A. Buying and Selling
III. Facilitating functions	A. Grading and Standardization B. Market financing C. Market Intelligence D. Risk bearing.

from five to 126 in the marketing journey of the products and services from farmers to final consumers. This functional approach served as the leading approach to the study of agricultural marketing from the early 1920s into the 1960s. The development of functional approach coincided with the development of agricultural marketing as a separate and unique activity within the business firm. Until the functions of marketing are recognized and isolated, agricultural marketing as a separate system with in the business firm could not be isolated. As such, with the recognition of marketing functions, a clear concept of agricultural marketing was developed. However, the dominant approach to the study of agricultural marketing at the present time is the Managerial approach. But, the approach that dominated the field of agricultural marketing for the longest time and many notable developments are witnessed are during the years of the domination of functional approach of marketing.

I. Physical Functions

This includes different activities like Packaging and Packing, Transportation, Processing and Storage that ensure handling, movement and physical change of the actual commodity itself. They are involved in solving the problems of 'when', 'what' and 'where' in marketing. Physical functions enable the actual flow of agricultural commodities through space and time from farmer to consumer and their transformation to a form desirable to the consumer.

II. Exchange Functions

This includes two important activities *viz.*, buying and selling and they ensure equilibrium price determination for the produce under consideration and also in transfer of title to the produce. Both buying and selling functions have their primary objectives of negotiating the price settlement between buyers and sellers. These functions involve finding a buyer or a seller, negotiating price and transferring ownership (but not necessarily physical transfer). These functions take place at the 'market', that is, the physical meeting point for buyers and sellers at the point of production or via some other means of communication.

III. Facilitating Functions

This includes important activities like grading and standardization, financing, market intelligence and risk bearing. These activities serve as catalyst for the smooth performance of the above physical and exchange functions. It is essential to note that, these activities are not directly involved in either physical handling of produce or exchange of title. But, in the modern era of agricultural marketing, these activities are of crucial significance and in the absence of these activities, it is difficult to imagine about smooth functioning of marketing system. These activities are popularly regarded as '*grease*' that makes the wheels of the marketing machine go round.

These three classes of functions create the marketing environment for agricultural commodities, whose elements are:

- ✰ Market and facilities - including all of the physical infrastructure that a market may depend on.
- ✰ Market information and intelligence - including informal and formal communication systems and standard weights and grades on which market information depends.
- ✰ Institutional environment - including the Government policy environment, regulations and supporting legislation.

Thus, it was found that, there are nine distinct agricultural marketing functions, grouped into three major classes. The identification of these functions is further believed to understand the role each participant in the market chain and the services rendered by each with regard to these functions. So, it is high time now, to discuss these important marketing functions in-detail.

I. Physical Functions

Physical functions address the problems of 'when', 'what' and 'where' in marketing. These are productive activities in their own right, certainly from the perspective of people performing the functions. These comprise packaging and packing, transportation, processing and storage and they are discussed here under.

A. Packaging and Packing

Though these two words are used interchangeably, but they are not synonyms. Packaging refers to the way the products are kept inside a box either individually or in sets using a wrapping material to protect them from any direct damage during transportation. In the words of Arens *'Packaging is the container for a product – encompassing the physical appearance of the container and including the design, color, shape, labeling and materials used'*. According to Olson and Jacoby, packaging can be defined quite simply as *'an extrinsic element of the product - an attribute that is related to the product, but does not form part of the physical product itself'*. It is mostly done to send the products in bulk. Thus, packaging is more than simply putting the product in a container or wrapping it up. It also involves the development of a container and the graphic design for a product. Packaging can be just as important as the product itself, when deciding what to choose. Well designed packaging can give a positive impression of the product and can encourage first-time consumers. As well as helping to preserve, inform, protect and promote a product, packaging can also create an image of luxury, sensuality and exclusiveness. When developing new packaging, the business firms take into account two most important of factors *viz.*, cost of packaging and quality packaging. This is because, consumers are prepared to pay more for quality packaging however, there is a limit as to how much consumers will pay. In these days of increasing community awareness of the environment and pollution, marketers need to give more emphasis on qualitative packaging material. Presently, recyclable and biodegradable packaging is deserving special attention. We know, the classification of marketing mix (given by McCarthy) comprises of 4Ps *viz.*, product, price, place and promotion. Some authorities want to include 'packaging' as fifth 'P' of marketing mix, keeping in view of its importance, but most continue to view it as a component of product strategy. Highlighting the importance of packaging, Coles and Beharrell stated that, *'it is the package that communicates more to the consumer than the actual product, at the point of purchase where the consumer decides.'* This statement though quoted in the context of consumer food products, it also has validity with respect to agricultural inputs. For example, when the farmer purchases a fertilizer bag or a pesticide bottle, it is the package that conveys about the features, benefits, applications and quality.

i. Levels of Packaging

There are three levels of packaging:

- ✰ ***Primary packaging:*** It is the material that first envelops the product and holds it. This usually is the smallest unit of distribution or use and is the package which is in direct contact with the contents. Examples include, cushioning, envelopes, plastic bags, plastic bottles, skin pack, tin cans, wrappers etc.
- ✰ ***Secondary packaging*** It is outside the primary packaging and is used to group primary packages together. Examples include, boxes, cartons, shrink wrap etc.

☆ ***Tertiary packaging***: It is meant for bulk handling and distant transport. Examples include, bales, barrel, crate, container etc.

These broad categories can be somewhat arbitrary. For example, depending on the use, a shrink wrap can serve as primary packaging, when applied directly to the product.

Packing refers to placing the already packaged products into a casing so that, they look attractive to consumers and arrive safely and securely into markets from where they are purchased. Thus, packaging is a part of packing. The packed products are often loaded and unloaded several times during transportation and can be roughly handled, dropped or exposed to potentially damaging weather. Packing also refers to all materials, procedures and methods used to prepare, present, handle, store, conserve and transport a product from the factory or packaging plant to the end consumer.

The good example to compare packaging and packing is that, mangoes, apples etc., are kept in straw (packaging) and these packaged mangoes are kept in wooden boxes (packing) for distant transport. Similarly, when we buy a product say, soap from the market, each soap is packaged inside a paper material. However, when such soaps as bundles are transported from the manufacturer to the wholesale shop or a retail shop, they are packed properly in cardboard boxes so that, they are transported safely. Similarly, biscuits, chocolates etc., are packed inside small cartons. However, there are some products for which the word package is used instead of packing such as packaged milk, packaged drinking water etc.

ii. Criteria for Selection of Packaging Material

The packaging and packing of agricultural commodities and their processed products assume greater significance, as they are an integral part of consumers' life and this function directly influences the safety, freshness and overall shelf life of the products. As these commodities are highly perishable, the farmers and other intermediaries are bound to take utmost care of their products after production until the time they are sold because, the spoilage of food adversely influences their business as well as wastage of time and labour. Since, there are different types of agricultural commodities and products, there are different ways of packaging and packing. In choosing packaging materials, it is essential to consider what product protection is needed, (*e.g.*, light, crushing, dehydration, oxygen). The protection offered by a package is determined by the nature of the packaging material and the package construction. Packages may be flexible (paper, foil, plastics) and rigid or semi-rigid (*e.g.*, cans, glass, some plastic containers). In general, the packaging material must be consumer friendly and for this, it should have the following features:

☆ protect the quality and quantity of the produce
☆ prevent spoilage during transportation and storage
☆ should not undergo chemical reactions with the product
☆ ensure that the product remains in perfect condition until it reaches the end consumer
☆ suitable according to climatic and environmental conditions
☆ compatible with the product, processing and storage conditions
☆ should have sufficient protective strength
☆ easy handling of the products during transportation
☆ be convenient and suit to the needs of the consumers
☆ be attractive for display

- ☆ provide information about quality, variety, date of expiry, weight and price etc.
- ☆ be cheap, clean, attractive and readily available
- ☆ be environment friendly and biodegradable
- ☆ be free from adverse chemicals
- ☆ conform to the requirements as laid down under Prevention of Food Adulteration Act (PFA) Standards as amended from time to time
- ☆ should provide technical information about the product and guide the consumers regarding how to use and preserve the products
- ☆ meets current legal requirements
- ☆ should be convenient to stack
- ☆ should be able to prevent spillage during transit and storage
- ☆ should be cost-effective
- ☆ should be helpful in reducing the Mc by reducing the handling and retailing cost
- ☆ should be reusable
- ☆ should be tamper proof
- ☆ should be easy to open
- ☆ should be convenient (sizes, resealable etc.).

It is important to note that, not one food package available today meets all of these criteria. It's up to the firm to decide, which are most important for the product under consideration and which can be compromised.

iii. Need for Packaging and Packing

They are essential because,

- ☆ To ensure protection to the products from shocks, vibration, compression, temperature etc.
- ☆ To provide barrier of protection from oxygen, water vapour, dust, light etc. However, some packages contain desiccants or Oxygen absorbers to help extend shelf life. Modified atmospheres or controlled atmospheres are also maintained in some food packages.
- ☆ To facilitate easy handling and transportation. For example, rasam powder packets, sambar powder packets, pencil box etc., ensure easy handling. Similarly, selling sealed one-litre-bottles of milk is better than asking the consumers to bring their own bottles to fill themselves.
- ☆ Some products need to be informed to the consumers about their technical ingredients so that, it guides them how to use, how to recycle or dispose of the product.
- ☆ It will encourage potential buyers to purchase the product.
- ☆ Packages may bear authentication seals to help indicate the significance of the products. For example, AGMARK label is indicated on the packed material and the presence of this label indicate that, the product is meant for export from India.
- ☆ Facilitate easy handling of the products. If the product is bulky (say, wheat), for convenient sake, it can be divided into packages of suitable size for individual households.

iv. Different Materials Employed in Packaging and Packing of Agricultural Commodities

The following are the important materials that are commonly used in packaging and packing of different agricultural commodities and their relative merits and demerits are shown through Table 5.3.

Table 5.3: Relative advantages and disadvantages of different materials used in packaging and packing of different agricultural commodities.

Item	*Advantages*	*Disadvantages*
Paper	◆ Low cost and readily available Easy to store ◆ Can be recycled ◆ Can be coated or laminated to improve impermeability to liquids, gases and vapors ◆ Opaque and excludes light ◆ Printable	◆ Poor barrier to liquids, oils, fats ◆ High water absorption capacity(hygroscopic)
Glass	◆ Does not react with food ◆ Can be transparent or colored to protect from light ◆ Clean, pure and hygienic material ◆ Water proof for normal purposes ◆ Resists internal pressure and high temperatures ◆ Can be stacked without crushing ◆ Consumers can see the contents and the product's appearance	◆ Brittle ◆ Heavy ◆ Printing not common (expensive) ◆ Less safety to product (due to breakage)
Plastic	◆ Minimum chemical interaction between the container and food ◆ Light weight, flexible and versatile ◆ Good mechanical resistance ◆ Can be recycled	◆ Permeable by gas and radiation ◆ Thermo stability can be a problem ◆ Printability depends on material used
Metal (tin, steel or aluminum cans)	◆ Non-toxic ◆ Strong coated or plated, so that they won't react with the food ◆ Good mechanical resistance ◆ Minimum chemical interaction between metal container and food ◆ Stable and impervious ◆ Good opacity (blocks light)	◆ Expensive ◆ Require expensive sealing equipment ◆ May have microscopic pinholes allowing gas exchange with environment ◆ Transportation and storage cost are higher due to greater volume ◆ Come in standardized forms
Wood	◆ Strong resistance to shocks, water and moisture ◆ Can be reused ◆ Allow air circulation around the product	◆ High cost ◆ Inconsistent resistance ◆ Requires more space for storing

Agricultural produce (fresh) are normally packed in wooden boxes and loaded on pallets. These pallets are an important element in export packing. Pallets can be made of wood, plastic or metal. To select the most suitable pallet, consider the weight of the load, climate, durability, local availability, cost and acceptance in the destination market. The benefits of using pallets include reduced use of labor in storage, shipping and distribution processes; reduced damages to products and increased utilization of storage space. Table 5.4 shows the most frequently used pallet sizes as per the International Organization for Standardization (ISO) standards.

- The products must be contained before they can be moved from one place to another. To function successfully, the package must contain the product. This containment function of packaging makes a huge contribution to protecting the environment. A better packaging help to maintain the quality of the product and reachability of the product in the consumer's hand without spillages. It gives better image to the firm.
- Packaging food makes it safer and less vulnerable to contamination.
- It provides convenience in handling during transportation and storage.
- Packaging helps create a positive product image in the minds of the consumers.
- The attractive packaging design of the product enables the consumers to easily identify and differentiate it from the competition. In this context, not only for processed products, even fresh produce is increasingly being packaged as a mechanism for market differentiation and consolidation. Distinctive product packaging, be it in the form of shape, size, colouring materials, print etc., can help in the positioning of a product in the consumers' minds. Suppliers of fresh produce, such as fruits, find that, it is difficult to effectively brand the product without packaging. Packaging in the form of attractively shaped and brightly coloured packages can enhance the product's appeal. Moreover, the quality of the packaging is often used by the prospective purchaser as an indication of the quality of the contents. So, packaging innovations will substantially increase sales volumes.
- A proper, standardized package of standard weight and suitable quality will bring a higher return to the farmer.
- Packaging design is capable of contributing to the improved performance of the supply chain in a variety of ways. By altering the shape and dimensions of the packaging, more quantity or number of products can be displayed on retailers' shelves (Ex. changing from bottle to card board boxes (packets) helps maximize the use of limited retailing space, since far more packets can be placed on a given area of retail display space than can bottles). Further, the manufacturers are preferring to go for small packs of the products, as they ensure the following advantages:
  - They help the consumers to pick the products at required quantity and at affordable price.
  - Individual use products like cashew, dry fruits, jam, biscuits, Rasna powder etc., are bought in smaller size.
  - Small packets are easy to display and they increase the visual appeal.
  - Small packs are convenient to retailer to do his business.
- Packaging and packing reduces bulkiness of the products.
- Packaging and packing helps in checking adulteration (purposefully spoiling the quality of produce).
- Packing is the only communicator with the consumer at the point of purchase, as it communicates what it sells to the consumer. This is especially true in case of international trade, where consumers of different spoken languages are involved in transacting and consuming the products. Here, packaging protects the interests of consumers and further provides useful information like quantity, grade, price, size, expiry date, time since packaging, colour etc.

- The packaging is often the most relevant element of a trademark and conduces to advertising or communication.
- The packaged material sometimes helps in improving the quality of the products. Ex. straw hastens the ripening of mangoes.
- Food lasts longer and stays fresh longer when it is packaged successfully. Unpackaged food can quickly become dry, mouldy or spoiled.
- Proper packaging of produce implies the product is hygiene.
- By proper packing along with labeling will help the consumer how to use, preserve the products.
- By scientific packaging of product, the storage capacity of the produce can be increased.
- Packaging plays a key role in brand promotion and management. It is of great importance in the final choice of the consumer because, it directly involves convenience, appeal, information and branding.
- *'Packaging is not a gimmick when it works'*. An innovative pack design can help to set a brand apart from its competitors. The marketing world is full of examples of brands that have used packaging to carve a unique position in the market place. The design of the pack itself can act as an incentive for purchase. For example, a strong, sturdy mineral water bottle might be chosen over its competitors, not for its content, but rather for its ability to be reused on future occasions.
- Packaging in different serving sizes can extend a product into new target markets or help to overcome cost barriers. In developing markets such as India, the pack size can mean the difference between the success or failure of a brand in the informal sector. Smaller packages are usually priced at a lower absolute level – making the product more readily affordable to a greater proportion of the population. Some examples of success in this regard include jams, tea packets, coffee packets, biscuits, wheat atta etc., which have increased the penetration of their brands substantially. The popularity of smaller packs for analgesics has proven that *'good things really do come in small packages'*. Where smaller packages are not available, entrepreneurial individuals often buy the product in bulk and transfer the same into smaller non-branded packaging for resale, which completely nullifies all the branding benefits of the original pack. On the other hand, larger packs can extend the category to a more social environment. For example, the large-sized fruit juice packs expanded the category from individual and home consumption to social and catering purposes. So, the larger size means that, the cost per volume is cheaper and more affordable for the masses.
- The paramount concern of packaging is the reachability of the product without any damage.
- Packaging is a very important marketing strategy to popularize the product among the consumers. Sometimes, packaging is so important that, its cost is more than the product itself in order to lure the consumers to buy it. It is logical to say, attractive packaging is crucial in order to get the first time buyers to buy the product. Of course, attractive packaging does not mean neglecting the quality of the product, rather it converts the first time buyers into loyal customers in purchasing the product on continuous basis.
- Packaging has a crucial (positive) impact on the efficiency of transport, handling and storage of products. Packaging should, therefore, be designed to permit easy handling, space-saving storage and stowage of products. The strength of the package required for stowing goods on

We know, most produce and products of agriculture need packing. However, there are some exceptions like grains and some larger fruits such as, jack fruit, coconut, pumpkins, watermelons etc., and these are transported in bulk. Similarly, leafy vegetables are also transported in bulk, as packaging is difficult for them, since they are not uniform in their length. So, for the examples quoted above, there is no packaging cost. However, for other produce and products like rice, wheat, pulses, edible oils, mangoes, sapotas, grapes, apples, sugar, jams, jellies, ice creams etc., packaging is inevitable, as its importance is already discussed in the earlier pages. At farmers level, rice, wheat, pulses etc., are packed in jute or gunny bags and the same may be used right through the marketing chain. However, in case of other produce like mangoes, sapotas, grapes, apples etc., more complex and expensive packaging such as plastic crates will be used and normally trader takes the responsibility regarding the use of these packaging materials. Unlike food grains, these fruits may be packed and repacked several times on the way between producer and consumer, depending upon the length of the marketing channel. The farmer generally goes for less expensive packaging like use of gunny bags for packaging fruits, but the traders or intermediaries in the marketing channel may transfer the same to wooden box or plastic crate during the course of transactions. Since, the retailers, who act as consumers' personal representatives may go small size packaging like plastic covers for convenient handling by the consumers. These different types of packaging by different marketing agencies involve costs and they are essential to compute, as they form part of the total marketing costs incurred in moving the produce from the farmers to the ultimate consumers. It is essential to note that, the costs incurred for transporting empty package or containers back to the beginning of the marketing chain should also be considered in computing the total packaging costs. On the other hand, if the trader owns a vehicle and if he transports the produce to the market, then his cost of returning the containers is negligible. The following example highlights the methodology generally followed in computation of total packaging costs in the marketing channel:

Assume that, a wholesaler is transporting 25 kgs mangoes in a wooden box to a processor and the same box can be used for 10 trips. The box cost is ₹ 20 and the costs incurred towards repairs and maintenance of it is ₹ 5 and each time the empty box is transported back to the wholesaler costs at ₹ 2. Then, the packaging cost per trip is given by the formula:

[(original cost + repairs and maintenance cost)/no. of trips] + transport cost of containers when empty

= [(20+5)/10] + 2 = ₹ 4.50 per 25 kg

= 4.50/25 = ₹ 0.18/kg

vi. Cost-benefit Analysis of Packaging

Assume that, a farmer cultivated a leafy vegetable crop and he is transacting 6 kgs output/day. Assume the market price of the commodity is ₹ 30/kg. The packaging costs incurred by the farmer is ₹ 5/kg and other marketing costs amounts to ₹ 5/kg. The impact of packaging can be studied by analyzing two situations *viz.*, Net revenue with packaging and Net revenue without packaging.

Net Revenue with Packaging

It is evident from the given data that, the net revenue/kg is (Price – Total marketing costs) = 30 - (5+5) = ₹ 20/kg. So, the Net revenue with packaging for transacting 6 kgs output is ₹ 120.

Net Revenue without Packaging

Without packaging the losses vary, as does the selling price. The same is exemplified through the Table 5.5. It is evident from the table that, it is better for the farmer to go for packaging of the produce considering Cases A, B and C, unlike Case D, where the net returns due to unpackaging (₹ 124.20) is more than net returns due to packaging of produce (₹ 120). So, to avoid extra packaging costs, under Case D, the farmer should not go for packaging of produce.

Table 5.5: Net revenue to the farmer: Packaging vis-à-vis Non-packaging of produce.

Item	*Net Returns for Farmer under Different Cases (in ₹)*			
	A	*B*	*C*	*D*
Net revenue with packaging (6kgs x (30-10))	120.00	120.00	120.00	120.00
Net revenue without packaging under different cases				
◆ Losses 10 per cent and Market price is ₹ 26/kg	110.40			
◆ Losses 10 per cent and Market price is ₹ 27/kg		115.80		
◆ Losses 5 per cent and Market price is ₹ 26/kg			118.20	
◆ Losses 5 per cent and Market price is ₹ 27/kg				124.20
Whether the farmer should go for packaging of produce	**YES**	**YES**	**YES**	**NO**

Note: The net revenue to the farmer for unpackaged sales is calculated as follows (for Case A):

= [Output transacted x (Price – Other marketing costs)] – [Price x (Output transacted x Loss of produce due to unpackaging)]

= [6 × (26-5)] – [26 × (6 × 0.1)] =₹ 110.4.

vii. Importance of Packaging and Packing

This function offers the following advantages:

- ☆ It provides physical protection to the products from any damage due to shock, vibration, compression, temperature etc., during storage, transportation and other marketing operations. Perishable produce is protected from excessive moisture, heat or cold, ultra violet light, pathological, physiological and/or mechanical damage in transit, storage or when awaiting purchase. In the case of horticultural produce, the packaging assumes greater significance. Some commodities are highly sensitive to ethylene gas and hence, need to avoid gas build-up in transit (*e.g.* avocado). In these cases, the packaging must allow for effective air ventilation. A package may also have to protect the produce from moisture loss. Polyethylene liners, usually with perforations to allow gaseous exchange, are used for some commodities. Other commodities have special treatments, which must be taken into account when designing the package. For example, sulphur dioxide treatment of grapes and in-pack use of ethylene absorbents.
- ☆ Small objects are typically grouped together in one package for transport and handling efficiency. Alternatively, bulk commodities (such as wheat) can be divided into packages that are a more suitable size for individual households.
- ☆ Information on how to use, transport, recycle or dispose of the package or product is often contained on the package or label.
- ☆ Packaging that cannot be re-closed or gets physically damaged (shows signs of opening) is helpful in the prevention of theft. Packages also provide opportunities to include anti-theft devices.

Table 5.4: Dimensions of most frequently used pallets as per ISO standard sizes.

ISO Standard Sizes			
Millimetres		*Equivalent Inches*	
Width	*Length*	*Width*	*Length*
800	1000	31.5	39.4
800	1200	31.5	47.3
1000	1200	39.4	47.3
1200	1600	47.3	63.0
1200	1800	47.3	70.9

Boxes and wooden pallets are required to comply with International Phyto-sanitary Measure No.15 (IPSM 15), which establishes that, all wooden packaging must be specially treated to prevent insects from entering the load, and that wooden packaging must be marked with a seal of compliance for this treatment. This regulation was accepted by the Interim Commission for Phyto-sanitary Measures of the International Plant Protection Convention (IPPC) in March 2002, amended in Annexure 1 of 2006. Two types of processes were approved for treating the wood *viz.*, heat treatment and fumigation with methyl bromide. Once treated, the packaging should the bear the seal as shown below (Figure 5.1)

- ✰ The symbol on the left is the logo of the IPPC.
- ✰ 'XX' refers to the ISO code of the country, where the wood was treated.

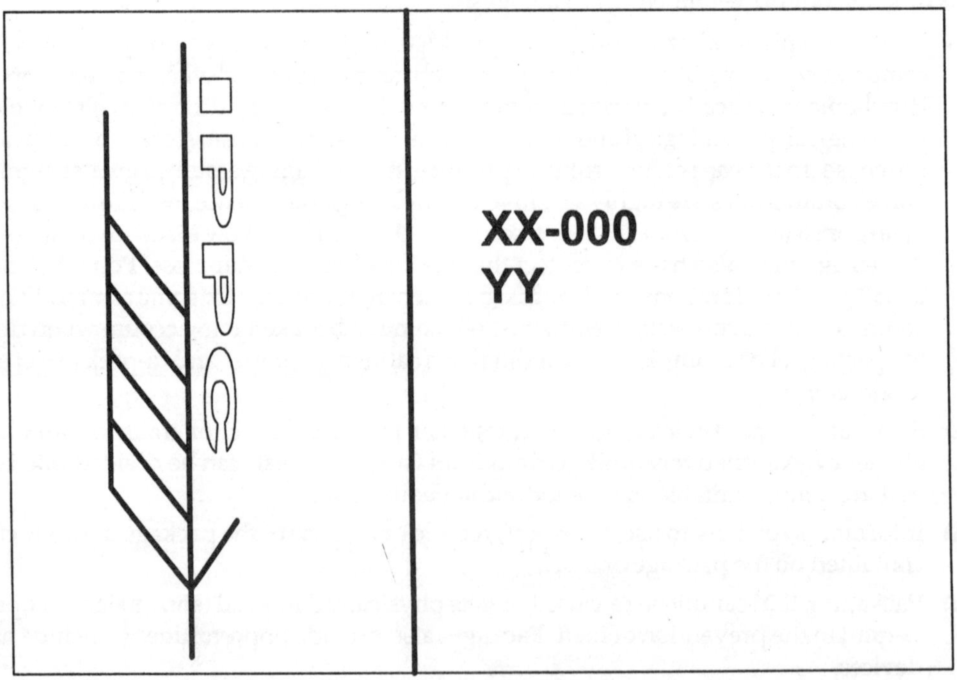

Figure 5.1: IPPC, ISO and NPPO recommendations on export packaging.

- ☆ '000' is the number assigned by the National Plant Protection Organization (NPPO) to the producer of the packaging.
- ☆ 'YY' refers to the treatment used (heat treatment, Methyl Bromide fumigation).

In the recent days, Poly Ethylene Terephalate (PET) bottles and jars have been extensively used to package carbonated beverages and fruit juices. With further introduction of an oxygen barrier polymer, its use has been extended to oxygen-sensitive products such as, beers. Likewise, plastic non-returnable packaging also gained momentum, but its usage is kept under check by the Government, as it is not environmentally friendly. Another way of restoring the quality of the agri-products is by resorting to *Sous-vide* vacuum packaging system wherein, certain measures like controlling temperature and/or humidity, removal of ethylene etc., are followed in the cases of fresh fruit and vegetables, meat, salad and vegetable products. A very recent innovation has been Active packaging and this type of packaging seeks to go beyond simply extending the shelf-life of fresh and pre-cooked foods and conserve quality attributes like colour, flavour, nutrients etc. Another form of 'Active packaging' is to put a '*time-temperature*' indicator on the pack to warn consumers that, when the product is in a certain condition it should not be eaten. Thus, packaging technology continues to change in tune with the changes in consumers' preferences and advances in marketing technologies and this guides the marketing managers to be vigilant all the time in the marketing process.

v. Packaging Cost

In transacting the product with special reference to packaging and packing function, the most important consideration is the packaging cost and it includes the following:

- ☆ ***Material cost***: It refers to the cost incurred on the quality and quantity of packing material employed in this function.
- ☆ ***Storage and handling cost of empty packages:*** It refers to the cost incurred on the handling and stocking charges on bulky packaging materials that are to be employed for packing the products.
- ☆ ***Packaging operation costs:*** It refers to the costs incurred on the operations like cleaning the produce, sizing of packaging material suiting to the products under consideration for packaging, labour costs incurred for packaging, labeling etc.
- ☆ ***Storage of filled packages:*** It refers to the costs incurred on the storage of packaged goods till their disposal in the market.
- ☆ ***Transportation cost of filled packages:*** It refers to the costs incurred due to movement of packaged products from seller to the buyers by various means like sea, air etc.
- ☆ ***Loss and damage cost:*** It is the cost indirectly incurred due to quantity loss of packaging material by various risk factors and physical damages during operation, transportation delivery etc.
- ☆ ***Insurance cost:*** It is the cost incurred by the agency for insuring depending on the vulnerability of packages.
- ☆ ***Obsolescence cost:*** It is the cost incurred due to innovative and inventive changes in the packaging materials, packages and labels.
- ☆ ***Package developmental costs:*** It is related to several costs like evaluation cost, pilot test cost, field testing cost, consumer research cost, feedback cost, final trial cost etc., about the packaging and packing materials.

top of each other demonstrates the close relationship between the loading and transport function and the protective function.

- ☆ In categories in which, children are the end consumers, appealing packaging can be a means of driving brand choice. Research has found that, 'pester power' can come from an attraction to packaging and as a result, packaging can heavily influence mothers' choices. Examples for the products like milk powder, nutritive products like Horlicks, Complan, Protinules, Boost, Maggie, Pediasure etc., the firms goers for attractive packaging, so as to gain popularity among the parents.
- ☆ In the recent days of marketing, manufacturers are also aware of the potential of adding to a product's utility by providing reusable packaging. These refill packs benefits the consumers in terms of price and also the packaging material should be reusable. For example, plastic containers are often preferred to cardboard cartons by consumers because, these can be used to carry and/or store other food items or liquids. Similarly, many farmers demand for fertilizers packed in High-Density PolyEthylene (HDPE) and Low-Density PolyEthylene (LDPE) sacks, as they feel that, they get sacks free of cost by purchasing fertilizers.
- ☆ Since, consumers are price conscious, the price of the product can be kept low, by low unit packaging. This is a common strategy adopted by many firms marketing in rural areas. For example, small chilli powder packets, tea packets, biscuit packets etc.
- ☆ The design of the product package is as important as the product itself. This is because the outer package creates the identity of the product, it is how customers will recognize the product.
- ☆ Promotional material placed on the packaging is intended to attract the potential purchaser's attention and to have a positive impact upon the purchasing decision. This information further guides the consumers regarding the use of the product. Examples are the nutritional details mentioned on noodles packets, biscuit packets, wheat atta packets etc.
- ☆ Packaging is especially important in certain industry, where future sales may be based largely on the quality, integrity and performance of a company's previous delivery.
- ☆ Professional packaging will determine a first purchase, while the quality of the product will bring repeat business.
- ☆ The packed products facilitate 'labeling' about the product and it ensures the following advantages:
  - ❖ Facilitates identification of a product
  - ❖ Indicate the grade of the product
  - ❖ Describe source of product, its content and major features
  - ❖ Guide the consumer in how to use the product, expiry date, technical ingredients etc.
  - ❖ Serve as a promotional tool.
- ☆ Packaging like branding helps identify a firm's product. For example, tea powder of different firms can basically be identified by the packaging rather than by the tea powder inside the packets.
- ☆ When the packing is attractive, it compels the consumer to pick up the article and, after reading what is mentioned on the package, to purchase it. Thus packaging can even perform the selling function.

viii. Criticisms of Packaging and Packing

This marketing function is subjected to the following criticisms:

- Some packages are difficult to open and such products are not even preferred by the consumers.
- Attractive packaging is a cost on the part of firms, but this burden is transferred on to the consumers in the form of high price of the product.
- In case of deceptive packaging and packing material, it may undergo chemical reactions with the stored produce and thereby, affects the human health. For example, in restaurants, hot food stuffs are often packed in a plastic material and it may undergo chemical reactions with the food products.
- Changing the package of an established product is not advisable because, consumers may either not buy the newly designed package or may buy the new package, but ignore the old package. In such case, the dealer may have a hard time getting rid of the old product.
- Product packaging is often viewed as wasteful and an unnecessary expense for which the consumer, ultimately, pays.

ix. Indian Institute of Packaging (IIP)

The IIP was established in the year 1966 jointly by the Ministry of Commerce, Government of India and the Packaging and Allied Industries in the country. The objectives set for the Institute was to bring about and sustain appropriate improvements in the standards of packaging particularly for exports. Its head office is located at Mumbai and branch offices are located at Chennai, Kolkata, Delhi and Hyderabad. The objectives of IIP include,

- To stimulate consciousness of good packaging
- To undertake R&D in packaging and package design
- To recommend standards for packages, packaging materials, processes and equipment
- To test, evaluate and certify packages, packaging systems, packaging materials, equipment and processes
- To provide consultancy services on packaging problems
- To study packaging for export commodity-wise and country-wise and suggest measures to authorities concerned for effecting improvements
- To provide short term and long term training in packaging technology for specified personnel employed in industry, commerce and Government
- To promote and stimulate action in universities and colleges, to provide education in packaging technology
- To analyze the implication of a sound policy on packaging on domestic production, domestic consumption and domestic prices
- To study methods and procedures of incentives appropriate for development of packaging
- To analyze the institutional framework most conducive to the promotion of packaging
- To study the packaging requirements with special reference to transport by road, water, rail and air
- To study the impact of packaging on marketing

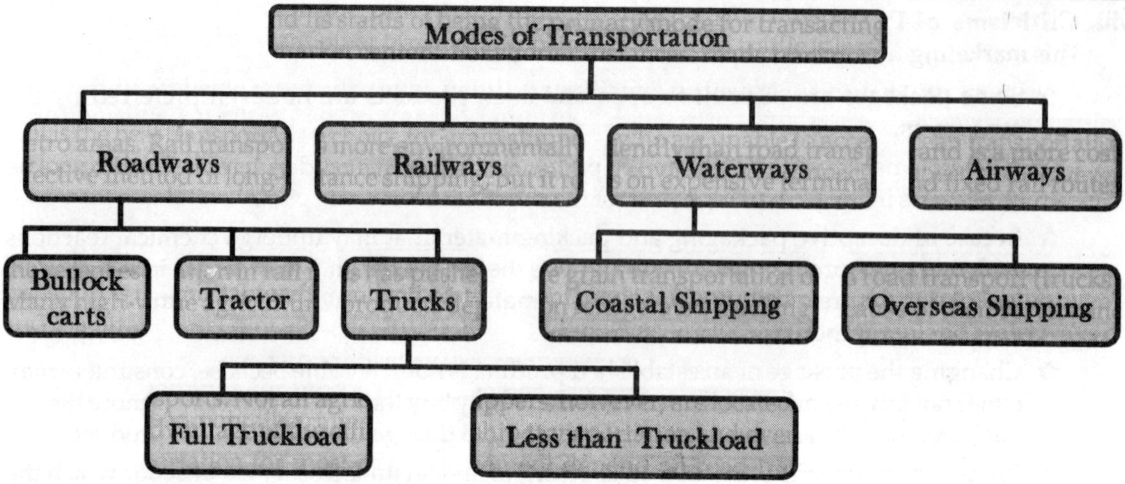

Figure 5.2: Different modes of transportation of agricultural produce.

Table 5.7: Comparative picture across different modes of transportation.

Mode	*Speed*	*Frequency of Shipments*	*Availability in different Locations*	*Flexibility in Handling*	*Cost*	*Nature of Produce*	*Meant for*
Road	Fast	Very high	Very high	Average	High	Durable and semi-durable goods	Domestic trade
Rail	Faster than road and water	Limited	Limited	High	Cheaper than road transport	Durable and semi-durable goods	Domestic trade
Water	Slowest	Very low	Very limited	Very high	Lowest	Durable goods	Export-Import trade
Air	Fastest	Average	Average	Limited	Very high	Highly perishable goods	Export-Import trade

of the perishables is still in primitive stage *i.e.*, more often stuffed in gunny bags or tightly packed in the baskets and staked one above the other, the quality of the road decide the extent of the damage to the produce. The recent developments in packaging has no doubt provided protection to the produce especially for fruits like grapes, apples etc., but majority of fruits are still outside the purview of proper packaging. So, compared to other modes of transport, road transport is much more important, as it is to be availed by the farmers or traders first and then followed by other modes of transport say, rail or ship or airlines for transporting the produce to distant areas. So, public investment in rural roads by increasing rural connectivity, can have a significant impact on access to markets by farmers, the development of supply chains and overall marketing efficiency in addition to other beneficial impacts on farmers. The earlier studies revealed that, the investments on strengthening of link roads contributed 36-68 per cent reduction in transportation costs, 27 per cent increase in agricultural wages, five per cent decrease in fertilizer costs and four per cent increase in output prices.

ii. Importance of Road Transport

When compared to other modes of transport, road transport has numerous advantages and they are as follows:

- ☆ Road transport requires much less capital investment, when compared to other modes of transport.

- ☆ Road transport provides door to door or warehouse to warehouse services, unlike other modes of transport.
- ☆ As discussed earlier, road transport is essential for transporting the agricultural produce from remote villages to the market centers. Such remote villages are not served by other modes of transport like rail, water or air transport.
- ☆ Road transport is more flexible over other modes of transport, as the routes and timings can be easily adjusted and even changed suiting to individual farmer's requirements without much inconvenience.
- ☆ Road transport has more advantage in transporting the agricultural produce to shorter distances and in right time, unlike other modes of transport. So, there is no chance of delays in transit of goods, as frequent loading and unloading of goods can be avoided. That is, the produce can be loaded directly into a road vehicle and transported straight to the place of destination. Further, due to avoidance of frequent loadings and unloadings, damage to the produce can be effectively checked and goods also require less packaging and packing.
- ☆ Road transport is more accessible to small and marginal farmers, unlike other modes of transport.
- ☆ The movement of goods must begin and ultimately ends by road transport. So, road transport acts as a feeder to the other modes of transport such as railway, ships and airways.
- ☆ Another advantage of road transport is that, big businessmen can afford to have their own motor vehicles and initiate their own road services to market their products without causing any delay in transacting the produce.

iii. Limitations of Road Transport

In spite of various merits, road transport has the following limitations:

- ☆ Road transport is not as reliable as rail transport, as it is not suitable during rainy or flood season.
- ☆ Road transport is not as safe as rail transport, as there may be chances of accidents and breakdown.
- ☆ It is unsuitable for long distances and bulky goods
- ☆ It is comparatively slow with reference to rail or air transports.
- ☆ The rates charged for road transportation frequently fluctuate.

iv. Reasons for Popularization of Bullock Transport in India

Among different means of road transport of agricultural produce, transportation by bullocks is more common in India on account of the following factors:

- ☆ It is cheaper and easily available conveyance
- ☆ Transportation costs are low
- ☆ Bullock carts can be easily designed and prepared by the local village artisans
- ☆ It can be used on any type of roads *i.e.*, on kuccha roads and pucca roads
- ☆ It facilitates multipurpose transport *i.e.*, transportation of both inputs and output.

Thus, each transportation mode has advantages and disadvantages for agricultural shippers. Some modes are expensive while others are cheap; others are fast whereas others take long to travel. The supply chain for agricultural produce often depends on multiple modes of transportation, each with its own price dynamics. The road transport has important advantages in its flexibility of service,

Table 5.6–*Contd...*

Nature of Transport	*Means of Transport*	*Advantages*	*Disadvantages*
		◆ Help reduce packaging costs because, some products require less packaging ◆ Can make rapid deliveries of large amounts of goods ◆ Monitoring of goods over transit is made easy due to ease of communication and locating the exact location There are two basic types of land transport through trucks *viz.*, Full truckload and Less than truckload. Full-truckload service is usually meant between farm gate and wholesaler's godowns. While less-than-truckload service is meant to carry produce from different individual farmers to a common location, from where they are redistributed	
Railways (Rail transport rates are based on distance, volume of container, sensitivity to damage, perishability of the product etc.)	–	◆ Suitable for carrying larger quantities of produce ◆ Suitable for long distances compared to any other means of road transport ◆ Comparatively cheaper and safer mode of transport ◆ Refrigerated transport also available ◆ This is a major mode of transport in terms of capacity, only second to ocean transport ◆ This mode of transport is relatively fast compared to land and water modes of transports. It is most economical especially, if the containers are fully loaded for a shipper on a regular basis ◆ It is less prone to interruption by poor weather	◆ Frequency of shipments is low ◆ Access to shipments at different locations is low ◆ Non-suitable for shorter distances ◆ High maintenance expenditure ◆ It might be unreliable, where the destination of the goods is not covered by the rail network and therefore, there is need to change the mode of transport
Water transport (river transport, canal transport and sea transport). Water transport rates are based on distance, volume of container, quantity of produce, etc. It is divided into Coastal and Overseas shippings.	This transport uses a loading device called 'container', which can be defined as a permanent,	◆ This is the oldest and cheapest mode of transport ◆ Suitable for carrying large quantities of produce especially for export and import trade ◆ Refrigerated transport also available ◆ The use of containers ensures easy handling of produce and thus saves labours and loading and unloading charges. Further, less	◆ Slowest mode of transport ◆ Vulnerable to bad weather like severe winter, cyclonic storms etc ◆ Less punctual, when compared with road or air transport ◆ Limited area of operation ◆ Buyers located far from port city must add rail or truck, which increases cost ◆ Cannot be used in the transportation of perishable goods

Contd...

Table 5.6–*Contd...*

Nature of Transport	*Means of Transport*	*Advantages*	*Disadvantages*
Coastal shipping provides service within the nation, while Overseas shipping is mainly used in the case of international trade	reusable, non-disposable shipping conveyance.	packaging is needed. These cost savings promote competitiveness in the international transit and also raises the service quality	◆ It is usually hard to monitor the location of goods being transported, when the goods are under transit
Air transport (Air transport rates are based on distance, volume of container, sensitivity to damage, perishability of the product etc.)		◆ Air transport is generally used, when delivery time is a key factor in product competitiveness ◆ This is the fastest mode of transport for long distances ◆ It is often used to ship highly perishable and high-value products ◆ This mode of transport is generally employed for export-import trade and in domestic trade especially for highly perishable goods like flowers etc ◆ Quick, reliable transit eliminates need for extensive warehouse accommodation and reduce risk of stockpiling, obsolescence, deterioration and capital tied up in warehouse and stock provision ◆ Reduces inventory expenses and storage costs	◆ This mode of transport can take less weight per shipment than other modes of transport ◆ It is the most expensive mode of transport ◆ This mode of transport involves certain technical limitations like maximum weight allowed, strength of cargo-hold floor, door size, flight autonomy etc ◆ Though this mode of transport is preferred for highly perishable and valuable goods due to its ability to move much faster over longer distances, it does lack the environment control and transfer ease of the land and sea transports. Also, during the flight, the cargo is stored in a 15°C–20°C environment, but close to 80 per cent of the time the package is exposed to exterior weather while waiting to be loaded onto the plane or being moved to and from the airfield. This is troubling considering the value of the food and the importance placed behind quality and freshness ◆ Low carrying capacity ◆ Vulnerable to adverse climatic conditions like fog, cyclonic storms etc ◆ There exist freight delays and cancellation in case of crisis such as weather problems ◆ High duties levied on goods delivered through this transport.

Of all the above means of transportation, the most crucial aspect is that, at first, the farmers must be able to move their produce from farm gate to the nearby market or distribution points. But, in India, this is more often ignored. So, the strengthening of link roads between the villages and nearby distribution centers is most essential, as it helps the farmers in transporting the agricultural produce without loss of time, so as to protect the quality. This is especially true in case of perishables. These farm roads and other road network also influence the quality of the perishable produce. Since, handling

☆ To undertake advertising, publicity and exhibition to promote packaging consciousness

☆ To provide and disseminate information relating to packaging.

x. Standardization of Packaging and Packing

As discussed above, since there are different types of agricultural commodities and products, the type of packaging and packing varies across them. However, certain standards must be followed in designing packaging and packing of commodities, as in the recent times, there is increased concern over issues like harmful affects of pollution, adverse affects of plastic material, use of more recyclable and biodegradable materials, restoration of ecological balance etc. Though for each commodity, the market is unofficial, but nevertheless rigid standards for packaging must be strictly followed. Further, the buyers of the products are heterogenous and thereby, it is essential to standardize this marketing function.

The above discussion infers that, packaging and packing marketing function assumes crucial significance both in domestic and international trades. Since, the standards of packaging and packing are fixed at the international level, it is essential on the part of India to rise up to these standards, so as to enjoy the fruits of international trade. This is because, goods damaged in transit or arriving at the destination in an unacceptable condition tarnishes the reputation of the firm or country in the market, besides colossal wastage of scarce economic resources. So, by strengthening the packaging and packing function, besides increasing export competitiveness, it also ensures the availability of both quality and hygiene products to the consumers.

B. Transportation

It is a necessary function of marketing because, most of the markets are geographically separated from the areas of production. It enhances the economic value through creation of place utility to the agricultural produce. That is, this function make the goods to move from a surplus area (production point) to a deficit area (consumption areas), where it is considered as a proper place for their marketing. So, transportation refers to the physical movement of goods from one place to another. To be more appropriate, it refers to an activity, which involves physical movement of goods from the place of production to the place of consumption. Goods in the sense, may be raw materials or resources, agricultural produce, processed products, finishing goods etc. Since agricultural produce is bulky and perishable unlike industrial products and they are produced or harvested over geographically distributed areas that are at a considerable distance from the central markets of major consumption centers, transportation is regarded as an indispensable marketing function. Further, time element has a direct impact on their quality and therefore, effective transportation ensures efficient marketing of agricultural produce. Various transportation methods may be employed in moving the produce depending upon the demand, cost involved, urgency, nature of the goods (perishable or durable) etc. So, appropriate transportation methods must be selected to maintain the quality of the produce. For example, milk and flowers must be provided with refrigerated transport facilities to keep them fresh at the time, when they are offered to the customers. However, for bulky and durable goods, the railways and waterways are the best. So, agriculture industry relies heavily on a consistent and dependable transportation system and any collapse of the transportation system will be a detriment to its survival. Hence, transportation function is often regarded as an indispensable marketing function and life blood of agri-business.

i. Different Modes of Transportation of Agricultural Produce

The means of transport most commonly used both for the domestic and international trades of agricultural produce and products include land, sea, air and rail (Figure 5.2). In general, road and rail transport are normally used for domestic market and air and sea transports are meant for export markets. Sometimes, the term 'Multimodal transport' is widely used, if more than one means of transport is used in moving the produce and this is more common in international trade. The different transportation modes followed for agricultural produce and products are discussed through Table 5.6 along with their advantages and disadvantages and their comparative picture is shown through Table 5.7.

Table 5.6: Different modes of transportation of agricultural produce and products.

Nature of Transport	*Means of Transport*	*Advantages*	*Disadvantages*
Road (Road transport rates are based on distance, product quantity, weight, volume, sensitivity to damage etc.)	Bullock carts	◆ Most popular and commonest mode for movement of agricultural produce to the assembling markets as well as to the distribution centers ◆ Suitable for transporting small quantity of produce ◆ Cheap and easily available ◆ Easy transport for short distance ◆ Easily manufactured by village artisans ◆ Easily operated on *kaccha* road, muddy or sandy path	◆ Poor maintenance of the existing road network ◆ Drastic increase in road user costs ◆ Road network may not be in good condition ◆ Inadequate road network to link villages and wholesale markets ◆ Delays due to traffic or equipment breakdown ◆ Subject to size and weight restrictions ◆ Traffic congestion may cause delay of delivery of produce
	Tractor	◆ Carry larger quantity of produce than bullock carts in less time ◆ Suitable to transport produce in primary assembling markets even in the absence of proper *pucca* roads connecting the villages and market ◆ Farmer can employ the same for multipurpose use	◆ There may be breakage of fragile goods as a result of careless driving ◆ Traffic regulations on speed may cause delays
	Trucks	◆ It is the most convenient means of transport throughout the country for longer distances and for transporting bulk quantity ◆ Comparatively easily available ◆ Quick movement ◆ Convenient during loading and unloading ◆ Provide door to door delivery of products ◆ Safe transport ◆ Timely delivery of goods ◆ Greater flexibility of travel schedules compared to other means of transport ◆ Suiting to the different varieties of produce or products, wide varieties of vehicles can be afforded ◆ Can be combined with other modes of transport	

Contd...

its competitive nature and its status of being the primary mode for transacting agricultural produce from remote villages to market centers. For shorter distances, roads transport is less expensive than rail or ship transport. On the other hand, long-distance road transport is not as efficient or environmentally friendly as other transportation modes and road traffic is less welcome in congested metro areas. Rail transport is more environmentally friendly than road transport and is a more cost-effective method of long-distance shipping, but it relies on expensive terminals and fixed rail routes. Rail is the best transportation choice for grain shippers, who are unable to access ocean transportation for long distance movements, both from the economic and environmental impact perspectives. However, the recent escalation in rail rates has pushed more grain transportation on to road transport (trucks). Many high-value agricultural products depend on refrigerated trucking because of their service and rapid delivery. Ocean transportation depends on interior transportation. Where available, ocean transport is the most fuel-efficient and cost-efficient mode for long-distance transport followed by rail and road transports. Not all agricultural shippers, however, are located near the inland waterways or able to take advantage of the efficiency of ship transportation and even grain shippers who use ship or rail transportation for most of the movement depend on road transport to get the grain to the elevator, ship port or rail terminal.

v. Means of Transport Used at different Stages of Agricultural Marketing

The following are the general means of transportation followed at different stages of marketing of agricultural produce (Table 5.8):

Table 5.8: Means of transportation used at different stages of agricultural marketing in India.

Stages of Marketing	*Agencies Involved*	*Means of Transport Used*
From farm gate to the village market or primary market.	Farmers and Local/Village traders	By bullock or truck or tractor
From primary market to secondary wholesale market and processor	Traders/Processors	By trucks, railways.
From wholesale markets and processor to retailer	Processors/Wholesalers/Retailers	By trucks, railways, tractor.
From retailer to consumer	Consumers	By hand and any other means of transport used by the consumer.
For export and import	Exporters and Importers	By air or ship

However, the modern agricultural production depends upon a complete transportation system that includes all major modes as discussed above and each operating in conjunction with each other to complement and supplement any deficiencies in the quick transportation of produce from one destination to another. The modern agricultural transportation has two specific effects. Firstly, it permits consumers to leave agricultural production areas and migrate to urban areas for new employment opportunities thereby facilitating the growth of manufacturing and other industries and second, it allows the production of specific agricultural commodities in accordance with the principle of comparative advantage. We know, Indian agriculture is strikingly diverse and requires a wide range of transportation services. Some products such as rice, wheat, pulses, sugar etc, are bulky commodities, while others, such as fresh fruits, vegetables, eggs, fish, meats etc., are highly perishable items, which require different transportation requirements. Also, the movement of highly perishable goods like milk, flowers etc., requires specialized handling and equipments. Further, modern agriculture is agri-business oriented and is input-intensive. So, it requires a wide variety of inputs and they also require a wide variety of transportation services. In the liberalized trade regime, the farmers and traders demand for cost-effective transportation services to take advantage of opportunities of the 'new' agriculture in both the domestic and international markets. So, this led to the development multimodal transport (also called Combined transport), where transportation of produce is done under a single contract, but performed with at least two different means of transport. This multimodal transportation is used to minimize expenses, choose optimal logistic itineraries, save time for delivery of the product etc. Under this transportation, the professional personnel will take care of the cargo during all the process of the transportation, closely monitor and coordinate all the counteragents involved in the process so that, it ensures reliable, timely and safe delivery of the products.

vi. Transportation Costs of Agricultural Produce

a. Factors Influencing Transportation Costs of Agricultural Produce

It is estimated that, on an average, transportation costs accounts for 50 per cent of the total cost of marketing and 30 per cent of the Consumer's Purchase Price (CPP) in having the commodity. Considering its importance, it is essential to analyze the major factors that contribute to higher transportation costs of the agricultural produce and the same is discussed here under.

- ✰ *Distance:* This is the major factor that influences the transportation costs of agricultural produce and it varies directly. That is, if the distance of transport increases, the transportation costs also increases.
- ✰ *Quantity:* The transportation cost per unit of produce will decrease with increase in the volume of the produce.
- ✰ *Mode of transportation :* The transportation costs of agricultural produce further depends upon the mode of transportation used. For bullock carts, the cost of transportation is low and for lorries and trucks, the transportation costs increases. In general, transportation costs increases with increase in quickness of moving the produce *i.e.,* if the produce is to be transported quickly, the transportation costs increases. So, transportation costs are highest through air mode of transport and lowest for ship mode of transport.
- ✰ *Nature of produce:* The transportation cost per unit of produce will be high, if the commodity is highly perishable, inflammable, fragile etc.
- ✰ *Risks associated:* If more risks are associated both in terms of quantity and quality of produce, the transportation costs will be high and *vice versa.*
- ✰ *Availability of return journey consignment:* If the transporting agency enjoys return journey consignment, then the transportation cost per unit of produce will be low.
- ✰ *Weight-for-volume ratio:* The produce, which have a low weight-for-volume ratio (for example green peppers) should be costed at a higher per kilogram cost than produce, which is heavier in relation to its volume (say, cucumbers).

b. Computation of Transportation Costs

As mentioned earlier, transportation is an indispensable marketing function and hence, both farmers and traders incur transport costs in moving the produce down the marketing channel towards the ultimate consumer. The computation of transportation costs is very easy, if a farmer or trader makes a direct payment to the truck owner for transporting his produce on per weight basis or per piece basis. However, the computation of transportation costs is not straight forward, as discussed under the following conditions:

- ✰ *Case 1-When the farmer or trader owns and operates his own vehicle:* In such case, it is essential to include opportunity costs like missed transportation charges from other farmers' produce and missed opportunity wage earning in the village.

- ☆ ***Case 2-If different types of agricultural produce are filled in the same truck:*** In this case, it is quite difficult to compute the transportation costs on per piece basis or per unit weight basis. This is because, transportation costs per piece basis or per unit weight basis will be computed by the truckers, only after assuming that, the entire truck was completely filled that too with the same commodity. So, when different commodities are loaded into the same truck, it is quite difficult to compute the transportation costs and generally the truckers over value the transportation costs. But, if the farmers or traders join as groups (co-operative marketing) and fill the truck with the same commodity, it is quite easy to compute the transportation costs on per piece basis or per weight basis.
- ☆ ***Case 3-If the farmer or trader owns a vehicle:*** In such case, the calculation becomes more complicated. This is because, truck owners are often accused of overcharging the transporting costs by including the following items:
  - ❖ wages paid to the driver and his assistant
  - ❖ cost of fuel, repairs and maintenance
  - ❖ cost of licenses, road tax, insurance and other necessary payments
  - ❖ costs incurred such as tolls or bribes paid at official or unofficial road blocks and charges for entering a market
  - ❖ depreciation costs of the truck (When roads are not in good condition, trucks may last only a few years and thus, depreciation will be a major cost.)
  - ❖ annual installment due of the loan taken from the bank in purchasing the truck

  In such case, if one farmer or trader uses his own truck and other farmers or traders who hire the other trucks, then the hiring costs on 'per piece basis' or 'per weight basis' paid by the latter can be used as a 'best guess' of the computation of transportation costs. However, assuming a normal case, when a farmer engages a truck on hired basis, then the transportation costs are computed considering the following items:
  - ❖ Total volume of the container
  - ❖ Cost incurred, if the total container is loaded
  - ❖ Actual space occupied by the loaded commodity
  - ❖ Weight of the commodity loaded into the container

The following formulae can be employed to compute the transportation costs per container and per unit weight basis:

Transportation costs per container is given by

[Hiring cost of total container/(Total volume of container/Volume occupied by the commodity in the container)]

Transportation cost per unit weight of produce is given by

[Transportation costs per container/Total weight of produce loaded in the container]

Example

To exemplify these formulae, assume that, a farmer hired a truck from a trader, which has a container of space of $50m^3$ and that costs ₹ 1000. The farmer loaded 10 kgs bhendi and 15 kgs tomato into the truck and they occupied $0.5m^3$ and $0.6m^3$ space respectively in the container. Then, the transportation costs for these commodities can be computed as follows:

Commodity bhendi

Transportation costs per container is given by:

=[Hiring cost of total container/(Total volume of container/Volume occupied by the commodity in the container)]

= [1000/(50/0.5)] = ₹ 10/container

Transportation cost per unit weight of produce (bhendi) is given by:

[Transportation costs per container/Total weight of produce loaded in the container]

= [10/10] = ₹ 1/kg

Commodity tomato

Transportation costs per container is given by

=[Hiring cost of total container/(Total volume of container/Volume occupied by the commodity in the container)]

= [1000/(50/0.6)] = ₹ 12/container

Transportation cost per unit weight of produce (tomato) is given by:

[Transportation costs per container/Total weight of produce loaded in the container]

= [10/12] = ₹ 0.83/kg

vii. Selecting the Mode of Transport

In selecting a mode of transport, it is important to determine how the qualities of each mode will influence the shipment. The following aspects should be taken into account:

- *Speed of transport:* Among different means of land transport, railways is the quickest and efficient mode of transport compared to road transport. But, for shorter distances, road transport is meritorious over rail transport. Though railways can carry much greater loads than road transport, but the loading and unloading of the wagons at the stations take much time compared to road transport. So, for long distance transport, railways is superior to road transport because, its carrying capacity is high and its speed of delivery is also quick. Compared to these two modes of transport, air transport is the quickest mode thereby, it is more suitable in transporting perishable produce and is mainly meant for export-import trade. Among all the modes of transport, water transport is the slowest form of transport. So, this mode is not suitable for perishable products unless proper arrangements for cold storage and refrigeration are made in the ships.
- *Cost of service:* This is one of the most important criteria in deciding the mode of transport availed by the farmers or traders in the marketing process. It is essential to note that, the cheapest mode of transport is not necessarily the shortest one neither is the quickest route always the shortest. So, transportation at low cost can be enjoyed by the marketing agency by selecting the right mode and by having full loads in both the directions. Among different modes of transport, rail transport is cheaper over longer distances for transporting cheap and bulky agricultural produce. For shorter distances however, rail transport is not economical because, much fuel or electricity is wasted, the trains cannot attain high speed and much delay is caused due to loading and unloading operations at the terminals. For shorter distances, road transport is economical and ideal, as the expenses are small, more flexible suiting to the marketing agencies requirements both in terms of quantity of produce transported and time of availment of service and loading and unloading expenses are also

quite small. If the time of delivery of produce is not a constraint and if the distances of transport are long, water transport is the best option, as it is cheaper than other modes of transport and also suitable for bulky non-perishable produce. Among all forms of transport, air transport is the costliest because of tremendous cost of fuel and this mode of transport is especially useful, when highly perishable produce like milk, flowers etc., are to be transported to very long distances in quickest possible time.

- ☆ ***Regularity of service:*** This criterion is very important considering the perishability of agricultural produce. Among different modes of transport, railway transport is least affected by weather conditions and it follows a regular schedule. But, the demerit is that, it is not suitable for shorter distances and the farmers especially from the remote villages do not have access to it. So, in general, majority of the farmers will avail road transport and that too depends mostly on bullock carts for the several advantages that are discussed earlier. Even road transport also follows a regular schedule and renders a dependable service, but its service is dependent upon weather conditions. Both water and air transports are much affected by adverse weather conditions compared to road and rail transports and hence, their regularity of service is not guaranteed.
- ☆ ***Flexibility of service:*** Compared to other modes of transport, road transport will offer more flexibility in service in the sense that, produce can be picked from any point at the desired time in the country. Though rail transport offers regularity of service, it is not that much flexible like road transport, as it is available only between two stations and that too it is not accessible to the farmers and traders residing in the remote villages, which have no access to railway lines. So, the goods and persons are required to reach the station first and then are able get this service, unlike road transport which offers door to door service and is especially suitable in those areas, where railway transport cannot be made available. On this ground of flexibility of service, both air and water transports are considered inflexible because, they offer service only between two points and hence, they are usually meant for export-import trade.
- ☆ ***Perishability of produce:*** The choice of the mode of transport by the marketing agencies depend on the nature of a produce, whether perishable or non perishable. For the perishable goods, the agencies will choose the fastest mode of transport (air transport) and discard the slow methods of transportation. For the non-perishable goods, the agencies will use other modes.
- ☆ ***Bulk or light weight goods:*** The mode of transport to be used for bulky goods will not be air transport, the reason being that it is very expensive. The bulky goods will usually be transported by road, rail or shipment. For light weight goods, the mode of transport will be air transport.
- ☆ ***Custom and excise duties:*** The imposition of tariffs by the Governments of the importing countries will also determine the mode of transport to be used by the exporting firms. Some transportation modes experience high taxation levels (air transport) and therefore, most industries avoid them as a result.
- ☆ ***Source and destination of goods:*** The source of the good and their destination will also determine the mode of transport to be chosen by the firms. In general, for short distance transportation, the mode of transport to be used will be road transport, but if the distance is long, then the mode used is rail, air or even the use of water transport.

Besides above, the marketing agencies should also consider other criteria like value of product, shipping cost, travel time, frequency of transportation service, reliability of transit time, traceability (possibility of monitoring the shipment), availability of intermodal connections, ability to serve both the origin and destination businesses etc., in selecting the right mode of transport. The above discussion infers that, each mode of transport offers advantages and disadvantages. The specific needs of each marketing agency and each product will, in the final analysis, determine, which mode of transport (or combination of modes of transport) best meets their needs. Although various shipping arrangements are available, it is recommended that, a freight forwarder or a customs agent be hired to help select the right mode of transport and to take care of all necessary shipping procedures, as they are much more experienced in transporting the produce.

viii. Importance of Transportation of Agricultural Produce

The words of President John Kennedy, '*it is not the wealth of a nation that builds roads, but the roads that build the wealth of a nation.*' highlight the importance of transportation of goods in general. The same is true in case of Indian agricultural sector and the present Economists are also of the opinion that, continued and reasonable investment in transportation is essential to the sustained economic health of the country. Further, with the advent of globalization and trade liberalization, the efficient transportation system ensures competitive trade in the international market. The economic effects of good transportation are as follow:

- ☆ Economists are of the opinion that, investment made on transportation increases productivity, which enhances competitiveness and brings prosperity to the farmers and traders depending on agricultural sector (Figure 5.3). This is because, transportation ensures in-time delivery of raw materials and products and this significantly reduces the inventory costs for many businesses. This change was possible only because, (processing) firms felt comfortable enough with the reliability of the transportation system, otherwise they have to shut down their production lines, if raw materials failed to arrive as planned. Further, the processing firms shifted their focus from 'push' distribution system to 'pull' distribution system because, 'pull' system depends upon reliable transportation services and advanced communication

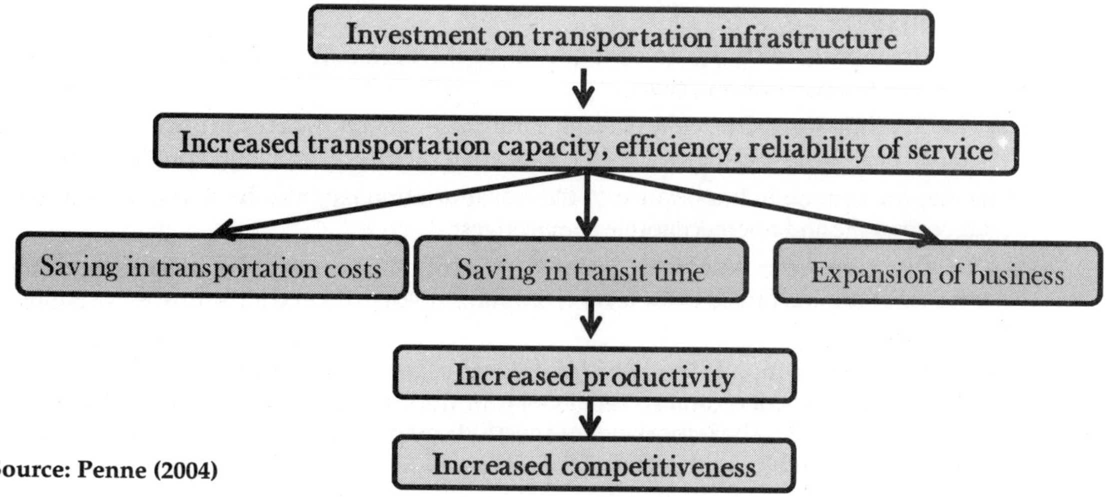

Source: Penne (2004)

Figure 5.3: Transportation enhances competitiveness of agricultural commodities.

services to pull raw produce from the farmers or traders as they are needed. The earlier 'push' distribution system pushed the products from the processors to retailers, who then had to deal with selling them. The shift from 'push' to 'pull' distribution system ensure further reduction in inventory costs and a more responsive retail system. So, the commodities that are demanded by the consumers are kept in stores by the retailers. This leads to production of commodities by the farmers and processors in accordance with the principle of comparative advantage and this ensures more competitiveness and prosperity to the farmers.

- Transportation ensures the assembly and distribution of inputs, raw agricultural produce and processed products from the place of production to the place where they are to be consumed. An efficient network of transport services encourages the movement of both labour and capital from one place to another place.
- The movement of goods, labour and capital from one place to another leads to specialization and this ensures maximization of resource use efficiency and thereby, reduces cost of production. Thus, an economic system makes the best use of resources through good transport system.
- The transportation of goods from one place, where there is surplus to another place, where is scarcity leads to price stabilization in the economy. It also helps in solving the problem of scarcity of goods (and factors also) in different regions of the country.
- Transportation creates place utility to the products, as it ensures the movement of products from the place of production to the places of consumption. Thus, transportation benefits the consumers in offering a wide range of products of their interest.
- It is essential to note that, transportation has a crucial role to play in economic development of a country (Figure 5.4). More specifically, adequate, reliable and economic transport is essential for the social and economic development of farming community. The direct impact of transportation on the well-being of farming community is through three effects *viz.*, lowering of production costs, increase in profits to the farmers and encouragement of investment. However, from a general standpoint, the economic impacts of transportation can be direct and indirect:
  - Direct impacts relates to accessibility change, where transport expands markets and enables to save time and costs.
  - Indirect impacts relate to the economic multiplier effects, where the price of inputs and output decreases and the availability of different varieties of inputs and products increase.

  Further, the economic importance of the transportation can also be assessed from both macroeconomic and microeconomic perspectives:

  - At the macro-economic level (the importance of transportation for a whole economy), transportation is linked to the level of output, employment and income within a national economy.
  - At the micro-economic level (the importance of transportation for specific parts of the economy, say agricultural sector alone) transportation is linked to farmers, marketing agencies and Mc. The importance of specific transport activities and infrastructure can thus be assessed for each sector of the economy.

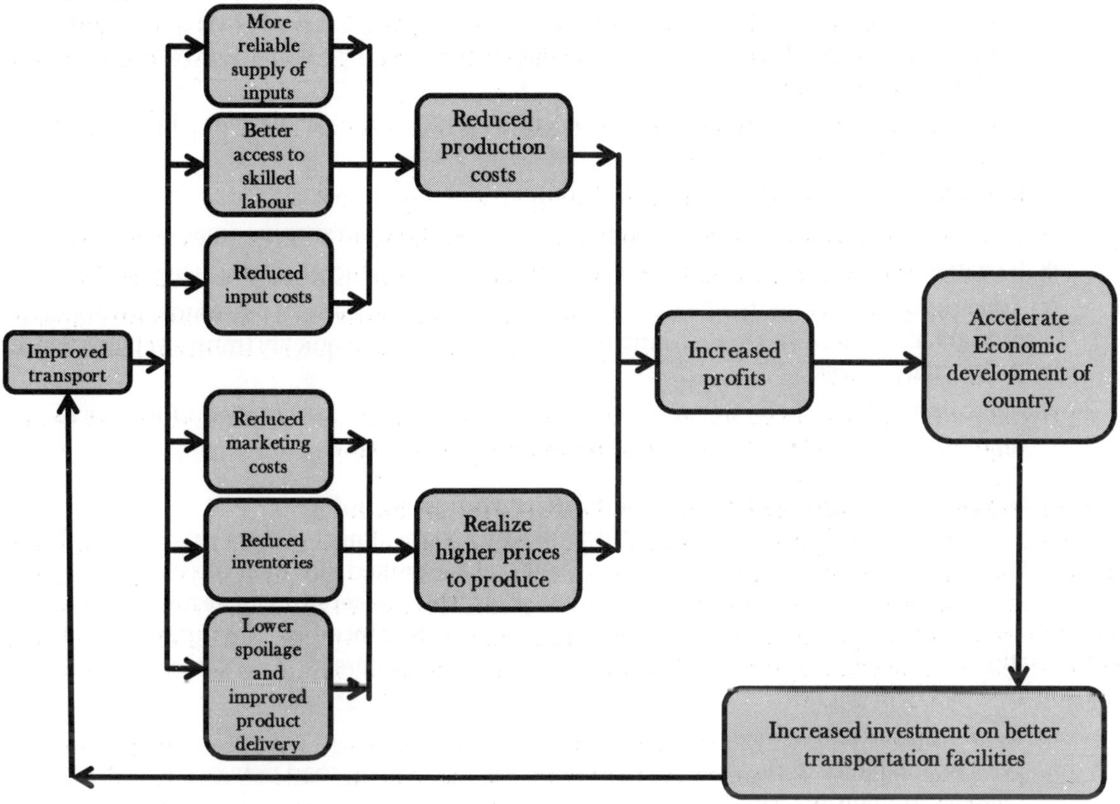

Figure 5.4: Relation between improved transportation of agricultural produce and economic development of the country.

- As commodities can be quickly transported from one place to another, local producers cannot charge prices at their own will. This discourages monopoly and encourages competition. Thus, transportation intensifies competition, which, in turn, reduces prices.
- A well developed transportation network ensures the availability of raw material in time and thereby, reduces the cost of production of processed products and hence, their prices also decrease in the market.
- Markets for agricultural produce expand both on the domestic and international fronts with the expansion of transportation network. For example, through the expansion of road, rail, air and water transport, the links between different parts of the country and with other countries of the world will be strengthened. Thus, transportation helps in promoting foreign trade of the country.
- Transportation function helps to create employment opportunities to many workers and also leads to performance of other functions like storage, processing etc.
- Means of transport are the source of advertisement also thereby, they also help in expanding agricultural education.

- Since transportation reduces the distance across the places, it leads to the availability of different types of products of preferred qualities to the consumers thereby, promotes standard of living among the people.
- It helps in the growth of industries, whose products require quick marketing like vegetables, flowers, milk, fish etc.
- It increases the demand for products through widening of market
- It helps in stabilization of prices by moving commodities from surplus areas to deficit areas.
- It enables the consumers to enjoy the benefits of many products not produced locally.
- Transportation enables the farmers to face natural calamities such as famine, earthquake, drought, floods etc., and in such emergencies, the farmers can quickly transport the produce to the safety places.
- The commodities can be transported to places, where there is scarcity and the prices are high from places where there is surplus and the prices are low.

ix. Constraints in the different Modes of Transportation in India

When it comes to transportation, agricultural producers are almost always price takers. Their transportation options for moving their produce to market are limited, but their buyers (processors, wholesalers etc) often have many sources from which to select. This market structure results in producers paying most of the transportation costs, which directly affects their incomes. The different modes of transportation employed for agricultural produce and products suffers from certain constraints and they are discussed here under.

- ***Road transport:*** The important constraints include insufficient network of major bridges, poor maintenance of the existing road network, inadequate road network to link interior villages to main market centres, encroachment on road reserves, inadequate financial resources from the Government and lack of private sector investment, lack of modern machinery and equipment to undertake an effective road maintenance, lack of sufficient cost recovery mechanisms etc. Road transport (truck) rates are more volatile than rail or ocean rates. They tend to fluctuate more frequently because, fuel prices are highly volatile. Thus, when fuel prices were increased, truck rates also increase rapidly. Further, because of the competitive nature of the trucking sector, some trucking companies' profit margins were squeezed and others went out of business or declared bankruptcy.
- ***Rail transport:*** The important constraints include less connectivity to interior villages, high freight rates, demands huge investments from the Government etc.
- ***Water transport:*** The important constraints include,
    - Since India enjoys comparative advantage with reference to several agricultural commodities and products, their export demand increases in the international market. But, fewer containers being available in locations suitable for export can result in lost sales and unreliable service to overseas buyers.
    - The increased demand for vessel capacity and container supplies can push up freight rates.
    - Many ports are working at or near capacity and expanding them is problematic because of their urban locations.

- ☆ ***Air transport:*** The exports of agro-products both in fresh form and as processed food products by air are expensive. There are three main reasons, why air transport costs are so high in India. In general, fuel costs are usually 10-15 per cent of airline operating costs, but in India, they are nearly 30 per cent. This is because, the average price of Aviation Turbine Fuel (ATF) in India is nearly 190 per cent higher than the international price. The excise and sales taxes constitute nearly 45 per cent of the ATF price. The fact that, only Government-owned oil companies supply the ATF and there is lack of competition in the fuel prices. Furthermore, the airport charges in India are 78 per cent higher than the international average charges. All these reasons cumulatively influences the higher air transportation charges compared to other modes of transport.
- ☆ There is also lack of multimodal transport facilities to link production areas with major markets.
- ☆ Transportation needs should be viewed from a system stand point. But, the Government oversees each mode of transportation *i.e.* road, rail, water and air separately rather than as a single interlocking system of transportation. There is lack of systematic planning and designing of multi-modal transportation system in rural areas. Thus, the Indian transportation system is a 'system of systems' made up of different transportation networks that need to work together. The system's statutory and regulatory framework is largely modal-based, providing with rules and regulations focused on singular modal solutions even though many of the challenges are associated with a multi-modal transportation system.
- ☆ The transportation system is fluid *i.e.,* when one mode is disrupted, freight shifts to other modes, pushing up their cost with the additional demand. For example, when ocean traffic was halted by cyclonic storms, freights will get shifted to other modes like rail and road immediately. Ocean port disruptions typically require ships and cargo to be redirected to other ports resulting in a significant logistical burden and transportation expense. It also adversely affects the transportation infrastructure. All these results in escalation of freight rates in other modes of transport due to sudden increase in demand and this also reduces the speed of transport.
- ☆ Since demand for Indian agricultural exports remains strong, the need for available containers to move the products is essential. The use of near-port distribution centers by large importers has reduced the number of containers available to agricultural exporters at interior locations. Exporters rely on import containers to supply an available container pool. When imports are down, exporters at inland locations are unable to find a sufficient number of containers.

x. Suggestions to Improve Transportation of Agricultural Produce in India

The following are the important suggestions that are to be considered to improve the transportation of agricultural produce:

- ☆ Recognize transportation as an investment rather than as a cost.
- ☆ Investments should be done more strategically, as all link roads, local roads, bypasses, railroads and ports are not equal. Investments in some will yield much greater benefits to the economy than investments in others.
- ☆ Think beyond jurisdictional boundaries, since freight does not stop at national borders. It does little good for one State to have a wonderful transportation system, if the next has bottlenecks.

- ✰ Plan and adopt multi-modal transportation strategies, so as to meet the demands with regard to transportation of agricultural produce as well as demands for environmental protection.
- ✰ Air transportation costs are much higher compared to other modes of transport. For example, transportation of mangoes by air is costlier by 10 times, as compared to sea in most of the Asian markets, especially in the West Asian markets. Thus, export of produce in bulk by sea is the best alternative. However, in order to preserve quality of produce during sea transportation, there is need for development of protocols for management of different varieties of agricultural produce during voyage time, through use of alternative techniques of refrigeration and controlled/modified atmosphere management.
- ✰ There is immense need in upgrading port facilities to facilitate transport and handling of agricultural produce.
- ✰ The Government should provide incentives for the purchase of new vehicles meant for transporting agricultural produce so that, transportation network will get strengthened.
- ✰ It is essential to plan for collaboration among different institutions to address the specific transportation needs of the agricultural sector and provide incentives to farmers for development of farm access roads.
- ✰ Since, road has become the dominant mode of transport for agricultural produce, it demands huge amounts of investments from the Government. However, to meet the growing demand, a shift to more economically as well as environmentally suitable modes *i.e.*, waterways and rail is vital. In addition to a greater emphasis on rail and waterways, the right mode of transport has to be used. Ideally, waterways should be prioritized for long distances, rail for medium distances and roads for shorter stretches. Such a balanced modal approach would lower transportation costs, achieve greater efficiency and be more environment-friendly in transporting agricultural produce and products.
- ✰ Better scheduling and transfers increase transportation speed and reduce spoilage. This further leads to improvements in the agro-processing and packaging industries because, lower shipping costs make it feasible to ship produce to one large processing firm.
- ✰ The road infrastructure problems (lack of roads and/or poorly maintenance of roads) can sometimes be addressed at the community or group level. The farmers have self-help schemes whereby, one day a week or month is given over to community activities such as, road construction, road maintenance etc. Thus, the transportation problems could be partly solved by communities themselves. Once, farmers had shown their commitment, they would be able to lobby local government to prioritize road improvements in local government spending plans.

Above all, the transportation constraints of agricultural produce can be effectively checked by facilitating collective marketing so that, farmers can afford to use intermediate and more expensive forms of transport to transport greater volumes, access to more distant markets and benefit from economies of scale in transportation costs. Further, by strengthening linkages with traders or transporters, they are able to collect produce from farmers directly or atleast nearby collection centres. Figure 5.5 shows the cost-saving approach of transporting cane from the farmers to the sugar factory. If each and every farmer individually transport the produce to the sugar factory, they incur huge transportation cost per unit of produce. However, if all the farmers dispose their produce to the nearby loading station, from where it is lifted to sugar factory, then the transportation cost declines per unit of

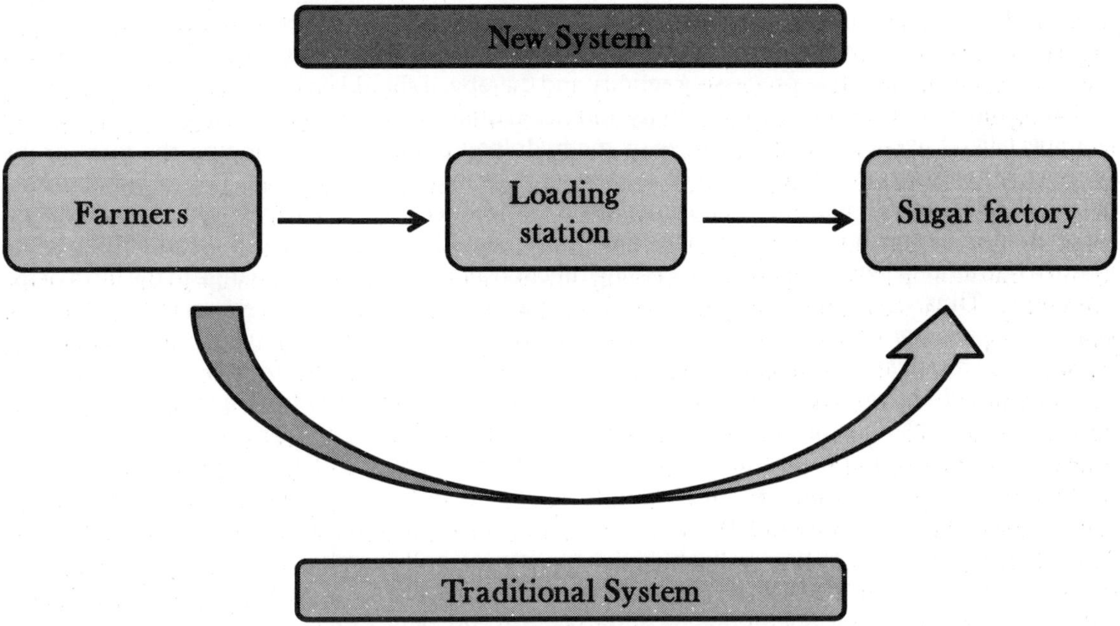

Figure 5.5: Loading station model for transporting sugarcane to sugar factory.

produce transported. This is because, transportation costs mainly depends upon the distance and not on the quantity of produce occupied in the container. Ideally, the loading station is a market trading spot for all sugarcane growers, and particularly for small farmers. The farmers transport sugarcane from their farms to the station in their own vehicles and rely on family labour. The sugar factory makes arrangement to collect the sugarcane from the loading station and arranges onward transportation to the processing plant. On a collective basis, farmers incur a meager amount for the transportation of cane from the loading station to the sugar factory. Under this new scheme, both farmers and truck drivers save time and costs, since they no longer have to wait in line to deliver their product. Under the former traditional system, farmers deliver their produce directly from their farms to the sugar factory by bulk carriers and had to bear the whole transportation cost themselves. So, this new system of transport ensures the following advantages to the farmers:

- It help to maintain steady supply of cane to the sugar factory
- It enables farmers to specialize in sugarcane, as there is assured market, and
- Small farms owned and operated by family members can rely on their own labour for cutting, loading and transporting cane up to loading station and thereby, save still more transportation costs.

C. Agro-processing

We know, most agricultural produce is not fit for direct consumption by the consumers, when it is first harvested. Rather, its form is to be changed, so as to make it fit for consumption. In this context, Kohls and Uhl opined that, processing function need not be recognized as a marketing function, as it simply involves change in form of the commodity. However, this activity is to be carried out by a specialized agency *i.e.*, 'processor' and this function adds value to the product. For example, sugarcane produced by the farmer has no market value and if the same is processed and converted into sugar or jaggery, its value in the market increases. Similarly, the conversion of paddy into rice, sunflower seed to sunflower oil etc., involves processing activity and thereby, it should be recognized as a specialized marketing function. Thus, processing activity adds form utility to the product and it is to be performed by a specialized agency and the process or methodology employed in changing the form of the commodity assumes greater significance in analyzing the marketing decisions. It is essential to note that, processing function will not alone add utility to the product, but other activities like transportation (place utility), storage (time utility), buying and selling activities (possession utility) add utility to the product. But, among these functions, processing function alone brings about change in the form of the commodity. Thus, agro-processing may be defined as *'an activity, which is performed to maintain or improve the quality or to change the form or characteristics of the agricultural product'*. It is, otherwise, defined as *'a set of techno-economic activities, applied to all the produces, originating from agricultural farm, livestock, aquacultural sources and forests for their conservation, handling and value-addition to make them usable as food, feed, fibre, fuel or industrial raw materials'*. Thus, processing converts the raw material and brings the produce nearer to the human consumption. It is concerned with the addition of value to the product by changing its form. So, processing may also be defined as a deliberate activity, which changes the form of a commodity. It converts farm products into a more usable and palatable form. Hence, the scope of the agro-processing industry encompasses all operations from the stage of harvest till the processed product reaches the end consumers in the desired form, packaging, quantity, quality and price. These processing operations are undertaken to add value to agricultural produce after their production. The main purpose of agricultural processing is to minimize the qualitative and quantitative deterioration of the produce after harvest. Important sector in agro-processing include food-processing sector and it deals with fruits and vegetables processing, grain processing, fish processing, milk processing, milling of jute, cotton etc., meat and poultry processing, packaged/convenience foods, alcoholic beverages and soft drinks etc. This food processing sector has experienced expansion during last five decades, starting with a handful of facilities, which were mainly operating at domestic/ cottage level.

Food processing aims to make food more digestible, nutritious and extend the shelf-life. Due to the seasonal variations, high levels of wastage or shortages can arise, if adequate measures are not taken to preserve and store the foods. Food processing covers all the processes that food items go through from the farm to the time it arrives on the consumer's plate. It includes basic cleaning, grading and packaging, as in case of fruits and vegetables and also alteration of the raw material to a stage just before the final preparation. Value addition processes to make RTE food like bakery products, instant foods, flavoured and health drinks etc., is also included in this definition. Processed food can be customized to suit the nutritious requirements of different age groups such as the elderly, pregnant women, infants, young children and athletes. Such foods are characterized by a balanced composition of energy suppliers in the form of fats, carbohydrates and proteins and by a combination of vitamins and minerals composed according to the current state of scientific knowledge. Food processing offers an opportunity for the creation of sustainable livelihoods and economic development for rural communities. The ever changing lifestyles, food habits and tastes of customers' globally have altered the dynamics of this industry. The world food production and consumption patterns are evolving with a change in the needs of the consumers. Increasing demand for ethnic and different foods from consumers across the world has redefined the market canvas for food processors across the world. With these changes, producers, processors, retailers and suppliers of food, world over, are reorienting their business plans to meet the new demands of the consumers. With a huge production base, India can easily become one of the leading food suppliers to the world, while at the same time serving the vast growing domestic market of over a billion people. India's large market size with growing incomes and changing life styles also creates incredible market opportunities for food producers, food processors, machinery makers, food technologists and service providers in this sector.

i. Food Processing in India

In India, 75 per cent of the total food processing units are in the unorganized sector (Figure 5.6). Out of this, 33 per cent belong to small scale industries and hence, huge investments and production of processed food products in tune with the SPS Standards are unlikely. Though organized sector constitute only 25 per cent, it is growing fast.

The size of global processed food industry is estimated to be valued around US $3.6 trillion and accounts for three-fourth of the global food sales. Despite its large size, only six per cent of processed foods are traded across borders compared to 16 per cent of major bulk agricultural commodities. Over 60 per cent of total retail processed food sales in the world are accounted by the U.S, EU and Japan taken together. Japan is the largest food processing market in the Asian region, though India and China are catching up fast and are likely to grow more rapidly. One of the most technically advanced food-processing industry globally is Australia, as the products produced are of international standards and at comparatively lower prices.

In the present day context of agri-business, food-processing is regarded as the sunrise sector of the Indian economy in view of its large potential for growth and likely socio-economic impact specifically on employment and income generation. Some estimates suggest that, 14 per cent of the total work force is engaged in food-processing sector directly or indirectly in developed countries and the same in India is only about three per cent thus, revealing the fact that, this sector is underdeveloped and vast potential remain untapped for employment. At present, the food-processing sector employs

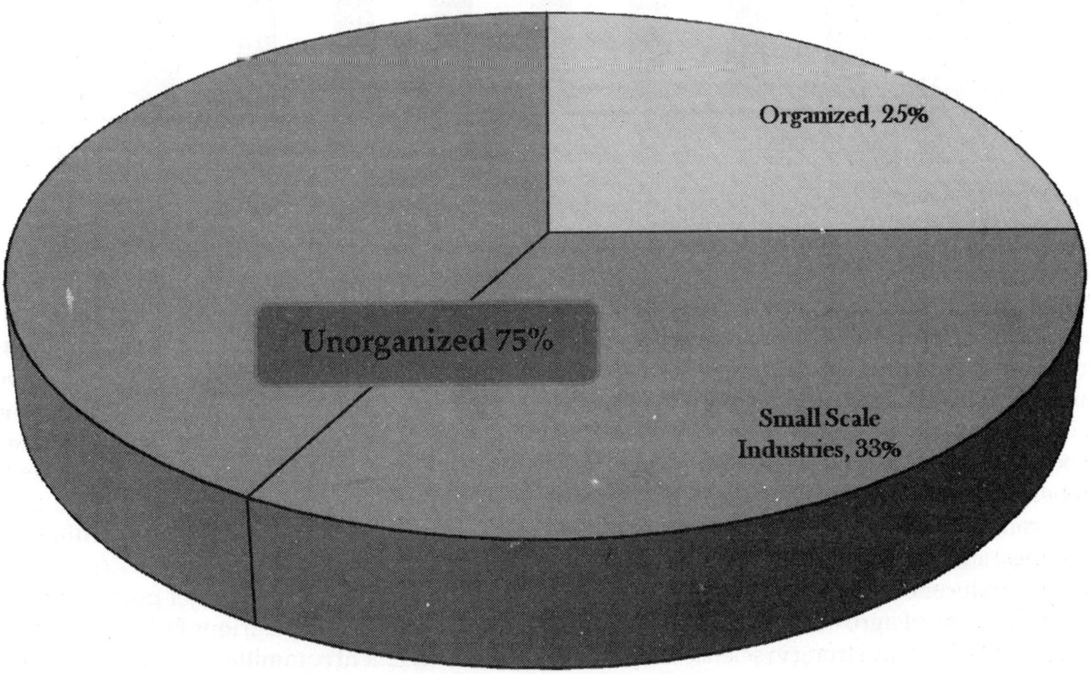

Figure 5.6: Structure of Indian food processing industry.

about 15 million people directly (Figure 5.7) and about 35 million people indirectly. In 2004-05, food-processing sector contributed about 14 per cent of manufacturing GDP with a share of ₹ 2,80,000 Crores. Of this, the unorganized sector accounted for more than 70 per cent of production in terms of volume and 50 per cent in terms of value.

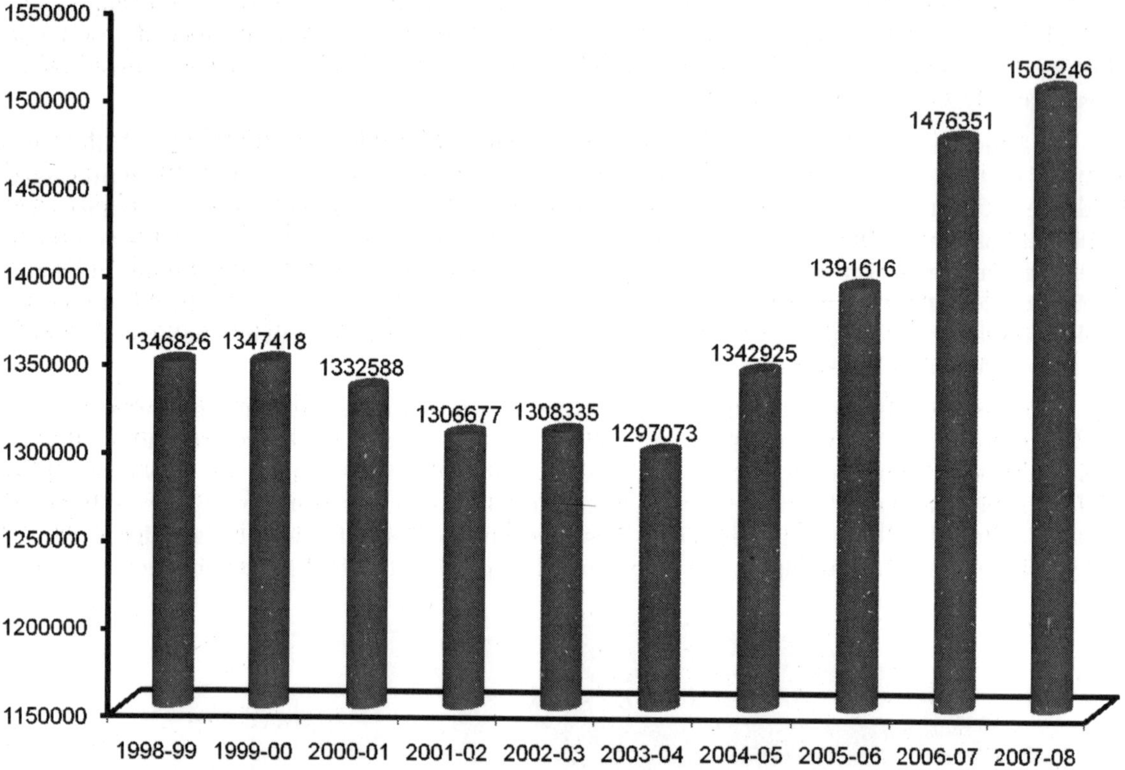

Figure 5.7: Number of persons employed under Registered food processing industries.

India is today the second largest producer of food in the world and has all the potential of becoming number one, if the goals and objectives framed under Rainbow revolution are achieved. Today (2009-10), India occupies a significant position in the world agriculture, as it has the second largest arable land (161 m. hectares) in the world, largest irrigated land (55 million hectares), largest producer of wheat (80.80 m tonnes, accounting for nearly 15 per cent of global wheat production), pulses (18.09 million tonnes, accounting for nearly 21 per cent of global pulses production) and milk (112.5 m tonnes, accounting for nearly 17 per cent of global milk production), largest producer and exporter of spices, second largest producer of rice (95.33 m tonnes, accounting for nearly 22 per cent of global rice production), sugarcane (339.17 m tonnes, accounting for nearly 21 per cent of the global sugarcane production, tea (accounting for nearly 28 per cent of the global tea production), fruits (71.52 m tonnes) and vegetables (133.74 m tonnes), largest exporter of the world's best basmati rice and third largest producer of coarse grains (42.22 m tonnes) etc. Added to these, there are other positive features like wide range of agro climatic regions and agri-produce, extremely industrious farmers, a country that is fundamentally strong in science and technology, a Government committed to Indian agriculture and an economy that is on the verge of double-digit growth. Yet the comprehensive outlook for Indian agriculture is far more complex than those statistics might suggest. The sector supports an estimated 65 per cent of the Indian population, but is also the most sluggish, having just extricated itself from a period of negative growth of -0.1 per cent in 2008-09 to rise to an unspectacular 0.4 per cent in 2009-10 with around 5.4 per cent expected for 2010-11. Adjusted for inflation, even this 5.4 per cent growth looks unexciting when compared to the growth rates in services and manufacturing. In 2010-11, agriculture accounts for 14.5 per cent of the country's GDP, compared to 51 per cent in the 1950s. Worse, India is amongst the world's largest wasters of food and faces a potential challenge to provide food security to its growing population in the light of increasing global food prices. But, due to poor handling of the produce, post-harvest losses are significantly high (Figure 5.8), resulting in a huge gap between gross production and the net availability to the consumer. The extent of post-harvest losses in agriculture range between 10 to 20 per cent in the entire supply chain from production to consumption (Table 5.9). Further, with the escalation of costs of agricultural inputs and stagnated crop yields during the past two decades has posed a worrisome picture in raising the cost of production of majority of the agricultural commodities. This escalation in the cost of production of commodities is at a faster pace compared to rise in their prices and this adversely affected their export competitiveness. So, farmers are about 15-20 per cent worse off, even after taking into account the gains in productivity.

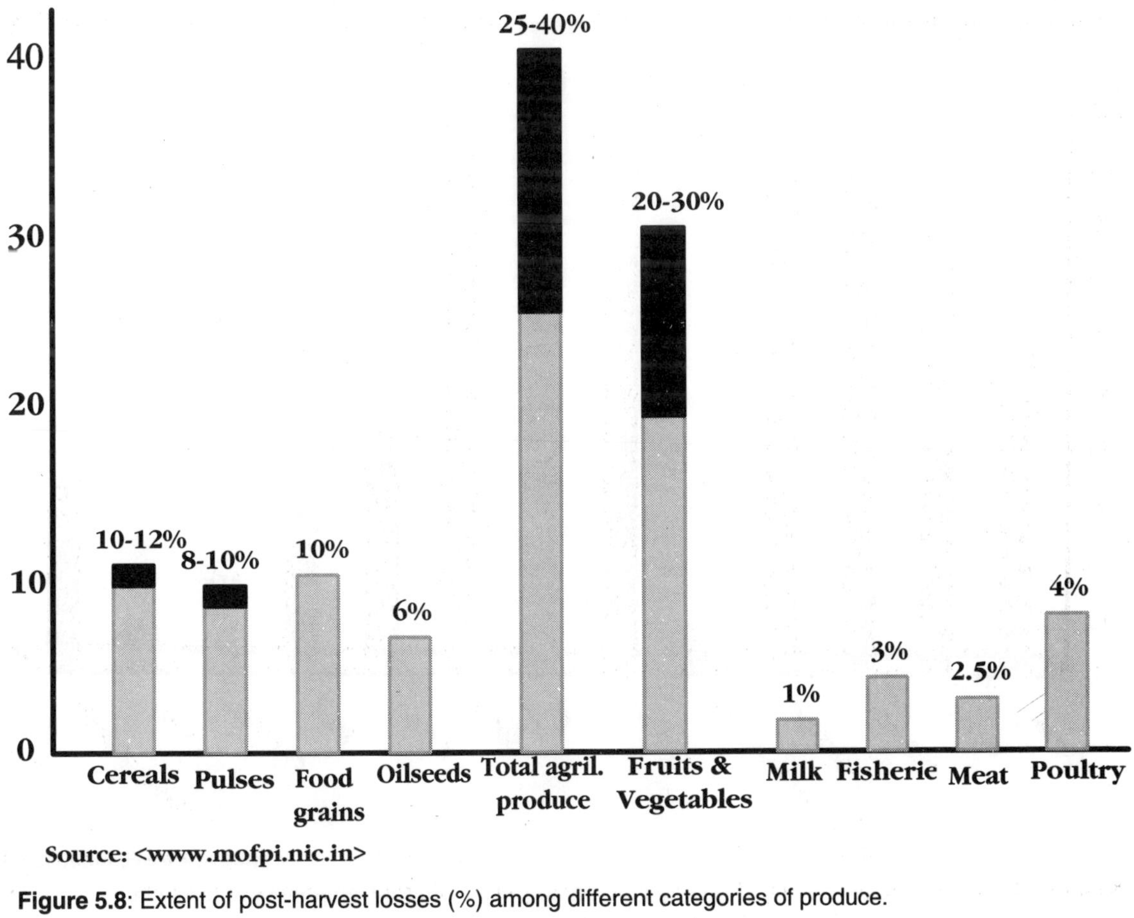

Source: <www.mofpi.nic.in>

Figure 5.8: Extent of post-harvest losses (%) among different categories of produce.

To overcome this set back, one major thrust area is promote agro-processing, as it increase the farmers' earnings through efficient and effective value addition of agricultural produce. But, the extent of processing of agricultural produce is far low in India, when compared to the developed nations with reference to various food items like food grains, fruits and vegetables, milk etc., and this is driven by the food habits of the population. The Indian population prefers fresh produce to processed products and this is the crucial factor for relatively less development of food-processing sector. The share of the processed products in India is low compared to that in the developed markets and they account for only 18 per cent of the total share (value), whereas non-processed foods account for nearly 50 per cent (value). The processing levels in India are very low and range from 2.2 per cent in fruits and vegetables to around 35 per cent in milk production (Figure 5.9). Across all segments, the level is around seven per cent and this is far low compared to developed countries and even below a developing country like Thailand (30 per cent). Thus, the extent of value addition due to processing to raw food material in India is only seven per cent while it is 23, 45 and 188 per cent in China, Philippines and UK respectively. Studies also revealed that, more than four dozen value-added products are produced from derivatives of paddy in a small country like Japan, which produces only two per cent of the total world production of paddy. India process only 2.2 per cent of fruits and vegetables as compared to 30 per cent in Thailand and 80 per cent in Malaysia. In India, the difference between price paid by the consumers for value added products and farmer's selling price of raw agricultural produce has been increasing rapidly. This implies that, through processing, there is more value addition to the agro-products. But,

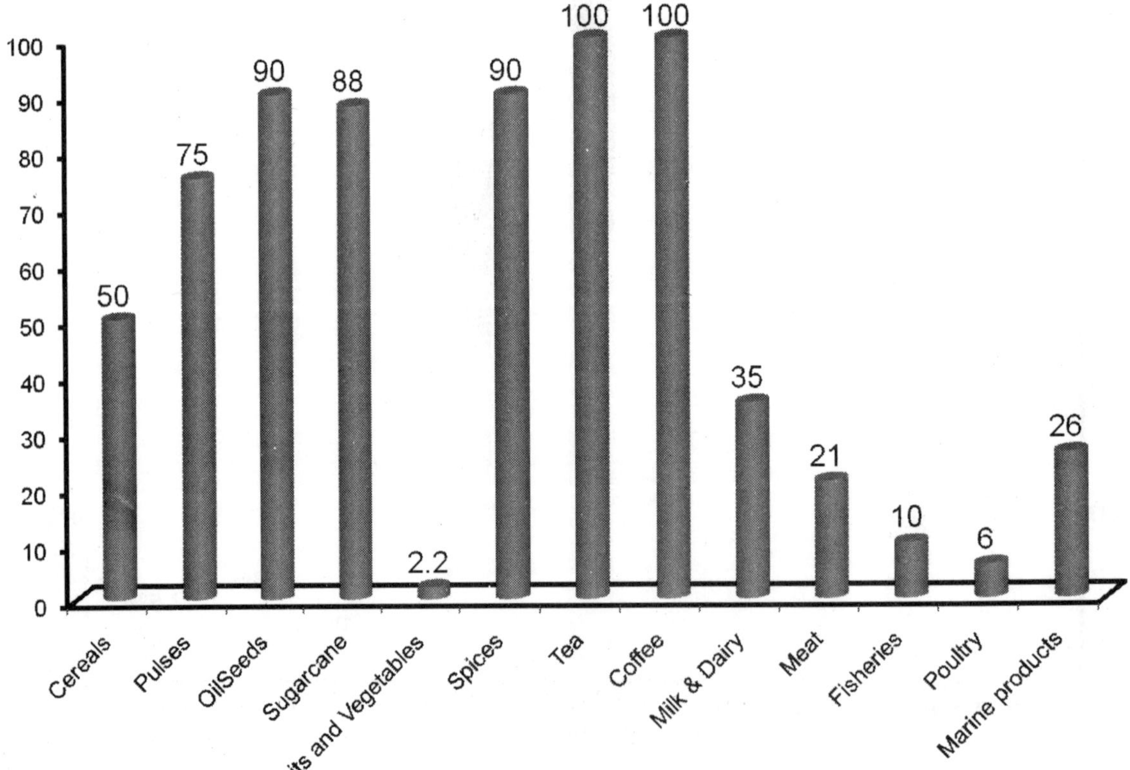

Source: <www.mofpi.nic.in>

Figure 5.9: Percentage of quantity of production processed of different categories of produce.

this activity (processing) is often understood only in the context of adding value to the product. However, in the modern days of agri-business, a new dimension from the consumer point of view is added to the existing understanding of value *i.e.*, how a consumer perceives the value delivered to him through a bundle of product services. So, this new approach of value addition is gaining more importance from the consumer point of view, as consumer orientation is required in all spheres of agricultural sector. In the context of value addition to agricultural products, there are three stages of processing as shown through Figure 5.10.

Table 5.9: Wastage levels of food produce in India.

Commodity	*Present Level of Production*			*Post-harvest Losses*		
	Quantity (m. tonnes)	*Average Price (₹/ton)*	*Value (₹ Crores)*	*Per cent*	*Quantity (m. tonnes)*	*Value (₹ Crores)*
Durables (cereals, pulses and oilseeds)	230	10000	230000	10	23	23000
Semi-perishables (potato, onion, sweet potato etc.)	40	3000	12000	15	6	1800
Perishables (fruits, milk, fish, eggs etc.)	210	15000	315000	20	42	63000
Total	480	11604	557000	14.8	71	87800

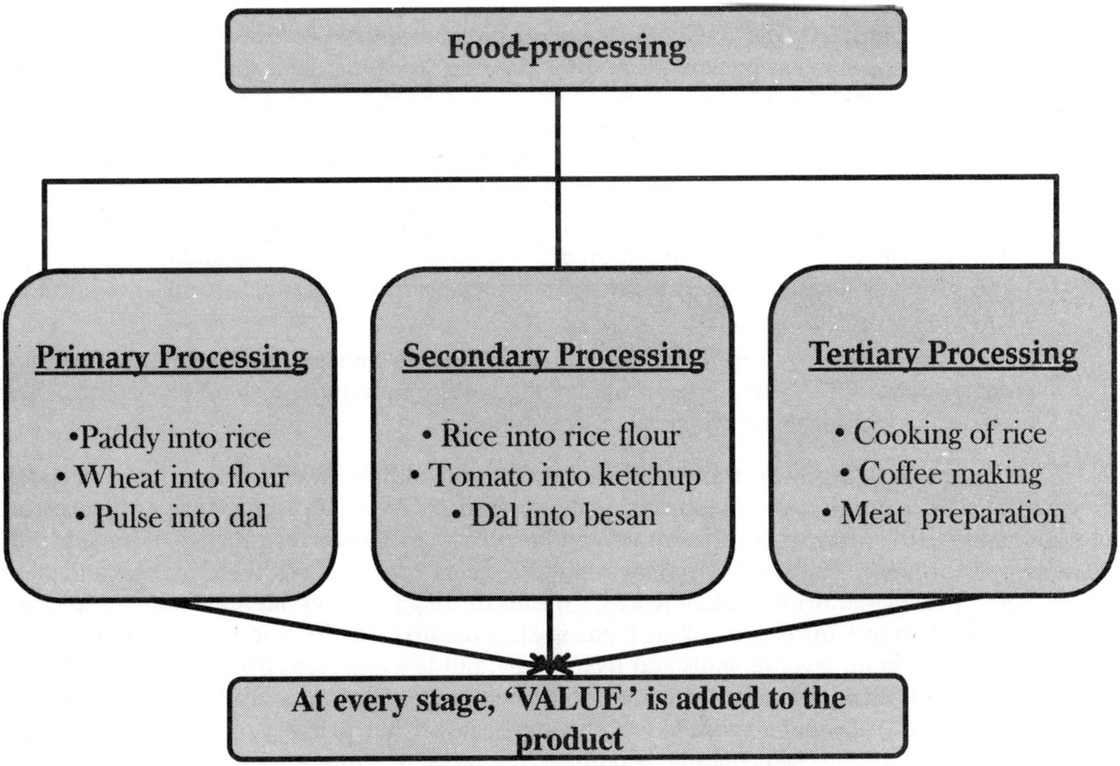

Figure 5.10: Different stages of food-processing adding value to the product.

☆ ***Primary Processing or Post-harvest Level Processing:*** It involves purification of raw agricultural produce by removing impurities, foreign matter, immature grain etc. Later the produce is subjected to various activities such as crop drying, shelling/threshing, milling, cleaning, grading and packaging. The milled produce is graded into different homogeneous lots. Later, each lot is eligible for secondary processing. These activities under primary processing are mainly carried out at the farm and they just transform the commodity into a slightly different form prior to storage, marketing or further processing. For example, conversion of paddy into rice falls under primary processing. Other examples include conversion of wheat into flour, pulse into dal etc. The following are the important conversion factors for different commodities under primary processing:

- Paddy into milled rice (conversion at 65–70 per cent, by-product is bran)
- Maize into maize meal (conversion at 65–85 per cent depending on quality of meal, by-product is bran)
- Green tea into black tea (conversion rate 28–32 per cent, no by-product)
- Cotton (kapas) into lint (conversion rate 30–35 per cent, by-product is cotton seed)
- Cherry coffee into green bean (conversion rate approximately 18 per cent, no by-product)
- Copra into coconut oil (conversion rate 60–65 per cent, by-product is copra cake)
- Soya beans into oil (conversion rate 18 per cent, by-product is soya meal) and
- Oil palm into palm oil (conversion rate 18–24 per cent, by-products are palm kernels and oil palm cake).

☆ ***Secondary processing:*** It involves processing of primary processed raw material into product, which is suitable for consumption after cooking, roasting, frying etc. Thus, secondary processing operations entail increasing nutritional or market value of the product and the physical form or appearance of the commodity is often totally changed from the original. For example, conversion of rice into rice flour falls under secondary processing. Other examples include, conversion of tomato into ketchup, fruits into jam, grinding groundnuts into peanut butter, pressing oil out of vegetable seeds, pressing juice out of fruit, making cheese out of milk etc. Depending on type of commodity, equipment needed for secondary processing is completely different from that used in primary processing or major adjustments/modifications need to be done to suit either.

☆ ***Tertiary processing or High end processing:*** It involves conversion of secondary processed material into RTE form. For example, cooking of rice falls under tertiary processing. Other examples include, meat preparation, tea making etc.

Thus, processing is done to make raw agricultural produce edible through primary and secondary processing and RTE through tertiary processing. It is essential to note that, at every stage of processing, value is added to the product. So, the term of food processing refers to the practice of transforming primary agricultural products into secondary or even tertiary commodities, which can be sold on the market for a much higher price. In other words, it means turning raw materials into more complex and expensive merchandise, instead of selling it cheap after the first step of production. Obviously, this would require a certain level of skills and technology, but it would also bring higher profits. The estimated value additions to the raw food materials through primary and secondary/tertiary processing in India are 75 and 25 per cent respectively. It, therefore, shows that, primary processing has a greater role to play in improving the economic benefits to the farmers. Thus, agricultural produce can be marketed in different forms as raw, primary processed, secondary processed and tertiary processed depending upon the nature of produce and consumption habits of the consumer. The farmers, in general, sell the raw agricultural produce in the market (marketed surplus), as they are in immediate requirement of liquid cash to meet their requirements. It is estimated that, on an average, the farmers market 52 and 56 per cent of paddy and wheat respectively in raw form and they are later processed by the wholesalers (also act as processors) or processors. Thus, in India, the processing function is generally performed at the wholesalers' or processors' level and not at the farmers' level. As mentioned earlier, this function is gaining wider significance because of the involvement of specialized marketing agency. Further, this function is gaining wider significance because of increasing importance for RTC (ready-mix idli, ready-mix dosa, pop corn, gulab jamun mix, sambar powder, rasam powder etc.) and RTE foods (rasagulla, ice creams, active pop corn, rasna powder, fruit juices etc). The markets for these processed commodities are consistently increasing with increasing demands by consumers. To meet their demand and to remain competitive, the processing firms must adopt low-cost improved technologies and improve market efficiency.

Farmers can use this value addition strategy (all the three stages of processing) to transform his unprofitable enterprise into a profitable one. Thus, these three stages of processing ensures the long-term survival of most small farms. Let us assume that, a coffee farmer in a village simply grows and harvests coffee cherry and then sells it 'as is' to a local processor at a price below the cost of production (say, due to glut in the market). So, the farmer can survive in the farm business in a short run, as long as the price covers only the AVC of production, if not Average Total Cost (ATC). So, the marketing strategy employed by the farmer *i.e.*, selling the produce immediately after harvest is a poor strategy because, the market price does not cover the ATC of production thereby, the farmer is not sustainable in coffee cultivation. However, considering the above different types of processing, the farmer has a range of options other than immediately selling coffee cherry to a local processor. The farmer can remove the cherry pulp, wash and dry the coffee beans to create 'parchment' (primary processing). The coffee parchment can then be either sold or stored for some time in this form. The farmer has now added value to the coffee, he initially harvested. So, the farmer gets extra income by selling the coffee parchment instead of selling the raw produce directly to the processor and this makes the farmer to sustain in the business, as now, price may cover the ATC of production. In the similar way, if all the coffee cultivating farmers join as a group, so as to enjoy the benefits of large scale marketable surplus, they can still boost the profits. Further, this farmers' group can themselves mill the parchment to form 'green beans'(secondary processing), which in turn can be either sold to roasters or roasted and packaged at the farm for retail sales as 'estate coffee'. The coffee grower can even add more value by opening the farm to visitors, who can tour the farm and learn about the whole process of growing, harvesting, pulping, drying, milling, grading, roasting, packaging and ultimately, tasting coffee (tertiary processing). Taking coffee to the agricultural tourism ('agtourism') level allows the entrepreneurial grower to operate a retail shop and to sell over the internet to those, who have already visited the farm. Once the farmer has added value all the way to the level of agtourism, he is selling more than just coffee *i.e.*, he is also selling a visitor experience.

ii. Commodity-wise Processing in India

There is a drastic increase in demand for the food processed products during the past six to seven years in the country. This is due to cumulative influences of various factors like significant increase in income and purchasing power of consumers, increasing propensity to spend, globalization, increased urbanization, liberalization of import policies, growth of organized retailing, impact of visual media, changing lifestyles and food habits of the people etc., and all these have opened the doors for the entry of imported food products from the international market. The recent global financial crisis has cast its shadow over the Indian economy. However, the Indian economy is expected to weather the storm much better than its compatriots, thanks to the well diversified and strong domestic economy and appropriate monetary and fiscal policy responses from the Central Government. Considering these meritorious aspects, food processing sector in India is gaining equal importance as agriculture or any other industry and is considered vital to the Indian economy, as it acts as a linkage between manufacturing industries and agriculture. It offers employment to around 13–15 million people in the country and contributes about 14 per cent to the GDP. In 2006–07, the food processing sector was estimated at US$82.6 billion (Table 5.10), growing at 13 per cent per annum. However, when compared to some of the developed countries, the Indian food processing sector is still in a nascent stage. For instance, the current level of fruits and vegetables processed is only 2.2 per cent and is far low against other developed and developing nations (Figure 5.13).

Table 5.10: Indian food sector and market size, 2006–07.

Item	*Value (US$ Billions)*
Food sector*	194.3
Food processing sector	82.6
Organized food processing sector	22.3

*: Includes fruits and vegetables, dairy products, fish and marine, meat and poultry, packaged foods, beer and wine.

Source: <www.mofpi.nic.in>

Further, in terms of employment generation, better price realization for farmers, reduction in agricultural wastage and improvement in quality, the food processing sector ranks way ahead of other sectors of the Indian economy. The multiplier for the food industry is reported to be much higher than other industries such as power and telecom. Therefore, the rapid growth in the Food Processing Industry (FPI) has stimulated the growth of domestic market and trade on the international front. In order to boost the food processing sector, the Government of India has introduced the concept of Mega food parks in the country. Catering to the entire range of food processing activities, food parks consist of groups of large, medium and small-scale food processing units. Facilities such as cold storages, warehousing, power and water supply are provided by the food parks thus, making it more affordable for small and medium scale units. The gap between the farmers and industries have also been closed in with the help of food parks by facilitating a firm buyback agreement between processing units located in the parks and the farmers. Various food parks are being set up by the Ministry of Food Processing Industries (MOFPI) across the states in the country such as Andhra Pradesh, Tamil Nadu, Maharashtra, Punjab and Jharkhand. With better market orientation, food processing units are enabled to be cost effective and also have auxiliary units such as packaging, transportation agencies etc. The FPI is dominated by the cottage industry sector, which accounts for 40 per cent of the total processed food market. The rest is shared by the small scale sector and the organized players, each accounting for almost 30 per cent of the market share. Small players in this industry have only a local or regional presence and use basic technologies for processing the food products.

The food processing sector is a highly fragmented industry, as it comprises of different sub-segments *viz.*, food grains, oilseeds, fruits and vegetables, milk and milk products, meat and poultry, marine products etc (Table 5.11). The most promising sub-segments includes: soft-drink bottling, confectionery manufacture, fishing, aquaculture, grain-milling and grain-based products, meat and poultry processing, alcoholic beverages, milk processing, tomato paste, fast food, RTE and RTC foods, food additives, flavours etc. Health food and health food supplement is another rapidly rising segment of this industry, which is gaining vast popularity amongst the health conscious consumers. A huge number of entrepreneurs in this industry are small in terms of their production and operations, and are largely concentrated in the unorganized segment. This segment accounts for more than 70 per cent of the output in terms of volume and 50 per cent in terms of value. Though the organized sector seems comparatively small, it is growing at a much faster pace.

Table 5.11: Different sub-segments of food processing sector and their relevant processed products.

Sub-segments	*Products*
Food grains	Flour, bakeries, starch glucose, cornflakes, malted foods, vermicelli, beer and malt extracts, grain based alcohol
Fruits and Vegetables	Pulps, Concentrates, Slices, Frozen and dehydrated products, Chips/Wafers etc.
Meat and Poultry	Frozen and canned products, Egg Powder
Fishery	Frozen and canned products
Dairy	Whole milk powder, Skimmed milk powder, Condensed milk, Flavoured milk, Ice-cream, Butter, Ghee, Cheese, Infant milk food, Malted milk food
Bakery Products	Bread, Biscuits, Cornflakes, Noodles etc.
Confectionery	Chocolates, Sugar candies, Bubble gum, Chewing gum
Beverages	Alcoholic and Non-Alcoholic (tea/coffee, carbonated soft drinks and non-carbonated soft drinks)
Packaged/Convenience Foods	RTE (jams, pickles/chutneys, ketchup, soup, snack foods) and RTC foods

Source: <www.mofpi.nic.in>

Table 5.12: Growth of Food Processing Sector (₹ Crores).

Activity	*2004-05*	*2005-06*	*2006-07*	*2007-08*	*2008-09*	*2009-10*	*CAG ₹* (Per cent)*
Manufacturing (Registered and Unregistered) under Food Processing sector							
Meat, Fish, Fruits, Vegetables and Oils	9,236	8,682	9,548	10,349	12,043	12,224	7.29
Dairy Products	3,509	4,342	4,319	4,608	5,419	4,762	6.66
Grain Mill Products	13,467	12,347	11,903	12,846	15,947	17,741	6.55
Other Food Products	14,722	17,794	20,895	22,522	25,775	23,664	10.71
Beverages	3,421	4,525	5,499	6,995	7,938	7,687	18.61
TOTAL	44,355	47,690	52,164	57,320	67,122	66,078	9.30
Per cent GDP Growth (FPIs)	–	7.52	9.38	9.88	17.1	–1.56	
GDP Agriculture	4,76,634	5,02,996	5,23,745	5,56,956	5,53,454	5,53,010	3.17
Fishing	27,152	28,749	30,650	32,427	33,561	35,215	5.34
GDP (Agriculture+Fishing)	5,03,786	5,31,745	5,54,395	5,89,383	5,87,015	5,88,225	3.29
Per cent GDP Growth (Agriculture+Fishing)	–	5.55	4.26	6.31	-0.40	0.21	

Note: * CAGRs: Compound Annual Growth Rates

Source: www.mofpi.nic.in

Growth in the food processing sector in the recent years has been significant as may be gauged from the Table 5.12. It may be seen from the table that, meat, fish, fruits and vegetables, oils, dairy products, grain mill products, beverages and other food products are the main items being processed. Beverages and other food products have exhibited the highest growth rates followed by meat, fish, fruits, vegetables and oils during 2004-05 to 2009-10. This is a positive development. It may also be seen that the growth in the number of registered processing units have been significant in grain mill products and beverages (Table 5.13). Further, it may be seen from Table 5.14 that, the Foreign Direct Investment (FDI) in food processing during 2000-11 was 1.78 per cent of the total FDI and the major sectors which attracted FDI were food processing and fermentation industries.

Table 5.13: Sector-wise number of registered processing units.

Year	*Processing of Meat, Fish, Fruits and Vegetables and Oils*	*Dairy Products*	*Grain Mill Products*	*Other Food Products*	*Beverages*	*Total*
1998-99	4241	737	12164	5682	1029	23853
1999-00	3819	795	12405	5810	1113	23942
2000-01	3740	735	12446	5985	1082	23988
2001-02	3454	865	12429	5688	1049	23485
2002-03	3284	769	12856	5899	1008	23816
2003-04	3352	912	12741	5757	1078	23840
2004-05	3484	927	13639	6093	1219	25362
2005-06	3549	1049	13893	6009	1225	25725
2006-07	3459	1015	13880	6245	1160	25759
2007-08	3667	1096	13805	6300	1351	26219

Source: www.mofpi.nic.in.

Table 5.14: Foreign direct investment in food processing sector from April 2000 to August 2011.

Sl.No.	*Sector*	*Amount of FDI Inflows*		*Percentage Share of Total FDI*
		(in ₹ Crores)	*(in US$ millions)*	
1.	Food Processing Industries	5872.16	1286.53	0.89
2.	Fermentation Industries	4269.92	979.74	0.65
3.	Vegetable Oils and Vanaspati	1103.22	238.72	0.17
4.	Tea and Coffee	446.61	99.38	
	Total Food Processing Sector	11691.91	2604.37	1.78
	Total FDI	658586.43	147088.13	100.00

Source: www.mofpi.nic.in.

The following are the products covered under food-processing sector:

a. ***Food grains:*** 1. Rice; 2. Wheat; 3. Maize; 4. Coarse cereals; 5. Pulses

b. ***Oilseeds***

c. ***Commercial crops:*** 1. Sugarcane; 2. Cotton
d. ***Fruits and Vegetables***
e. ***Milk***
f. ***Poultry and Meat***
g. ***Fish***

a. Food Grains

This is an important sector in the context of export earnings, as majority of the commodities like rice, wheat, maize, coarse cereals (ragi), pulses enjoy export competitiveness in the international market.

1. Rice Milling

Paddy production in India is about 95.33 m tonnes in 2009-10. The cleaned paddy, on an average, yields 22 per cent husk, six per cent bran and 72 per cent rice. Among the grains, rice is most processed grain. Paddy is milled into raw or parboiled rice and flaked rice. Puffed rice is also produced as snack food. Milling is generally done in hullers, shellers and modern rice mills. In the conventional paddy-rice system of harvesting at low moisture content, parboiling in masonry tanks, drying in sun and milling by hullers, the total yield of rice seldom exceeds 65 per cent with 20-30 per cent broken, when milled as raw, total yield of rice is around 68 per cent with 15-20 per cent broken as parboiled paddy. Moreover, the by-products cannot be used economically. Excessive breakage during milling in the conventional system reduces the total recovery of rice. The recent investigations have shown that, it is possible to increase the total out turn of better quality rice by 10 per cent with improved harvesting, parboiling, drying and milling technologies. A rice milling system can be a simple one or two step process, or a multi stage process. In a one step milling process, husk and bran removal are done in one pass and milled or white rice is produced directly out of paddy. In a two step process, removing husk and removing bran are done separately, and brown rice is produced as an intermediate product. In multistage milling, rice will undergo a number of different processing steps. Depending on whether the paddy is milled in the village for local consumption or for the marketing, rice milling systems can be classified into the categories *viz.*, village rice mills and commercial mills. Milling is a crucial step in post-production of rice. The basic objective of a rice milling system is to remove the husk and the bran layers, and produce an edible, white rice kernel that is sufficiently milled and free of impurities. Depending on the requirements of the consumer, the rice should have a minimum of broken kernels. Most rice varieties are composed of roughly 20 per cent rice *hull* or *husk*, 11 per cent *bran* layers and 69 per cent *starchy endosperm*, also referred to as the *total* milled rice. In an ideal milling process this will result in the following fractions: 20 per cent husk, 8-12 per cent bran depending on the milling degree and 68-72 per cent milled rice or white rice depending on the variety.

Modern rice mills (Table 5.15) process 65 per cent of paddy production and rest by huller/sheller mills. The recovery of whole grains in a traditional rice mill using steel hullers for dehusking is around 52-54 per cent, whereas in modern rice mills-rubber roll shellers for dehusking operation is around 62-64 per cent in raw and 66-68 per cent in parboiled paddy. The conversion ratio (*i.e.* recovery percentage of various final product and byproduct for every 100 kg feed of raw paddy) for these improved rice mills are: milled rice: 62-68 per cent, rice bran: 4-5 per cent, rice husk: 25 per cent and germ wastages: 2-8 per cent. Thus, there is a need to establish small capacity (150 kg/h) modern rice mills in villages.

Table 5.15: State-wise number of rice mills (as on 1-1-2002).

States/Union Territories (UTs)	*Hullers*	*Shellers*	*Hullers cum Shellers*	*Modern/Modernized Rice Mills*	*Total*
Andhra Pradesh	4609	1776	2364	12995	21744
Assam	431	14	2133	242	2820
Bihar	4749	63	9	51	4872
Gujarat	1890	159	67	1045	3161
Haryana	807	–	–	990	1797
Himachal Pradesh	890	1	3	222	1116
Jammu & Kashmir	–	–	–	–	–
Karnataka	9131	462	1103	3674	14370
Kerala	13664	–	13	2533	16210
Madhya Pradesh	3918	201	262	1761	6142
Maharashtra	8199	273	541	1759	10772
Manipur	838	–	–	660	1498
Meghalaya	252	–	8	–	260
Nagaland	–	–	–	–	–
Odisha	6398	125	289	552	7364
Punjab	4416	442	–	1965	6823
Rajasthan	152	2	6	193	353
Sikkim	17	–	–	–	17
Tamil Nadu	13684	448	1324	3922	19378
Tripura	1030	6	8	1	1045
Uttar Pradesh	5707	562	150	1415	7834
West Bengal	9554	3	72	926	10555
Andaman & Nicobar Islands	116	–	–	–	116
Arunachal Pradesh	–	–	–	–	–
Chandigarh	16	–	–	11	27
Dadra & Nagar Haveli	8	1	–	–	9
Delhi	2	–	–	27	29
Goa	670	–	25	56	751
Mizoram	–	–	–	–	–
Pondicherry	138	–	8	88	234
Lakshadweep	–	–	–	–	–
India	91287	4538	8385	35088	139298

Source: <www.indiastat.com>

2. Wheat Milling

India produces 80.80 m tonnes of wheat. It is mainly consumed in the form of atta, suji, maida and baking flour. Most of the wheat is milled in small capacity disc mills to produce wheat flour (atta).

Maida-the refined wheat flour, used in the bakery industry, could be produced in capital intensive and sophisticated roller flour mills (Table 5.16). Small scale bakery industry is often confronted with the problem of availability of right quality of maida. Burr mills (chakkis) are very common for milling of wheat in rural and urban areas. Uttar Pradesh ranks first in having large number of roller flour mills followed by Maharashtra, Karnataka, Tamil Nadu and Andhra Pradesh (Table 5.17). Central Food Technological Research Institute (CFTRI), Mysore has now developed a simple mini wheat mill which simultaneously produces bakery flour, maida, atta and suji. At present, flours made by the roller mills are sold to institutional buyers like defense, hotels etc., and the household purchase is limited to only one per cent due to absence of open policy for sale of wheat flour through PDS. Sale of soya blended and branded wheat flour is likely to increase due to better quality flour and thus scope of organized wheat milling will increase in future. Bran is available as a bye product, which is used as animal feed. Wheat is processed for flour, refined flour, grits, and whole-wheat flour. Roller flour mills process over 50 per cent of wheat production with a milling capacity of 8-10 million tonnes each.

Table 5.16: Number of wheat processing mills in India (1993, 2004 and 2010).

Type	*Number*		*Processing Capacity*	*Projection Number*
	1993	*2004*	*(Million Tonnes)*	*2010*
Roller Mills	750	750	10	1000
Chakki	266000	266000	80	300000

Source: <www.indiastat.com>

Table 5.17: State-wise number of roller flour mills in India (As on 1.1.2002).

States/UTs	*Number*	*States/UTs*	*Number*
Andhra Pradesh	58	Assam	40
Bihar	58	Chandigarh	5
Delhi	18	Goa	2
Gujarat	34	Haryana	47
Himachal Pradesh	15	Jammu & Kashmir	19
Karnataka	60	Kerala	27
Madhya Pradesh	44	Maharashtra	65
Manipur	1	Meghalaya	2
Nagaland	4	Odisha	19
Pondicherry	3	Punjab	48
Rajasthan	10	Sikkim	3
Tamil Nadu	60	Tripura	2
Uttar Pradesh	127	West Bengal	49
	India	820	

Source: <www.indiastat.com>

3. Maize

Maize is usually processed by two distinct processes, namely wet milling and dry milling. Maize is generally processed to manufacture corn starch by wet milling method the world over. The by-products of starch manufacture, like corn oil, corn steep liquor, gluten etc. are the important value added products. The average recovery of various products and co-products of maize during the wet milling include Starch (60-62 per cent), Gluten (8-9 per cent), Germ (6-7 per cent) and Husk (22-24 per cent). The different products that result from dry milling include, Grits (40 per cent share), Coarse meal (20 per cent share), Germ (14 per cent share), fine meal (10 per cent share), flour (5 per cent share) and Hominy feed (10 per cent share). The technology has been standardized by CFTRI, Mysore. Dry milling units have a crushing capacity of 10 M Tonnes/day and such units can be set up any where depending upon the availability of raw material, power and suitable land.

4. Coarse Cereals

Production of coarse cereals has risen from 15.4 to 42.22 m tonnes between 1950-51 and 2009-10. The growth has not been as rapid as in case of rice and wheat. It is because of low profitability of these crops for farmers, till 1950s, milling is done by manual chakkis or water mills. By the year 1998-99, power operated equipment were available for milling. Technology has also been developed for production of value-added products from coarse cereals such as extruded snacks developed from ragi at CFTRI, Mysore, ragi based snacks at UAS, Bangalore and IIT Kharagpur, sorghum-soybean fortified foods at IIT Kharagpur etc. The trends indicate that, coarse cereals are now increasingly used as cattle feed, specialty/occasional foods, and industrial products such as starches. Among coarse cereals, ragi is very important in meeting the nutritive diet requirements of the consumers. Ragi can be milled by wet conditioning. It can be steamed followed by milling in a hammer or plate mill or a roller flour mill. Compared to other millets, ragi is most suitable from the stand point of product quality and enzyme release for malting. The malted ragi flour can be used along with germinated green gram flour to formulate a high calorie dense weaning food having excellent nutritional qualities. Ragi flour can also be used with milk beverages. Parboiling of ragi helps in the improving its quality by eliminating its slimy texture. Flour from puffed ragi has good flavour and can be used in snacks and supplementary foods in South India. The following (Table 5.18) are the major players in cereals industry in India:

Table 5.18: Major players in cereals industry in India.

Company	*Brands*	*Products*
KRBL	India Gate, Lion, Doon, Bemisal, Nur Jahan, Rice king, Taj Mahal	Rice
Kohinoor foods	Kohinoor	Rice, Convenience Food
LT Overseas	Daawat, Heritage, Orange, Josh, Apsara	Rice, Wheat
Lakshmi Energy and Foods	Lakshmi Foods	Rice, Wheat
Usher Agro	Rasoi Raa	Rice, Other cereals
REI Agro	Kasauti, Real Magic, Hansraj, Rain Drop	Rice

Source: Nisha (2008).

5. Pulses

Pulses are the major sources of protein for the vegetarians in India. In a total world production of 60 million tonnes of pulses, India, China, Brazil, Turkey and Mexico account for 2/3 rd of the output. India produced 8.4 m tonnes of pulses in the year 1950-51. The production grew to 18.09 million tonnes by the year 2009-10. Starting with nearly 500 *dhal* mills in the country in 1950-51, there were about 20000 *dhal* mills in the year 2004. In the area of milling of pulses, CFTRI developed a *dhal* mill that has the advantage of not being dependent on natural sun shine. It involves subjecting the pulse grain to high temperature (120 degree Celsius) for short time and the dehusking by carborundum rollers resulting in higher *dhal* recovery. For small entrepreneurs in rural areas, *dhal* mills have been designed at Central Institute of Agricultural Engineering (CIAE), Bhopal; PDKV, Akola; IIPR, Kanpur, TNAU, Coimbatore; GBPUA&T, Pantnagar and CFTRI, Mysore. These units in specific regions have gained popularity, as these are low investment machines which can be owned and operated with low risk. In a number of *dhal* mills, improved machinery including cleaners, graders, magnetic separators, washers, driers, polishers, colour sorters and packaging systems are being used. With complete phasing out of hand operated *dhal* chakkis commonly used during 1950s, the technology has turned fully mechanized and more-and-more urban based. There is a need to evolve more efficient machines and processes for pre-treatment of the grain, dehusking, sorting, polishing and packaging in order to improve Dhal recovery and consume less energy. Also, there is a need for product diversification and development of technology for quick cooking and RTE dhal.

Dhal (split pulse) milling is the third largest processing industry in India, after rice and wheat milling. Dhal recovery potential is 83-85 per cent, but at present, it is 68-70 per cent in conventional mills and 72-78 per cent in modern dhal mills. The mills are processing more than 10.5 million tonnes of dhal. By-products are generally used as cattle feed. In India, there are two conventional pulses milling methods *viz.*, wet milling method and dry milling method. The latter is more popular and used in commercial mills. However, there is no common processing method for all types of pulses. But, both wet milling and dry milling processes involves two basic steps *viz.*, precon-ditioning of pulses by alternate wetting and sun drying for loosening husk and subsequent milling by dehusking and splitting of the grains into two cotyledons followed by aspiration and size separation using suitable machines. Complete dehusking and splitting of pulses are seldom achieved particularly in cases of certain pulses like red gram, black gram and green gram. Of them, red gram is the most difficult pulse to dehusk and split. Only about 40 to 50 per cent red gram grains are dehusked and split in the first pass of pre-conditioning and milling. As sun drying is practiced, the traditional method is not only weather dependent, but also it requires a large drying yard to match with the milling capacity. As a result, it takes three to seven days for complete processing of a batch of 20 to 30 tonnes of pulses into dhals. Moreover, milling losses are also quite high in the traditional method of milling of pulses. In general, the raw pulses may contain 2 to 5 per cent impurities (foreign materials), some insect infested grains and extra moisture. Though the clean pulses contain about 10-15 per cent and 2-5 per cent germs, the yield of dhals commercial *dhal* mills varies from 68-75 per cent. It may be noted that, the average potential yields of common *dhals* vary from 85 to 89 per cent. These milling losses in the commercial pulses mills can be attributed to small brokens and fine powders found during scoring and simultaneous dehusking and splitting operations.

b. Oilseeds

India is the fifth largest producer of oilseeds in the world, behind US, China, Brazil, and Argentina. India is one of the world's largest edible oil economies with over 15,000 oil mills, 689 solvent extraction units, 251 Vanaspati plants and over 1,000 refineries, employing more than one million people. In the year 1950-51, the country produced 5.2 m tonnes of oilseeds. Production by the year 2009-10 had increased to 24.88 m tonnes. The OTM has helped in increasing the oilseeds production. On an average, oilseeds contain 40 per cent oil and 20 per cent protein. Present utilization of total oilseed production in India is seven per cent as seed, eight per cent direct food uses and 85 per cent for oil extraction. Nutritionally, the edible oil is the richest source of energy (9 kcal/g) and provides essential fatty acids and transports fat-soluble vitamins in human body. India being deficient in oils has to import 40 per cent of its consumption requirements. One of the ways of augmenting edible oil supply is to reduce post-harvest losses of oilseeds and enhance oil extraction efficiency. In the year 1950-51, most of the oilseeds were crushed by bullock operated oil ghanies. But, they resulted in high volume of edible oil left in the cake (10-15 per cent oil with cake). This oil extraction through bullock operated oil ghanies has been a cottage level activity in the country. However, the mechanical power ghanis have now been introduced in the country, but they also leave 8-12 per cent oil with cake. These ghanis are phasing out in favor of power operated oil expellers. The single and the double chamber expellers of this type are capable of extracting most of the oil except 6-8 per cent, which is left with the cake. The introduction of high capacity mechanical expellers and solvent extraction technology has brought in modernization of oil extraction. Small capacity oil expellers are also developed, which could be installed in rural areas for promoting agri-business and that might provide more employment. Like pulses, oil seeds are also rich in protein and in addition, they contain a high level of fat. Hence, they are not only good sources of protein, but are concentrated source of energy. They do not contain an appreciable amount of carbohydrate, but contain high level of B-vitamins. Groundnuts are particularly rich in thiamine and nicotinic acid. Oil seeds produced in the country are mostly used for oil extraction. In India oilseeds like groundnut, mustard, safflower, are more commonly grown for edible oil purpose. The oilseeds are cleaned, dehusked and oil is extracted by mechanical pressing, screw pressing, prepress solvent extraction and direct solvent extraction. After solvent extraction, it is treated with dilute sodium hydroxide at pH 8. The proteins are separated by filtration and washing with water and then used for different purposes.

Oil Extraction

- ☆ ***Groundnut oil:*** Groundnuts resemble pulses in general nutritive value, except that they are rich in fat. Groundnuts are boiled or roasted and consumed. But, the chief product is the oil, which can be used either as cooking oil or for making margarine and soap. The secondary product is the residue or cake left after the expression of the oil. It is also purified and used in supplementary mix. The major proteins of groundnut are arachin and conarachine, which is lacking in lysine and methionine and its quality can be improved by either adding these amino-acids or foods rich in the same.
- ☆ ***Sunflower oil:*** Both solvent extraction process and mechanical expression methods are in use for the extraction of oil from the sunflower seeds. In the extraction of oil, 60 to 70 per cent of the oil from the meal is recovered. First, the seeds are cleaned to remove all foreign materials including stones, dust etc. Then, dehulling of the cleaned seed is done to reduce the fiber content and increase the oil and the protein contents in the dehulled kernel. The kernels are then allowed to pass through the roller mills to increase the oil recovery efficiency. The rolled flakes are then cooked by heat treatment for further increase in oil recovery. The oil is extracted by pressing the cooked flakes in the expellers. The oil is then filtered to remove the seed particles. The sizes of the deoiled cakes are reduced and then these are bagged for storage. The crude oil is refined for the production of edible grade oil.
- ☆ ***Soya bean oil:*** Soyabeans have value due to their high quality oil and protein. Historically, soya protein products account for about two-thirds of the value of soybeans with the balance coming from the oil fraction. For this reason, many refer to soybeans as a protein crop as opposed to an oil crop. The protein fraction of soyabeans finds its way into soyabean meal as a livestock feed ingredient, as soya protein ingredients for the food industry and for

industrial uses (paper coatings, adhesives etc.). Soya oil has found many food uses due to its excellent nutritional qualities, widespread availability, economic value and wide-use functionality. Soya oil is also an important ingredient for industrial products such as paints, plastics, lubricants and biofuels. Other by-products of soybean processing include soybean hulls - mainly used in animal feeds and as a source of soya fiber and soya lecithin phospholipids – a nutritional supplement and functional emulsifying food ingredient. In the 1930s, soya beans were mechanically processed using hydraulic or screw presses (often known in the industry as expellers), which squeezed out the oil of the heated or cooked soyabeans. In the late 1940s and early 1950s, most of the industry converted to the solvent-extraction process, which removes more oil from the soybean. Today, more than 99 per cent of the processing capacity is using the solvent extraction process. Several of these plants process more than 3,000 tonnes per day. Soyabean Processing and Utilization Centre at CIAE, Bhopal has developed a number of processing technologies and equipments for extraction of soyabean oil. These are being promoted in rural/urban areas for agribusiness and nutrition security and so far over 300 cottage scale units have been established throughout the country.

c. Commercial Crops

1. Sugarcane

More than 50 per cent sugarcane is estimated to be processed by sugar mills and the balance by jaggery (Gur) and Khandsari industries. Although, the efficiency of jaggery (Gur) and Khandsari sector is low compared to sugar mills, but these units provide more employment opportunities to rural work force hence, requires considerable attention. As compared to mill made sugar, establishment of jaggery and khandsari units have low capital requirements. Sugarcane cultivation is done on around 4 million hectares of land in India and its production has fluctuated between 230-300 million tonnes in past several years. The 250 billion sugar industry has about 450 sugar mills in India with an average installed capacity of 18 million tonnes. India is the largest consumer of sugar and second largest producer in the world. Ninety per cent of total sugarcane and sugar production in the country is with Maharashtra and Uttar Pradesh alone, accounting for 60 per cent of Indian's total sugar production. As against an average annual rise of 2.5 per cent in world sugar production during the past ten years, global sugar consumption has grown by about two per cent per annum, while in India the consumption has been higher at about 3.5 per cent per annum. Uttar Pradesh is the largest sugarcane producing state in the country (Table 5.19). It contributes 44 per cent to the country's total sugarcane production and is also the largest producer of sugar in the country. Gur and Khandsari are among the major agro processing industries found in rural sector of our country. Nearly 50 per cent of total sugarcane produced in the country is used for manufacture of about eight million tonnes of gur, which is known as the most nutritious agent among all sweeteners (Madan, 2004). Even today, this sector utilizes about 45 to 50 per cent of the total sugarcane grown in the country and provides employment to about 2.5 million people. It is therefore, imperative to expand this sector because, it provides higher food value jaggery and khandsari at lower cost, it boosts-up the rural economic system, the transportation cost of raw material is low and there is no need of highly technical machinery and labour (Baboo and Solomon, 1995). Jaggery and Khandsari, which still dominate in the preparation of Indian cookies, are of ancient origin. Gur making plants are generally tiny in size, and the machineries of the plants are fabricated by local artisans or engineering workshops.Regarding sugar production, the process of separating sugar from the sugarcane is accomplished through two steps: sugar mill crushing and sugar refinery extraction. Sugarcane is initially processed into raw sugar at mills nearer to the cane fields because, cane is bulky and relatively expensive to transport and it must be processed as soon as possible to minimize sugar deterioration. The raw sugar is then shipped to refineries to produce refined sugar. The final products of refining include powdered, granulated and brown sugar, which is sugar that contains some molasses. Maharashtra ranks first (143) in having largest number of sugar mills followed by Uttar Pradesh (Tables 5.20 to 5.22).

Table 5.19: State-wise production and consumption of sugarcane by sugar mills in India (Lakh MT).

States	*2004-05*	*2005-06*	*2006-07*	*2007-08*	*2008-09*	*2009-10*
A. Production						
Andhra Pradesh	157	177	217	203	154	117
Bihar	41	43	60	39	50	50
Gujarat	146	146	156	152	155	134
Karnataka	143	183	287	262	233	291
Maharashtra	205	389	786	884	606	566
Madhya Pradesh	21	24	28	32	30	24
Punjab	52	49	60	67	47	37
Rajasthan	3	5	6	6	4	3
Tamil Nadu	234	351	411	381	328	318
Uttar Pradesh	1187	1255	1339	1247	1090	1087
Uttarakhand	64	61	61	77	56	58
India	2371	2812	3555	3482	2850	2778
B. Consumption by sugar mills						
Andhra Pradesh	92	123	173	132	60	55
Bihar	26	44	52	36	24	27
Gujarat	74	108	134	128	94	113
Karnataka	103	179	252	267	161	240
Maharashtra	195	446	798	761	400	614
Madhya Pradesh	7	10	19	18	6	9
Punjab	32	37	51	58	26	21
Tamil Nadu	115	232	275	230	166	143
Uttar Pradesh	514	608	895	747	455	567
Uttarakhand	40	45	56	41	24	32
India	1248	1887	2792	2499	1450	1855

Source: <www.indiastat.com>

2. Cotton

The cotton gin is, where cotton fiber is separated from the cotton seed. The first step in the ginning process is, when the cotton is vacuumed into tubes that carry it to a dryer to reduce moisture and improve the fiber quality. Then it runs through cleaning equipment to remove leaf trash, sticks and other foreign matter. Ginning is accomplished by one of two methods. Cotton varieties with shorter

Table 5.20: Selected state-wise number of working sugar mills in India (2006-2007 to 2009-2010).

States/UTs	2006-2007	2007-2008	2008-2009	2009-2010
Punjab	16	16	16	15
Haryana	14	14	15	14
Rajasthan	1	1	1	1
Uttarakhand	10	10	10	10
Uttar Pradesh	133	132	132	128
Madhya Pradesh	8	8	9	11
Chhattisgarh	1	1		
Gujarat	18	18	18	18
Maharashtra	163	172	147	143
Bihar	9	8	9	9
Assam	0	0	–	–
Odisha	5	5	–	–
West Bengal	1	1	–	–
Nagaland	0	0	–	–
Andhra Pradesh	38	38	35	35
Karnataka	47	51	50	54
Tamil Nadu	39	39	38	42
Pondicherry	2	2	–	–
Kerala	0	0	–	–
Goa	1	1	–	–
India	504	516	488	490

Source: <www.indiastat.com>

Table 5.21: Selected state-wise installed capacity of sugar mills in India (Lakh MT).

Year	States									India
	Andhra Pradesh	Bihar	Gujarat	Karna-taka	Mahara shtra	Madhya Pradesh	Punjab	Tamil Nadu	Uttar Pradesh	
1995-96	6.79	4.12	8.54	8.37	39.87	0.99	6.25	12.63	34.79	128.46
1996-97	6.79	4.12	8.54	8.37	39.87	0.99	6.25	12.63	34.79	128.46
1997-98	7.49	4.01	8.96	8.37	43.00	0.99	6.25	12.73	36.39	134.59
1998-99	8.35	4.27	9.60	8.93	45.86	0.99	6.25	13.12	38.99	142.25
1999-00	8.05	4.05	10.00	9.00	40.23	0.80	6.25	14.63	38.99	146.28
2000-01	8.05	4.05	10.00	9.20	47.17	0.80	5.25	14.63	38.99	148.72
2001-02	8.05	4.05	10.00	9.20	47.17	0.80	5.25	14.63	38.99	148.90
2002-03	8.12	3.19	9.64	12.84	65.73	1.45	6.49	15.24	40.60	176.85
2003-04	9.08	4.84	10.71	14.46	69.35	1.54	6.84	15.24	44.11	188.03
2004-05	9.08	4.85	10.71	14.60	69.78	1.54	6.84	15.24	45.37	189.81
2009-10	9.73	4.85	10.79	20.48	75.30	2.50	6.84	19.03	74.27	236.25

Source: <www.indiastat.com>

Table 5.22: Selected state-wise number of mills and production of sugar in India.

States	*1995-96*	*1996-97*	*1997-98*	*1998-99*	*1999-00*	*2000-01*	*2001-02*
Manufacturing units (Nos)							
Andhra Pradesh	33	34	35	36	35	35	35
Bihar	19	19	11	11	10	10	10
Gujarat	16	17	18	17	18	17	15
Karnataka	29	28	28	33	34	35	36
Maharashtra	109	105	95	119	123	137	135
Madhya Pradesh	8	7	7	7	7	8	8
Punjab	21	20	21	21	21	22	22
Rajasthan	3	3	3	3	2	2	2
Tamil Nadu	36	36	37	38	39	37	36
Uttar Pradesh	116	117	120	115	109	100	100
India	416	412	400	400	423	436	434
Production of sugar (000' MT)							
Andhra Pradesh	859	772	782	1113	1182	–	–
Bihar	382	362	299	258	368	–	–
Gujarat	1130	967	890	1025	1141	–	–
Karnataka	1263	871	959	1372	1577	–	–
Maharashtra	5394	3445	3847	5337	6503	–	–
Madhya Pradesh	125	86	68	82	103	–	–
Punjab	633	613	331	311	421	–	–
Rajasthan	31	24	29	23	14	–	–
Tamil Nadu	1671	1083	1266	1266	1734	–	–
Uttar Pradesh	4379	4083	3922	3729	4556	–	–
India	16451	12905	12855	15538	18200	–	–

Source: <www.indiastat.com>

staple or fiber length are ginned with saw gins. This process involves the use of circular saws that grip the fibers and pull them through narrow slots. The seeds are too large to pass through these openings, resulting in the fibers being pulled away from the seed. Long fiber cottons must be ginned in a roller gin because, saw gins can damage their delicate fibers. The roller gin was invented in India centuries ago and this concept is still used in modern gins. A roller gin uses a rough roller to grab the fiber and pull it under a rotating bar with gaps too small for the seed to pass. The raw fiber, now called lint, makes its way through another series of pipes to a press, where it is compressed into bales (lint packaged for market). Tamil Nadu (803) ranks first in having large number of cotton/man made fiber textile mills in India followed by Maharashtra and Andhra Pradesh (Tables 5.23 to 5.25).

Table 5.23: State-wise number of cotton/man made fiber textile mills functioning in India (As on 31.05.2011).

States/UTs	*Number of Mills*	*States/UTs*	*Number of Mills*
Andhra Pradesh	119	Assam	0
Bihar	0	Chhattisgarh	0
Dadra and Nagar Haveli	11	Daman and Diu	0
Goa	1	Gujarat	44
Haryana	31	Himachal Pradesh	16
Jammu and Kashmir	1	Jharkhand	1
Karnataka	21	Kerala	20
Madhya Pradesh	38	Maharashtra	165
Manipur	0	Odisha	1
Pondicherry	10	Punjab	80
Rajasthan	38	Tamil Nadu	803
Uttar Pradesh	20	Uttarakhand	7
West Bengal	16		
	India	1443	

Source: <www.indiastat.com>

Table 5.24: Growth of cotton/man-made fiber textile mills in India.

March End	*Number of Mills*			*Spindles Installed (million)*	*Rotors Installed (Thousand)*	*Looms Installed (Thousand)*
	Spinning	*Composite*	*Total*			
1995	1148	268	1416	30.7	180	139
1996	1294	275	1569	31.75	226	132
1997	1438	281	1719	33.15	276	124
1998	1504	278	1782	33.88	313	124
1999	1543	281	1824	34.72	383	123
2000	1565	285	1850	35.1	392	123
2001	1565	281	1846	35.53	394	123
2002	1579	281	1860	35.75	409	123
2003	1599	276	1875	36.1	379	119
2004	1564	223	1787	34.02	383	88
2005	1566	223	1789	34.24	385	86
2006	1570	210	1780	34.14	395	73
2007	1608	200	1808	35.61	448	69
2008	1597	176	1773	35.01	461	56
2009	1653	177	1830	37.03	485	57
2010	1673	180	1853	37.68	494	57
2011	1757	183	1940	–	–	–

Source: <www.indiastat.com>

Table 5.25: Capacity utilization of cotton/man-made fiber textile mills in India.

Year	*Installed Spindles Nos. (in Million)*	*Installed Rotors Nos. (in Thousand)*	*Percentage Utilization*	*Installed Looms Nos. (in Thousand)*	*Percentage Utilization*
1995-96	31.75	226	86	132	52
1996-97	34.59	276	86	124	56
1997-98	35.39	340	86	124	54
1998-99	36.67	434	79	123	51
1999-00	37.08	444	83	123	51
2000-01	37.91	454	85	123	47
2001-02	38.33	480	82	123	42
2002-03	39.03	468	80	119	41
2003-04	37.03	482	83	88	53
2004-05	37.47	500	87	86	58
2005-06	37.51	520	88	73	61
2006-07	39.5	601	89	69	62
2007-08	39.07	621	90	56	61
2008-09	41.34	570	88	56	55
2009-10 (P)	42.04	674	90	57	62

Source: <www.indiastat.com>

d. Fruits and Vegetables

India is the world's second largest producer of fruits and vegetables. Joint efforts of R&D institutions, farmers, Government agencies etc. has resulted in India emerging as a major producer of fruits and vegetables in the world. It has potential to grow all types of temperate, sub-tropical and tropical fruits and vegetables because of varied agro-climatic diversity. The total production of fruits is 71.52 m tonnes and vegetables is 133.74 m tonnes in the year 2009-10. The horticultural crops in the country presently cover 13.6 million hectares of arable land, *i.e.* seven per cent of the gross cropped area and contribute 18-20 per cent of the gross value of India's agricultural output. India accounts for 11 and 17 per cent share in global productions of fruits and vegetables respectively. On an average, the post-harvest losses are estimated to the extent of 20-30 per cent due to lack of proper harvesting, processing, transportation and storage facilities, which is valued at ₹ 230 billion (Figure 5.11). The growth in post harvest sector has not kept pace with the production of fruits and vegetables. The processed products from fruits and vegetables include beverages, jams, jellies, candies, preserves, canned fruits and vegetables, dehydrated fruits and vegetables, pickles, soup mixes, sauces and ketchups. Products that have growing demand, especially in the Middle East countries include pickles, chutneys, fruit pulps, canned fruits and vegetables, concentrated pulps and juices, dehydrated vegetables and frozen fruits and vegetables. The new arrivals in this segment are vegetable curries, canned mushroom and mushroom products, dried fruits and vegetables and fruit juice concentrates. The primary processing of fruits and vegetables is almost 100 per cent in all the countries, but the extent of secondary processing varies significantly across the countries and it is as low as 1.3 per cent in India (Figure 5.12). People generally prefer fresh fruits and vegetables in India due to abundance of seasonal fruits throughout the year available at low price. The production of pickles and chutneys has

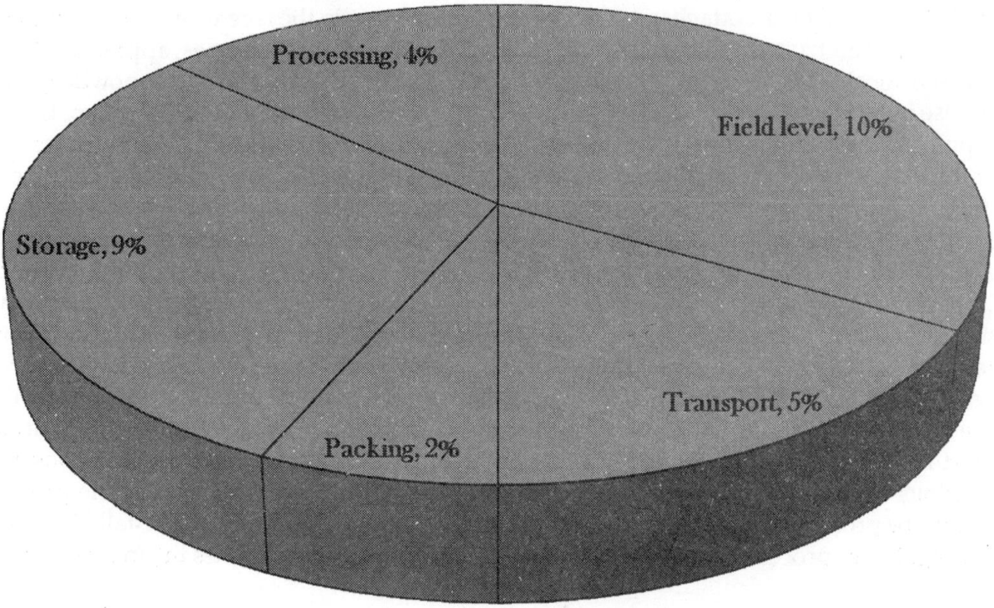

Figure 5.11: Average post-harvest losses (%) of fruits and vegetables at various stages of marketing.

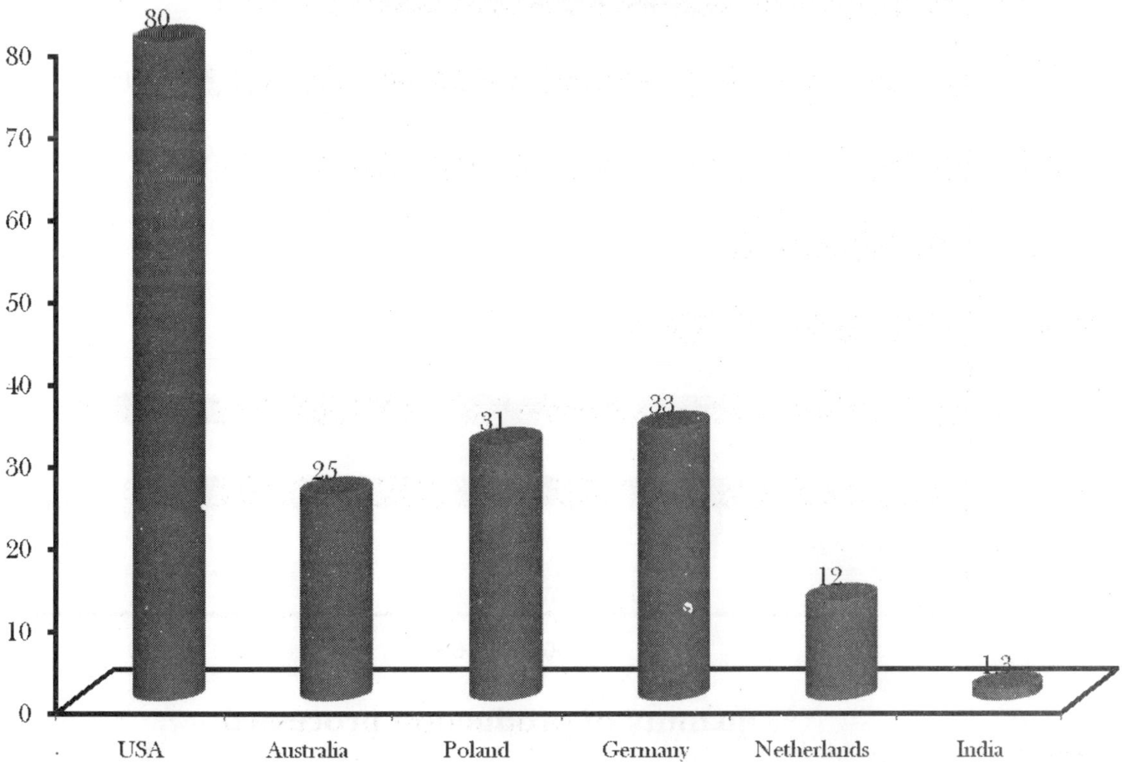

Source: <www.mofpi.nic.in>

Figure 5.12: Level of secondary processing (%) in fruits and vegetables across different countries.

traditionally been rural level cottage industrial activity. However, in the recent years, processed foods in the form of canned fruits such as pineapple, mango slices and pulps, grapes, apple, peaches etc., have increased considerably. The uses of fruits in the form of concentrated juice, dry powder, jam and jelly have also increased. The percentage production of processed fruits and vegetables are fruit juice and fruit pulp (27), jams and jellies (10), pickles (12), ready to serve beverages (13), synthetic syrups (8), squashes (4), tomato products (4), canned vegetables (4) and others (18). Presently, a mere 2.2 per cent of total fruits and vegetables production are processed, even as the country ranks second in the world in terms of their production. This is comparatively low when compared to other countries (Figure 5.13) like Thailand (30 per cent), Brazil (70 per cent), USA and UK (65 per cent), Philippines (78 per cent), China (23 per cent), Malaysia (83 per cent) and South Africa (80 per cent). Even, within the country, the processing level of fruits and vegetables (2.2 per cent of total production) is comparatively low when compared to milk (35 per cent), marine products (26 per cent), meat (21 per cent) and poultry products (6 per cent) and these processing standards are significantly lower by international standards (Figure 5.9). More importantly, the lack of processing and storage of fruits and vegetables results in huge per cent of post-harvest losses and these vary across the commodities (Table 5.26) and these losses also vary at various stages of marketing process (Figure 5.11). So, the fruits and vegetable processing industry must be promoted for minimization of post-harvest losses. No doubt, the installed capacity of fruits and vegetables processing industry has doubled from 1.1 million tonnes in January 1993 to 2.1

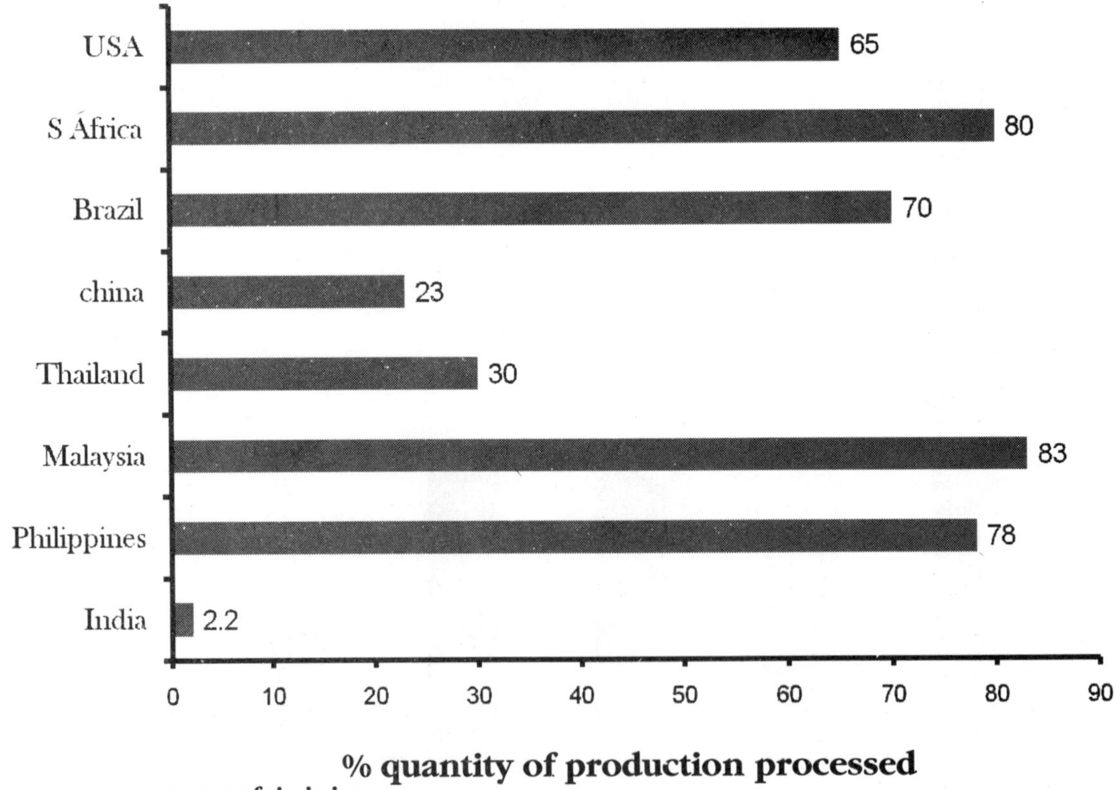

Source: <www.mofpi.nic.in>

Figure 5.13: Percentage of quantity of production of fruits and vegetables processed in different countries.

Table 5.26: Extent of post-harvest losses (per cent) of selected fruits and vegetables at All-India level.

Sl.No.	*Item*	*Per cent of Post-harvest Losses*
1.	Banana	20-80
2.	Mango	17-36
3.	Citrus	20-95
4.	Guava	10-15
5.	Papaya	40-100
6.	Apple	14
7.	Grape	20-25
8.	Tomato	5-50
9.	Onion	25

Source: National Horticultural Board, 2004.

million tonnes in 2006, but the fruits and vegetable processing industry is rather fragmented. A large number of units are in household and small-scale sector, having low capacities of up to 250 tonnes per annum. From the year 2000 onwards, the industry has seen a significant growth (Tables 5.27 and 5.28) in ready-to-serve beverages, pulps and fruit juices, dehydrated and frozen fruits and vegetable products, pickles, processed mushrooms and curried vegetables, and units engaged in these segments are export oriented. This processing industry has been facing problems of low capacity utilization, technological obsolescence and marketing. It has to work under

Table 5.28: Fruits and Vegetables processing licensed industries

Year	*Number*
1990	3629
1991	3846
1992	3925
1993	4057
1994	4132
2004	5198
2010 (Projected)	6200

Source: www.indiastat.com

Table 5.27: State-wise number of fruits and vegetables processing units and their installed capacity in India (As on 01.04.2007).

State/UTs	*Number*	*Installed Capacity (m tonnes)*
Maharashtra	865	283430
Goa	38	8180
Madhya Pradesh	95	40590
Chhattisgarh	4	420
Gujarat	234	120870
Dadra & Nagar Haveli	6	480
Assam	44	8260
Arunachal Pradesh	2	60
Manipur	14	2510
Meghalaya	10	390
Mizoram	3	885
Nagaland	7	1920
Sikkim	3	770
Tripura	6	2770
Delhi	181	43510
Chandigarh	15	4620
Rajasthan	111	20100
Haryana	176	101540
Jammu & Kashmir	102	20330
Punjab	240	121410
Himachal Pradesh	130	91625
Uttar Pradesh	497	138780
Uttrakhand	108	29510
Andaman & Nicobar Islands	3	200
Bihar	39	12440
Jharkhand	23	4160
Odisha	17	11310
West Bengal	267	109341
Andhra Pradesh	237	887658
Karnataka	247	208904
Kerala	368	41340
Pondicherry	9	1680
Tamil Nadu	432	360826
India	4533	2680819

Source: <www.indiastat.com>

the constraints of high fluctuations in raw material quality, fluctuations in market prices, poor technology for handling and storage, inadequate R&D support for product development, high cost of energy and uncertainty in availability of adequate quantity for processing purposes, inadequate and expensive cold chain facilities and varying requirement of processing conditions from one material to another. Future R&D has to focus on the issues of economically producing value-added products and product diversification, besides the issues mentioned above.

Major Players in Fruits and Vegetables Processing

The following are the major players (Table 5.29) in fruits and vegetables processing sub-sector in India

Table 5.29: Major players in Fruits and Vegetables processing sub-sector.

Company	*Brands*	*Products*
Hul	Kissan, Knorr, Annapurna	Jams, Ketchups, Wheat Flour, Fruit Beverages, Soups
Dabur India	Real, Real Activ, Coolers	Fruit Beverages
Mother Dairy (Safal)	Safal	Frozen processed fruits and vegetables, Jam, Pickel
Temptation Foods	Pure Temptation	IQF fruits and vegetables
Godreg F&B Capital Foods	Private Label	Fruit Juices, Fresh Fruits & Vegetables (Retail), Frozen Foods
Mafco	Mafco	Frozen Fruits and Vegetables
Priya Foods	Priya	Pickels, Fruit Juices
MTR Foods	Mtr	Frozen Foods, Pickels, Spices & Masala
Allana Cold Storage	Allana	Frozen Foods

Source: Nisha (2008).

e. Milk

India is with highest livestock populations in the world, it accounts 50 per cent of the buffaloes and 20 per cent of the world's cattle population, most of which are milch cows and milch buffaloes. India has a large livestock population base constituting 278 million livestock including 180.5 million cattle, 82.8 million buffaloes, four million sheep and 9.2 million goats. The livestock population is projected to increase to 322 million by the year 2015. India's dairy industry is considered as one of the most successful development industry in the post-Independence era. India has emerged as the largest milk producing country in the world with present level of annual milk production estimated as 112.5 million tonnes in the year 2009-10. The expected production level is 135 million tonnes by the year 2015. Annual milk production in India has grown more than six times since independence. The average annual growth rate in the production of milk in recent years has been close to four per cent. Even though the level of per capita availability at 263 gram/day for India in 2009-10 is much lower than that in developed countries, it is well above the developing country average. India's milk production will continue to grow at about three per cent per annum in spite of difficulties due to stagnant livestock herd size and shortage of fodder. During 1993-2005, the dairy industry recorded an annual growth of four per cent, which is almost three times the average growth rate of the dairy industry in the world. Due to increasing population, per capita availability of milk will increase by only about 1.5 per cent per annum. For an economy growing at about six per cent per annum, this increase in availability will be grossly inadequate. The Eleventh FYP envisages an overall growth of 6-7 per cent per annum for the sector.

India has a unique pattern of production, processing and marketing/consumption of milk, which is not comparable with any large milk producing country. Approximately 70 million rural households in the country are engaged in milk production. Over 11 million farmers are organized into about 0.1 million Village Dairy Co-operative Societies. About 80 per cent of milk produced in the country is still handled in the unorganized sector and only the remaining 20 per cent is equally shared by co-operatives and private dairies. About 35 per cent of milk produced in India is processed (Table 5.30). The organized sector (large scale dairy plants) processes about 13 million tonnes annually, while the unorganized sector processes about 22 million tonnes per annum, while remaining is either consumed at farm level or sold as fresh, non-pasteurized milk through unorganized channels. In an organized dairy industry, dairy co-operatives account for the major share of processed liquid milk marketed in India. As in the case of fruits and vegetables, the level of primary processing of milk across the developed countries is almost 100 per cent, but the extent of secondary processing (Figure 5.14) differs significantly and it is quite low in India (27 per cent). Milk is processed and marketed by 170 Milk Producers' Co-operative Unions, which include 15 State Co-operative Milk Marketing Federations. Over the years, several brands have been created by co-operatives like Amul (Gujarat), Vijaya (Andhra Pradesh), Verka (Punjab), Saras (Rajasthan). Nandini (Karnataka), Milma (Kerala) and Gokul (Kolhapur). The milk surplus states in India are Uttar Pradesh, Punjab, Haryana, Rajasthan, Gujarat, Maharashtra, Andhra Pradesh, Karnataka and Tamil Nadu and the manufacturing of processed milk products is very much concentrated in these states due to the availability of milk in huge quantity. According to the MOFPI, exports of dairy products have been growing at the rate of 25 per cent per annum in terms of quantity and 28 per cent in terms of value since 2001. Significant investment opportunities exist for the manufacturing of value-added milk products like milk powder, packaged milk, butter, ghee, cheese and ready-to-drink milk products.

Table 5.30: Levels of processing (per cent) of food produce in India.

Products	*Level of Processing (Per cent)*		
	Organized	*Unorganized*	*Total*
Fruits and vegetables	1.4	0.8	2.2
Milk and milk products	13	22	35
Buffalo meat	21		21
Poultry	6		6
Marine products	8		8

Processed Dairy Products

☆ ***Cheese:*** The organized cheese market including its variants like processed cheese, mozzarella, cheese spreads, flavored and spiced cheese, is valued at around ₹ 4.5 billion. Processed cheese at 60 per cent of the overall market is ₹ 2.7 billion. The next most popular variant is cheese spread claiming a share of around 30 per cent of the total processed cheese market. The market is primarily an urban phenomenon and is known to be growing at around 15 per cent. The market for cheese cubes, slices and tins is growing. The flavored cheese segment has been constantly declining. Gujarat Co-operative Milk Marketing Federation (GCMMF) with the Amul brand continues to be the main operator in the branded cheese market in India. It pioneered the market for processed, branded cheese. GCMMF has developed the technology to make cheese from buffalo milk, while world over, it is made from cow milk.

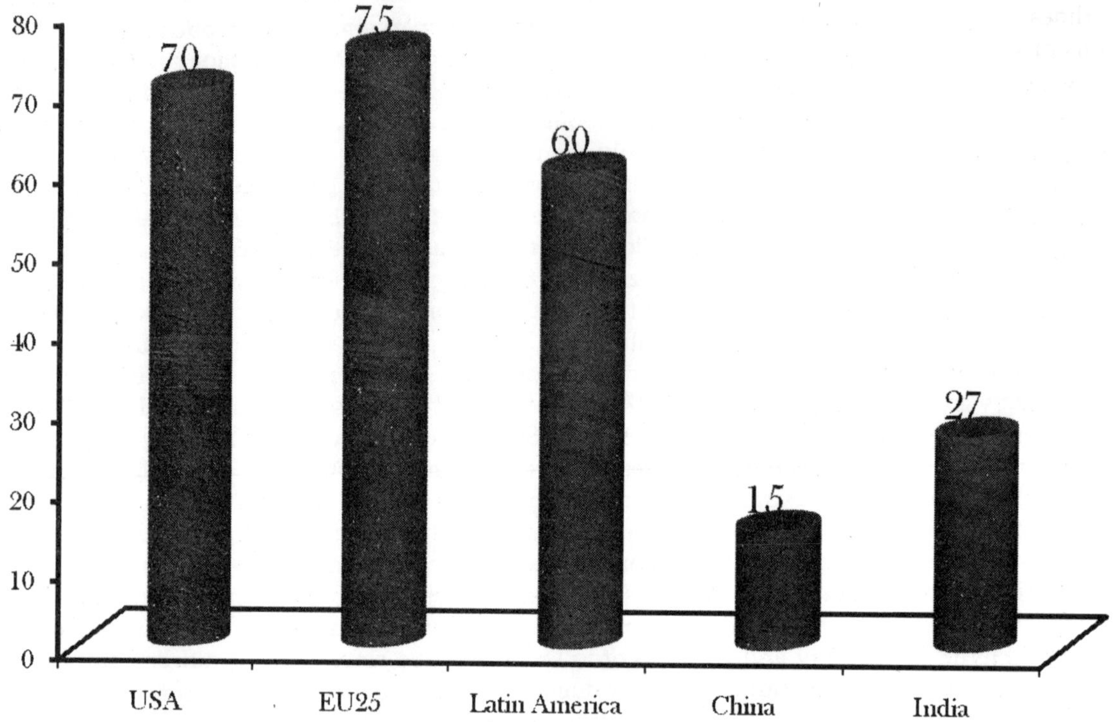

Source: <www.mofpi.nic.in>

Figure 5.14: Level of secondary processing (per cent) in milk across different countries.

Britannia Industries joined the fray in the cheese market in mid-1990s through an arrangement with Dynamix Dairy Industries.

☆ ***Processed cheese:*** It is a food product made from normal cheese. Many flavours, colours, and textures of processed cheese exist. Processed cheese has several technical advantages over unprocessed cheese, including extended shelf-life, resistance to separation when cooked, and uniformity of product. Its production also enjoys significant economic advantages over traditional cheese making processes, most often through the ability to incorporate any of a wide variety of less expensive ingredients.

☆ ***Dairy whiteners:*** The market for dairy whiteners (commercially knows as beverage milk powders and condensed milk) and creamers is around ₹ 3,000 millions. Apart from MNCs like Nestle and companies like Britannia, the Indian enterprises have also made perceptible progress. Names like Amul, Sapan, Vijaya, Mohan, Parag and several others have been seen in the market place with their whiteners. These are available mostly in pouches, tetra packs and in the near future, may be in miniportion cups.

Since, India is a major consumer of tea and coffee, it would be a very large market, if only the price was not a constraint. In addition to domestic consumption, the whiteners/creamers find a high level of institutional acceptance, especially by railways, hotels and restaurants, airlines, hospitals and nursing homes and corporate offices. The institutional market can be tapped first, in particular, the airlines, railways and hotels. The penetration can then be extended to the household sector. Milk and milk products account for a significant 17 per cent of India's total expenditure on food. The potential for exports, especially to neighbouring countries and the countries in the Middle East, the Gulf and Africa, also exist and could be exploited.

Major Players in Dairy Industry

The packaged milk segment is dominated by the dairy co-operatives. GCMMF is the largest player. All other local dairy co-operatives have their local brands (For example, Gokul, Warana in Maharashtra, Saras in Rajasthan, Verka in Punjab, Vijaya in Andhra Pradesh, Aavin in Tamil Nadu etc). Other private players include J.K. Dairy, Heritage Foods, Indiana Dairy, Dairy Specialties etc. The following (Table 5.31) are the major players in dairy industry in India:

Table 5.31: Major players in dairy industry in India.

Company	*Brands*	*Major Products*
Nestle India	Milk Maid, Cerelac, Lactogen, Milo, Everyday	Sweetened Condensed Milk, Malted foods, Milk Powder and Dairy Whitener
Milk Food	Milk Food	Butter, Ghee, Milk Powder, Ice Cream and other Milk Products
Glaxo Smith Kline Consumer Healthcare	Horlicks, Boost, Maltova, Viva	Malted Milkfood, Ghee, Butter, Powdered Milk, Milk Fluid and other Milk Based Baby Foods
Kwality Dairy (India)	Indana, Kream Kountry	Skimmed Milk Powder, Whole Milk Powder, Dairy Milk Whitener, Ghee
GCMMF	Amul	Milk, Butter, Cheese, Ghee, Ice Cream and other Milk Products
H.J. Heinz	Farex, Complan, Glactose, Bonniemix, Vitamilk	Infant Milkfood, Malted Milkfood
Heritage Foods	Heritage	Milk, Curd, Ghee, Buttor Milk
Britannia	Milkman	Flavoured Milk, Cheese, Milk Powder, Ghee
Cadbury	Bournvita	Malted Food
Mother Dairy	Mother Dairy	Milk, Ice Cream, Milk Products

Source: Nisha (2008).

f. Poultry and Meat

India ranks first in world's cattle population, 50 per cent of buffalo population and one-sixth of total goat population of the world. There is vast scope to set up modern slaughter facilities and cold store chains in meat and poultry processing sector. The market has not been tapped for RTE and semi processed meat products in the domestic market as well as for exports to neighboring countries especially to the Middle East.

Since 1995, production of meat and its products has been significantly growing at a rate of four per cent per annum. Presently, the processing level of buffalo meat is estimated at 21 per cent, poultry is estimated at six per cent, while marine products are estimated at eight per cent. But, only about one per cent of the total meat is converted into value added products like sausages, ham, bacon, kababs, meatballs, etc. Processing of meat is licensed under the Meat Food Products Order, 1973. Presently, the country has more than 3,600 slaughterhouses, nine modern abattoirs and 180 meat-processing units licensed under the Meat Products Order. India's meat industry took a step forward, when the

Government of India launched a National Meat and Poultry Processing Board (NMPPB), an autonomous body on 19th February 2009. This Board serves as a 'single window service provider' for producers/manufacturers and exporters of meat and meat products, in order to promote and regulate the meat industry, as a whole. The Board will also help industry to utilize slaughterhouse waste materials, set up quality control laboratories for meat and meat products, promote meat manufacturers to adopt Good Manufacturing Practices (GMP), Hazard Analysis and Critical Control Points (HACCP) systems, ISO-9001: 2000 standards, help industry to create and disseminate data, train workers and technicians and work as a central and National hub to address meat-related issues.

Approximately, 70 per cent of the Indian population consumes meat and/or poultry products. Of the total production, the percentage of meat produced from different sources are: cattle and buffaloes - 40 per cent; sheep and goat - 26 per cent; poultry 17 per cent; pig - 16 per cent; and others - one per cent. However, slaughter rates in relation to the population of the animals are: 1.45 per cent in case of cattle; 3.45 per cent in case of buffaloes; 32.5 per cent in case of sheep; 35.58 per cent in case of of goat; 26.21 per cent in case of pigs. India's meat production is just two per cent of the world's total. Despite the potential of its large livestock population, India's meat industry has not taken its due share. Though India has acquired number one status in the world contributing around 15 per cent of global milk production in 2007-08, meat production is comparatively low at just two per cent of the world's total. The processing status is one of the major key issues of concern in the meat sector of India. Quality and hygiene levels are very low along with tremendous waste of meat, contamination/deterioration of meat and there are concerns over animal welfare. Currently, about one per cent of the meat produced is converted into value added products and most meat is purchased by the consumers in the country in the fresh/frozen form and converted into various meat products at homes, restaurants, etc. The maximum conversion takes place only in case of pork products.

Poultry industry is also among the faster growing sectors rising at a rate of eight per cent per year. There are eight integrated poultry processing units in the country, which of course hold a significant share in the industry. It is believed that, the Indian poultry industry is 5,000 years old, but it is only in the last four decades that it has begun to witness remarkable growth from backyard to poultry industry. India is the fifth largest producer of eggs and ninth largest producer of poultry meat in the world. The poultry meat production is gaining ground in the country. It is estimated that, over ₹ 1000 crores of capital was employed in the poultry processing in the country. Compared with meat, poultry industry has registered significant growth. The organized sector of poultry industry contributes nearly 70 per cent of total output, while the remaining 30 per cent comes from the unorganized sector. The broiler industry is well dominated in southern states in our country with nearly 60 to 70 per cent of total output coming from these states. The layer industry once again is represented more in southern states especially, Andhra Pradesh, Tamil Nadu and Maharashtra, which produce nearly 70 per cent of the country's total egg output. Three-quarter of India's egg production is consumed by just 25 per cent of the population living in urban and semi-urban areas. Presently, there are only five egg powder plants in India, which is considered insufficient in view of growing export demand for different kind of powder - whole egg, yolk and albumen. The years 2008 and 2009 witnessed a major set-back for the chicken industry which was on the growth graph in the past 10 to 15 years due to the bird flu influenza. Presently, 100 per cent of FDI is permitted in the food processing sector. FDI in food retailing covering dairy, poultry, marine, vegetables and fruits might help the entire FPI to grow. Although India has abundant raw materials, a vast network of manufacturing facilities all over the country, and a large and skilled labour force, further development is hampered by poor infrastructural facilities, inadequately developed links between R&D labs and industry and a lack of awareness about hygienic practices. As a result, the Indian meat and poultry market stands in a low position in respect to the global markets. The Ministry is exploring opportunities like setting of mega food parks, modern abattoirs, improvised processing units, providing added incentive to develop existing projects while taking care of the changing consumption patterns. Further, the meat consumers are becoming more and more aware and demanding for meat and poultry products processed in clean and sanitary environment and 'convenience items' such as, semi cooked, RTE, RTC meat food products. So, this separate body (NMPPB) will cater to these upcoming issues. It also creates awareness regarding utilization of animal waste materials for processing into by-products.

Indian consumers prefers to buy freshly cut meat, rather than processed or frozen meat. A mere six per cent of production of poultry meat is sold in processed form. Of this, only about one per cent undergoes processing into value added products (RTE/RTC). Processing of large animals is largely for the purpose of exports. This is because of consumers' preference for fresh meat. The total meat processing capacity in India is over one million tonnes per annum of which, 40-50 per cent is utilized. The total processed meat production in India is likely to double in the next ten years and has a huge potential with the growing number of fast food outlets in the country. With the rise in per capita incomes and busy lifestyles, the demand for processed meat products, which can be quickly cooked, has been rising. Most of the production of meat and meat products continues to be in the unorganized sector. Branded products like Venky's and Godrej's Real Chicken are, however, becoming popular in the domestic market. The following (Table 5.32) are the major players in the poultry and meat industry.

Table 5.32: Major players in the poultry and meat industry in India.

Company	*Brands*	*Products*
Arambagh Hatcheries	Arambagh	Meat, Poultry
Hind industries	Sibaco, Eatco	Frozen buffalo meat, Frozen sheep and goat meat
Venkatechwara Hatcheries	Venky's	Poultry products
Alkabeer Exports Limited	Alkabeer	Frozen buffalo meat, chilled
ASF Sea Food	ASF Sea Foods	Sea food
Bell foods	Bell Foods	Marine foods
Frigo Refico Allana	Allana	Frozen buffalo and other meat
Godrej Agrovet	Real Good Chicken	Poultry products
MAFCO, Mumbai	MAFCO	Pork and other meat products

Source: Nisha (2008).

7. *Fish*

India is third largest producer of fish (7.85 million tonnes in the year 2009-10) in the world. Categorically India's potential for fishes, from both inland and marine resources, is supplemented by the 8,000 km coastline, 3 m. ha. of reservoirs, 50,600 sq km of continental shelf area, 1.4 m. ha of brackish water and 2.2 mn sq km of exclusive economic zone. In India, nearly 10 million people, living in 4,000 coastal villages and more number of interior villages depends on fisheries sector. With regards to infrastructure facilities for processing of marine products, there are more than 375 freezing units and 504 frozen storage facilities for safe storage with a capacity of 138,229 tonnes, besides there are 11 surimi units, 473 pre-processing centres and 236 other storages. Processed fish products for export include conventional block frozen products, individual quick frozen products, minced fish products like fish sausage, cakes, cutlets, pastes, surimi, texturised products and dry fish etc. Exports of marine products have been inconsistent and on a declining trend, which can be owed to the adverse market conditions prevailing in the European and American markets. The anti-dumping procedure initiated by the US Government has affected India's shrimp exports to the US. The marine products exports account for approximately 1.1 per cent of the total exports from India.

Fish processing in India is done almost entirely for export purposes. At present, India has freezing units, cold stores, ice plants, canning units and fish meal plants. Capacity of most of these processing and storage units is small, when compared to the facilities in fish processing industry in technologically advanced countries. The total fish processing and storage facilities in India grossly is inadequate compared to the potential for fish production and processing. Inland fisheries need low cost palletized feeds and special containers to transport fingerlings and fish. More rearing ponds are needed. Techniques to reduce seepage loss of water have to be introduced. Obsolete fishing gear needs replacement with better gear. Extensive network of refrigerated handling, transport, storage and retailing, better fish waste and by-product use need to be developed.

Processing of fish into canned and frozen forms is carried out almost entirely for the export market. It is widely felt that, India's substantial fishery resources are under-utilized and there is tremendous potential to increase the output of this sector. The potential could be gauged by the fact that, against fish production potential in the exclusive economic zone of 3.9 million tonnes, actual catch is to the tune of 2.87 m tonnes. Harvesting from inland sources is around 2.7 m tonnes. In the last five years, there was substantial investment in fisheries to the tune of ₹ 3,000 crores of which foreign investments were of the order of ₹ 700 crores. The following (Table 5.33) are the major players in the marine and fish industry.

Table 5.33: Major players in the marine and fish industry in India.

Companies	*Key Products*
Allanasons	Pomfrets, Seer Fish,Squids,Prawns and Cuttle Fish
ASF Sea foods	Seafood
Bells Foods Marine Division	Crab, Cuttlefish, Shrimps, Squid,Fish Octopus
Deep Sea Products	Marine Products
IFB Agro Pvt. Ltd.	Pomfrets,Crabs,Prawns and Sea Food
Sea Sparkle	Octopus, Squid, Crabs and Tuna
Sumero	Pomfrets, Crabs, Prawns and Sea Food

Besides above sub-segments, the food-processing industries also deals with consumer foods. The consumer food industry mainly consists of RTE and RTC food products (Table 5.34), salted snacks, chips, cocoa based products, bakery products, biscuits, soft drinks etc. The bakery industry is among the few processed food segments, whose production has been increasing consistently in the country in the past few years. Products of bakery include bread, biscuits, pastries, cakes, buns, rusk etc. This activity is mostly concentrated in the unorganized sector. Bread and biscuits constitute the largest segment of consumer foods with an annual production of around 4.00 million tonnes. Bread manufacturing is reserved for the small-scale sector. Out of the total production of bread, 40 per cent is produced in the organized sector and remaining 60 per cent in the unorganized sector. In the production of biscuits, the share of unorganized sector is about 80 per cent. In the basket of consumer goods, RTE foods market in India is expected to reach ₹ 2,900 crores by 2015 from ₹ 128 crores (in 2006). The factors contributing to this growth would be changes like cold chain development, streamlining of taxation, economies of scale on the supply side, coupled with increasing disposable incomes, diminishing culinary skills and the rising need for convenience on the demand side. However, the RTE foods market in India has remained under-penetrated owing to factors like consumers' penchant for freshness, low affordability and the Indian housewife's preference for home cooked food. So, this RTE foods industry in India has not experienced significant growth due to inadequate demand arising from low household incomes and consumer preference for fresh and home-cooked food. There is thus, a huge untapped market opportunity arising due to rapid demographic shifts in income, urbanization and proportion of urban working women in India. The industry needs to concentrate on broadening the market and increasing penetration amongst Indian consumers.

Table 5.34: Major players of RTE and RTC foods in India.

Company	*Brands*	*Products*
Dabur India	Hommade	RTE and RTC Foods
Priya Foods	Priya	Instant Mixes, Puries,Pulihora Paste, RTE
Capital Foods	Ching's Secret, Smith & Jones	Cooking Paste, Sauce & Ketchups
Haldirams	Haldirams	Packaged Bhel Puri Chats, Chana Masala, Samosa, Pakoras
ITC	Aashirvaad Atta, Bingo, Kitchens of India	RTE and RTC Foods
MTR	MTR	Indian Curries, Gravies and Rice
Satnam Overseas Ltd	Kohinoor	RTE Indian Delicacies
Gits	Gits	Variety of Ethnic Indian Cuisine

Souroo: Nicha (2008).

Table 5.35 summarizes the above discussion in presenting the important value added products and by-products of processing of major agricultural commodities.

iii. SWOT Analysis for Food-Processing Industries in India

It is essential to note that, changing lifestyles, food habits, organized food retail and urbanization are the key factors that influence the strengthening of the food processing sector in the economy. It is a known experience that, the consumption pattern of consumers in India has drastically changed towards fruits, vegetables, meats and dairy products from cereals and pulses. This shift implies a need to diversify the food production base to match the changing consumption preferences and offers wide scope for producing a wide range of processed products. In the recent period, there is a drastic shift towards to RTE and RTC products and organic foods. These shifts in the consumption pattern offers a wide range of opportunities to the agro-processing industries to excel in the market economy. In this context, to assess the potentiality of success of agro-processing industries in the country, Strengths (S), Weaknesses (W), Opportunities (O) and Threats (T) (SWOT) analysis is done with reference to the Indian food-processing sector.

Strengths

- ✰ Social acceptability of processed products.
- ✰ Domestic and export competitiveness for food-processed products.

Table 5.35: Value added food products and their by-products

Sl.No.	Commodity	Value Added Products		By-products of Processing	Recent Products, Process and Technologies
		Domestic Market	Export Market		
1	Paddy	Packaged and branded rice (including Basmati rice)	Packaged and branded Basmati rice	Rice-husk particle boards, bran oil, straw board paper	☆ Fully automatic modern rice mills ☆ Partially cooked/quick cooking rice ☆ Breakfast cereals and value-added products ☆ Attractive packaging and branding
2.	Wheat	Fortified atta, bread, biscuits, vermicelli, noodles, pizza etc.	Fortified atta, biscuits	Straw board paper	☆ Fully automatic roller flour mills ☆ Whole bran wheat flour ☆ Fortified wheat flour ☆ Attractive packaging and branding ☆ Large number of backed products ☆ Automatic chapatti making machines
3.	Pulses	Branded and processed	—	—	☆ Automatic processing units for pulses with driers, colour sorters and packaging unit ☆ Attractive consumer packaging with branding ☆ Cold storing of processed pulses ☆ Snack foods and other value added products
4.	Oilseeds	Double filtered refined branded consumer packs	Oils, cakes	Flour	☆ Production of full fat soy flour/enzyme active soy flour for bakery and fortified foods ☆ RTE snack foods
5.	Sugarcane	Sugar, jaggery	Sugar, jaggery	Molasses, paper Boards etc.	
6.	Cotton	Textiles	Textiles, Cake of cotton seed, oil of cotton seed, cotton lint	Cattle feed	
7.	Fruits	Slices, juices, squashes, jams, jelly, fruit based drinks, pickles	Pulp, slices, juices, squashes, jams, jelly, fruit based drinks and concentrates, pickles	—	
8.	Fish	Fish cutlets, meat, fish burger, prawn kababs, fish pickles etc.	Processed fish and fish products, RTE consumer packs	Fish oil, fish meal	

- Government recognized this sector as one of the priority sectors for economic development and makes increased allocations by the Government towards strengthening this sector (Table 5.36 and 5.37)
- Strengthens both backward and forward linkages of marketing process.

Table 5.36: Number of FPI's in India.

Year	*No. of FPI*
2004-05	1139
2005-06	704
2006-07	644

Table 5.37: State-wise number of food processing units in India.

States/UTs	*2004-05*	*2005-06*	*2006-07*
Jammu & Kashmir	2	1	5
Himachal Pradesh	3	4	11
Punjab	21	14	21
Uttarakhand	2	5	15
Haryana	7	3	16
Delhi	1	2	2
Rajasthan	3	10	26
Uttar Pradesh	23	21	36
Bihar	1	1	2
Mizoram	0	0	0
Nagaland	0	1	1
Tripura	0	0	1
Meghalaya	1	1	2
Assam	8	4	12
West Bengal	13	9	22
Jharkhand	2	3	1
Odisha	1	4	3
Chhattisgarh	1	3	6
Madhya Pradesh	4	10	8
Gujarat	7	10	15
Maharashtra	24	43	83
Andhra Pradesh	26	24	41
Karnataka	15	13	26
Goa	1	1	0
Tamil Nadu	11	23	35
Pondicherry	1	1	2
Kerala	10	11	30
Manipur	0	0	4
India	188	222	426

- ☆ Food processing sector generates significant employment.
- ☆ India is the world's largest producer of milk, spices and second largest producer of fruits, vegetables and oilseeds, largest livestock population and this enable the country to earn significant amount of foreign exchange.
- ☆ India's relatively inexpensive, but skilled labour force

Weaknesses

- ☆ Demands huge amount of initial investment and also working capital
- ☆ Fast changing technology makes the purchased equipment technically obsolete
- ☆ Demands more skilled labour in executing the operations
- ☆ Weak linkages between R&D and processing industry
- ☆ Lack of adequate quality control and testing methods as per international standards thereby, the quality standards of processed products are not in tune with the SPS standards imposed by the importing (developed) countries
- ☆ Long and fragmented supply chain with independent players
- ☆ Seasonal availability of raw material and its high cost during lean marketing season
- ☆ Socio-cultural factors such as preference for fresh and cooked food, variation in food habits across the country, easy availability of raw materials for cooking, preference for consumption of food at home etc.
- ☆ Inadequate infrastructure for sorting, grading, packing etc.
- ☆ Low production volumes from FPI due to low utilization of installed capacities
- ☆ Food safety and quality standards are not stringent
- ☆ Poor distribution net work.

Opportunities

- ☆ Diversified agro-ecological conditions in the country are suitable to produce different types of produce and this offers vast potential for food processing activities to produce a wide range of processed products
- ☆ Facilitate integration of developments in contemporary technologies such as electronics, material science, computer, bio-technology etc., thereby, ensures vast scope for rapid improvement and progress
- ☆ With the advent of globalization and trade liberalization, the opportunities were greatly increased for the food-processing industries for boosting the exports of processed products (Figure 5.15). This leads to earning of more amount of foreign exchange, increase in investments and thereby, employment opportunities
- ☆ Setting of Special Economic Zones (SEZ) and Food parks will further increase the marketing opportunities for processed products
- ☆ Rising income levels and changing consumption patterns of the people
- ☆ Favourable demographic profile and changing lifestyles
- ☆ Encouraging FDI into the country (Table 5.38 and Figure 5.16)
- ☆ Contract farming gaining more importance

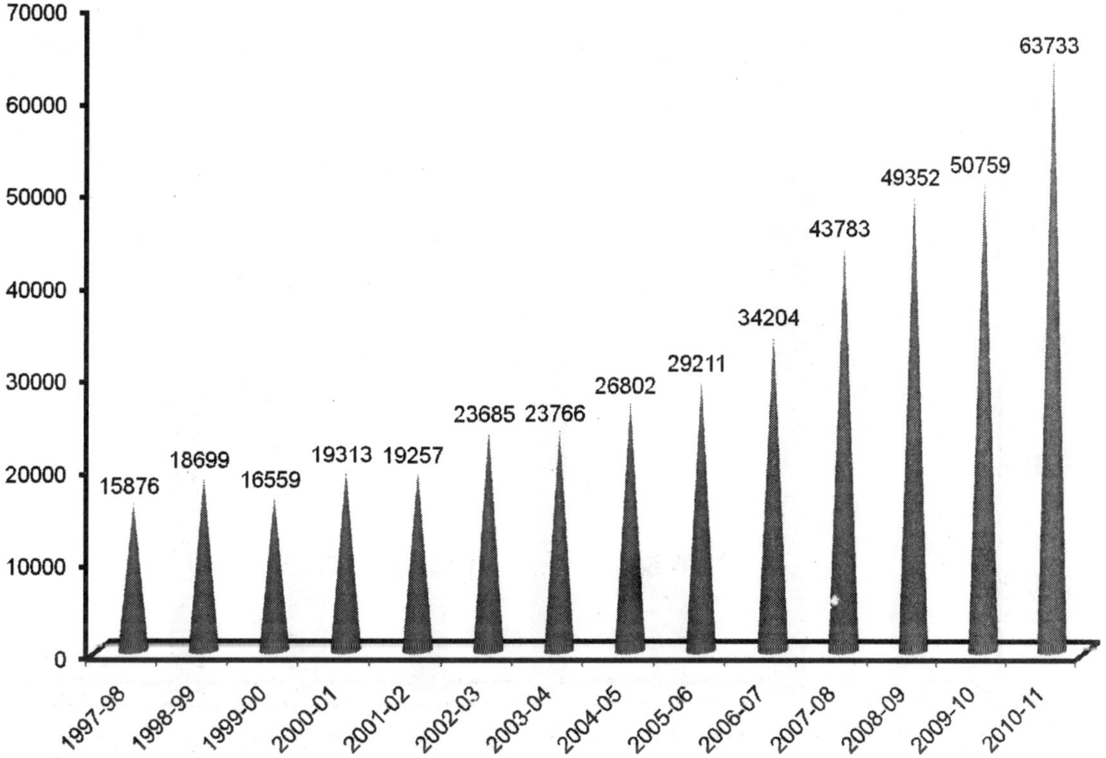

Figure 5.15: Value of exports (Rs. Crores) of processed food products.

Table 5.38: FDI in food processing sector (April to November, 2011).

Sector	*FDI Inflows (₹ Crores)*	*Per cent of Total FDI Inflows*
FPI	6207.32	0.89
Fermentation industries	4422.74	0.66
Vegetable oils and Vanaspathi	1133.51	0.16
Tea and Coffee	446.61	0.07
Total food processing sector	12210.18	1.77
Total	686151.29	100.00

Source: <www.mofpi.nic.in>

- ☆ Increases investments on infrastructure through Public Private Partnership (PPP)
- ☆ Promotes integrated cold chain and Food Safety Management Systems
- ☆ Food products today are the single largest component of household consumption expenditure. The current food consumption in India is estimated at ₹ 8,60,000 Crores (Table 5.39). Processed food account for ₹ 4,60,000 Crores and share of primary processed food (includes packed fruits and vegetables, packed milk etc) is at ₹ 2,80,000 Crores.

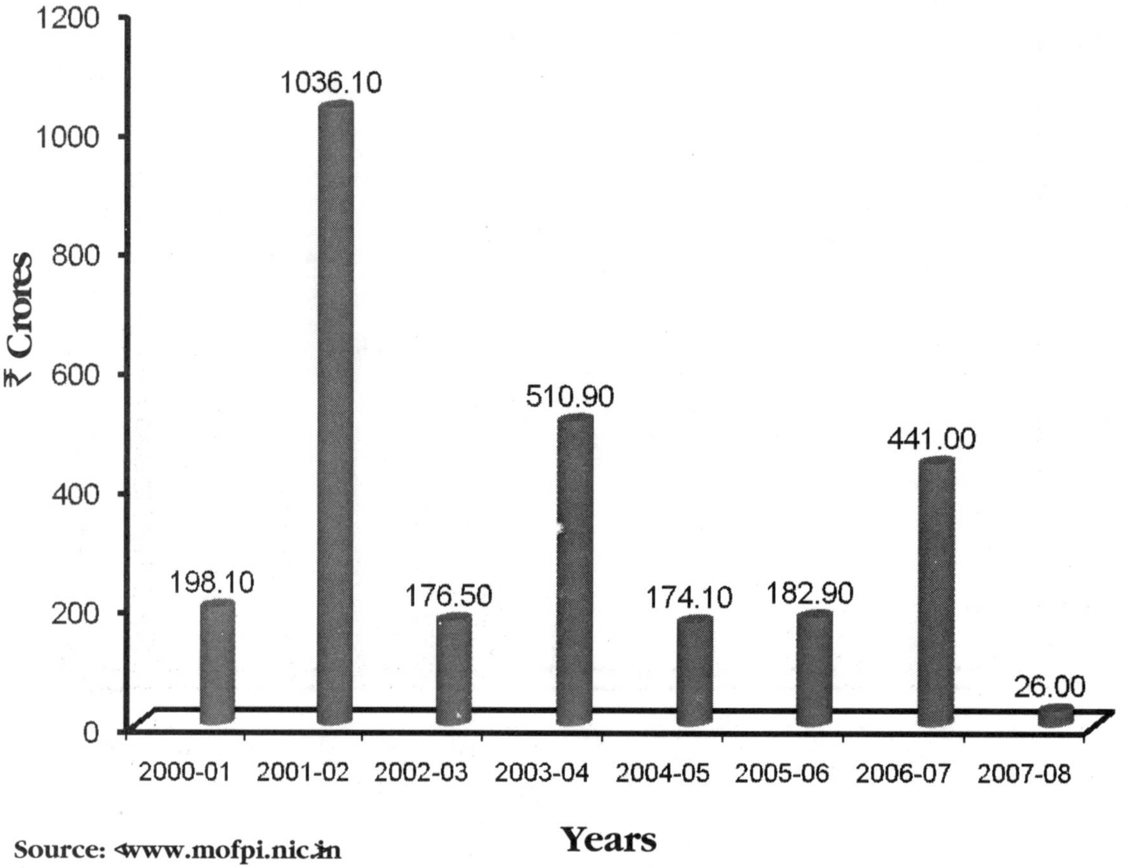

Source: <www.mofpi.nic.in>

Figure 5.16: FDI flows into the food-processing sector in India.

Total 5.39: Estimated total food market in 2003-04 at current prices.

Description	*Market Size (₹ Crores)*
Total food consumption*	8,60,000
Processed foods**	4,60,000
Primary processed food (include packed fruits and vegetables, packed milk, unbranded edible oil, milled rice, flour, tea, coffee, sugar, pulses, spices and salt etc)	2,80,000
Value- added foods (includes processed fruits and vegetables-juices jams, pickles, squashes, concentrate; processed dairy products-ghee, paneer, cheese , butter, ethnic Indian products; branded edible oil, breads, biscuits, snack foods, pasta based foods, processed meat, poultry and marine products, Confectionary and chocolates, alcoholic beverages, spirits wine, aerated and malted beverages etc).	1,80,000

Note: * Excluding consumption of alcoholic beverages and out of home consumption.

** Excluding out of home consumption, but including alcoholic beverages and processing inunorganized sectors in dairy (halwais) and grain milling(chakkis).

- The significant shift in India's demographic profile in favour of a younger population, increasing disposable (surplus) incomes and changing socio-economic environment, is set to change the food consumption pattern in favour of processed foods, which are convenient, hygienic and of consistent quality.
- FDI in food processing sector is quite encouraging (Tale 5.14).

Threats

- Stiff competition from the developed nations, as their processing technology is far superior than in India
- If trained manpower migrates to other industries and other professions for want of better working conditions and payments, it will adversely affect the prospects of the industry and may even lead to shortage of manpower in the processing sector
- Low affordability and cultural preferences towards fresh food may affect the expansion of processing industry
- High prices of the processed products compared to fresh produce
- Prospects of FPI fluctuate in accordance with the Government's policies
- Tariff and non-tariff barriers place restrictions for trade in the international market.

iv. Principle of Establishment of Food-Processing Unit

The basic criterion or underlying principle of establishment of an food-processing unit is availability of surplus raw agricultural produce, which can be estimated by the following equation (Bharti *et al*):

$$S = Pt - (Un + L)$$

where, S = Surplus produce available to a food-processing unit, Pt = Total production of a commodity, Un= Total utilization of raw produce by the population to meet their various needs, L = Total losses during various marketing operations.

The correct assessment of quantum of raw material availability is the basic step for consideration of the establishment of a food-processing unit in a particular area. If there exists similar processing units in that area dealing with the same raw material, then it is essential to consider their requirements also. Further, correct assessment of demand for the processed products to be produced is also essential to analyze the business prospects of FPI.

v. Different Institutions Related to Food-Processing in India

The following are the important institutions engaged in the promotion of food-processing in the country:

a. MOFPI

The MOFPI, set up in July 1988, is the main central agency of the Government responsible for developing a strong and vibrant food processing sector; with a view to create increased job opportunities in rural areas, enable the farmers to reap benefit from modern technology, create surplus for exports and stimulating demand for processed food. In the era of economic liberalization where the private, public and co-operative sectors are to play their rightful role in development of food processing sector, the Ministry acts as a catalyst for bringing in greater investment into this sector, guiding and helping the industry in a proper direction, encouraging exports and creating a conducive environment for the healthy growth of the FPI. With this overall objective, the MOFPI aims at:

- Better utilization and value addition of agricultural produce for enhancement of income of farmers
- Minimizing wastage at all stages in the food processing chain by the development of infrastructure for storage, transportation and processing of agro-food produce
- Induction of modern technology into the FPIs from both domestic and external sources
- To encourage R&D in food processing for product and process development and improved packaging
- To provide policy support, promotional initiatives and physical facilities to promote value added exports.

The MOFPI has been supporting a range of initiatives for the growth of FPI. However, the response to some of the schemes/programmes of MOFPI has not been encouraging. MOFPI, therefore, took the decision to critically review its existing programmes so that, they reflect changing economic realities. It was also realized that, a paradigm shift is required in implementation of the various schemes to achieve the targets under the Vision 2015. MOFPI constituted following five sub-groups on the various sub sectors of FPIs:

- Sub-group on Meat, Poultry, Fisheries and Dairy
- Sub-group on Fruits and Vegetables, Wine and Beer
- Sub-group on Infrastructure
- Sub-group on Taxation and Finance
- Sub-group on R&D, Human Resource Development (HRD), Labs and Standards

For rapid growth and sustainability of food-processing sector, it is essential that, the industry requires availability of matching number of skilled manpower. So, realizing this importance, MOFPI provides financial assistance for creation of infrastructural facilities for running degree/diploma courses in food processing sector, entrepreneurship development programme, setting up of Food Processing Training Centres (FPTC) and training at recognized National/State Level Institutions sponsored by MOFPI. For promoting research work in the sector, Ministry provides financial assistance R&D in processed food sector. The financial assistance is given as grant-in-aid to various Universities, IITs, Central/State Government Institutions, Public Funded Organizations etc. MOFPI also undertake several schemes (Table 5.40) to promote demand driven R&D work in the field of food processing sector for product and process development, improved packaging, value addition and leading to innovative products and processes with commercial value.

To ensure steady and better price to the farming community as well as availability of commodities in processed form to the consumer throughout the year, Food Processing Scheme is promoted, which reduces the wastage of agricultural produce to a great extent. Department of Agricultural Marketing and Agri-Business is the Nodal Agency at the State level for the MOFPI, Government of India. MOFPI aims at technology upgradation, establishment/modernization of FPIs with effect from April 2007 through Banks/Financial institutions. Being a State Nodal Agency, different project proposals under Entrepreneur development programmes, contract farming, Food processing training centre, Seminars and Exhibition on food processing, infrastructure development, setting up of Food testing laboratory and Food park etc., are being received, scrutinized and forwarded to the MOFPI, Government of India.

Table 5.40: Schemes offered by the MOFPI to promote food processing sector schemes.

Schemes	*Features*
Income Tax Relief	☆ Tax deduction for 100 per cent of profit for five year and 25 per cent of profits for the next five years in case of new agro-processing industries set up to process, preserve and package fruits and vegetables.
Excise Tax Relief	☆ Exemption limit of excise duty for Small Scale Industry raised from ₹ 1 crore to ₹ 1.5 crores
	☆ Excise duty on reefer vans (refrigerated motor vehicles) reduced from 16 per cent to 8 per cent and on dairy machinery has been fully exempt for promotion of dairy processing industries
	☆ Excise duty on condensed milk, ice cream, preparations of meat, fish and poultry, pectins, pasta and yeast, soya bari (food supplements) and RTE packaged foods, on vanaspati fully exempt.
	☆ Excise duty on ready-to-eat packaged foods and instant food mixes like dosa and idli mixes reduced from 16 per cent to 8 per cent.
Customs Duty Relief	☆ Custom duty reduced from 7.5 per cent to 5 per cent on food processing machinery, 20 per cent to 10 per cent on refrigerated vans, 65 per cent to 50 per cent on sunflower oil (crude), 75 per cent to 60 per cent on sunflower oil (refined)
	☆ Customs duty on packaging machines reduced from 15 per cent to 5 per cent.
Scheme for Infrastructure	☆ Food Parks, Packaging Centres, Modernizing Abattoirs, Integrated Cold Chain facilities, Irradiation facilities, Value added centers – at an estimated budget of ₹ 2613 Crores
Development	☆ Scheme for Technology Up-gradation/Establishment/ Modernization of FPIs at an outlay of ₹ 600 Crores
Scheme for Quality Assurance, Codex standards and R&D	☆ Food safety and quality assurance mechanisms, Strengthening Codex cell, Continuous R&D, Setting up/up gradation of quality control laboratories, Promotional activities such as participation in exhibition/ fairs/supporting seminars/workshops/ studies and surveys, Generic advertisement, Preparation of short films and publicity material for different events at an expense of ₹ 250 Crores
Scheme for Human Resource Development	☆ Setting up of Food Processing Training Centres, imparting training to update skills, Entrepreneurship Development Programme, facilitating Universities/Institutions for running degree/diploma courses at an expense of ₹ 65 Crores
Scheme for strengthening of institutions	☆ ₹ 325 Crores is allotted for strengthening of institutions, setting up Wine Board and Meat and Poultry Processing Board

National Mission on Food Processing (NMFP)

MOFPI proposed to launch a new Centrally Sponsored Scheme: NMFP during 12th Five Year Plan to be implemented through States. The basic objective of NMFP is decentralization of implementation of Ministry's schemes, which will lead to substantial participation of State Governments. The proposed NMFP contemplates establishment of a National Mission as well as corresponding Missions in the State and District level. The proposed NMFP is likely to improve significantly to the Ministry's outreach in terms of planning, supervision and monitoring of assistance through this Scheme. This scheme is proposed to be implemented as a new centrally sponsored scheme in the ratio of 75:25 by Government of India and State Government.

Objectives of NMFP

- ☆ To augment the capacity of food processors working in unorganized sector and upscale their operations through capital infusion, technology transfer, skill up-gradation and handholding support
- ☆ To support established self help groups working in food processing sector

- ☆ Capacity development and skill upgradation through institutional training to ensure sustainable employment opportunities and also to reduce the gap in requirement and availability of skilled manpower in food processing sector
- ☆ To ensure the standards of food safety and hygiene to the globally accepted norms
- ☆ To facilitate FPIs to adopt HACCP and ISO certification norms
- ☆ To augment farm gate infrastructure, supply chain logistic, storage and processing capacity
- ☆ To provide better support system to the organized food processing sector.

Major Programmes/Schemes to be Covered under NMFP

- ☆ Scheme for Technology Up-gradation/Setting up/Modernization/Expansion of FPI;s
- ☆ Scheme for Supporting Cold Chain Facilities and Reefer Vehicles for other than Horticultural Products
- ☆ Scheme for HRD
  - ❖ Creation of Infrastructure Facilities for Running Degree/Diploma/Certificate Courses in Food Processing technology
  - ❖ Entrepreneurship Development Programme
  - ❖ FPTC
- ☆ Scheme for Promotional Activities
  - ❖ Organizing Seminar/Workshops
  - ❖ Conducting Studies/Surveys
  - ❖ Support to Exhibitions/Fairs
  - ❖ Advertisement and Publicity.

The Government of India also proposed to start agro-processing units with farmers' and private sector participation. On behalf of farmers, 25 per cent of the share capital contribution will be borne by the Government through the Department of Agricultural Marketing and Agri-Business and the remaining 75 per cent of the share will be borne by the private entrepreneurs.

b. Agricultural and Processed Food Products Export Development Authority (APEDA)

The APEDA was established by the Government of India under the Agricultural and Processed Food Products Export Development Authority Act passed by the Parliament in December, 1985. The Act (2 of 1986) came into effect from 13th February, 1986. The Authority replaced the Processed Food Export Promotion Council (PFEPC). APEDA has marked its presence in almost all agro potential states of India and has been providing services to agri-export community through its head office located at New Delhi, five Regional offices at Mumbai, Kolkata, Bangalore, Hyderabad and Guwahati and 13 Virtual offices. The Virtual offices have been established in association with respective State Governments/agencies. Basic information about APEDA, its functions, registration and financial assistance schemes etc., is being made available to entrepreneurs/prospective exporters by these virtual offices. This authority looks after the following functions:

- ☆ Development of industries relating to the scheduled products for export by way of providing financial assistance or otherwise for undertaking surveys and feasibility studies, participation in enquiry capital through joint ventures and other reliefs and subsidy schemes;

- ✰ Registration of persons as exporters of the scheduled products on payment of such fees as may be prescribed;
- ✰ Fixing of standards and specifications for the scheduled products for the purpose of exports;
- ✰ Carrying out inspection of meat and meat products in slaughter houses, processing plants, storage premises, conveyances or other places where such products are kept or handled for the purpose of ensuring the quality of such products;
- ✰ Improving of packaging of the Scheduled products;
- ✰ Improving of marketing of the Scheduled products outside India;
- ✰ Promotion of export oriented production and development of the Scheduled products;
- ✰ Collection of statistics from the owners of factories or establishments engaged in the production, processing, packaging, marketing or export of the scheduled products or from such other persons as may be prescribed on any matter relating to the scheduled products and publication of the statistics so collected or of any portions thereof or extracts there from;
- ✰ Training in various aspects of the industries connected with the scheduled products.

APEDA is mandated with the responsibility of export promotion and development of various products like, fruits, vegetables and their products, meat and meat products, poultry and poultry products, dairy products, biscuits and bakery products, honey, jaggery and sugar products, cereal and cereal products, groundnuts, peanuts and walnuts, floriculture and floriculture products etc.

c. *Marine Products Export Development Authority (MPEDA)*

The Marine Products Export Development Authority (MPEDA) was constituted in 1972 under the Marine Products Export Development Authority Act 1972 (No.13 of 1972). The role envisaged for the MPEDA under the statute is comprehensive - covering fisheries of all kinds, increasing exports, specifying standards, processing, marketing, extension and training in various aspects of the industry. MPEDA functions under the Ministry of Commerce, Government of India and acts as a coordinating agency with different Central and State Government establishments engaged in fishery production and allied activities. The Seafood Industry of India has come a long way and today seafood is exported to nearly 100 countries from India. The various activities of MPEDA include:

- ✰ Registration of infrastructure facilities for seafood Export trade
- ✰ Collection and dissemination of trade information
- ✰ Projection of Indian marine products in overseas markets by participation in overseas fairs and organizing international seafood fairs in India
- ✰ Implementation of development measures vital to the industry like distribution of insulated fish boxes, putting up fish landing platforms, improvement of peeling sheds, modernization of industry such as upgrading of plate freezers, generator sets, ice making machineries, quality control laboratory etc.
- ✰ Promotion of aquaculture for production of shrimp and prawn for export
- ✰ Promotion of value added seafoods
- ✰ Promotion of Tuna fishery
- ✰ Implementation of organic farming
- ✰ Conservation management.

d. NMPPB

The Government of India launched a NMPPB on 19th February 2009 and is located at New Delhi. It is an autonomous body and would initially be funded by the Government of India for two years and would be managed by the industry itself. The Board will have 19 Members including CEO of the Board. The mandate of the Board is to oversee the growth and further promotion of Meat and Poultry Processing sector in the country. This has been considered necessary because of a felt need for a focused intervention in meat and poultry sector. This industry driven institution has been launched to work as a National hub for addressing all key issues related to Meat and Poultry processing sector for the systematic and proper development of this sector. The Board would serve as a single window service provider for producers/manufacturers and exporters of meat and meat products, for promoting and regulating the meat industry as a whole and it would result in large number of employment opportunities. It would also help the rural farmers for increasing their income The Board will address issues related to production of hygienic, safe and wholesome meat and meat products, including:

- ☆ Lay down policies for healthy development of the meat and poultry sector
- ☆ Evolve standards of hygiene and quality
- ☆ Harmonization of domestic standards with international standards
- ☆ Address current environmental pollution issues
- ☆ Organize capacity building efforts
- ☆ Capacity-building to address human resource development issues
- ☆ Develop uniform and effective national meat quality testing systems and laboratories
- ☆ Provide credible consultancy services to the Meat and Poultry Processing Industries/for the Abattoir/Slaughter houses modernization/setting up all over the country.

The Board will also help industry to utilize slaughterhouse waste materials, set up quality control laboratories for meat and meat products, promote meat manufacturers to adopt GMP, HACCP systems, ISO-9001: 2000 standards, help industry to create and disseminate data, train workers and technicians and work as a central and National hub to address meat and poultry processing related issues so that more hygienic and whole some meat and poultry products are available to the consumers.

e. CFTRI

Established in 1950 as the third National Laboratory of Council of Scientific and Industrial Research (CSIR), CFTRI today stands out among the largest and most diversified technology laboratories in the world. The major R&D programmes of the Institute relate to development of food products and processes for optimal utilization of the country's agricultural produce, technology packages for convenient and ready -to-eat foods, processes for value added products from plantation produce, protocols for post-harvest handling and transportation of fruits and vegetable for export and inland transportation, methods of packaging of food materials and processed food products for extended shelf life, techniques for crop/grain protection, technologies for eliminating/reducing post-harvest losses of perishables and durables, and modernization of the primary FPI through development of energy-efficient and cost-effective machinery and equipment.

The role of CFTRI has always been on a high pedestal in the areas of human resource development and R&D partnerships cherished through the international linkages with institutions such as the Institute of Food Technologists (USA), UN University (Tokyo), European Economic Commission (Belgium), National Science Foundation (USA) and many more. CFTRI is an ISO 9001:2008 and ISO

14001:2004 organization and NABL accredited for chemical and biological testing of samples. These unique features of CFTRI are of importance in building an everlasting partnership with industries and agencies for excellence in science and technology, sponsoring consultative research and business tie-ups for long term strategic research as well as unparalleled technology development in the area of food science and technology. This vibrant scenario has consolidated CFTRI to evolve its capabilities and to position itself as a global R&D hub with a firm commitment in reaching its R&D to people through adaptable, affordable and appropriate technologies. The basic objectives of CFTRI include:

- Generate and apply knowledge of food science and food technology for optimal conservation and utilization of the nation's food resources
- Integrate scientific and technological knowledge into conventional and traditional systems and practices, and local and regional realities
- Add value and utility to agro-resources through R&D and contribute to sustained development, food security and food safety
- Aid and promote the development of food industry through inter-disciplinary, innovative and state-of-the-art solutions
- Set national standards for food quality and spread food quality consciousness all around
- Sustain leadership in long-term strategic research and technology development
- Integrate the food supply chain from the cultivator to the consumer so that cultivators get optimal returns from processing, and consumers get the food that they want, when they want, where they want, in whatever form they want and at affordable cost
- Build and bolster bonds with nodal agencies from the global to the grassroots level, particularly in the area of multi-level human resources development
- Develop new knowledge continuously, to address contemporary challenges and answer future emergencies

f. National Fish Processing Development Board (NFPDB)

Fish processing in India is done almost entirely for export. Open sun dried fish and fish meal are the only major exceptions. At present India has – freezing units, - cold stores, - ice plants, - canning units and – fish meal plants. Capacity of most of these processing and storage units is very low when compared to the facilities in fish processing industry in technologically advanced countries. The total fish processing and storage facility in India is grossly inadequate compared to the potential for fish production and processing. Barely five per cent of India's seafood exports are in processed form. Most exports are in the form of frozen fish. Also, the Indian brand does not exist in markets of advanced countries. In fact, more than 60 per cent of India's exports to south-east Asia are re-exported after processing. In view of the above facts about the vast potential for fish processing industry to cater to the domestic market and supplement the existing export of processed sea food, it is necessary to constitute a NFPDB.

NFPDB will be an autonomous development agency promoted by the Government of India under the MOFPI, New Delhi. The NFPDB will work as a facilitator to ensure the smooth and orderly development of the Indian fish meat/processed fish products sector. The important functions of NFPDB include:

- Promote, establish, manage Modern fish meat/fish products Processing Plants for fish meat/processed fish products production in India in furtherance of Government's policy to promote exports and domestic sale of fish meat/processed fish products with specific emphasis to improve the sanitary and hygienic conditions for their production.

- Initiate, facilitate rearing/production of highly productive fishes emphasizing backward linkages for fish meat/processed fish products production.
- Promote, establish, manage fish markets, to promote domestic sale of fish meat/processed fish products.
- Promote safe and hygienic production of fish meat and processed fish products at par with the international standards including SPS requirements, transparency in the international and national standards, and global trade without any discrimination.
- Promote fish/aquaculture health requirements in synchronization with OIE guidelines.
- Regulate the production of fish meat and processed fish products in the country and promote a single window service to producers, manufacturers and exporters of fish meat and processed fish products.
- Promote and train fish meat producers, manufacturers, entrepreneurs to adopt GMP, HACCP, and ISO-9001:2000 in fish meat/processed fish products production.
- Create facilities to utilize the waste material from fish markets/processing houses to prepare fish byproducts with value addition.
- To work as a dynamic, progressive and financially viable nodal and autonomous development agency promoted by the MOFPI, Government of India for the development of fish meat/processed fish products sector.
- To work as a facilitator to ensure smooth and orderly development of the fish meat/processed fish products industry in the country for taking comprehensive measures leading to "quality" production of fish meat/processed fish products and value added products for the domestic and export market.
- To ensure, as and when required, the necessary participation of government agencies and R&D institutions, Agricultural-Universities, fish meat/fish product producers, exporters and other entrepreneurs.
- To undertake national level planning and policy measures for the development of the fish meat/processed fish products industry
- To help the existing research organizations to carry out required applied research in production of fish meat/processed fish products and help in processing and marketing of fish meat/processed fish products for the overall benefit of fish meat/processed fish products industry of India.
- To promote the co-operative efforts among the fish growers, and processors of fish meat/processed fish products and encourage contract farming and buy-back arrangements.
- To undertake research and development work with the assistance of the existing institutions on the new technology for production of highly productive fish/aquaculture for fish meat production and for modernization of process technologies for fish meat/processed fish products.
- To conduct training programs with the help of existing and new training institutions for the development of skilled personnel for production, processing and export of fish meat/processed fish products.

- To help the producers, processors and exporters to achieve the desired "quality" standards for fish meat/processed fish products and fish byproducts.
- To assist the fish meat/processed fish products industry to develop appropriate extension/ education programs for farmer-producers, meat processors and meat consumers in the country.
- To establish co-ordination and liaison with the State Agricultural Universities (SAUs), R&D institutes, central and state government departments, other relevant institutes, development boards, private sector industry etc.
- To suggest suitable changes/modifications in existing policies and other regulations for taxation/excise on fish meat/processed fish products and fish by -products and take suitable measures to implement the policies.
- To suggest suitable measures for collection of required statistical/market data on fish meat production, processing, domestic and international markets and prices and establish Market Intelligence Service for the stakeholders in the fish meat/processed fish products industry.
- To assist exporters to develop export of fish meat/processed fish products and fish byproducts and if necessary carry out domestic and international market surveys to seek the potentials available in new markets for fish meat/processed fish products.

g. Indian Institute of Crop Processing Technology (IICPT)

The IICPT is a National Institute with International Repute in the field of crop processing technology. It serves as a world class R&D and Educational Institute located as under the MOFPI, Government of India. The mandatory activities of IICPT are teaching, research and outreach activities in post harvest processing, preservation and value addition of agricultural and horticultural produces. IICPT and its scientists are experts in their own fields of research. IICPT has created in its main campus at Thanjavur in Tamil Nadu with world class research laboratories for conducting research in different areas of food processing technologies.

h. Indian Grape Processing Board (IGPB)

The Union Government in 2009 gave its approval for the establishment of the IGPB at Pune, Maharashtra which is close to the principal grape growing and processing areas in the country. The following are the basic functions and objectives of the IGPB:

- To focus on R&D, Extension, and Quality upgradation, market research and information, domestic and international promotion of Indian wine.
- To foster sustainable development of Indian Wine Industry
- To formulate a vision and action plan for the growth of Indian Wine Sector including research and development for quality upgradation in new technologies/processes.

During two years of its existence, the Board has focused on the promotion of 'Wines of India' in the domestic as well as international market by participating in important and relevant exhibitions, fairs, consumer awareness and training programmes, undertaking advocacy work with the various state governments/central ministries on various issues related to taxes/levies and promotion aspects.

i. National Institute of Food Technology, Entrepreneurship and Management (NIFTEM)

NIFTEM is the brainchild of the MOFPI. MOFPI in its Vision document-2015, envisaged creation of a world-class institution to cater to the various stakeholders such as entrepreneurs, food processing industry, exporters, policy makers, Government and existing institution. NIFTEM will work actively in assisting in setting up food standards, businesses incubation and can also include knowledge sharing *i.e.*, information/knowledge sharing. It would also be an apex institution in the field of food technology and management, networking and coordinating with other institutions in the same field. NIFTEM campus is located at Haryana and it serves as an international centre of Excellence with state of the art infrastructure and world class man-power for research and training. The major objectives of NIFTEM include:

- Developing world class managerial talent and entrepreneurship with advanced knowhow in food science and technology.
- Providing intellectual backing for regulations which will govern food safety and quality and at the same time foster innovation.
- Functioning as a knowledge repository on various aspects of food processing such as product information, production and processing technology, market trends, safety and quality standards, management practices among others.
- Facilitating business incubation services with its ultra modern pilot plans for fruit and vegetables, dairy, meat and grain processing.
- Promoting co-operation and networking among existing institutions within India and as well as with international bodies.

NIFTEM will have various 'theme centres' with the aim of promoting thought leadership and innovation–driven practices in the key sector of food processing with strong sectoral/functional focus are planned. Each theme centre will have world-class manpower and infrastructure for research and training. The theme centres would benefit through long term associations with prominent international institutions, through a faculty exchange programme or joint research initiatives in various food processing areas such as dairy, seafood and wine. Sector focused theme centres will ensure a comprehensive supply chain approach. The theme centres planned are dairy, cereal based products, animal protein, fruits and vegetable based foods, beverages and confectionery. The cross sectoral theme centres would work closely with the other theme centres. These include management studies/ business incubation, trade/knowledge management/networking, packaging and engineering, food standards, testing and safety, biotech, nutrition and culinary science. Another critical role of NIFTEM would be to provide adequate operational independence to the food standards testing and safety department as this will be used as a national referral laboratory. NIFTEM would actively collaborate with the world's best institutions to promote development of path breaking technologies, which are globally relevant, Exposure of these institutions to global trends and industry needs will support in developing NIFTEM as an international centre of excellence. Some areas of collaboration would be course design/content and faculty exchange, expertise in research and development in various food sectors, business incubation, food standards, testing and safety, consumer research and information sharing and assist in setting up of the themes centres. Developing and strengthening research programmes (both fundamentals and applied) and their linkages with industry and end consumers.

vi. Major Players in Indian Food Processing Sector

As far as food production is concerned, India holds second position next to China and FPI is considered as one of the largest industries in India in terms of expected growth, export, consumption and production. This industry offers important synergies and linkages between agriculture and industry. Initially, this sector was confined to transportation, packaging and food preservation, which includes pickling, drying, curdling, salting, etc. However, over the years with emerging technologies and new markets, the scope of this sector has widened considerably. The list of top players (MNCs and major domestic players) in the processing sector and the products dealt by major players are shown through Tables 5.41 and 5.42 respectively.

Table 5.41: Leading players in food-processing sector in India.

Major MNCs	*Major Indian Companies*	*Indian MNCs likely to Enter*
Nestle, Pepsi, Coke, Kellogs,Unilever, Perfetti, Glaxo Smithkline, Heinz, Wyeth, Ajinomoto, Nissinmet, Walmart.	ITC, Dabur, Britannia, Parle, Amul, Haldiram, Godrej etc.	Reliance, Bharti Group,Tatas, Wipro etc.

Source: Nisha (2008).

Table 5.42: Major players in the food processing sector in India and products.

Company	*Products*
Agro Tech Foods	Edible oils, Wheat flour, Dried green peas, Packaged food, Microwave popcorn, Snacks (Potato and corn chips)
Britannia Industries Ltd	Biscuits and Bakery products, UHT Milk, Milk Products (cheese, butter, flavoured milk
Cadbury India Ltd	Chocolates, Confectionery, Malted drinks, Cocoa powder, Hard boiled toffees
Dabur Foods	Friut juices, Cooking pastes, Candies, Coconut milk, Lemon drink, Chilli powder
Gilts Food Products Pvt Ltd	Sweet mix, Snack mixes, Meal mix, Pure ghee, Dairy whitner, Milk powder
Glaxo Smithkline Beecham	Malted milk food, Biscuits
Godrej Foods Ltd	Fruit juices, Tomato puree, Cooking medium, Bakery Fats
GCMMF	Ice creams, Butter, Ghee, Cheese, Milk powder, Infant milk powder, Traditional Indian sweets, Chocolates, Curd
Haldirams Marketing Pvt Ltd	Snack foods, Traditional Indian Sweets, Syrups, Vermicelli
Heinz India Ltd	Malted milk food, Tomato ketchup
Hindustan Unilever Ltd	Ice creams, Atta, Salt, Tea/coffee, Bread, Jam and friut juice, Biscuits, Soups
ITC Ltd	RTE foods, Snack foods, Confectionary, Biscuits
Kellogg's India Ltd	Breakfast cereal
Mother Dairy Fruit and Vegetable Ltd	Ice creams, Butter, Cheese, Milk powder, Traditonal Indian sweets, Chocolates, Ghee, Processed fruits and vegetables
Marico Industries Ltd	Cooking medium, Fruit jams
MTR Foods Ltd	Conceinience foods, Ice creams, Snack foods, Pickles, Readymade mixes, Spice and masala powders, Frozen foods, Pickles
Nestle India Ltd	Chocolates, sugar Confectionery, Malted drinks, milk powder, instant coffee, tea, noodles and sauces, UHT and Condensed Milk, Yoghurt
Parle Products Ltd	Biscuits, toffees
Parle Agro	Fruit pulp, mineral water, fruit drinks
Pepsi Foods India Ltd	Soft drinks, Snack foods, Fruit juices
Venkateshwara Hatcheries Ltd	Poultry products, olive oil pastes premium cookies

Information Pertaining to the Important Players in the Indian Food Processing Sector

ITC Limited

ITC made its entry into the branded and packaged foods business in August 2001 with the launch of the Kitchens of India brand. A more broad-based entry has been made since June 2002 with brand launches in the Confectionery, Staples and Snack Foods segments. ITC has over the last 100 years established a very close business relationship with the farming community in India and is currently in the process of enhancing the Indian farmer's ability to link to global markets, through the e-Choupal initiative, and produce the quality demanded by its customers. This long-standing relationship is being leveraged in sourcing best quality agricultural produce for ITC's Foods business.

Parle Products Private Limited

Parle Products has been in the FPI for more than 80 years now and they are popular in the field of manufacture of confectionery and biscuits. Their brand name Parle symbolizes great taste, nutrition and quality. They are manufacturers of snacks, sweets and biscuits in some of the popular brand names.

Agro Tech Foods Limited

Agro Tech Foods Limited is a public limited company, which is engaged in the marketing of food ingredients and food items to institutional customers and consumers. Some of the popular brand name of products dealt by them in the FPI are Crystal, Healthy World, Sundrop Peanut Butter, Sundrop Snack Break and Sundrop.

Amul

Amul products are being used by households in India right from the year 1946 and some of the popular products are Amulya, Amul Milk, Nutramul, Amul Ice Cream, Amul Shirkhand, Amul Chocolates, Amul Cheese, Amulspray, Amul Ghee, Amul Milk Powder and Amul Butter. They are leaders in the dairy product production in the FPI in India.

Perfetti VanMelle

Perfetti VanMelle is the manufacturer of some of the popular brands like Mentos, Happydent, Chocoliebe, Chlormit, Center Fresh, Big Babol and Alpenliebe in the Indian FPI. They are operating from their corporation office in the city of Gurgaon and sales offices at Mumbai, Kolkata and Bangalore.

Cadbury India Limited

Cadbury India Limited is operating in five different categories in the FPI namely candy, gum, biscuits, beverages and confectionery. As far as their chocolate production is concerned, they are the undisputed all-time favourite among children and some of their key brands include Oreo, Tang, Eclairs, Halls, Celebrations, Bournville, Perk, 5 Star and Cadbury dairy milk.

Pepsico India

Pepsico came into existence in India in the year 1989 and they hold the fourth position in the largest consumer product manufacturers in India. They are the owners of three state-of-the-art food plants in West Bengal, Maharashtra and Punjab. They are the proud manufacturers of some of the popular brand in food section namely Aliva, Uncle Chipps, Quaker Oats, Lehar Namkeen, Lay's, Kurkure and Cheetos.

Nestle India Private Limited

In the FPI, Nestle India Private Limited are engaged in the manufacture of food items in different categories like chocolates and confectionery, beverages, prepared dishes and cooking aids and milk products and nutritional products.

Dabur' s Foods Division

Dabur's Foods division pioneered the concept of packaged fruit juices in India way back in 1997. Realising the potential of processed and packaged food industry in India (4th largest in the world), it successfully established brands in a market, where the barrier to accept packaged food is high, and the fledging processed food industry was in a nascent stage in India. Today, Dabur's Foods division has kept its promise not only to the Indian consumers but to the world by providing them the very best products in packaged food class. It also provides a range of leading brands to choose from - Réal Fruit Juice, Réal Active Fruit and Vegetable Juice, Burrst Fruit Beverage, Hommade, Lemoneez and Capsico.

Britannia Industries Limited

Britannia Industries Limited is an Indian food-products corporation based in Kolkata, India. It is famous for its *Britannia* and *Tiger* brands of biscuit, which are popular throughout the country. Britannia has an estimated 38 per cent market share. The Company's principal activity is the manufacture and sale of biscuits, bread, rusk, cakes and dairy products.

Haldiram

Haldiram's is one of India's largest sweets and snacks manufacturers, based in Delhi, India. It was founded in 1937 by Gangabisenji Agrawal, as retail sweets and namkeens shop in Bikaner, Rajasthan, India. It deals with various food products like Namkeens, Sweets, Cookies, Sherbets (Sharbat), Minute Khana (Microwaveable Food), Papads, Pani Puri, Bhel Puri, Chips (Boletos, Takatak, Whoopies), Royal Temptations (Packaged Dry Fruits), Gujarati Snacks, South-Indian Snacks etc. In additional to these packaged products, Haldiram has its own outlets where it sells sweets and eatables. It also has a range of restaurants in and around Delhi and the National Capital Region. Haldiram's products are exported to several countries worldwide including Pakistan, Bangladesh Canada, Australia, Srilanka, Singapore, Malaysia, South Africa, Indonesia, Qatar, Saudi Arabia, Hong Kong, Japan, Kenya, Libya, South Korea, Nigeria, Mauritius, United Kingdom, United Arab Emirates, Zambia and Bahrain. Their products are sold in U.S. supermarkets in the foreign foods/import section.

Godrej Agrovet Limited (GAVL)

GAVL is a diversified agribusiness company dedicated to improving the productivity of Indian farmers by innovating products and services that sustainably increase crop and livestock yields. With FY 2011-12 sales of ₹ 2439 crore, GAVL has interests in animal feed, oil palm plantations, agri-inputs and poultry. The Animal Feed business is India's largest, producing over 9,57,000 tons annually of high-quality feed and cutting-edge nutrition products for dairy cattle, poultry and aquaculture. GAVL has also entered Bangladesh through a joint venture with the Advanced Chemical Industries Ltd. (ACI) Group, where it is rapidly becoming a key player in animal feed and poultry breeding. The Oil Palm Plantation business has over 44900 hectares of smallholder cultivation across Andhra Pradesh, Goa, Gujarat, Odisha, Tamil Nadu, Karnataka and Maharashtra. With potential cultivation of over two lakh hectares, Godrej Agrovet is positioned to significantly reduce India's crippling dependence on imported edible oil. The Agri-Inputs business is a niche player in innovative agrochemicals, with strong market share in plant growth promoters, soil conditioners, and cotton herbicides. The Poultry business, best known for the Real Good Chicken and Yummiez brands, is now a joint venture with Tyson Foods and has 45 per cent market share in processed poultry.

vii. Importance of Agro-Processing

The processing of agricultural produce is essential because, very few farm produce like milk, eggs, fruits and vegetables are consumed directly in the form in which they are obtained by the farmer. All other products such as paddy, wheat, sugarcane etc., have to be processed into a consumable form.

This processing function is important, both for the farmers and for consumers, as it increases the total revenue to the farmer by adding value to the product and makes it possible for the consumer to have food products in the form desired by him for consumption. Indian FPI is widely recognized as a 'sunrise industry' considering the following advantages:

- Food-processing involves value addition to agricultural and horticultural produce and changes raw produce into edible and palatable forms. The value added by processing to the total value produced at the farm level varies from product to product. It is nearly seven per cent for rice and wheat, about 79 per cent for cotton and 86 per cent for tea, five per cent each for fruits and vegetables, buffalo meat and poultry and 10 per cent each for dairy and marine products. It is generally higher for commercial crops than for food crops.
- Food processing benefits all the sections of the society. It helps the:
  - Farmers - get higher yield, better revenues and lower the risks drastically,
  - Consumers - have access to a greater variety and new products at affordable prices,
  - Economy - gets benefitted with new business opportunities for the entrepreneurs and the work force gets employment.
- Processing function also enhances shelf life of processed products. The processed products like jams, jellies, sauces, juices, butter, ghee etc., will have more shelf-life compared to raw agricultural and horticultural produce.
- Processing satisfies the needs of consumers at a lower cost. If it is done at the door of the consumer, it is more costly than if it is done by a firm on a large scale. Processing saves the time of the consumers and relieves them of the difficulties and botheration experienced in processing.
- It is estimated that, the present levels of post-production losses are to the tune of 5-15 per cent in durables, 20-30 per cent in semi-perishables and 30-50 per cent in perishables. It is also estimated that, the post-harvest losses of food items is to the tune of ₹ 75000 crores per annum. A substantial amount of these losses could be prevented, if scientific agro-processing units with strong forward and backward linkages have been developed.
- Farmers will have more bargaining power in offseason sales and seasonal high fluctuations of prices will also be reduced, if 'form utility' of products is created by processing activity.
- Setting up processing facilities is a first step towards stimulating both consumer demand for the processed products and an adequate supply of the raw material, while the provision of transport, power and other infrastructure facilities required for processing industries also benefits agricultural production. The development of this industry and allied industries provides a better atmosphere for technical progress and the acceptance of new ideas in farming itself.
- India has varied agro-climatic conditions, it has a wide-ranging and large raw material base suitable for food-processing industries. It has a vast coastline of 8000 km, vast marine land with 10 major ports. As discussed earlier, India enjoys significant position in the world in terms of productions of milk, fruits and vegetables livestock (largest number), food grains, fish, poultry eggs etc. Thus, food-processing industry is significant for India's development because, it has important link and synergy with industry and agriculture. Though the food-processing industry is mainly unorganized with 75 per cent of the processing units and only 25 per cent is in the organized category, it is growing fast. The food production is

expected to double in the next decade and the consumption of processed food products is expected to grow at a much faster pace. This growth will benefit the economy in terms of increasing agricultural yields, create employment and raise the standard of living of various associated people. Rising consumer affluence and economic liberalization is opening up new opportunities in the sector.

- Processing of food commodities increases food security in four major ways *viz.*, by reducing post-harvest losses, by extending the shelf-life of food, by adding value to commodities and by improving the quality and safety of foods through establishing appropriate certification, traceability systems and harmonization of standards. However, in the country, the vast majority of the R&D work undertaken to improve food security has concentrated only on increased food production. The role that food-processing can play have not received sufficient attention.
- Agro-processing results in 'sideway linkages' or linkages which derive from the use of the by-products or waste products of the main industrial activity. For example, animal feed industries can utilize several agro-industrial products, such as blood, carcass and bone meal, while agricultural wastes can be used as fuel, paper pulp or fertilizer.
- Food processing sector generates significant employment. The multiplier effect of investment in FPI on employment generation is 2.5 times than in other industrial sectors, higher than any other sector. Even within FPI, the employment intensity is significantly higher in the unorganized sector as compared to the organized sector for the same level of investment. This arrests the migration and reduce social problems in cities (mitigation of congestion in cities). However, the employment generated by agro-processing industries in developing countries is far less than in developed countries. Some estimates suggest that, in developed countries, up to 14 per cent of the total work force is engaged in agro-processing sector directly or indirectly. However, in India, only about three per cent of the work force finds employment in this sector revealing its underdeveloped state and vast untapped potential for employment. Properly developed agro-processing sector can make India a major player at the global level for marketing and supply of processed food, feed and a wide range of other plant and animal products.
- Food Safety and Standards will be developed in tune with the latest developments of International Standards and this contributes more foreign exchange in the international market.
- Agro-processing industry provides the crucial farm-industry linkages, which help to add value to the produce and increase the net income to the farmers. Thus, agro-processing sector have a potentially important role in the economic development of the country - directly as a source of income and employment and indirectly for their backward linkages with agriculture. The latter is especially critical for the commercialization of Indian agriculture, which is widely recognized as important for adoption of modern farming technologies, improving agricultural productivity and incomes and eventually for rural poverty reduction. Not only is food-processing industry expected to raise incomes, but also improve nutritional levels of the consumers. It can contribute to improved nutrition indirectly through generating income to purchase a more varied and nutritious diet and directly through producing food products that contribute to a nutritious diet. For example, the consumption of processed products like jams, ragi malt, Horlicks lite etc., enrich the consumers with more nutrients. Similarly, meat products are a rich source of protein to the consumers.

- Processing of agricultural produce results in low volume and this will save the storage and transportation costs, as processed products require less space compared to raw produce. Further, contemporary consumers prefer to purchase the value-added processed products for various reasons *viz.*, saving of time to cook food, less storage space in small urban houses and better presentability.
- Processing is a necessary step in conversion, value addition and loss prevention of agricultural produce. The processing of agricultural commodities at the village level will check the post-harvest losses at significant level.
- If the surplus agricultural produce can be processed by processing firms, the demand for the agricultural produce will remain higher and this helps the farmers get high prices even in times of surplus production.
- Globalization brings both challenges and opportunities to the agro-processing sector in the country. We know, globalization reduces the magnitude of tariff escalation in the developed countries and this, in turn, provides easy market access for the agro-processed products of the developing countries. Today, agro-industries have more market opportunities than under a protected situation and can select the most profitable market for their products. In this regard, processing activity must be encouraged, as the processed products enjoy high export competitiveness in the international market.
- Processing of agricultural commodities is closely linked to product perishability. The perishability of raw agricultural produce can be greatly reduced by the processing activity, as it extends the potential shelf life of the processed product and may facilitate easy transportation. Even though some processed products like butter, ice cream etc., demand refrigerated transport and higher transportation costs, these can be counterbalanced by the considerable value added due to processing.
- The production of food-processed products increases demand for other inter-industrial inputs (preparation of food items like Maggie noodles, Kissan jam etc). The agro-processing industries require agricultural raw materials as input, which in turn stimulates the demand for other industrial outputs such as chemical fertilizers, insecticides and improved seeds. Similarly, the production of fertilizers requires a greater volume of inputs from other industries. This process continues for many rounds. Thus, an improvement or increase in agro-processing industries generates a greater degree of multiplier effects.
- The value-added contribution of agro-processing industries is also in line with the level of development and the extent of urbanization of the country. To some extent, the contribution of agro-processing also depends upon the types of agricultural produce available. If the country produces and exports a greater proportion of cereals such as rice, wheat and maize, the value added by the grain mills will be relatively high. This, in turn, increases the contribution of the agro-processing industries to the national economy.
- The strengthening of agro-processing sector generates the demand for more agricultural output resulting into more intensive agriculture. Increasing intensity of agriculture would result in more employment generation in rural areas. Apart from generating employment in agriculture, this sector will directly generate employment opportunities within sector and more opportunities in service sector. Thus, the processing sector ensures for all round economic development. Keeping this in view, the Government of India has established MOFPI for strengthening this sector at national level, and also changed APMC Act so that, raw materials for this sector are easily available.

- ☆ When considering food security and poverty alleviation, one aspect that is frequently neglected is the cultural significance of food products. All cultures have traditional food products that they value for more than just the nutritional role. Small-scale agro processing helps to maintain production of these products.
- ☆ Agro-processing encompasses the development and use of appropriate equipment and technology. These technologies are designed to reduce drudgery, save time and to make the processing of foods less laborious.Above all, the overall potential of agro-processing is huge, as it can increase the value of crops and yield higher returns, expand marketing opportunities, improve livelihoods of people, improve palatability of commodities, enhance food security, overcome seasonality and perishability constraints and empower women, who are often involved in agro-processing.

viii. Constraints in the Agro-Processing of Commodities in India

Inadequate attention on the agro-processing sector in the past put both the producer and the consumer at a disadvantage and it also hurt the economy of the country. The FPI has very low processing level *i.e.*, only 2.2 per cent for fruits and vegetables, around 35 per cent in milk, 21 per cent in meat and 6 per cent in poultry products, which is significantly lower by international standards. The processing of food produce is around 40 per cent in China, 30 per cent in Thailand, 70 per cent in Brazil, 78 per cent in the Philippines and 80 per cent in Malaysia. Value addition to agriculture produce in India is just 20 per cent with wastage estimated to be valued at around US $13 billion. The present levels of post-production losses are 5-15 per cent in durables, 20-30 per cent in semi-perishables and 30-50 per cent in perishables. A substantial amount of these losses could be prevented, if appropriate agro-processing centers having backward linkage with farmers to ensure constant supply of quality raw materials are established and operated. The following are the important constraints or problems inhibiting the growth of the agro-processing industries:

- ☆ Limited availability of varieties suitable for processing and this affects processing in two ways. Firstly, it causes low processing volumes. For instance, in India only about 2.2 per cent of total fruit and vegetables grown are processed and is far lower compared to developed countries and even other developing countries. Secondly, it results in high processing costs. For instance, to produce one tonne of orange juice in India, 14 tonnes of raw material is needed, whereas only about 10 tonnes is needed in other countries.
- ☆ The raw agricultural produce is highly perishable and requires a reasonable storage and transportation system, which does not exist on scientific lines in India. The lack of these facilities causes huge annual wastage. Government plans to upgrade cold storage facilities continue to be on the back burner. The existing facilities are highly inadequate to keep pace with rising production of agricultural commodities in the country thereby, adversely affecting the adequate and timely availability of raw material to agro-processing industries. Further, absence of effective backward linkages between the processing industries and farmers cause mismatches between industry production schedules and the supply of raw materials. This also leads to less utilization of installed capacity of processing plants.
- ☆ Processing of agricultural produce is concentrated mainly in unorganized and tiny sectors, where economies of scale are not possible. Attainment of international acceptable quality products in these small units is difficult.
- ☆ The technology employed by the majority of the processing units (especially, small scale units) are not technologically upgraded, and they employ obsolete technology in processing,

packaging and distribution of the processed products. Such products are less competitive in the international market.

- Weather disturbances such as drought, floods etc., in the country severely affect supply and raise prices of the raw material to the processing industries.
- The capital requirement is very high both for starting and operating the processing firms. For example, the fruit (mango) processing firm runs only during the season of any particular fruit, which means that, the production (mango juice) done during the fruit season has to be sold during the entire year. Inventory requirements are also high due to storage requirements of material, like bottles required for nationwide marketing. Such factors lead to large working capital requirements for the firms. The financial sector is reluctant to provide loans to a new and untested industry like fruit processing. Further, majority of the rural agro-processing industries are small in size and practice traditional technologies. The production levels are usually limited and their marketing growth is usually slow. This slows down the returns and it hampers the re-investment and expansion of the industry. More often, these factors also do not favour financial institutions in providing loans needed.
- Inadequate support services and collaboration between research stations, training institutions, private sector consultants, small enterprise advisors, research institutions and engineering workshops.
- Erratic supply and increased cost of fuel coupled with frequent power cuts affect the functioning of processing firms
- The technology available in India with reference to agro-processing equipment sector is not much advanced when compared to the developed countries. In India, the major thrust in R&D in the agro-processing sector has been on the processing of agricultural produce rather on developing equipments in this sector. Most of the technologies available in the equipment sector, which could be considered as globally competitive fall in the category of pre-harvest technologies rather than on post-harvest technologies. Hence, it is high time now, to strengthen R&D to focus on upgrading the processing equipment sector considering the high priority for processed products in the liberalized trade regime.
- The level of food processing in India is very low, as by and large, Indian consumers prefer fresh foods to processed products, unlike the trend in developed countries. This is because, many Indian consumers perceive that, processed food reduces food value and that processing and packaging increases their cost.
- There is a need for appropriate technologies for the processing of many crops. This is particularly the case for the traditional products. Most research in processing sector is on technologies suitable for medium or large-scale production with little effort directed at small-scale agro-processing.
- The export market information available to many major firms in the processing industry was not adequate. This results in putting many limitations on the exporters, when exploring the new markets.
- There is a serious dearth of trained human resource in food processing activities. The public universities relating to food science and technology are loathed, rigid and reluctant to re-shape, redesign courses embracing the changing needs of the economy. Further, not much research work has so far been done in this country, particularly on processing and packaging development. Lot more of empirical research and development in this field, including

development of marketing systems for the up-coming processed products is essential. Since not enough was done on R&D, India is lacking proper intermediate technologies to give momentum for the processing activities. Processing revolution requires appropriate production and handling technologies for many agricultural produces in India.

- The raw materials are of sub-standard grades, because of inconsistency in supply and availability. These also usually affect the production flow and the marketing set-up.
- Proper hygiene and sanitary practices are often neglected.
- Low price-elasticity of demand for processed food products.
- The processing methodology and equipment is often crude and has been used over many years. Efforts to mechanize, improve the efficiency and modernize the operations have been insufficient. Production efficiency in many of the industries is therefore, low due to the various laborious operations.
- Quality product development was not given due attention and the output was only expanded to meet growing demands and changing consumption habits.
- Lack of awareness of the importance of product presentation. Many good quality products do not sell well because of poor packaging and presentation.
- Lack of managerial and marketing know-how leads to poor promotion, marketing and distribution of processed products, which results in slow production and low returns.
- Small food industries have been known to indiscriminately use chemicals such as preservatives, stabilizers, colouring materials and non-nutritive sweeteners in food preparations.
- Different varieties of agricultural crops have been developed primarily for increasing yields and they lack the quality characters required for making processed products.
- The operating capacity of agro processing enterprises is very low. It is mainly because, the investments are made without taking the future market conditions into consideration thereby, leading to poor marketing and sales. Further, most enterprises run on a small scale and they have limited ability to do product development by themselves or adopt technology for improving quality. Many firms experience difficulties in coping rapidly with market changes owing to lack of new technology, development of new products and improvement of processing technology.
- The share of raw material cost in production of processed products' cost is relatively higher, compared to other costs. This is because, at times of harvesting season of a crop, the raw material cost is low, but it is not easy to obtain the required quantity of necessary material (that too at lower prices) steadily throughout the year.
- Many of the small and medium-sized FPIs are engaged in primary processing and not in secondary or tertiary processing, which reduces the scope for value addition.
- Most of the food processing enterprises have difficulties in securing professional workers and procuring labour during the busy season. Due to the seasonal variations of agricultural production, the labour demand increases suddenly at certain periods and high dependency on aged female labourers also hampers productivity. Low operating capacity and the heavy burden of high costs for raw material and personnel are the primary elements that increase the production cost, which subsequently causes inactivity in sales.

- A time lag exists between current technologies in the agro-processing firms and educational practices in the institutions. So, inadequacy of in-depth technical understanding and lack of management skills in the work force restricts innovation and consistent performance. The physical working environment for many agro-processing firms is generally uncomfortable and the compensation incentives and labour practices are not always at par with other labour intensive industries. The seasonal nature of production and processing compounds difficulties and insecurities for workers.
- The small and unorganized sector accounts for 75 per cent of the total FPI. There are only a few organized large industrial houses, which make their presence felt in the food-processing arena in India. The FPI's were neglected during previous planning periods. The liberalized economy, poses a threat to the small and unorganized sector from a marketing point of view. When products from other countries are available in the Indian market at lower prices, the local small processing firms are facing severe threat in marketing their products so that, some of them had to close down in a competitive business environment.
- The multiple and complicated tax regimes have made the FPI uncompetitive. The tax levels imposed on processed foods in India are among the highest in the world (Tables 5.43 and 5.44). Few, if any, other countries in the world impose excise duty on locally processed food. There is also a distinction between branded and un-branded products for taxation in India. Excise duty (0-24 per cent), sales tax (8-28 per cent), octroi tax, mandi and samiti tax, taxes of the local bodies for entry and exit increase the tax burden on the product from 27 to 35 per cent. India is also one of the few countries to levy tax on the machinery and equipment for FPI. Because of these various taxes and duties, the product becomes very highly priced and cannot compete both in the domestic and international markets consequently, the processing firms have to leave the market. The investments in food processing involve huge risk and yield low returns. Investments are further depressed, despite the advantage of offering hygiene and quality products, as the price sensitive Indian consumers has stayed away from the high priced packaged foodstuffs.

Table 5.43: Excise and Sales tax structure on processed food products across the countries.

Country	*Excise Tax*	*Sales Tax*
India	0-24	8-28
UK	Nil	Nil
Ireland	Nil	Nil
Thailand	Nil	7
USA	Nil	Nil
Malaysia	Nil	Nil

Source: <www.mofpi.nic.in>

The prices of processed food products in the country are forced to remain high owing to the imposition of high import duties (Tables 5.45 and 5.46) ranging from 20 per cent to over 60 per cent. India, traditionally has maintained high tariffs and import restrictions on food items. The various duties levied on imported processed food products into India include Basic duty, Additional duty of

Table 5.44: Tax incidence on different processed food products in India.

Product	*Tax Incidence (Per cent)*
Biscuits	20.8
Condensed milk	26.9
Jams	14.6

Source: <www.mofpi.nic.in>

customs (Countervailing duty (CVD)), Education Cess and Special additional duty. Apart from these four duties, additional levies such as antidumping duties and safeguard duties may also be applied. It may be noted that, at present, additional duty of customs and special additional duty are nil for most food products.

Table 5.45: Import duties on select food products.

Sl.No.	*Category*	*Basic Duty (%)*	*CVD*	*Cess (%)*	*Additional Duty (%)*	*Total Customs Duty (%)*
1.	**Fruits**					
	✩ Apples	50	0	3	0	52.02
	✩ Oranges, grapes, pineapples, guavas, mangoes	40	0	3	0	41.61
	✩ Pears and quinces	35	0	3	0	36.41
	✩ Other fruits	30	0	3	0	31.21
2.	**Vegetables**					
	✩ Potatoes	30	0	3	4	36.50
	✩ Other vegetables	30	0	3	0	31.21
3.	**Fresh and processed meat**					
	✩ Fresh, chilled and frozen meat of bovine, swine, sheep, goats	30	0	3	4	36.50
	✩ Sausages and similar products	100	0	3	4	111.15
4.	**Fresh and processed fish**					
	✩ Fresh	30	0	3	0	31.21
	✩ Processed	30	0	3	4	36.50
5.	**Dairy products**					
	✩ Skimmed milk, milk food for babies	60	0	3	4	68.95
	✩ Yogurt, butter milk, cheese and curd	30	0	3	0	31.21
	✩ Butter, dairy spreads, ghee, butter oil, whey, ice cream	30	0	3	4	36.50
6.	**Preparations of cereals, flour, starch of milk and pastry products**					
	✩ Pasta (cooked or uncooked)	30	0	3	4	36.50
7.	**Confectionary (chocolate and sugar)**	30	10.5	3	8	48.09
8.	**Processed fruits and vegetables**					
	✩ Juices, jams, jellies, marmalade, sauce, soups	30	0	3	4	36.50
9.	**Other food products**					
	✩ Natural honey	60	0	3	4	68.96
	✩ Olive oil					
	✩ Virgin	0	0	0	0	0
	✩ Edible Grade	0	0	0	0	0

Source: www.mofpi.nic.in.

Table 5.46: Customs duty on select imported food products.

Sl.No.	Category	Customs Duty (Per cent)
1.	**Fruits**	
	☆ Bananas	31.21
	☆ Dates, figs, pineapples, avocados, guavas, mangoes	41.61
	☆ Oranges	41.61
	☆ Grapes	41.61
	☆ Melons	31.21
	☆ Apples	52.02
	☆ Pears and qunaces	36.41
	☆ Apricots, cherries, peaches, plums	31.21
	☆ Strawberry, Raspberry, Kiwi fruit, mulberry	31.21
	☆ Pomegranate, Tamarind, Litchi, Chico	31.21
2.	**Vegetables**	
	☆ Potatoes	36.50
	☆ Tomatoes	31.21
	☆ Onions, shallots, garlic, leeks	31.21
	☆ Cabbages, Cauliflowers, Kohirata	31.21
	☆ Lettuce, chicory	31.21
	☆ Carrots, turnips, salad beetroot, radish, celeriac	31.21
	☆ Cucumbers, ghernics	31.21
	☆ Peas, beans, other leguminous vegetables	31.21
	☆ Asparagus, Aubergines, Mushrooms, Olives, Green chilly, Pumpkin, Spinach	31.21
	☆ Green pepper	31.21
3.	**Fresh and processed meat**	
	☆ Fresh, chilled and frozen meat of bovine, swine, sheeep, goats	36.50
	☆ Sausages and similar products	111.15
4.	**Fresh and processed fish**	
	☆ Fresh, chilled and frozen	31.21
	☆ Processed marine products (crab, shrimp, prawn, lobster, squid, octopus, tuna, culltefish, oysters, salmon, herrings)	36.50
5.	**Dairy products**	
	☆ Skimmed milk, milk for babies	68.95
	☆ Yogurt, Butter milk	31.21
	☆ Butter, Dairy spreads, Ghee, Butter oil	36.50
	☆ Whey	36.50
	☆ Cheese and curd	31.21
	☆ Ice cream	36.50

Contd...

Table 2.1–*Contd...*

Sl.No.	*Category*	*Customs Duty (Per cent)*
6.	**Preparation of cereals, flour, starch or milk and pastry products**	
	✩ Pasta (cooked or uncooked	36.50
	✩ Corn flakes	48.09
	✩ Sweet biscuits	59.19
	✩ Pastries and cakes	13.00
7.	**Fruit juices**	41.91
8.	**Confectionary**	
	✩ Chocolate confectionary	48.09
	✩ Sugar confectionary	48.09
9.	**Processed fruits and vegetables**	
	✩ Jams, jellies, marmalade	36.50
	✩ Sauce (soya, tomato, mustard, chilli, mayonnaise)	36.50
10.	**Other food products**	
	✩ Natural honey	68.96
	✩ Olive oil	0.00
	✩ Virgin	0.00
	✩ Edible grade	0.00

Source: www.mofpi.nic.in.

- ✩ The existence of multifarious laws and multiple authorities is another hurdle. Many of the laws were framed some 50 years ago like PFA Act, Meat Food Products Order, Essential Commodities Act etc., which were meant to serve the purpose of food safety and to prevent adulteration. The plethora of laws covering food and processed items, involving multiple agencies, at present, hinders growth and makes administration of the processed food industry difficult.
- ✩ Almost all the FPIs, particularly the small and medium ones, are in the hands of the private sector, which do not have access to latest technology, because of the prohibitive cost of technology. Even in the Information Technology (IT) age, many Indian agro-processing industries do not have a database, one of the major hindrances in setting up new industries.
- ✩ The MOFPI was demoted to the status of a Department under the Ministry of Agriculture in October 1999. This diluted the focus attached to the sector by the Government.
- ✩ Majority of exports from India in the food processing sector are in the primary processed foods category. Nearly 75 per cent of this industry is unorganized. Hence, the sector is not able to reap the benefits of economies of scale. Also there is lack of investment in technology in cold storage, transportation and supply chain systems. This has curtailed productivity, reflecting on the slow growth in competitiveness even though the food industry is firmly propelled by rapidly increasing consumption expenditure.
- ✩ Despite a gradual increase in the demand for processed food products, the market is still relatively small, as the demand for such products is mainly from the higher-income groups only.

- Lack of consistent supplies of brands is another major issue faced by retailers in India. Very few importers deal with particular brands for a long period of time. Most of them, buy products that are fast moving and shift to other brands or products in a short span leaving the Indian consumers to remain one time buyers of most imported products.
- A number of imported food products do not adhere to the basic guidelines of the PFA Act such as, stating the vegetarian and non-vegetarian symbols on the package, spelling out the manufacturing and expiry dates, not giving directions for use and other information in English and so on.
- Infrastructure challenges such as lack of sufficient cold storages, poor development of ports and related infrastructure in India is a hindrance to the import of food products, especially the ones with a shorter shelf life.

ix. Challenges in Agro-Processing

India is predominantly an agricultural economy with 60-65 per cent of her population living in villages and earns their livelihood through agriculture and allied activities. Rural population of India was 91 per cent in 1901 and may decline to 50 per cent by 2020. Population of India by 2011-12 is estimated to be 1221 millions with an annual compound growth rate of 1.80 per cent. Rural people migrate to urban areas for employment and better amenities, as such opportunities are not adequately available in rural areas. However, these could be created through selective mechanization of agriculture and appropriate post-harvest processing and value addition to the agricultural produce in the production catchment. To keep pace with the present population growth and consumption pattern, food grain requirement of India has been estimated to be 300 m tonnes by 2025. But, the post-harvest losses are at significant level, 5-15 per cent in durables, 20-30 per cent in semi-perishables and 30-40 per cent in perishables. About 50 per cent of these post-harvested losses can be prevented by using appropriate post-harvest approaches. Here lies the importance of agro-processing, as it ensures presenting the product in near natural form with added convenience. Considering these setbacks in the Indian agricultural sector, the following are the challenges for strengthening agro-processing:

- Postharvest losses account for 25-40 per cent of total agricultural production
- Due to inefficient management of raw material availability, even the installed capacity of agro-processing industries is not fully exploited.
- Inadequate cold chain facilities
- The local FPI face increasing competition from developed nations and even from other developing nations like Pakistan, Thailand etc. The market shares of Indian processed products are insignificant in the international market and they face stiff competition from established, large supply-ing countries, stringent SPS barriers or high protectionism. This calls for upgrading of processing technologies suiting to the importers' needs.
- The SPS standards prescribed for processed products by the developed nations are too difficult for India to maintain and further, these standards vary across the member countries
- Frequent failure or interruption of power supply
- Unequipped processing laboratories
- Lack of adequate trained manpower
- Lack of coordination links among different stakeholders of agro-processing sector
- The recent slowdown in the Indian economy is of serious concern, as the growth rate has been tending down for the last three years. Presently, agricultural growth has dropped to 0.9 per cent from the average of 3.9 per cent in eighties and 3.3 per cent in the nineties. So, it

is high time now, to strengthen agro-processing sector and this should start from agricultural research institutions and laboratories to help farmers in scientific cultivation, particularly to produce quality raw material that are best suitable for production and export of agro-processed products. This further calls for more and more mechanization and automation in processing and subsequent functions and utilization of agri-wastes generated during processing. As world markets are undergoing a major transformation, there is immense need to modernize the processing mills in the country, so as to enjoy export competitiveness. In this context, closer farmer-processor relationship is essential and this should be in a win-win proposition, with reduction of intermediaries.

- There is a convergence of thought at all levels that, the strengthening of agro-processing function in the production catchments ensures more employment and income to the beneficiaries. This further helps in stabilizing the market prices through removing vulnerabilities of perishables by transforming them into semi-perishables or durables. But, the present scenario of declining public investments, weak public-private linkages and lack of coordination between them etc., impede the development processes towards agro-processing. Besides these, there are also production and productivity related constraints that impede acceleration in agro-processing. Many of the crop varieties are tailored for table consumption purposes, when used for processing the product recoveries tend to be low, much lower than the varieties available with the developed countries.
- The most important challenges among others in the food processing sector include avoidance of the significant wastage at every level and in value addition. A study undertaken by the Central Institute of Post Harvest Engineering and Technology (CIPHET) has calculated the wastage in various produce (Table 5.47).

Table 5.47: Wastage in the various commodity groups.

Commodity Group	*Cumulative Wastage (%)*
Cereals	3.9 to 6.0
Pulses	4.3 to 6.1
Oilseeds	6.0
Fruits & vegetables	5.8 to 18.0
Milk	0.8
Fisheries	2.9
Meat	2.3
Poultry	3.7

Source: CIPHET, Ludhiana.

One of the most critical aspects in the value chain, wastage happens at all levels in the value chain. The following (Table 5.48) are the different factors responsible for wastage at different levels of post harvest management and value addition.

Table 5.48: Wastage in agricultural commodities.

Agency	*Factors Responsible for Wastage*	*Agency*	*Factors Responsible for Wastage*
Farmer	Improper harvesting techniques Poor packaging Poor transportation	Wholesaler	Moisture loss, poor handling Poor transportation, multiple handling Poor storage
Retailer	Poor handling Poor transportation Handling by customers Moisture loss		

Source: CIPHET, Ludhiana.

x. Suggestions to Strengthen Agro-Processing Sector in India

India accounts for less than 1.5 per cent of international food trade despite being one of the world's major food producers, which indicates huge potential for both investors and exporters. With rapid increase in the per capita income and purchasing power along with increased urbanization, improved standards of living, there lies a large untapped opportunity to cater to 1000 million domestic consumers. It is estimated that, 300 million upper and middle classes consume processed food. But, the FPI in India has traditionally been confined mainly to a few commodities such as sugar, edible oils, tea, coffee and spices. In recent times, this sector has been expanding and diversifying into new commodities such as fruits, vegetables, meat (poultry), dairy products etc. So, it is essential to strengthen this sector and the following are the important areas:

- ✰ Strengthening both backward and forward linkages of processing function. The strengthening of backward linkage include,
  - ❖ Establishment of a sustained and lasting linkage between the farmers and the processors regarding efficient utilization of existing infrastructure.
  - ❖ Mechanism to reduce the gap between the farm gate price of agricultural produce and the final price paid by the processor.
  - ❖ Development of futures markets in the best interest of both the farmers and the processors such that, it ensures price stability to the farmer and a sustained supply of raw material to the processor.

  The strengthening of forward linkage include,
  - ❖ Establishment of a strong linkage between the processor and the market to effect cost economies by elimination of avoidable intermediaries.
  - ❖ Establishment of other requisite market infrastructure that ensures prompt and reliable supply of quality products to the consumers.
- ✰ Opportunities presented by contract farming must be exploited.
- ✰ Exploitation of increasing health and safety awareness of the Indian consumer - this would pave the way for value added products on a health platform.
- ✰ Enhance investments in supply chain in order to minimize wastage of raw material and ensure efficient transportation of processed products.
- ✰ In order to produce high value processed products that will be well-received by consumers, it is necessary that, product development be based on the concept of 'meeting consumer needs' rather than 'from available local products'.
- ✰ Since the developed varieties of agricultural crops lack the quality characters required for making processed products, R&D programmes must focus on developing new varieties for specific end-use and processing qualities.
- ✰ There should be strengthening of extension services to the farmers and co-operatives in the areas of post harvest management of agro-produce to encourage creation of pre-processing facilities near the farms like washing, fumigation, packaging etc. There should be efforts to encourage setting up of agro-processing facilities as close to the area of production as possible to avoid wastage and reduce transportation cost.
- ✰ In the modern days of agri-business, food processing firms can be able to access raw material supply units around the world through the internet. So, information dissemination about availability of raw materials, latest processing technologies, literature on modern machinery,

contact details of suppliers etc., is essential. Hence, a Technology Bank could be set up for this purpose and this would enable development and adoption of modern technologies in agro-processing. This could be facilitated by a mechanism helping quick transfer of technologies to field through a net work of R&D institutions. It is heartening to say that, significant increase has taken place after early fifties in the number of institutions engaged in agro processing research. The SAUs are offering degree, post-graduate and doctoral programmes. Among R&D organizations in the area of agro-processing, ICAR has established several institutes with some component of Post Harvest Technology, CSIR has established laboratories, SAUs and IITs have various research programmes. Some of the leading Government funded R&D Institutes include: CFTRI, Mysore; CIPHET, Ludhiana; IARI, New Delhi; NDRI, Karnal; DFRL, Mysore; CIAE, Bhopal; IIT, Kharagpur; GPBUA&T, Pantnagar; IGMRI, Hapur; TNAU, Coimbatore; PAU, Ludhiana; GAU, Anand; RAU, Udaipur; BCKV, Kalyani; OTRI, Anantpur; PPRC, Thanjavur; MERADO, Ludhiana; MPKV, Rahuri; ILRI, Ranchi; IVRI, Izatnagar; NIRJAFT, Kolkata; CIRCOT, Mumbai; IISR, Lucknow; IGFRI, Jhansi; KVIC, Mumbai; HBTI, Kanpur and PHT Institute, Pune. The ICAR also executed various All India Coordinated Research Projects in various important areas like post-harvest technology, processing, handling, and storage of commodities, application of plastics in agriculture etc.

- Government should selectively support agro-processing firms according to their market performance. That is, the Government should facilitate in providing all the requisite processing technology on the modern lines in these selective firms, so as to enable them enjoy large scale economies.
- Policies of the agro-processing firms must be market-oriented, rather than production-oriented. This is because, when the strategies are production oriented, many firms have experienced difficulties at times of sales expansion. Each product in the market enjoys a definite life cycle, as the tastes and preferences of the consumers are fast changing. In order to meet consumer demand, the processing firms should continuously monitor consumer's preferences and the market situation.
- Government should support agro-processing firms in terms of technical aspects and management skills because, most of the firms do not have specialists to develop new products and improve their quality. Also the entrepreneurs have limited business experiences and management ability. So, it is essential to promote effective linkages between various stakeholders of agro-processing sector development.
- For the proficient enterprises, continuous support for operating funds is necessary. The financial assistance at the startup of the business is not enough to continue the business because, it is only for providing facilities. Therefore, additional assistance is needed for successful operation. If the enterprise is proficient and has potential then, Government may provide loans for operating costs such as raw material procurement costs and wages. Government could set up a special soft loan programme to support all activities involved with agro-processing development. International funding may even be encouraged to flow into this sector.
- There should be a contract between farmers' associations or farmers and processing firm, to stabilize the management system. Farmers should have a common understanding that, they all will profit from the system, when they co-operate. Through this arrangement, the processing firm can provide bulk-purchased seeds, fertilizers and pesticides to the farmers at lower cost and at standardized quality. The farmers' supply raw materials according to

the contract for the processing firm. The processing firm will then share the profit with farmers from the product sale. In this way a strong management system can be established.– Processing of agricultural raw material generates a sizable amount of utilizable by-products, commonly termed as 'waste'. Experimental protocols for converting these into usable coproducts are available. These need to be developed into commercially viable technologies. These ventures would be a success, if main product manufacturing is undertaken in consonance with the by-product processing. Government regulations and policies must be suitably amended to facilitate this.

☆ Establishment of basic infrastructure such as roads, water supply and power is recognized as essential for the growth of agro-processing industries. Selected post-harvest facilities like cool stores and drying facilities as well as farm-to-factory roads are to be strengthened for the proper delivery of raw materials. In order to address the problem of infrastructure in food processing sector, the Government has implemented the scheme for infrastructure development comprising the following components:

  ❖ ***Food Park Scheme***: Since small and medium scale entrepreneurs are shortage of capital in investing in food processing sector, the Government started establishing 'food parks' in different parts of the country. These food parks provide common facilities like cold storage, food testing and analysis lab, effluent treatment plant, common processing facilities, packaging center, power supply, water supply, seminar/conference/training facilities etc., to different stakeholders of food processing sector.

  ❖ ***Packaging Centres:*** The scheme aims to provide facilities for packaging, which may help in enhancement of shelf life of food products and make them internationally acceptable.

  ❖ ***Integrated Cold Chain Facility:*** The scheme is intended to improve viability of cold storage and enhance cold storage capacity.

  ❖ ***Value Added Centre:*** The scheme is intended to provide value addition to extend shelf-life and to facilitate traceability.

  ❖ ***Irradiation Facilities:*** The scheme aims at enhancing shelf life of the food product through irradiation techniques by preventing infestation like in flour, sprouting and change in chemical composition of the product (as in potato).

  ❖ ***Modernized Abattoir:*** The scheme aims at scientific and hygienic slaughter, causing least pain to the cattle and ensuring better byproduct utilization.

☆ The CFTRI, Mysore, a premier institute in the field of technology development in food sector, and other institutes like APEDA, MPEDA, NFPDB etc., are making constant endeavors to popularize and propagate traditional food technologies. It is high time now, these institutes should focus on generating cost-effective processing technologies in tune with the latest developments in the developed nations. This will exert significant impact on the social and economic structure of the rural areas. In this context, it is even essential to call for PPPs to strengthen the food-processing sector.

☆ Comprehensive processing units for developing products as well as by-products should be set up. For example, wheat flour mills can earn added benefits by undertaking wheat germ and oil production in consonance. Rice processing units should simultaneously handle de-husking and processing of bran for food, feed, oil and biochemicals, as is being done in many other countries. One such unit has recently come up in Punjab.

- Policy and legislation must be amended to enable the processors to procure raw produce from the farmers directly. This ensures twin benefits *viz.*, the processing firms get the raw produce in time and by procuring the produce at the announced support price by the processors, distress sales at the farmers' level can be effectively checked. Further, the burden on Government's exchequer can be reduced by involving processors to procure the raw produce at the announced support prices. Since marketing of processed products is more remunerative than sale of raw commodities, farmer - processor linkages are needed to add value as per demands of the consumers.
- Separate Boards should be established commodity-wise (which enjoy both domestic and export competitiveness) and they should focus on all the activities right from R&D, production scenario for quality output, post-harvest handling to processing them and marketing the processed products including image building both in the domestic and international markets. Strategy formulation at the State and National level, development of requisite infrastructure and technology, updating the products suiting to changing tastes and preferences of the consumers etc., should also be the role of these boards.
- The New Generation Co-operatives model should be adopted, as it binds both farmers and processors to honour commitments and agreements. The same concept should be adopted even in the existing co-operative units. Further, contract farming involving small farmers should be promoted rather than involving a few big farmers.
- Agriculture must diversify in favour of high-value enterprises. The emphasis should be on production of high value commodities like, fruits, vegetables and fish with enhanced quality and specific nutritional and processing characteristics, rather than increasing production *per se* as in the past. Pricing policies also need to be changed, as linking these with the quality of the produce or a product is the basis for fixing per unit price, just as fat content in milk; higher protein quality/quantity in wheat; better aroma or cooking quality in rice and shelf life of fruits and vegetables.
- The Town and Village Enterprises (TVEs) model of China is an excellent example for involving surplus rural labor in industrial activity by providing them alternative work at their doorstep. This can be considered for adoption with suitable modifications, for primary processing of the agricultural produce and effectively linked to urban units for secondary processing, product development and marketing.
- Location of food-processing units should be strategically placed depending upon the raw material availability, labour, product utilization and domestic and/or export marketing. It should be nurtured to evolve on a natural course after the initial nucleation, as has been done for the IT industry.
- Realizing its importance, the food-processing industry has been identified as a focus area for development and has been included in the priority-lending sector. So, the Government of India has to formulate policies and plans for the improvement of food processing infrastructure, including upgrading technology and the enforcement of quality standards and activating the domestic market with a focus on exports. The Government should establish a National bank, on the lines of NABARD, to lend credit to FPIs. This will ensure speedy disbursals of the funds to food processing sector. Also, the State Governments should play a catalytic role in partnership with banks, financial institutions and technical and management institutions so that, small and unorganized players become globally competitive. The following are the few policy guidelines:

- ❖ Agro-processing industry is included in the priority-lending sector. Most have been exempted from the provisions of industrial licensing under Industries (Development and Regulation) Act, 1951.
- ❖ Where foreign investment is concerned, automatic approval for even 100 per cent equity is available for the majority of processed foods.
- ❖ The expenditure outlay of Ministry/Department of Food Processing has been substantially increased during the successive FYPs.

☆ In terms of policy support, the MOFPI has taken the following initiatives

Regulation and Control

- ❖ FDI up to 100 per cent is permitted under the automatic route in the food infrastructure (food park, cold chain/warehousing).
- ❖ Most of the food-processing industries with the exception of beer and alcoholic drinks and items reserved for small scale sector, like vinegar, bread, and bakery have been exempted from the provisions of industrial licensing under Industries (Development and Regulation) Act, 1951.
- ❖ Automatic approval to FDI up to 100 per cent equity in FPIs sector is available excluding alcoholic beverages and a few reserved items.
- ❖ FDI up to 100 per cent is permitted on the automatic route for distillation and brewing of alcohol subject to licensing by the appropriate authority.
- ❖ No industrial license is required for almost all of the food and agro processing industries except for some items like beer, potable alcohol and wines, cane sugar etc.

Fiscal policy and Taxation

- ❖ Fruits and vegetables, and dairy machineries are completely exempted from central excise duty. Central excise duty on preparation of meat, poultry and fish, pectin, pats and yeast is also completely exempt.
- ❖ Quantity restrictions on all food products have been removed. Peak rate of customs duty has been reduced from 30 to 25 per cent (excluding agricultural and dairy products) and duty structure on designated items has been rationalized.
- ❖ Customs duty on refrigerated goods transport vehicles has been reduced form 20 to 10 per cent.
- ❖ Excise Duty of 16 per cent on dairy machinery has been fully waived off and excise duty on meat, poultry and fish products has been reduced from 16 per cent to 8 per cent

Export promotion

- ❖ Food-processing industry is one of the thrust areas identified for exports. Free Trade Zones (FTZ) and Export Processing Zones (EPZ) have been set up with all infrastructures. Also, setting up of 100 per cent Export Oriented Units (EOU) is encouraged in other areas. They may import free of duty all types of goods, including capital foods.
- ❖ Capital goods, including spares up to 20 per cent of the Cost, Insurance and Freight (CIF) value of the capital goods may be imported at a concessional rate of customs duty. Export linked duty free imports are also allowed.
- ❖ 50 per cent of the production of EPZ/FTZ and 100 per cent EOU units are saleable in domestic tariff area.

- All profits from export sales are completely free from corporate taxes.
- Setting up of 60 agri-zones for end-to-end development for export of specific product from geographically contiguous areas.
- 53 food parks approved to enable small and medium food and beverage units to set up and to use capital intensive common facilities such as cold storage, warehouse, quality control labs, effluent treatment plant, etc.

Above all, an integrated approach is essential to strengthen the agro-processing sector in India. There should be active involvement of all the stakeholders in the strengthening of agro-processing sector. The various stakeholders include R&D organizations, Universities, private sector, NGOs, farmers' organizations etc. The Universities and Research organizations should educate and train the personnel and develop need based technologies for updating the processing technologies. Since public investment is on the decline in the post-liberalization era, there is a greater need to enhance investments from the private sector. To encourage private sector in the establishment of agro-processing industries, the Government of India, MOFPI has formulated two models of Food Park for food-processing industries apart from other incentives. They include, Industrial Estate Model under which, the entrepreneurs are encouraged to set up agro processing industries in well-developed, well laid out industrial plots. The developer of the food park also provide common facilities, like water supply, storage, laboratories, warehousing facilities, common effluent treatment plant and other facilities. Another model, which is being tried out in Punjab State, where all common facilities are located in a hub with agro-processing units functions in a radius of 50 kms. In this second model, the initial capital outlay is less and small agro-processing units and farmers can utilize the services at a reasonable cost. Besides private sector, even NGOs and farmers' associations should actively involve in investing in the processing sector in terms of establishment of critical infrastructure such as cold storage, grading and cleaning facilities. For this, the banking sector should liberalize the credit disbursement, as agro-processing sector is recognized as one of the priority sectors for economic development of the country. The NGOs need to continue to be active with their catalytic role, in activities relating to capacity building, awareness building, empowerment etc. The farmers associations should focus on production (including planning for production) and marketing suiting to the changes in consumers' tastes and preferences. This will also be helpful for reducing cost of production because of economies of scale of operation. This will widen the market supplies and empower the producers to be able to bargain with the corporate anchor industries (Figure 5.17). These desirable changes will increase food processing in India from an existing low level of two per cent up to a double digit figure. It would require an investment of ₹ 14 trillion in the food processing sector. This investment would generate direct employment for about 7.7 million persons and indirect employment for about 30 million. This could also reduce food wastage worth ₹ 800 billion. Considering these desirable changes in the food processing sector, MOFPI in its Vision 2015 document has rightly estimated that, the size of processed food sector to treble, processing level of perishable to increase from six per cent to 20 per cent, value addition to increase from 20 per cent to 35 per cent and India's share in global food trade to increase from 1.5 per cent to three per cent. Table 5.49 shows the other projections that can be achieved through integrated planning in food processing sector by 2014-15. The Government's focus towards FPI as a priority sector will ensure policies to support investment in this sector and attract more FDI. India with its vast pool of natural resources and growing technical knowledge base has strong comparative advantages over other nations. According to CII estimates, food-processing sector has the potential of attracting US $33 billion of investment in 10 years and generate employment of nine million person-days. Thus, the food-processing sector in India is clearly an attractive sector for investment and offers significant growth potential to investors.

Regulations by Government	Stakeholders	Resource endowment	Market structure	Demand conditions
• Tax exemptions for new entrants into the industry • FDI up to 100 per cent is permitted under the automatic route in the food infrastructure	• Agricultural Universities •Research stations •Farmers associations •NGOs • Private sector	India is the second largest producer of food next to China, world's third-largest producer of food grains, the second - largest producer of fruits and vegetables and the largest producer of milk; it also has the largest number of livestock	• Highly fragmented industry • Insignificant FDI levels • High prices of processed products •The industry is mainly unorganized with 75 per cent of the processing units in the unorganized category.	• Rise in per capita income of consumers • Rapid industrialization • Increasing preferences for ready-to-eat and ready-to-cook processed foods

FOOD-PROCESSING SECTOR IN INDIA

Figure 5.17: Factors influencing growth of food-processing sector in India.

Table 5.49: Future growth of food processing sector in India.

Items	*2003-04 ($billion)*	*2014-15 ($billion)*
Total Food Consumption	205	—
Processed foods	126	274
Primary processed food	79	136
Value Added Food	48	138
Share of value added products in food production (%)	16	50

Source: www.mofpi.nic.in

xi. Plan Schemes Under Food-Processing Sector

1. Scheme for Infrastructure Development

The creation of integrated and holistic infrastructure is extremely important for the food processing sector. Towards achieving this end, the MOFPI had launched new schemes in eleventh FYP for the creation of modern enabling infrastructure that can facilitate the growth of food processing and an integrated cold chain mechanism for handling perishable produce. The MOFPI initiatives include a launch of the following schemes for strengthening infrastructure in food processing sector:

A. Mega Food Parks Scheme (MFPS)

The MFPS is the flagship programme of the MOFPI during the eleventh Five Year Plan. The Scheme aims to accelerate the growth of the FPI in the country by facilitating establishment of strong food processing infrastructure backed by an efficient supply chain. The MFPS provides for a capital grant of 50 per cent of the project cost in difficult and Institute for Transportation and Development Policy (*ITDP*) notified areas (with a ceiling of ₹ 50 crores). The grant shall be utilized towards creation of common infrastructure in Central Processing Centre (CPC) and Primary Processing Centres (PPCs) in the park. Such facilities are expected to complement the processing activities of the units proposed to be set up at the CPC in the Park. Each Mega Food Park may take about 30-36 months to be completed. Out of 30 Mega Food Parks proposed during the eleventh five year plan, the Ministry has taken up 15 projects under the Scheme so far. Among these, final approval has been accorded to 15 Mega Food Parks in the states of Andhra Pradesh, Punjab, Jharkhand, Assam, West Bengal, Uttarakhand, Tamil Nadu, Maharashtra, Gujarat, Uttar Pradesh, Madhya Pradesh, Tripura, Odisha, Bihar and Karnataka. The total assistance from the Government to these projects is estimated at ₹ 750 crore.

B. Scheme for Cold Chain, Value Addition and Preservation Infrastructure

The Task Force on Cold Chain set up by the Ministry of Agriculture has identified a huge gap of 9 to 10 million tonnes of cold storage capacity in the country. The MOFPI through its Scheme for Cold Chain, Value Addition, and Preservation Infrastructure has been successfully addressing the above issue. The Scheme was approved in 2008 with an objective to provide integrated and complete cold chain, value addition and preservation infrastructure facilities without any break, for perishables from the farm gate to the consumer. The assistance under the Scheme includes financial assistance (grant-in-aid) of 50 per cent of the total cost of plant and machinery and technical civil works in General areas and 75 per cent for the NE region and difficult areas subject to a maximum of ₹ 10 crore. In the first phase, the Ministry has approved 10 integrated cold chain projects in 2008-09, which are already being implemented in different parts of the country. Out of the 10 projects, 8 have started commercial operation. Substantive value addition, reduction in wastage and enhancement in farmers' income is evident from concurrent evaluation of the projects. In the second phase, 39 projects have been approved. The approved projects envisage creating a cold chain capacity of about 1.60 lakh metric tonnes. Taking note of the high demand and the gap in the requirement of cold storage, processing, preservation and cold logistics facilities in India, the Ministry is planning to upscale the Scheme and the Planning Commission has already accorded an 'In-principle' approval for the same.

C. Construction and Modernization of Abattoirs

The Ministry has approved 10 projects in first phase which are at various stages of progress. Two projects have been completed. A proposal for up-scaling the scheme is under consideration.

2. Scheme for Technology Upgradation, Establishment, Modernization of FPIs

Under the Scheme for Technology Upgradation, Establishment, Modernization of FPIs, financial assistance is provided in the form of grants-in-aid for the setting up of new food processing units as well as Technological Upgradation and Expansion of existing units in the country. The Ministry extends financial assistance in the form of grant-in-aid to entrepreneurs at 25 per cent of the cost of Plant and Machinery and Technical Civil Works subject to a maximum of ₹ 50 lakhs in general areas or 33.33 per cent subject to a maximum of ₹ 75 lakhs in difficult terrains. The implementation process of the Scheme has been made more transparent and decentralized from 2007 onwards.

3. Quality Assurance, Codex Standards and Research and Development and Promotional Activities

In the global market today, quality and food safety gives a competitive edge, which is an important factor for the enterprises producing processed foods and providing services. Apart from domestic standards for food products, processes and management practices, Codex prescribes international standards for safety and quality of food as well as codes of good manufacturing practices, which are accepted worldwide. Further, equal emphasis is required to be accorded to R&D activities for the development of innovative products, cost effective processes and efficient technologies for the food processing sectors. The scheme for food safety, codex and R&D has been successful in making a dent in this area in the country. The following are the components of Food Safety Schemes:

- ☆ Setting up/Up gradation of Food Testing Laboratories.
- ☆ Implementation of HACCP/ISO/GMP/Safety Management System in food processing units.
- ☆ Research and Development in food processing sector.
- ☆ Promotional activities including advertisement and publicity.

xii. Emphasis for Food-Processing in 12th FYP

India's food-processing sector, still at a nascent stage, has a share of around six per cent of GDP. Its sales turnover is around ₹ 2.8 trillion and it employs around 13 million people. Availability of raw materials, changing lifestyles and favourable policies has given a considerable push to the sector's growth. The Indian food processing sector holds considerable significance, because of the linkages and synergies that it promotes between the two key sectors of the economy, *i.e.*, primary sector (agriculture) and secondary sector (manufacturing). Strengthening this link is of critical importance to improve the value of agricultural produce, ensure remunerative prices to farmers and at the same time create favourable demand for Indian agricultural products in the international market. The growth of food processing sector has nearly doubled to 13.7 per cent during the last four years and the MOFPI has set a growth target of 20 per cent by 2015. A dominant segment of the food industry, food processing is estimated to be worth ₹ 2784 billion with a 32 per cent share. Estimates by the MOFPI indicate that, by 2012-13, the size of the industry will be ₹ 5400 billion (Figure 5.18), around 40 per cent of the food industry. The following are the important aims formulated for the strengthening of food-processing sector by the MOFPI in the XII FYP:

- ☆ Providing greater assurance about safety and quality of food to consumers
- ☆ Enhancing the competitiveness of FPI in both domestic as well as international markets
- ☆ Making the sector attractive for both domestic and foreign investors
- ☆ Achieving integration of the food processing infrastructure from farm to market

xiii. Regulatory Framework for Food-Processing Sector in India

Different Laws and Acts govern the food processing sector in India and they aim at verifying the quality of food products that are made available to the consumers. This system of food regulation is considered to be one of the best in the world. The laws/regulations prescribe various standards regarding food additives, contaminants, food colours, preservatives and labeling. In order to rationalize the multiplicity of food laws, a Group of Ministers was recently set up to suggest legislative and other changes to formulate a modern, integrated food law, which will be a single reference point in relation to the regulation of food products. The food laws in India are enforced by the Director General of Health Services, Ministry of Health and Family Welfare, Government of India. There are different laws that govern the food-processing sector in India. They include,

- ☆ ***PFA Act, 1954 and Rules (Ministry of Health and Family Welfare):*** Covers specifications related to food colour, preservatives, pesticide residues, packaging and labeling, and regulation of sales.
- ☆ ***The Standards of Weights and Measures Act, 1976 and Standards of Weights and Measures (Packaged Commodities) Rules, 1977:*** Designed to establish fair trade practices with respect to packaged commodities. Agriculture Produce (Grading and Marking) Act (Ministry of Rural Development).

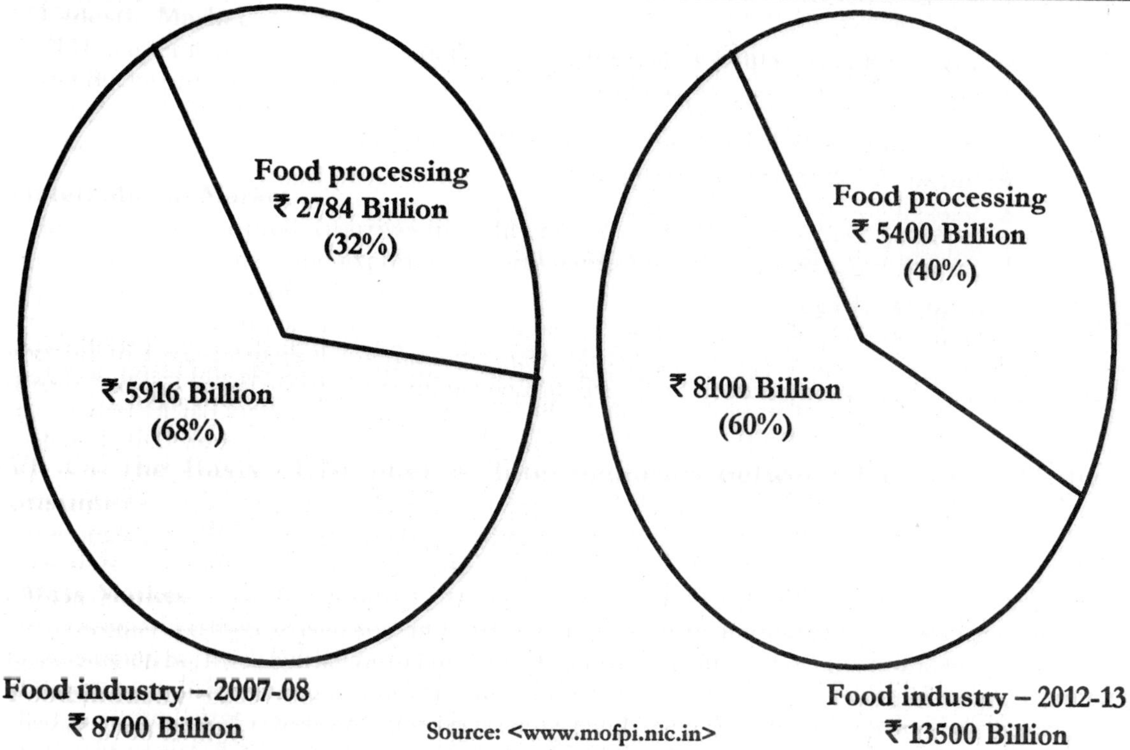

Figure 5.18: Share of food processing in total Indian food industry (2007-08 vs 2012-13).

- ☆ ***Agriculture Produce (Grading and Marking) Act, 1937*** (Ministry of Rural Development).
- ☆ ***Essential Commodities Act, 1955*** (Ministry of Food and Consumer Affairs): to ensure availability of essential commodities to the consumers and to protect them from exploitation by unscrupulous traders.
- ☆ ***Fruit Products Order (FPO), 1995:*** Specifications and quality control requirements regarding the production and marketing of processed fruits and vegetables, sweetened aerated water, vinegar, and synethic syrups.
- ☆ ***Meat Food Products Order, 1973:*** Administers the permissible quantity of heavy metals, preservatives, and insecticide residues for meat products.
- ☆ ***Milk and Milk Products Order, 1992:*** Regulates the production, distribution, and supply of milk products; establishes sanitary requirements for dairies, machinery and premises; and sets quality control standards for milk and milk products.
- ☆ ***The Food Safety and Standards Act, 2006:*** In August 2006, the Government of India had passed a new legislation Food Safety and Standards Act. The Act proposes establishment of a new authority, the Food Safety and Standards Authority, reorganization of scientific support pertaining to the food chain through the establishment of an independent risk assessment body and a new Food Law, merging eight separate Acts.
    - ❖ The Infant Milk Substitutes, Feeding Bottles and Infant Foods (Regulation of Production, Supply and Distribution) Act, 1992 and Rules 1993.

- The Insecticide Act, 1968.
- Export (Quality Control and Inspection) Act, 1963.
- Environment Protection Act, 1986.
- Pollution Control (Ministry of Environment and Forests).
- Bureau of Indian Standards Act, 1986.
- Vegetable Oils Products Control Order, 1947
- Solvent Extracted-Oil (Control) Order, 1967.

xiv. Future Outlook of FPIs

The above discussion infers that, FPI has evolved into a modern industry from a traditional, small-scale production system. This industry today caters to the diverse needs and tastes of Indian consumers. As discussed earlier, rapid change in the lifestyle of the consumers particularly urban dwellers, rise in disposable incomes, increase in the number of working women, growth of nuclear and double income families, explosion of the information and communication sector, adequacy of raw materials production, increase in literacy rate, rapid growth in urbanization, rising per capita incomes etc., are responsible for drastic increase in the demand for processed food in India. Further, in the context of trade liberalization, the Indian consumer is witnessing a sea change in his food habits. This offers great marketing opportunities in the days to come. The issues that deserve attention are the proper infrastructure and Government support. Presently, a lack of proper standardization of the agricultural output and indiscriminate use of pesticides and chemical fertilizers makes the processed foods unacceptable in the international markets. If the Indian FPI has to match with the international standards, a thorough development of the technology, both at the production of agricultural products and the processing stages is essential. The processing technologies must be dynamic (updated) in tune with various changes like change in technology, change in capital-labour price ratio, change in managerial skills, change in the demand for product quality in the market and change in the volume to be processed. No doubt, the potential is huge, but what is required is the will power to match with the very best in the world. Keeping these aspects in view, MOFPI has finalized the Vision 2015 Document emphasizing the crucial role of FPIs to exploit both the domestic and international markets.

D. Storage

We know, agricultural produce is characterized by its perishability and seasonality, but the demand for their consumption is regular. Hence, it is essential to store or stock the produce, keeping in view of the demand requirements from the consumers. For example, sugarcane produced by the farmers is processed in sugar factories and then the sugar is made available to the consumers or to firms engaged in producing chocolates, sweets etc. We know, sugarcane is produced once in a year, but there is a need to supply the same to the sugar factories on continuous basis for their viability. So, here comes the importance of 'storage' of sugarcane produce by the farmers and sugar factories in their godowns. These sugar factories purchase raw sugarcane at times of peak harvesting season (generally the prices are low) and stock the produce in their godowns and utilize the same throughout the year for running the sugar factory. Again, after production of sugar, it requires some time for sale or distribution. Thus, the need for storage arises both for raw material as well as for finished products. Thus, storage involves proper arrangement for preserving goods from the time of their production or purchase till the actual use.

i. Need for Storage of Agricultural Commodities

Storage of agricultural commodities is necessary due the following reasons:

- ☆ ***Seasonality of production:*** As discussed above, since agricultural production is seasonal, but consumption is regular, storage of commodities or products is essential for their regularity of supply to the needy consumers.
- ☆ ***Seasonal demand:*** For some commodities like water melon, demand arises only during summer season and not/very low in winter and rainy seasons. So, to make them available to the consumers in plenty during summer season, their production is to be done throughout the year and store the produce in a godown to make them available at the time of consumption.
- ☆ ***Large-scale production:*** In the modern days of agri-business, the production of commodities or processed products (sugar) is done to meet both the existing as well as future demand of the products. Further, the entrepreneurs produce the goods in large quantities, so as to enjoy the benefits of economies of large scale. So the products, which are produced on a large scale, need to be stored properly till they are cleared by sales.
- ☆ ***Quick supply:*** The production of agricultural commodities is carried out in accordance with the principle of comparative advantage. So, some commodities are produced at some locations only (basmati rice, apples, shrimp etc), but they are to be made available throughout the country. So, it is essential to stock these commodities near the place of consumption so that, without making any delay these commodities are made available to the consumers at the time of their need.
- ☆ ***Continuous production:*** For production of agricultural commodities, the farmers employ several resources like fertilizers, pesticides, weedicides etc. At times of sowing season, the prices of these resources are too high and hence, the farmers purchase them during off-season at low prices, stock them and employ the same during the crop production programmes.
- ☆ ***Price stabilization:*** Storage of agricultural commodities or products ensures price stabilization. For example, at times of peak marketing season, the supply of commodities into the market increases thereby, their prices fall drastically in the market. So, by storing the commodities at times of peak marketing season, their supply into the market decreases and prices will get stabilized. Similarly, at times of scarcity of supply of commodities, their prices drastically increase in the market. So, by destocking these commodities, their supply into the market increases and this leads gradual fall in the prices. Thus, by maintaining a balance of supply of goods, storage function leads to price stabilization.
- ☆ ***Spoilage of produce:*** After harvesting the produce, it is subjected to spoilage by various agencies like:
  - ❖ ***Biotic agents:*** Insects, rodents, birds, fungi, mites, bacteria
  - ❖ ***Abiotic agents:*** Moisture and temperature

It is estimated that, storage losses due to insects is highest (2.55 per cent) followed by rodents (2.50 per cent), birds (0.85 per cent) and moisture (0.68 per cent). So, the total post harvest losses accounts to 9.33 per cent. Further, while performing other functions like threshing, transport and processing, the losses amounts to 1.68, 0.15 and 0.92 per cent respectively. Besides these losses, there is also degradation in quality, loss in weight, poor germinability, discolouration and finally poor marketability. Therefore, it is important to protect grains from loss by providing scientific storage.

- ✩ ***Perishability:*** Considering the perishability of agricultural produce, it is essential to have scientific storage facilities, so as to preserve it for a longer period of time. Examples include fruits, vegetables, milk, fish, eggs etc.

Besides these, storage of agricultural produce is also essential in view of regular supply of goods in all seasons, to guard against delayed shipments, to adjust the demand and supply position of the products in the market and to promote the orderly markets.

ii. Requirements of Good Storage Structures

A good storage structure should satisfy the following requirements:

- ✩ It should provide maximum possible protection from both biotic and abiotic agents
- ✩ It should permit aeration and fumigation when required
- ✩ It should provide the necessary facility for inspection, disinfection, loading, unloading, cleaning and reconditioning
- ✩ It should facilitate proper cleaning and should be self cleaning, if it is silo
- ✩ It should be economical on unit storage cost basis
- ✩ It should be elevated and away from moist places in the house
- ✩ It should be airtight
- ✩ The area surrounding the structure should be clean to minimize insect breeding
- ✩ It should be plastered with an impervious clay layer to avoid termite attack or attack by other insects.

Best Storage Performance

To ensure best storage performance of a commodity,

- ✩ The produce must be thoroughly cleaned and graded
- ✩ The produce must be dried to the safe storage moisture level of 10-12 per cent for food grains and 7-9 per cent for oil seeds (on wet basis) for a safe storage period of 6-12 months.
- ✩ Storage structures should to be properly repaired, cleaned and disinfected
- ✩ Structures should bear the load of grain stored and do not permit contact/exchange with outside humid air
- ✩ Structures should be constructed in the coolest part of the house/farm.

iii. Steps Necessary for Good Storage of Agricultural Produce

The quantitative and qualitative losses in storage may be on account of the activities of both biotic and abiotic agents. Damage by insect pests results in the food grain becoming weevilled, causing losses not only in weight but also in food value. Leakage of water through roof, and dampness in the structure through floors and side walls make the grain deteriorate in many ways. One of the several effects is excessive oxidation, which causes a rise in temperature of the grain, including its heating and cake formation. Moisture also encourages fungus, mould and termite growth, at times rendering the entire grain stock unfit for human consumption. Beside this, rats cause serious loss by eating or breaking the grains into pieces. The losses in quantity also occur in threshing yards, processing plants and during handling. It is estimated that, annually 10 per cent of the food grains are lost on account of losses during the post-harvest operations. Further, due to inadequate and unscientific methods of storage, the losses have been estimated to range between 1.5 to 5 per cent. It is roughly estimated that, about 2/3rd of the total grain produced is retained by the farmers for their food and seeds. The remaining 1/3rd quantity is considered as marketable surplus. The total storage capacity available with various agencies of the Government is of the order of about 10 per cent. So, it is obviously quite inadequate and need of storage facilities should be increased considerably in order to reduce losses. The following steps are to be followed for ensuring good storage of agricultural produce:

Before Storage

- ☆ Checking for leakage of rain water and sufficiency of drainage facilities
- ☆ Hygiene maintenance of storage area
- ☆ Assessment of capacity of the storage facility
- ☆ Pesticidal treatment
- ☆ Security and fire fighting arrangements
- ☆ Repairs to available storage equipment

After Receipt of Produce

- ☆ Inspection for variety and quality of produce
- ☆ Inspection about the moisture content in the produce and any infestation on the produce and assess the nature of infestation,
- ☆ Check the quantity of produce received

During Storage

- ☆ Maintenance of hygiene
- ☆ Ensure proper aeration in the storage area
- ☆ Checking for leakage after rains
- ☆ Inspection for insects, rats and mites at fortnightly intervals
- ☆ Pesticidal treatments necessarily based on observations

iv. Classification of Storage Structures

About 65-70 per cent of the total food grains produced in India is retained by farmers for their self-consumption or meeting their other financial requirements. The food grains at farm level are stored in traditional as well as in modern storage structures. A large number of traditional rural grain storage structures are found in India. These structures are made of locally available cheaper materials like mud, bamboo, wood, paddy and wheat straw and stones. Food grains are stored in bulk in these storage structures, which are neither rodent proof nor moisture proof. These structures cannot protect the food grains from insects, pests, fungi and rodents. There are estimates that, substantial quantity of food grains (about 6 to 10 per cent of total production) are damaged in these storage receptacles due to moisture, insects, rodents and fungi. The storage structures designed for storing agricultural produce can be conveniently classified based on several criteria (Figure 5.19).

1. Based on the Location of Storage Structure

According to this criterion, there are two types of storage structures and they are discussed here under.

- ☆ ***Underground storage structures:*** Holding grain in bulk in underground is an age old method of rural storage. Wheat, paddy, sorghum etc., can be stored underground for a period of two

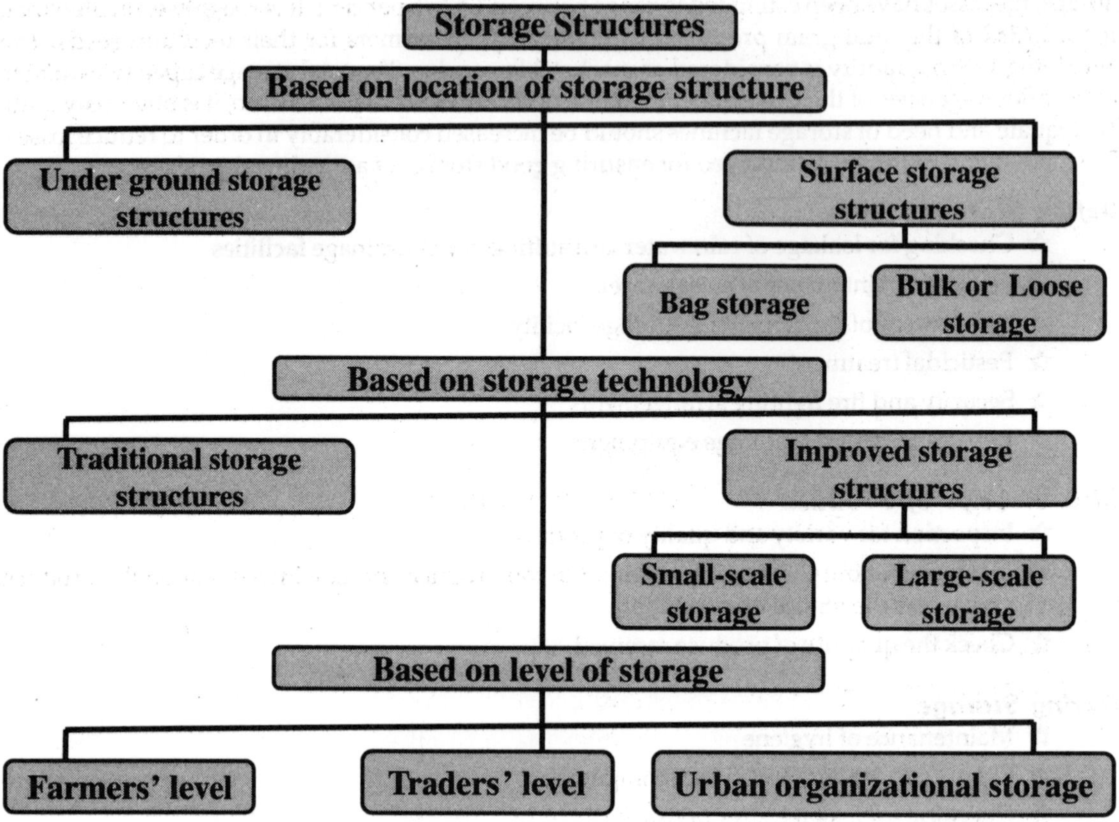

Figure 5.19: Classification of storage structures.

years. These structures are simple underground dig-outs up to a depth of five meters varying in sizes to hold from a small quantity up to 50 tonnes. Sometimes, the pits are lined with brick or concrete so that, moisture from walls and bottom does not damage the grain. At the time of filling grain in the pit, a layer of straw is placed on all sides. After the pit is filled, straw is spread over the grain and then topped with a layer of soil. The structures may be circular or rectangular in shape. The capacity varies with the volume and nature of produce to be stored. The underground storage structures are common in dry parts of Karnataka, Uttar Pradesh, Bihar, Madhya Pradesh, Maharashtra and Andhra Pradesh. These storage structures are safer from threats from various external sources of damage, such as theft, rain or wind and they can also be temporarily utilized for some other purposes with minor adjustments. The depletion of oxygen in the underground storage structures discourages infestation. This underground storage is cheaper over surface storage structures. However, this underground structure is not suitable for high rainfall and high water-table areas, as seepage of moisture results in mould growth and grain damage by rodent is possible. Further, the grain stored underground have poor appearance and musty smell. Hence, different types of surface storage structures are used to store the produce.

☆ ***Surface storage structures:*** They can be sub-divided into Bag storage and Bulk or Loose storage. Food grains are generally stored either in bags or in bulk. A combined system of bag-cum-bulk storage is also practiced in some parts of the country. In villages, the bulk storage system is more common than the storage in bags, which is considered to be a practicable method of storing grains in the Government godowns as well as in trade.

- ***Bag storage:*** The size of a bag used for grain storage is large enough to contain about 90-100 kg of grain. These bags are made of jute fiber and are commonly known as gunny bags. In bag storage, the produce is stored either inside a warehouse or in the open and then covered by tarpaulin sheets. These structures are easy to handle at times of loading, transportation and unloading the produce. Further, it is easier to keep separate lots with identification marks on the bags. Even if the produce was infested, it can be treated easily. Each bag contains a definite quantity which can be bought, sold or dispatched without difficulty. In the tropical regions, the grain is stored in bags. Storage in bags requires considerable labour, but the minimum investment is enough on permanent structures and equipment. The storage in bags has the advantage of being short-term storage. Bag storage can be done under a roof of Galvanized Iron sheets, a plastic covering where grain is intended for very early onward movement. Usually no control measures against insects are needed for short-term storage. If bag storage produce is intended for long time, the control measures have to be taken against insect pests.
- ***Bulk or loose or open storage:*** In this, the produce is stored in concrete and/or metal bins. The exposed peripheral surface area per unit weight of grain is less. Consequently, the danger of damage from external sources is reduced. Further, pest infestation is less because of almost airtight conditions in the deeper layers. When compared to bag storage, bulk storage is more efficient because,
  - it reduces congestion at the depots by not allowing for bagged produce to be dumped all over the depot yard
  - reduces handling costs
  - saves expenditure on bags, tarpaulin and fumigation and lowers storage losses
  - no difficulty in loading and unloading of grain
  - easy inspection (save labour and time)
  - low running costs
  - less labour is involved in loading and unloading and there is no need of investment in purchasing gunny bags.
  - efficient and effective fumigation operation
  - less land area requirement
  - complete control of aeration
  - possible to store the grain for longer periods
  - possible to mechanize all operations
  - possible to store moist grain for short periods
  - the insect infestation is also lower over bag storage

But, bulk storage is disadvantageous over bag storage in the sense that, its initial investment is high, it is inflexible in terms of not being easily expandable to cope with changes in intake and off-take levels, it relies heavily on an efficient transport system and it needs skilled manpower to run and maintain the entire system. But, the bag storage ideally overcomes the above problems associated with a bulk system. But, bag storage itself is not free from the disadvantages such as, higher quality losses in storage due to insect pests and rodents, frequent expenditure on the purchase of bags and tarpaulin sheets.

Sometimes, the farmers use mud (Mud bins) and paddy straw (Straw bins) to construct surface storage structures. Both these are neither rodent-proof, nor secure from fungal and insect attack. On average, out of a total six per cent loss of food grains in such storage structures, about half is due to rodents, and half to insects and fungi.

- ☆ ***Mud bins or Kachha Kothi:*** The mud bins are made of unburnt clay mixed with straw with 1 to 3 inches thick wall and are oval, rectangular or circular. A small hole is provided at the base for taking out the grain and a larger hole is provided at the top for filling it with grain. Both the inlet and outlet holes are plugged, while grain is stored. The bottom portion of the wall section and floor is made in the form of rings, which can be put one over the other, so as to make the structure of desired capacity. Both the sides of the structure are plastered with mud and cow dung. These mud bins are available in Uttar Pradesh, Haryana, Bihar and Madhya Pradesh and are used for the storage of wheat, paddy, rice, maize and pulses. However, the food grains stored in mud bins can be easily damaged by insects, rodents and moisture.
- ☆ ***Paddy straw bins:*** For storing paddy in humid zones, dried plants are used for making temporary structures, which after being filled with grain are further reinforced from outside by winding paddy straw ropes around the whole structure. Each structure holds 2 to 6 quintals of grain. These structures generally store 1-6 quintals of paddy grain. Sometimes, palm leaves are also incorporated in these structures to provide them extra strength and safety from heavy rains, which are prevalent in these regions. Such type of structures is used in Jharkhand, West Bengal and Odisha.

2. Based on the Storage Technology Employed

Based on this criterion, the commonly used grain storage structures in India can be classified into Traditional storage structures and Improved storage structures.

a. Traditional Storage Structures

In villages, agricultural produce is stored in traditional structures of different shapes and sizes with capacity varying from 2 to 5 tonnes. Various R&D organizations in India have identified some proven, age-old structures for stocking agricultural produce and the important structures are discussed below.

- ☆ ***Earthen pots and gourds:*** These storage structures are common in the dry tropics and are useful for storing small quantities. They should be kept under a shelter. By treating the pots and gourds (the hard, dried outside cases of certain fruits or vegetables) with varnish, paint or linseed oil, and sealing the lids with mud or cow dung, an airtight form of storage is obtained. These structures are suitable for storing of cereals, pulses, oilseeds etc., for about a year period.
- ☆ ***Baskets:*** These structures are also suitable in the dry tropics and they are raised above the ground floor, as they can be damaged by the rodents. They also do not give enough protection against insects, but this can be improved by applying mud, clay or cow dung to the inside and outside the structure. To make the basket airtight, the basket should be plastered. This

structure facilitates to apply pesticides by dusting inside the box or mixing the same with grain. This structure is suitable for storing cereals, pulses, oilseeds etc., and they can be stored up to 6 - 9 months.

- ☆ ***Jute sacks:*** These sacks are especially suitable for the dry tropics. These sacks should be placed above the ground floor, as they absorb the moisture quickly. So, these structures should be elevated and preferably they have to be placed on plastic sheets or on wooden pallets. Further, these sacks should not be stacked against the walls, as insects can get into the grain from the walls. While stacking the sacks, space should be left between the stacks so that, air moves freely and this also permits to use fumigants. But, the main disadvantage is that, these sacks do not give much natural protection against insects, rodents, fungi and moisture. Further, they are easily damaged during transport and handling. This structure is useful for storing cereals, pulses, oilseeds, potatoes and the produce can be stored up to a year.
- ☆ ***Plastic bags:*** These are suitable for storage of agricultural produce both in humid and dry tropics. It is essential that, before stocking the produce in these bags, it must be well-dried because, during storage further drying is not possible. Like jute bags, these plastic bags also do not offer much protection against rodents thereby, special attention is required. Further, due attention is required that the stocked produce with its sharp ends should not damage the plastic bags. Since, plastic becomes weak or brittle after continued exposure to ultra violet light; they should be frequently replaced for every 2-3 years. This structure is suitable for storage of cereals, pulses, oilseeds, copra etc.

b. Improved Storage Structures

The Government of India has made efforts to promote improved storage facilities at the farm level. Various State Agricultural Universities (SAUs) and Research centers functioning under ICAR have designed improved storage structures suiting to handle different quantities of agricultural produce. The improved storage structures may be small-scale and large-scale storage structures.

i. Small-scale Storage Structures

- ☆ ***PAU bin*** : This is a galvanized metal iron structure. It's capacity ranges from 1.5 to 15 quintals. It was designed by Punjab Agricultural University.
- ☆ ***Pusa bin:*** Developed by the Indian Agricultural Research Institute. A typical 'Pusa' bin has a foundation of bricks, compacted earth, or stabilized earth. A polyethylene sheet is laid on this, followed by a concrete slab floor 10 cm thick. An internal wall of the desired height (usually 1.5 to 2 metres) is constructed of bricks or compacted earth, with a sheet of polyethylene wrapped around it. This sheet is heat-sealed to the basal sheet, and the external wall is then erected. During the construction of the wall an outlet pipe is built into its base. The concrete slab roof is supported by a wooden frame and, like the floor, is constructed of two layers separated by a polyethylene sheet. During its construction, a man-hole measuring 60 x 60 cm is built into one corner. The Pusa bin has been widely adopted in India, and has been demonstrated in some African countries. It gives good results when loaded with well dried grain. The grain and seed both remain safe in the bin for more than one year with proper precautions. It is suitable for the dry and wet tropics, provided it is protected from rain. It gives good protection against insects and rodents, especially if the bottom 50 cm of the outside walls and the bottom layer are made of fired bricks or concrete. This structure is suitable for storing cereals, pulses and oilseeds.

- ☆ ***Hapur tekka:*** This structure was designed by Indian Grain Storage Institute, Hapur, Uttar Pradesh. It is a cylindrical rubberized cloth structure supported by bamboo poles on a metal tube base, and has a small hole in the bottom through which grain can be removed.
- ☆ ***Coal tar Drum bin:*** This structure was developed and recommended by CIAE, Bhopal for grain storage at farm level. These are the used tar coal or bitumen drums, which are discarded after using the material in road construction works. These drums are heated by open fire to remove excess coal tar. A layer of bitumen remains inside the drum which serves as insulator and protects the galvanized iron sheet. A lid is fabricated by local artisans and its capacity is about 1.5 quintals. These are popular among farmers in Madhya Pradesh and Chhattisgarh.
- ☆ ***Domestic Hapur Metal Bins:*** These domestic grain bins have been developed by Indian Grain Storage Management and Research Institute (IGMRI) previously known as Indian Grain Storage Institute (IGSI) Hapur, Uttar Pradesh. These are made of galvanized iron sheet or aluminum sheet of 18 gauge and have covered inlet and outlets. Their capacity ranges from 2 to 10 quintals.

ii. Large Scale Storage Structures

- ☆ ***Cover and Plinth Storage (CAP):*** Another facet of the FCI's manifold activities is the provision of scientific storage for the millions of tonnes of food grains procured by it. In order to provide scientific storage of food grains, the FCI developed CAP storage structure. CAP storage is a term given to storage of food grains in the open with adequate precautions such as, rat and damp proof plinths, use of wooden dunnage and covering of stacks with specially fabricated polythene covers etc. It involves the construction of brick pillars to a height of 14" from the ground with grooves into which wooden crates are fixed for the stacking of bags of food grains. The structure can be fabricated in less than three weeks. It is an economical way of storage of food grains on a large scale and they can be maintained for 6-12 months. But, CAP storage is vulnerable to wind damage and the covers should be inspected frequently to detect damage. The system requires careful management, if severe losses are to be avoided. For example, only sound stock should be accepted at the plinth with a maximum of 14 per cent moisture content at 25°C ambient temperature, and regular opening of covers is necessary to allow some aeration. Careful quality control is achieved with regular sampling. The advantage of CAP is its low establishment cost, which is only one quarter the cost of godown storage.
- ☆ ***Pusa Cubicle:*** This is a room like structure (3.95 x 3.15 x 2.60 m), a modification of Pusa bin to provide large storage capacity of 24 tonnes on a platform of 3.73 m x 2.93 m x 0.07 m is made with unburnt bricks on a concrete floor (except 22 cm of outer sides with burnt bricks). A polyethylene sheet is placed on this platform and another platform of similar dimension is made with unburnt bricks. The 22 cm thick inner walls are constructed up to 2.6 m height. A wooden frame of 1.89m x 1.06 m for door is fixed in the front side of 3.95 m wall. The roof can be made by wooden beam placed at 15 cm distance and covered with unburnt bricks. This structure is air tight, moisture impervious, thermally insulated and rodent proof. The changes in weather conditions outside does not affect stored produce in this structure.
- ☆ ***Pusa Kothar:*** Storage is practiced in small compartments of a room (5.3 m x 2 m x 4 m) called kothar. The roof is constructed with the help of wooden poles and mud slabs, leaving near the front wall three filling holes each of 0.5 m x 0.5 m size. Two out lets of galvanized iron sheets of 15 cm diameter and 30 cm length are fitted at the floor wall on the front side. This

is a low cost storage structure and it saves the stored grains from insects and environmental hazards.

- ☆ ***Metal bins:*** Bins made of steel or Aluminum are used for storage of grains outside the house. These bins are fire and moisture proof. They have long durability and produced on commercial scale. The capacity ranges from 1 to 10 tonnes.
- ☆ ***Silos:*** Silos are huge bins made with steel/aluminium or concrete and they are usually circular in shape. A silo has facilities for loading and unloading grains. In these structures, the grains in bulk are unloaded on the conveyor belts and through mechanical operations, are carried to the storage structure. The storage capacity of each of these silos is around 25,000 tonnes.

3. Based on the Level of Storage

The grains are stored at three different levels, *viz.*, at the farmer's level (rural storage) trader's level and urban organizational storage. The urban organization storage uses modern facilities and structures like silos, warehouses and also undertakes periodical inspection and pesticidal treatment of grains for ensuring their quality during storage.

v. Different Storage Structures Employed for Stocking Agricultural Produce

Different types of storage structure will be employed for stocking the agricultural produce and they are discussed here under.

1. Farmers' Level

The most common means of storage of agricultural produce by the farmers are earthen pots, baskets, jute sacks, plastic bags, *kuthla* (cylindrical bin of mud and bricks mixed with straw or dung with a capacity of 1 quintal to 2 tonnes), *Katchi Kothi* (rectangular bin of mud bricks mixed with straw or dung with a capacity of 2 to 8 tonnes), *thekka*, a cylindrical rubberized cloth structure supported by bamboo poles on a metal tub base (with a capacity of 10-15 quintals) and the metal drums.

2. Rural Godowns

At rural level, farmers store the produce in their own houses and by using different structures. It is a known fact that, the marginal and small farmers are not economically sound to retain their produce till the market prices are favourable for sale. Considering the importance of rural storage in marketing of agricultural produce, in the year 1979, the Government of India launched a scheme called 'National Grid of Rural Godowns', with the main objective of creating a network of rural godowns in the States and UTs, primarily to meet the storage requirements of the small and marginal farmers. The other objectives of this scheme include,

- ☆ Prevention of distress sales of food grains and other agricultural commodities immediately after harvest
- ☆ Checking both quantity and quality losses arising at present, by storage in sub-standard places
- ☆ Reduction in pressure on transport system in post-harvest period
- ☆ Creation of employment opportunities in rural areas
- ☆ Helping the farmers in getting the loans against stored produce
- ☆ Helping in easy procurement of food grains by FCI

In the year 2001, the DMI, an attached organization of Ministry of Agriculture, Government of India initiated a Rural Godowns Scheme or Gramin Bhandaran Yojana (GBY) (a Central Sector Scheme for Construction of Rural Godowns) in collaboration with National Bank for Agriculture and Rural Development (NABARD) and National Co-operative Development Corporation (NCDC), to construct rural godowns with scientific facilities in rural areas and to establish a network of rural godowns in different States and UTs (Tables 5.50 to 5.52). These rural godowns are primarily meant for providing warehousing facilities to the farmers. The godowns are of 100 to 1000 m tonnes capacity. They are owned by FCI, Central Warehousing Corporation (CWC) and State Warehousing Corporations (SWCs), APMCs or co-operatives looking to the need for having storage structures or facilities for agricultural produce. The following are the important features of GBY:

- ✩ ***Subsidy:*** 33.33 per cent of the capital cost of the project in case of projects located in North – Eastern States, hilly areas and those belonging to Women Farmers/their self help groups/ co-operatives and SC/ST entrepreneurs and their Self-Help Groups (SHGs)/Co-operatives subject to a maximum ceiling on subsidy of ₹ 62.50 lakhs. No maximum ceiling on subsidy in the case of co-operatives assisted by NCDC. 25 per cent of the capital cost of the project to all categories of farmers (Other than Women Farmers), agriculture graduates, co-operatives and CWC/SWCs subject to a maximum ceiling on subsidy of ₹ 46.87 lakhs. No maximum ceiling on subsidy in the case of co-operatives assisted by NCDC. 15 per cent of the capital cost of the project to all other categories of individuals, companies and corporations etc., subject to a maximum ceiling on subsidy of ₹ 28.12 lakhs. 25 per cent of the capital cost of the project for renovation of godowns of co-operatives with assistance from NCDC.
- ✩ ***Eligibility:*** Individual farmers/group of companies/growers partnership proprietary firms, Non-Government Organizations (NGOs), SHGs, Co-operatives, Agro-processing Co-operative Societies, Marketing Boards and Agro processing corporations.
- ✩ ***Unit size/cost:*** Minimum size of godown 100 m tonnes or cost ₹ 2 lakhs and Maximum size of godown 10,000 m tonnes or cost ₹ 150 lakhs.

Objectives of GBY Scheme

They include:

- ✩ Creation of scientific storage capacity with allied facilities in rural areas to meet the requirements of farmers for storing farm produce, processed farm produce, agricultural inputs etc.
- ✩ Promotion of grading, standardization and quality control of agricultural produce to improve their marketability
- ✩ Strengthen agricultural marketing infrastructure in the country by paving the way for the introduction of a national system of warehouse receipts in the respect of agricultural commodities stored in such godowns
- ✩ Prevention of distress sale immediately after harvest by providing the facility of pledge financing and marketing credit and
- ✩ Reverse the declining trend of investment in the agriculture sector by encouraging the private and co-operative sectors to invest in the creation of storage infrastructure in the country.

Table 5.50: State-wise number of rural godowns in India.

State	Number of Rural Godowns	No.of Godowns Available per 1000 sq.km	Capacity	No. of Godowns per 1000 MT
Andhra Pradesh	310	1.19	1320000.00	58.38
Arunachal Pradesh	14	0.17	4000.00	13.97
Assam	2	0.03	10000.00	2.34
Bihar	334	3.55	33000.00	2.67
Chhattisgarh		0	0.00	0.00
Goa	0	0	0.00	0.00
Gujarat	97	0.49	90000.00	8.57
Haryana	217	4.91	1124000.00	67.93
Himachal Pradesh	40	0.72	3000.00	2.13
Jammu and Kashmir	0	0	0.00	0.00
Jharkhand		0	0.00	0.00
Karnataka	286	1.49	178000.00	14.25
Kerala	161	4.14	22000.00	36.67
Madhya Pradesh	332	1.08	601000.00	28.77
Maharashtra	38	0.12	132000.00	8.90
Manipur	9	0.40	1000.00	2.41
Meghalaya	19	0.85	2000.00	8.22
Mizoram	0	0	0.00	0.00
Nagaland	65	3.92	3000.00	5.12
Odisha	26	0.17	186000.00	24.54
Punjab	192	3.81	2464000.00	89.91
Rajasthan	22	0.06	2000.00	0.09
Sikkim	0	0	0.00	0.00
Tamil Nadu	106	0.82	16000.00	1.96
Tripura	0	0	0.00	0.00
Uttar Pradesh	92	0.39	453000.00	9.46
Uttarakhand		0	0.00	0.00
West Bengal	208	2.34	25000.00	1.48
All India	2570	0.79	6669000.00	25.44

Source: Compiled from the data obtained from DMI, MOA, GOI, Faridabad.

Table 5.51: Progress of Rural Godown Scheme (as on 31st March, 2011).

Agency	No.of Projects Sanctioned	Capacity in Lakh Tonnes	Subsidy Involved	Subsidy (Released)
NABARD	19715	244.37	862.48	649.83
NCDC (new construction)	2445	16.00	57.08	42.65
NCDC (renovation)	2546	22.90	6.14	3.97
Total	24706	283.27	925.69	696.45

Source: Compiled from the data obtained from DMI, MOA, GOI, Faridabad.

Table 5.52: State-wise progress of Rural Godowns Scheme (as on 31st March, 2011).

Sl.No.	State	Sanctioned by NABARD		Sanctioned by NCDC (New)		Sanctioned by NCDC (Renovation)		Total	
		No. of Projects	Capacity in Tonnes	No. of Projects	Capacity in Tonnes	No. of Projects	Capacity in Tonnes	No. of Projects	Capacity in Tonnes
1.	Andhra Pradesh	1027	3932050	25	38580	8	800	1060	3971430
2.	Arunachal Pradesh	1	945	0	0	0	0	1	945
3.	Assam	190	326459	1	650	0	0	191	327109
4.	Bihar	66	94941	659	69461	24	4800	749	169202
5.	Chhattisgarh	271	774071	77	258350	0	0	348	1032421
6.	Goa	3	290	0	0	0	0	3	290
7.	Gujarat	6427	1839229	64	176564	19	19600	6510	2035393
8.	Haryana	372	2253681	592	296933	525	274817	1489	2825431
9.	Himachal Pradesh	11	3122	45	5000	0	0	56	8122
10.	Jammu and Kashmir	4	6751	0	0	0	0	4	6751
11.	Jharkhand	6	8530	0	0	0	0	6	8530
12.	Karnataka	2409	1731985	164	129387	3	415	2576	1861787
13.	Kerala	38	33871	108	23280	0	0	146	57151
14.	Madhya Pradesh	1625	3563269	106	98300	64	75120	1795	3736689
15.	Maharashtra	2179	2731272	31	196880	53	132000	2263	3060152
16.	Meghalaya	8	11749	7	750	0	0	15	12499
17.	Mizoram	1	756	0	0	0	0	1	756
18.	Nagaland	2	250	0	0	0	0	2	250
19.	Odisha	292	548334	0	0	4	11000	296	559334
20.	Punjab	1238	2866002	18	3661	213	387066	1469	3256729
21.	Rajasthan	773	622835	96	41350	127	12700	996	676885
22.	Tamilnadu	159	492673	344	73500	943	120576	1446	686749
23.	Uttar Pradesh	264	1377197	15	115600	546	1230768	825	2723565
24.	Uttarakhand	122	265991	33	25250	4	18600	159	309841
25.	West Bengal	2227	950848	54	5550	13	1300	2294	957698
26.	UTs	0	0	0	0	0	0	0	0
27.	NAFED	0	0	05	30800	0	0	5	30800
28.	NCCF	0	0	01	10000	0	0	1	10000
	Total	19715	24437101	2445	1599846	2546	2289562	24706	28326509

Source: Compiled from the data obtained from DMI, MOA, GOI, Faridabad.

3. Mandi Godowns

Majority of the Indian farmers are small and marginal farmers thereby, they do not have the financial capability to retain their surplus produce till favourable prices prevail in the market. So, they are often compelled to sell their produce immediately after the harvest, when the prices are usually low. The only way out to overcome this problem is through creating infrastructure, which could provide them with safe and scientific storage of their produce and also make proper avenue for availing marketing finance against their stored produce. The creation of rural godowns coupled with the system of pledge finance is emerging as a befitting mechanism to benefit them in a great way. According to the RBI guidelines, loan/advances up to 75 per cent of the value of the produce stored in the godown can be advanced to farmers against pledge/hypothecation of agricultural produce (including warehouse receipts) subject to a ceiling of ₹ 1 lakh per borrower. Such loan shall be for a period of six months, which can be extended up to 12 months based on financing banks commercial judgment. The commercial/co-operative banks/RRBs provide credit to the farmers for the produce stored in the godown under this Pledge Loan Scheme. The banking institutions accept the godown receipt and sanction the pledge loan against hypothecation of produce, as per RBI guidelines. Farmers are given freedom to take back their produce once the pledge loan is repaid. Facility of pledge finance is extended to all farmers, whether they are the borrowing members of Primary Agricultural Credit Societies (PACS) or not and the District Central Co-operative Banks (DCCBs) directly finance individual farmers on the strength of the pledge. The major benefits of this pledge finance scheme include,

- ✰ Minimizes the farmers' dependence on the commission agents, as the pledge finance provides financial support to them immediately after harvest period.
- ✰ Participation of the farmers, irrespective of their land holding, helps in increasing the arrivals in market yards throughout the year.
- ✰ Gives a sense of security to the farmers even if their produce is not sold out in the market yard immediately.
- ✰ Increases the retention capacity of the small farmers to avoid distress sale.

Most of the States and UTs have enacted Agricultural Produce Market (Regulation) Act and as per this Act, the APMCs will provide godowns facilities in the market yard. At the time of keeping the produce in the APMC godown, a receipt is issued indicating the quality and weight of produce that is stored. The receipt is treated as negotiable instrument and eligible for pledge finance from the Scheduled Banks. The CWC and SWCs were also allowed to construct godowns in the market yards. Commission agents and traders also have their permanent storage godowns in the premises of market yards.

4. CWC

CWC was established during 1957. It is a premier warehousing agency in India and is the largest public warehouse operator in the country offering logistics services to a diverse group of clients. Presently (2010-11), CWC is operating 479 warehouses across the country with a storage capacity of 9.98 million tonnes providing warehousing services for more than 400 commodities including agricultural produce, industrial raw-materials, finished goods and variety of hygroscopic and perishable items. Warehousing activities of CWC include food grain warehouses, industrial warehousing, custom bonded warehouses, container freight stations, inland clearance depots and air cargo complexes. Apart from storage and handling, CWC also offers services in the area of clearing and forwarding, handling and transportation, procurement and distribution, disinfestation services, fumigation services and other ancillary activities. CWC also offers consultancy services/training for the construction of warehousing infrastructure to different agencies. During 2010-11, an additional storage capacity of 1.45 lakh m tonnes was created with a capital outlay of ₹ 65.00 crores for storing of food grains. So, the overall capacity utilization (occupancy) increased from 85 per cent to 88 per cent. The following are the important objectives of CWC:

- ☆ To meet the changing needs of agriculture, trade, industry and other sectors by providing scientific warehousing, logistic services and related infrastructural facilities.
- ☆ To assist in the implementation of the Warehousing (Development and Regulation) Act, 2007 with a view to expand credit through banking institutions against warehoused goods.
- ☆ Efficient human resource management in order to enhance productivity and customer's satisfaction and reduce losses during warehousing and distribution.
- ☆ To be a major player in providing pest control services
- ☆ To provide total quality service in the field of warehousing, logistic services and related activities with value addition to the customer's satisfaction.
- ☆ To acquire and build godowns and warehouses at suitable places in India or abroad as it thinks fit.
- ☆ To run warehouses for the storage of agricultural produce, seeds, manures, fertilizers, agricultural implements and notified commodities offered by individuals co-operative societies and other institutions.
- ☆ To arrange facilities for transport of agricultural produce, seeds, manures, fertilizers, agricultural implements and notified commodities to and from warehouses
- ☆ To subscribe to the share capital of SWCs.
- ☆ To act as an agent of the Government for the purposes of purchase, sale, storage and distribution of agricultural produce, seeds, manures, fertilizers agricultural implements and notified commodities.
- ☆ To enter into, with the previous approval of the Central Government, joint ventures with any Corporation established by or under any Central Act or any State Act or with any Company formed and registered under the Companies Act, 1956 including foreign company or through its subsidiary companies, for carrying out the purpose of this Act.

5. SWCs

Various States have set up their own warehouses in the country and first SWC was established in Bihar in 1956. The area of operation of the SWCs is district place of the state. The total share capital of the SWCs is contributed equally by the CWC and the concerned State Government. Thus, the SWCs are under the dual control of the State Government and the CWC.

In India, CWC and SWCs are the two Government controlled warehouses that provide agro-produce storage facility across the country. They provide a good networking across India because, the people can take delivery at the convenience of their nearest warehouse. The warehousing system also helps in the better integration of ready market for commodities across time and place thereby, facilitating better integration of spot and futures markets. This in turn improves efficiency of futures market in price discovery and price risk management functions. However, despite the storage facilities offered both by the Central and State Governments and Co-operatives, the farmers still retain and store about 60 per cent of total food grains production and use the same as seed, feed and payment of wages to labourers. So, only 40 per cent is marketable surplus and is handled by traders, co-operatives and Government agencies. Even with the advent of improved storage technology, the farmers are not utilizing them and they often resort to distress sales of produce, as they suffer from prior indebtedness and pressing family needs.

So, the scientific storage facilities offered by the Government *i.e.*, rural godowns, CWC and SWCs are being utilized mostly by the traders and less by the farmers. Further, the retentions made by the farmers in their own storage structures are not on scientific lines and hence, the extent of loss is on the high side. At farmers' level, post-harvest losses occur both in terms of quantity and quality. The produce is stored mostly in rooms as bulk storage, which is prone to extensive damage by fire accidents, rodents and insects and infected by diseases. Thus, the storage facilities are not on scientific lines and thus, 5-10 per cent losses in quantitative terms are incurred during storage period and these losses may still increase during rainy season.

6. FCI

Besides these two public sector agencies *i.e.*, CWC and SWCs, there is another agency *i.e.*, FCI, which is involved in large-scale storage of commodities in the country. While the FCI uses its warehouses mainly for storing food grains, the storage capacities with CWC and SWCs are used for the storage of food grains as well as other items. This FCI is the single largest agency in the country with a largest storage capacity meant for storing food grains. The FCI has the largest agricultural warehousing systems with over 24.33 million tonnes of storage capacity in over 1451 godowns located all over India. This includes owned as well as hired warehouses.

The FCI has a total of 31.61 m tonnes of storage capacity (Tables 5.53 and 5.54) including Covered and CAP both owned and hired with a capacity utilization of 91 per cent as on 1st April, 2011. Apart from storing in covered godowns, wheat is also stored in CAP which is also a scientific method of storage. Due to the increased procurement of food grains and to reduce the storage in CAP, the Government formulated a Scheme for Construction of Storage Godowns through private entrepreneurs, CWC and SWCs. Assessment of additional storage needs under the Scheme is based on the overall procurement/consumption and the storage space already available. For the consuming areas, storage capacity is to be created to meet four months requirement of PDS and other Welfare Schemes in a State. For the procurement areas, the highest stock levels in the last three years are considered to decide the storage capacity required.

Table 5.53: Storage capacity with FCI (m tonnes).

Capacity	*1st Apr. 2004*	*1st Apr. 2005*	*1st Apr. 2006*	*1st Apr. 2007*	*1st Apr. 2008*	*1st Apr. 2009*	*1st Apr. 2010*	*1st Apr. 2011*
Covered								
Owned	12.82	12.91	12.93	12.94	12.95	12.97	12.97	12.99
Hired	10.85	10.46	9.9	9.34	8.71	10.12	12.89	15.46
Total	23.67	23.37	22.83	22.28	21.66	23.09	25.86	28.45
CAP								
Owned	2.21	2.25	2.21	2.29	2.2	2.17	2.51	2.62
Hired	1.36	0.41	0.51	0.63	0.03	0.02	0.47	0.54
Total	3.57	2.66	2.72	2.92	2.23	2.19	2.98	3.16
Grand Total	27.24	27.03	25.55	25.2	23.89	25.28	28.84	31.61

Source: <fciweb.nic.in>

Table 5.54: State-wise storage capacity with the FCI, as on 30/09/2008) (Lakh m tonnes).

Region/U.T	*Covered*							*CAP*			
	FCI Owned	*Hired*						*Owned*	*Hired*	*Total CAP*	*Grand Total*
		State Govt.	*CWC*	*SWC*	*Pvt. Parties*	*Total Hired*	*Total Covered*				
East Zone											
Bihar	3.66	0.03	0.54	0.72	0.48	1.77	5.43	0	0	0	5.43
Jharkhand	0.66	0.02	0.13	0.18	0.2	0.53	1.19	0	0	0	1.19
Odisha	2.93	0	0.92	2.84	0.15	3.91	6.84	0	0	0	6.84
West Bengal	8.59	0.21	0.76	0	0.59	1.56	10.15	0	0	0	10.15
Sikkim	0.1	0.01	0	0	0	0.01	0.11	0	0	0	0.11
Total	15.94	0.27	2.35	3.74	1.42	7.78	23.7	0	0	0	23.72
NE Zone											
Assam	2.07	0	0.16	0.07	0.39	0.62	2.69	0	0	0	2.69
Arunachal Pradesh	0.18	0.02	0	0	0	0.02	0.2	0	0	0	0.2
Meghalaya	0.14	0	0.07	0.05	0	0.12	0.26	0	0	0	0.26
Mizoram	0.22	0.01	0	0	0	0.01	0.23	0	0	0	0.23
Tripura	0.27	0.05	0.16	0	0	0.21	0.48	0	0	0	0.48
Manipur	0.2	0	0	0	0	0	0.2	0	0	0	0.2
Nagaland	0.2	0	0.1	0	0	0.1	0.3	0	0	0	0.3
Total	3.28	0.8	0.49	0.12	0.39	1.08	4.36	0	0	0	4.36
North Zone											
Delhi	3.36	0	0	0	0	0	3.36	0.34	0	0.34	3.7
Haryana	7.68	4.06	1.88	3.97	2.55	12.46	20.14	3.17	0	3.17	23.31
Himachal Pradesh	0.14	0.06	0.05	0	0	0.11	0.25	0	0	0	0.25
J&K	1.03	0.16	0	0	0.09	0.25	1.28	0	0	0	1.28
Punjab	21.84	0.02	0.75	23.93	3.72	28.42	50.26	6.33	0	6.33	56.59
Chandigarh	0.4	0	0.38	0.2	0	0.58	0.98	0.08	0.01	0.09	1.07
Rajasthan	7.06	0	0.65	0.55	0.11	1.31	8.37	1.58	0.12	1.7	10.07
U.P	14.95	0.08	2.09	4.72	0.18	7.07	22.02	4.15	0.2	4.35	26.37
Uttarakhand	0.66	0.08	0.15	0.26	0.05	0.54	1.2	0.09	0.06	0.15	1.35
Total	57.12	4.46	5.95	33.6	6.7	50.7	108	15.7	0.39	16.1	124
South Zone											
Andhra Pradesh	12.66	0	1.37	17.6	0	18.97	31.63	1.97	0	1.97	33.6
Andman & Nicobar	0.07	0	0	0	0	0	0.07	0	0	0	0.07
Kerala	5.17	0	0	0	0	0	5.17	0.21	0	0.21	5.38
Karnataka	3.78	0	0.25	0.35	0	0.6	4.38	0.92	0	0.92	5.3
Tamil Nadu	5.8	0	1.33	0.38	0	1.71	7.51	0.58	0	0.58	8.09
Pondicherry	0.44	0	0	0.05	0	0.05	0.49	0.1	0	0.1	0.59
Total	27.92	0	2.95	18.4	0	21.3	49.3	3.78	0	3.78	53.03

Contd...

Table 5.54–*Contd...*

Region/U.T	*Covered*							*CAP*			
	FCI Owned	*Hired*						*Owned*	*Hired*	*Total CAP*	*Grand Total*
		State Govt.	*CWC*	*SWC*	*Pvt. Parties*	*Total Hired*	*Total Covered*				
West Zone											
Gujarat	5	0.29	0.73	0.12	0.06	1.2	6.2	0.3	0	0.3	6.5
Dadar Haveli	0	0	0	0	0	0	0	0	0	0	0
Lakshadweep	0	0	0	0	0	0	0	0	0	0	0
Maharashtra	11.77	0.13	1.54	1.59	0.63	3.89	15.66	1.18	0	1.18	16.84
Goa	0.15	0	0	0	0	0	0.15	0	0	0	0.15
Madhya Pradesh	3.37	0.38	1.71	1.59	1.33	4.96	8.33	0.24	0.02	0.24	8.57
Chhattisgarh	5.12	0.03	0.06	0.68	0.03	0.8	5.92	0	0.01	0	5.92
Total	25.41	0.83	4.04	3.93	2.05	10.9	36.3	1.72	0.39	1.72	37.98

Source: <fciweb.nic.in>

The higher production levels and drastic increase in production of major agricultural commodities (Tables 5.55 and 5.56) necessitates the strengthening of storage system in the country. India has total agri-storage capacity of around 91 million m tonnes (Figure 5.20) as on 31st March, 2010. Out of this, State agencies own 41 per cent of the capacity and the balance distributed among private entrepreneurs, co-operative societies, farmers etc. However, the State agencies use 66 per cent (60 million m tonnes) of India's total agri-storage capacity, which also includes hired capacity of 23 million m tonnes. The total State owned storage capacity of 37 million m tonnes is held through three public sector agencies *viz.* CWC, SWCs and FCI. With the average buffer stock norms for food grains of around 25 million m tonnes and peak of 32 million m tonnes in July every year, the storage capacity available with Government agencies is primarily used for keeping central stock of food grains for the buffer stock, PDS and other Government schemes. Due to insufficiency of covered storage capacity of food grains, which was around 47.55 million m tonnes as on March 31st, 2000 and due to progressively increasing Government stocks, considerable amount of food grains had to be stored under open CAP storage and to an extent in hired capacities, mainly during the year 2000 to 2003. But, with the falling procurement levels between 2002-03 and 2006-07, the additional capacities hired by FCI during higher procurement levels in 2000-01 and 2001-02 had to be released due to idle capacities. The years from 2007-08 to 2009-10, saw highest ever levels of procurement of food grains by Government agencies, resulting in severe strain on the available storage capacities. Higher MSP, better reach and consequent higher procurement helped ensure better and remunerative prices to farmers. However, it caused strain on available storage capacities with the State agencies involved in procurement. Before 2000, private sector interest in warehousing industry was low due to its capital intensive nature and high uncertainty over revenue generation. But, with the introduction of 'Construction of Rural Godowns Scheme' on 1st April, 2001 being implemented through NABARD and NCDC, the private sector interest in warehousing industry was substantially increased. The main objectives of this scheme include creation of scientific storage capacity with allied facilities in rural areas to meet various requirements of farmers for storing farm produce, processed farm produce, agricultural inputs etc., and prevention of distress sale by creating the facility of pledge loan and marketing credit. Under the original scheme, a subsidy @ 25 per cent of capital cost of the project was provided. In case of North Eastern States, hilly areas and SC/ST entrepreneurs, subsidy was provided @ 33.33 per cent of the capital cost of the project. The Scheme has since been modified with effect from 20th October 2004, to provide subsidy @ 25 per cent to farmers, agricultural graduates, co-operatives and CWC and SWCs. All other categories of individuals companies and corporations are now given subsidy @ 15 per cent of the project cost. The scheme has been made farmers' friendly by allowing subsidy for smaller godowns of 50 m tonnes size in general and of 25 m tonnes in hilly areas. Five lakh tonnes capacity to be created is reserved for small farmers.

Table 5.55: India's position in world agriculture in 2008.

Sl.No.	*Item*	*India*	*World*	*Per cent Share*	*India's Rank*	*Next to*
	1	*2*	*3*	*4*	*5*	*6*
1.	**Total Area** (Million Hectares)	329	13442	2.4	Seventh	Russian Federation, Canada, U.S.A., China, Brazil, Australia
	Land Area	297	13009	2.3	Seventh	Russian Federation, China, U.S.A., Canada, Brazil, Australia
	Arable Land	159	1411	11.3	Second	USA.
2.	**Total Population*** (Million)	1181	6750	17.5	Second	China
	Agriculture	583	2617	22.3	Second	China
3.	**Economically Active Population*** (Million)					
	Total	472	3178	14.9	Second	China
	Agriculture	262	1295	20.2	Second	China
4.	**Crop Production** (Million Tonnes)					
	(A) : Total Cereals	267	2521	10.6	Third	China, USA
	Wheat	79	683	11.5	Second	China
	Rice (Paddy)	148	686	21.6	Second	China
	Total Pulses	15	61	24.6	First	
	(B) : Oilseeds					
	Groundnut (in shell)	7	38	18.8	Second	China
	Rapeseed	6	58	10.3	Third	Canada, China
5.	**Fruits and Vegetables** (Million Tonnes)					
	(A) : Vegetables and Melons	90	932	9.7	Second	China
	(B) : Fruits excluding Melons	67	580	11.6	Second	China
	(C) : Potatoes	35	326	10.6	Second	China
	(D) : Onion (Dry)	14	73	18.5	Second	China
6.	**Commercial Crops** (Million Tonnes)					
	(A) : Sugarcane	348	1736	20.1	Second	Brazil
	(B) : Tea	0.81	3.9	20.7	Third	China, Turkey
	(C) : Coffee (green)	0.26	8.25	3.2	Seventh	Brazil, Vietnam, Colombia, Indonesia, Ethiopia, Mexico

Contd...

Table 5.55–*Contd...*

Sl.No.	*Item*	*India*	*World*	*Per cent Share*	*India's Rank*	*Next to*
	1	*2*	*3*	*4*	*5*	*6*
	(D) : Jute and Jute like Fibres	2.02	3.20	63.1	First	
	(E) : Cotton(lint)	3.77	22.85	16.5	Second	China
	(F) : Tobacco Leaves	0.52	6.88	7.6	Third	China, Brazil
7.	**Livestock** (Million Heads)					
	(A) : Cattle	175	1372	12.7	Second	Brazil
	(B) : Buffaloes	105	185	56.7	First	
	(C) : Camels	0.63	24.73	2.5	Tenth	Somalia, Sudan, Ethiopia, Niger, Mauritania, Kenya, Mali, Pakistan, Chad
	(D) : Sheep	65	1086	6.0	Third	China, Australia
	(E) : Goats	126	864	14.5	Second	China
	(F) : Chicken	584	18139.1	3.2	Fifth	China, USA, Indonesia, Brazil
8.	**Animal Products**					
	(A) : Total Milk (000 MT)	109000	694235.3	15.7	First	
	(B) : Eggs Total (000 MT)	3060	66103	4.6	Third	China, U.S.A.
	(C) : Total Meat (000 MT)	4353.1	277847.5	1.6	Fifth	China, U.S.A., Brazil, Germany
9.	**Implements** (Thousands numbers)**					
	Agricultural Tractors-in-use	3149	29320	10.7	Second	USA

* Estimated figure for 2008. ** Figure relates to 2007.

Source: <FAO, Regional Office for Asia and the Pacific, Bangkok>

Table 5.56: Compound growth rates of area, production and yield indices of principal crops during the selected reference periods (Base: TE 1981-82 = 100).

Crops	*1980-81 to 1989-90*			*1990-91 to 1999-00*			*2000-01 to 2011-12*		
	Area	*Production*	*Yield*	*Area*	*Production*	*Yield*	*Area*	*Production*	*Yield*
Rice	0.41	3.62	3.19	0.68	2.02	1.34	0.04	1.72	1.68
Wheat	0.46	3.57	3.1	1.72	3.57	1.83	1.22	2.37	1.14
Course cereals	–1.34	0.4	1.62	–2.12	–0.02	1.82	–0.75	3.01	4.39
Total pulses	–0.09	1.52	1.61	–0.6	0.59	0.93	1.7	3.47	1.91
Sugarcane	1.44	2.7	1.24	–0.07	2.73	1.05	1.37	1.96	0.58
Total oilseeds	1.51	5.2	2.43	–0.86	1.63	1.15	2.08	4.45	3.39
Total food grains	–0.23	2.85	2.74	–0.07	2.02	1.52	0.43	2.32	2.91

Source: <www.agricoop.nic.in>

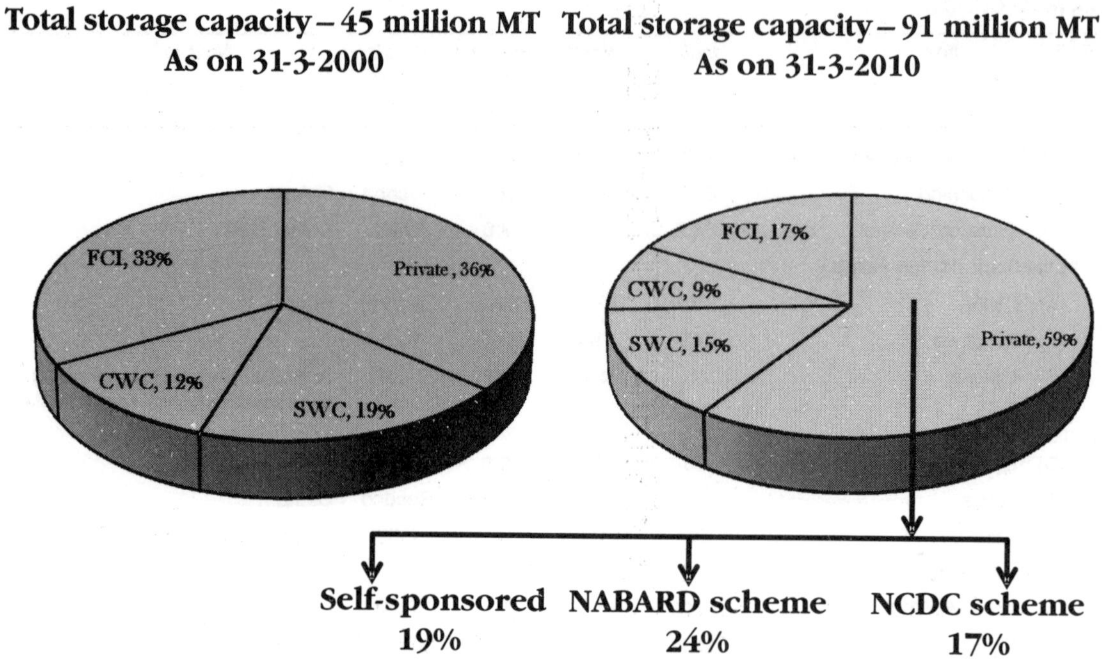

Figure 5.20: Total agri-storage capacity in India – Comparative view.

7. Co-operatives

To meet the increasing need for storage capacity, the NCDC encourages construction of scientific storage facilities by co-operatives, particularly at rural and market level. The NCDC was established by an Act of Parliament in 1963 as a statutory Corporation under the Ministry of Agriculture. This Corporation provides storage facilities to the farmers at cheaper rates, which reduces the storage cost. The co-operatives also provide pledge loan against the stored produce. The storage is more systematic and scientific than traditional storage. Financial assistance and subsidies are provided by Government organizations/banks to build co-operative storage.

Agriculture co-operatives have been playing a significant role in stepping up the growth of agricultural production in the country through supply of farm inputs and other technical services to the farmers. They also provide post-harvest services for procurement, grading, processing, storage and marketing of farmers' surplus produce. These co-operatives are rendering an equally useful service by maintaining the supply line and distribution of consumer articles including essential commodities in rural areas. To undertake these functions effectively, co-operatives require adequate captive storage capacity of their own at the village, mandi and rail head centres. To meet this requirement, the NCDC has been assisting primary co-operative societies working at the village level. Besides, the Corporation has also been assisting co-operative marketing societies functioning at the mandi, district, regional, state and national levels for construction of godowns of their own to create storage facility, to expand the existing capacity and also for renovation/upgradation of rural godowns under the Central Sector and Corporation Sponsored Scheme. The Government of India in the year 2002 approved to provide Credit Linked Subsidy for construction/renovation/expansion of rural godowns during the period of 10th FYP *i.e.* from 2002-07. Under this Scheme, the construction of rural godowns will be taken up by co-operative societies in addition to other eligible entrepreneurs. The primary societies are generally assisted for construction of rural godowns of 50 tonnes and 100 tonnes capacity and Marketing co-operatives in the range of 250 tonnes to 5000 tonnes capacity and in special cases, assistance is also provided for large sized godowns of 10000 tonnes capacity and above depending upon the potential. GBY, a Capital Investment Subsidy Scheme for Construction/Renovation/ Expansion of Rural Godowns (with a minimum capacity of 50 tonnes without any ceiling, but subsidy is available for maximum capacity of 10000 tonnes and limited to ₹ 37.50 lakhs, based on normative costs) was introduced by Government of India, Ministry of Agriculture and Co-operation, DMI. Under the scheme, 25 per cent subsidy (other than North Eastern States, Hilly area, SC/ST co-operatives) and 33.33 per cent subsidy for North Eastern States, Hilly area and SC/ST Co-operatives are available. Thus, NCDC, since its inception in 1962-63, has been making systematic and sustained efforts to assist in the creation of scientific storage facilities at various levels in co-operative sectors.

The NCDC also supports fruit and vegetable marketing and processing co-operatives. It is a unique organization, which not only plays a developmental role, but also provides financial assistance for creating infrastructure for marketing, processing and storage of agricultural produce in the Co-operative Sector. The details of schemes of the Corporation for providing financial assistance are listed below:

- ☆ ***Strengthening of share capital base of fruit and vegetable co-operatives***: Under this scheme, assistance up to ₹ 10 lakhs is provided for increasing the capital base of the beneficiary Co-operative societies for improving their marketing operations.
- ☆ ***Margin money assistance for raising working capital:*** Assistance is provided to enable the co-operative society to raise working capital from Banks for increasing business operations. This assistance is available to State Federations, District Level Federations and Commodity Marketing Societies.
- ☆ ***Assistance for purchase of transport vehicles:*** The Corporation provides assistance to the beneficiary societies for purchase of transport vehicles for moving stocks from growing areas to consuming centers.
- ☆ ***Construction of packing and grading sheds and godowns:*** For construction of godowns and sheds, assistance is provided to co-operatives at the primary and mandi level,
- ☆ ***Construction of cold storages and ice plants***: Assistance to the co-operatives is provided for setting-up of new ice plants and cold storages and also for expansion/modernization of the existing units.
- ☆ ***Creation of marketing infrastructure including retailing:*** Assistance is provided to create integrated marketing infrastructure including that of retail outlets.
- ☆ ***Setting up of processing units:*** Assistance is provided to Co-operatives for establishment of processing units of fruits and vegetables including that of other products like Ayurvedic medicines, honey, spices processing units etc.

Listed below are a few good working marketing co-operatives:

- ☆ Horticultural Producers' Co-operative Marketing and Processing Society Limited (HOPCOMS), Lalbagh, Bangalore
- ☆ MAHAGRAPES, Pune, Maharashtra
- ☆ Lahoul Potato Growers Co-operative Marketing Society Ltd., Manali (LPS), Himachal Pradesh
- ☆ Rau Co-operative Cold Storage, District. Indore, Madhya Pradesh.

- ☆ Uttar 24 Paraganas Krishi Samabya Himghar Samity Ltd., Distt. 24 Parganas, West Bengal.
- ☆ Mahua Co-operative Cold Storage Ltd. District, Vaishali, Bihar

8. Private Warehouses

With a view to increase the participation of private sector and development and regulation of warehousing industry, Government of India introduced 'The Warehousing (Development and Regulation) Act'* in September 2007. The main focus of this Act is to establish a regulated environment for issuance of Negotiable Warehouse Receipt (NWR) under Warehousing Development and Regulatory Authority (WDRA), which was constituted in October 2010 under the said Act. The functions of WDRA include registration and accreditation of warehouses intending to issue NWR and to improve the storage capacity in the country and also help producers and consumers get a better deal by cutting out the intermediaries and wastages. The important large national level private players involved in handling, storing and marketing of agricultural commodities, which have emerged over the last decade owing to the available capital subsidy include National Bulk Handling Corporation Ltd., National Collateral Management Services Ltd., Adani Agri Logistics, Star Agriware housing and Collateral Management Ltd., Shree Shubham Logistics Ltd., Ruchi Infrastructure Ltd., Guru Warehousing Corporation, Paras Warehousing and LTC Commercial. Despite the importance of agricultural warehousing in the economy, adequate steps to augment capacity and investment have not been taken in the past. The enactment and implementation of 'The Warehousing (Development and Regulation) Act', 2007 is, therefore, a very significant step, as it aims in bringing the desired paradigm shift in the way warehousing as an economic activity is being looked at or the way banks look at post harvest credit. Most significantly, it addresses one of the major woes of Indian farmer that of getting liquidity against the produce, thus giving him an option and opportunity to time his sale and realize better prices. Receipts issued by WDRA registered warehouses are negotiable and the holder will get due protection in the form of getting a pledge loan. This NWR system also brings desirable changes in agriculture credit disbursal system. This is because, the commercial banks have a mandate to lend at least 18 per cent of their credit to agriculture sector. However, almost all of it has been in pre-harvest phase. So, warehouse receipts as financial instruments, backed by the provisions of WDRA, will further increase lendings to priority sector and boost the returns to farmers. But, this warehouse receipt financing has not reached its true potential vis-à-vis other modes of funding to agriculture because of important factors like low credibility of warehouse management, perishability of agricultural produce etc. In some cases, the warehouse receipt can be negotiable or non-negotiable, according to the remark mentioned on it. If nothing is mentioned, receipt is considered as negotiable.

CWC and SWCs warehouses are generally located in big cities and medium-size towns, though SWCs have warehouses in small towns. As indicated above, co-operatives were to fulfill the rural warehousing role, but co-operatives have not performed as vigorously as envisaged, leaving a significant gap in the provision of services. Small private warehouses fill the gap in some villages. Scarcity of

---

* The Warehousing (Development and Regulation) Act, 2007 had been made effective from 25th October, 2010. As per the provisions of the Act, the Warehousing Development and Regulatory Authority (WDRA) was set up by the Government on 26th October, 2010. The main objectives of the Warehousing (Development and Regulation) Act, 2007 are to make provision for the development and regulation of warehouses, negotiability of warehouse receipts, establishment of a WDRA and related matters. The Negotiable Warehouse Receipts (NWRs) issued by the warehouses registered under this Act would help farmers to seek loans from banks against NWRs to avoid distress sale of agricultural produce. It will also be beneficial for a number of other stakeholders such as banks, financial Institutions, insurance companies, trade, commodities exchanges as well as consumers.

warehouses and increased food grains production and procurement has prompted the Centre to rope in private players to construct storage facilities and the Government has already approved a capacity of 12.76 million tonnes under private sector warehouses (Table 5.57). The Centre's move was crucial following the acute storage of warehouses over the last few years with the FCI of India procuring 20-25 million tonnes of wheat every year.

Table 5.57: State-wise storage capacity of private warehouses (2009-10).

State	*Storage Capacity (Tonnes)*
Andhra Pradesh	36,000
Bihar	300,000
Chhattisgarh	5,000
Gujarat	45,000
Haryana	3,880,000
Himachal Pradesh	142,550
J&K	361,690
Jharkhand	175,000
Karnataka	205,000
Kerala	15,000
Maharashtra	99,500
Punjab	7,125,000
Tamil Nadu	345,000
Uttarakhand	25,000
West Bengal	5,000
Total	12,764,740

Source: Ministry of Consumers Affairs, Food and Public Distribution.

Private Entrepreneurs Godown (PEG), 2008 Scheme

To meet the situation arising out of high procurement level of wheat and rice as a result of increase in MSP during last five years, the Government formulated a scheme for creation of additional storage capacity for food grains through private sector participation in 2008. The scheme is known as PEG-2008 Scheme. A state-wise mapping of existing capacities and analysis of additional requirements was undertaken based on certain criteria by State Level Committees and a High Level Committee of the FCI. Under the scheme, the FCI would now give a business guarantee of ten years for assured hiring. A capacity of 15.29 million m tonnes is proposed to be created in 19 States under the PEG scheme through private sector participation and CWC and SWCs. CWC and SWCs are constructing 5.31 lakh tonnes and 15.49 lakh tonnes storage capacity respectively under the scheme. 3.5 lakh tonnes storage capacity has already been constructed by CWC/SWCs. To make the scheme more attractive for private entrepreneurs, the guarantee period was increased from five years to seven years and subsequently to 10 years. The ceiling of rate fixed for hiring of godowns has been revised from ₹ 3.80 per quintal per month to ₹ 4.78 per quintal per month. In appropriate cases, the High Level Committee has been empowered to decide higher rates by recording reasons in writing. The Planning Commission is also keen for argumentation of storage facilities for the storage of food grains through PPP. There is a proposal for setting up modern silos of 2.0 million m tonnes in PPP mode. All these efforts will be beneficial for creating integrated modern warehousing capacities in the country. For meeting the capital expenditure on construction of silos, the private entrepreneurs would be eligible for Viability Gap Funding (VGF) under the existing VGF scheme, which allows grants of up to 20 per cent of capital cost on the basis of competitive bidding. The FCI would provide an additional VGF of up to 20 per cent of capital costs. For storage of wheat in these silos, the developer will be entitled to receive a recurring service charge provided he meets the required performance and maintenance standards.

9. Cold Storages

With a view to enhance shelf life of perishables like fruits, vegetables, milk etc., cold storages in the country have also been promoted. These perishable commodities require controlled temperature and humid conditions to enhance their shelf life (Tables 5.58 and 5.59). Presently, a total of 5274 cold storages (Table 5.60) are in the country with a total capacity of 24.31 million tonnes (2010-11). Most of these cold storage units are in the private sector (Table 5.61). Public and co-operative sector accounts for a very small capacity. The present storage capacity of cold stores is sufficient for only 12 per cent of the total production of fruits and vegetables. There are only two states (Manipur and Mizoram), where cold storage facilities are not available. On the other hand, states like Assam, Himachal Pradesh, Jammu and Kashmir, Kerala, Sikkim and Tamil Nadu have cold storage capacity available only for one per cent of their produce. There are only four states *i.e.*, Uttar Pradesh, Punjab, West Bengal and Rajasthan, which have more than All-India average capacity available for their produce. Uttar Pradesh (1589) ranks first in having highest number of cold storages followed by Maharashtra (466) and West Bengal (463). Presently, density of cold storage is about two per thousand sq. km of area. Looking to the available quantities of perishable products (fruits and vegetables), the cold storage capacity available in the country is inadequate and requires their promotion both in the production as well as consuming areas (Table 5.62 and Figures 5.21 and 5.22). The DMI provides consultancy and technical services to prospective entrepreneurs in construction, maintenance and operation of cold storages. It also trains

Table 5.58: Desired storage environment of fruits and vegetables in the cold storage.

Commodity	*Temperature (°C)*	*Relative Humidity (per cent)*
Apple	–1-3	90-98
Apricots	–0.5-0	90-95
Avocado	7-13	85-90
Asparagus	0-2	95-97
Beans, green	4-7	90-95
Beet root	0-2	95-97
Broccoli	0-2	90-95
Black berry	–0.5-0	95-97
Cabbage	0-2	90-95
Carrots	0-2	90-95
Cauliflower	0-2	90-95
Cherries	0.5-0	90-95
Cucumber	7-10	90-95
Brinjal	0-2	90-95
Grapes	–1-1	85-90
Lemons	4-15	86-88
Lettuce	0-1	95-98
Lime	3-10	85-90
Mango	11-18	85-90
Melon water	2-4	85-90
Orange	0-10	85-90
Peach	–1-1	88-92
Potato	1.5-4	90-94

Source: <www.ncdc.in>

managers and cold storage operators for efficient operation and management of cold storages/cool stores at farm level. It provides technical support in preparation of master plan for cold storage requirements at micro/macro level. Conducting seminars and problem oriented studies related to cold stores is a regular feature of the DMI. It acts as a nodal agency to promote cold storages in the country by coordinating R&D in cold storage.

Table 5.59: Requirements for successful storage of important fruits and vegetables.

Name of Fruit	*Storage Temp.*	*Cold Storage Life (Weeks)*	*Remarks*
Apple	32-35	7 to 26	Small fruits keep better than large ones
Apricot	30-32	2 to 4	–
Berry fruits	30-32	2 to 3	–
Bananas	55-60	2 to 3	–
Cherries	30-32	2 to 3	–
Guavas	47-50	3 to 4	–
Grapes	30-32	4 to 6	Fumigation with 1-2 per cent sulphur dioxide before storage reduces decay and storage life can be increased to 8-12 weeks
Lemon	50-55	9 to 13	Lemon for storage should be picked at the greenish yellow colour
Lime	50-55	4 to 5	–
Mangoes	45-50	4 to 6	–
Mandarin Oranges (Nagpur), Coorg (Assam)	40-42	9 to 13	Mandarin oranges can be kept satisfactorily at lower temperature if to be stored for less than 4 weeks
Oranges (Tight skinned)	36-39	13 to 17	Tight skinned oranges can also be kept at lower temperature if to be stored for less than 8 weeks
Pears	30-32	2 to 4	–
Peaches	30-32	2 to 4	–
Plums	30-32	2 to 4	If fruits is to be stored for jam making, the storage life can be extended by another two weeks.
Pine apple	50-55	3 to 4	–
Quinces	30-35	9 to 17	–

Source: <www.ncdc.in>

The first cold store in India was reported to have been established in Calcutta in 1892. However significant progress in the expansion of the cold storage industry in the country has been made only after independence. The regulation of cold storage started with the enactment of the Central Cold Storage Order 1964 under Section 3 of the Essential Commodity Act, 1955 to provide technical specifications and minimum norms for hygienic storage of edible commodities. The Agricultural Marketing Advisor to the Government of India is the Licensing Officer. The Order was applicable all over India. Subsequently, states like West Bengal, Uttar Pradesh, Haryana, Punjab etc., sought permission from the Central Government to enact their own independent legislation for regulation of cold storages in their respective states. These states wanted to include some provisions, which were not included in the Central Cold Storage Order such as compulsory insurance, compensation to the farmers, fixation of storage charges, power to requisition storage space, compulsory issue of receipts

Table 5.60: State-wise cold storage facilities in India (2010-11).

State/UTs	*No. of Cold Storages*	*Capacity (MT)*	*Cold Storage/ 1000 sq.km*	*Capacity Available/ 1000MT*
Andhra Pradesh	290	900606.00	1.12	0.05
Arunachal Pradesh	1	5000.00	0.01	0.02
Assam	24	88068.00	0.31	0.01
Bihar	246	1147041.00	2.61	0.06
Chhattisgarh	69	341885.00	0.51	0.07
Goa	29	7705.00	7.83	0.04
Gujarat	398	1267304.00	2.03	0.09
Haryana	244	393121.00	5.52	0.09
Himachal Pradesh	18	19858.00	0.32	0.01
Jammu and Kashmir	19	42869.00	0.09	0.01
Jharkhand	45	170148.00	0.56	0.04
Karnataka	170	407165.00	0.89	0.03
Kerala	193	58105.00	4.97	0.01
Madhya Pradesh	197	808052.00	0.64	0.11
Maharashtra	466	546748.00	1.51	0.03
Manipur	0	0.00	0.00	0.00
Meghalaya	3	3200.00	0.13	0.00
Mizoram	0	0.00	0.00	0.00
Nagaland	2	6150.00	0.12	0.03
Odisha	101	291039.00	0.65	0.03
Punjab	422	1345193.00	8.38	0.28
Rajasthan	110	324226.00	0.32	0.15
Sikkim	1	2000.00	0.14	0.01
Tamil Nadu	148	238536.00	1.14	0.01
Tripura	11	29450.00	1.05	0.04
Uttar Pradesh	1589	10118000.00	6.66	0.40
Uttarakhand	15	68499.00	0.28	0.04
West Bengal	463	5682000.00	5.22	0.23
All India	5274	24311968.00	1.62	0.12

Source: <www.ncdc.in>

in the prescribed format etc. Government of India, in the interest of the farmers, permitted these States to have their own cold storage legislations and also promulgated another cold storage order known as Central Cold Storage Order, 1980, wherein all the above provisions were included. DMI, Ministry of Agriculture and Government of India enforced this Central Cold Storage Order 1980. Government of India modified the Central Cold Storage Order during May 1997 in line with the policy of liberalization. At the same time, the States were also advised to modify their respective cold storage orders to remove

Table 5.61: Sector-wise distribution of cold storages as on 31/12/2009.

Sl.No.	State/UT	Private Sector		Co-operative Sector		Public Sector		Total No.	Total Capacity in m tonnes
		No.	Capacity	No.	Capacity	No.	Capacity		
1.	Andaman and Nicobar Islands (UT)	01	170	00	00	01	40	02	210
2.	Andhra Pradesh	269	885557	11	11598	10	3451	290	900606
3.	Arunachal Pradesh	01	5000	00	00	00	00	01	5000
4.	Assam	19	85948	01	1000	04	1120	24	88068
5.	Bihar	236	1100641	10	46400	00	00	246	1147041
6.	Chandigarh (UT)	05	11216	01	1000	00	00	06	12216
7.	Chhattisgarh	67	341815	01	29	01	41	69	341885
8.	Delhi	77	103277	02	5201	16	17680	95	126158
9.	Gujarat	372	1230198	21	30669	05	6437	398	1267304
10.	Goa	29	7705	00	00	00	00	29	7705
11.	Haryana	234	378319	04	3403	06	11399	244	393121
12.	Himachal Pradesh	09	12896	02	767	07	6195	18	19858
13.	Jammu and Kashmir	15	40689	03	2134	01	46	19	42869
14	Jharkhand	37	142733	08	27415	00	00	45	170148
15.	Kerala	176	55335	06	1080	11	1690	193	58105
16.	Karnataka	135	390882	18	6689	17	9594	170	407165
17.	Lakshadweep (UT)	00	00	00	00	01	15	01	15
18.	Maharashtra	379	507678	55	25346	32	13724	466	546748
19.	Madhya Pradesh	172	704270	20	101348	05	2434	197	808052
20.	Manipur	00	00	00	00	00	00	00	0
21.	Meghalaya	01	1200	00	00	02	2000	03	3200
22.	Mizoram	00	00	00	00	00	00	00	0
23.	Nagaland	01	5000	01	1150	00	00	02	6150
24.	Odisha	81	248739	16	38100	04	4200	101	291039
25.	Pondicherry(UT)	02	35	01	50	00	00	03	85
26.	Punjab	404	1306101	18	39092	00	00	422	1345193
27.	Rajasthan	100	320380	09	3832	01	14	110	324226
28.	Sikkim	00	00	00	00	01	2000	01	2000
29.	Tamilnadu	130	225712	13	7562	05	5262	148	238536
30.	Tripura	03	12750	01	5000	07	11700	11	29450
31.	Uttar Pradesh	1505	9842000	84	276000	00	00	1589	10118000
32.	Uttarakhand	12	60499	00	00	03	8000	15	68499
33.	West Bengal	413	5380000	50	302000	00	00	463	5682000
	TOTAL	4885	23406745	356	936865	140	107042	5381	24450652

Source: <www.ncdc.in>

Table 5.62: Commodity-wise distribution of cold storages as on 31/12/2009.

Sl.No.	State/UT	Potatoes		Multipurpose		Fruits and Vegetables		Meat and Fish		Milk and Milk Products		Others		Total No.	Total Capacity (m tonnes)
		No.	Capacity	No.	Capacity	No	Capacity	No	Capacity	No	Capacity	No.	Capacity		
1.	Andaman and Nicobar Islands (UT)	00	00	00	00	00	00	02	210	00	00	00	00	02	210
2.	Andhra Pradesh	00	00	212	839456	11	9843	44	26068	15	6838	08	18401	290	900606
3.	Arunachal Pradesh	00	00	01	5000	00	00	00	00	00	00	00	00	01	5000
4.	Assam	00	00	24	88068	00	00	00	00	00	00	00	00	24	88068
5.	Bihar	228	1069841	18	77200	00	00	00	00	00	00	00	00	246	1147041
6.	Chandigarh (UT)	01	1000	04	11131	00	00	01	85	00	00	00	00	06	12216
7	Chhattisgarh	14	39242	52	302543	00	00	00	00	03	100	00	00	69	341885
8.	Delhi	00	00	54	117373	01	53	04	1239	09	4757	27	2736	95	126158
9.	Gujarat	213	967000	97	265574	02	1063	60	22211	25	11440	01	16	398	1267304
10.	Goa	00	00	01	3633	00	00	26	4004	02	68	0	0	29	7705
11.	Haryana	172	225991	66	165787	00	00	00	00	06	1343	00	00	244	393121
12.	Himachal Pradesh	05	9748	05	3837	06	6100	01	78	01	95	00	00	18	19858
13.	Jammu and Kashmir	05	11281	12	31473	00	00	00	00	02	115	00	00	19	42869
14.	Jharkhand	08	27415	37	142733	00	00	00	00	00	00	00	00	45	170148
15.	Kerala	00	00	08	13200	00	00	178	43815	06	1080	01	10	193	58105
16.	Karnataka	09	16530	96	378244	09	3870	34	5141	15	3128	07	252	170	407165
17.	Lakshadweep (UT))	00	00	00	00	00	00	01	15	00	00	00	00	01	15
18.	Maharashtra	0	0	189	413885	115	47914	62	63263	70	18407	30	3279	466	546748
19.	Madhya Pradesh	114	564600	67	237737	04	2577	00	00	11	3125	01	13	197	808052
20.	Manipur	00	00	00	00	00	00	00	00	00	00	00	00	00	00
21.	Meghalaya	00	00	03	3200	00	00	00	00	00	00	00	00	03	3200
22.	Mizoram	00	00	00	00	00	00	00	00	00	00	00	00	00	00
23.	Nagaland	00	00	02	6150	00	00	00	00	00	00	00	00	02	6150
24.	Odisha	39	139630	35	130060	04	13737	22	6412	00	00	01	1200	101	291039

Contd...

Table 5.62–*Contd...*

Sl.No.	State/UT	Potatoes		Multipurpose		Fruits and Vegetables		Meat and Fish		Milk and Milk Products		Others		Total No.	Total Capacity (m tonnes)
		No.	Capacity	No.	Capacity	No	Capacity	No	Capacity	No	Capacity	No.	Capacity		
25.	Pondicherry(UT)	00	00	00	00	01	15	01	20	01	50	00	00	03	85
26.	Punjab	344	1097609	76	236994	00	00	00	00	02	10590	00	00	422	1345193
27.	Rajasthan	19	65896	79	256445	00	00	00	00	08	1781	04	104	110	324226
28.	Sikkim	00	00	01	2000	00	00	00	00	00	00	00	00	01	2000
29.	Tamil Nadu	00	00	67	213966	03	7675	59	11908	13	4512	06	475	148	238536
30.	Tripura	03	11000	08	18450	00	00	00	00	00	00	00	00	11	29450
31.	Uttar Pradesh	1286	8719533	294	1390021	04	3580	02	4027	02	801	01	38	1589	10118000
32	Uttrakhand	00	00	15	68499	00	00	00	00	00	00	00	00	15	68499
33.	West Bengal	402	5460000	61	222000	00	00	00	00	0	0	0	0	463	5682000
	TOTAL	2862	18426316	1584	5644659	160	96427	497	188496	191	68230	87	26524	5381	24450652

Source: <www.ncdc.in>

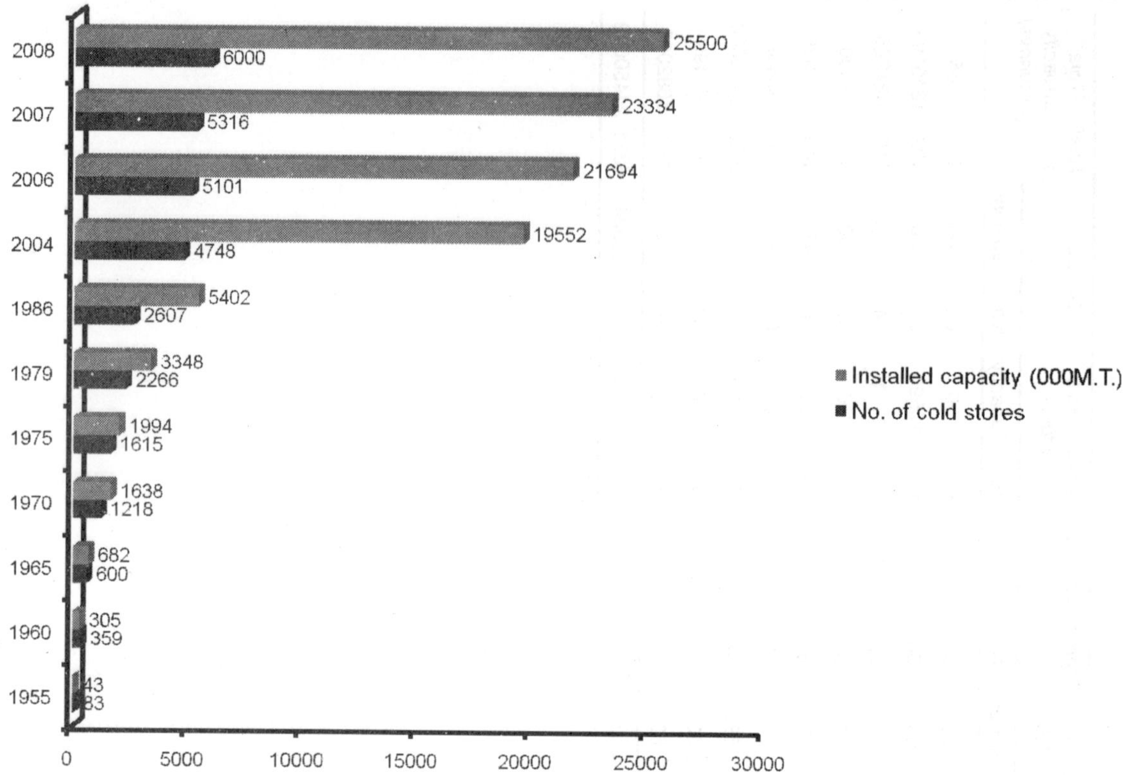

Figure 5.21: Growth in cold stores in India.

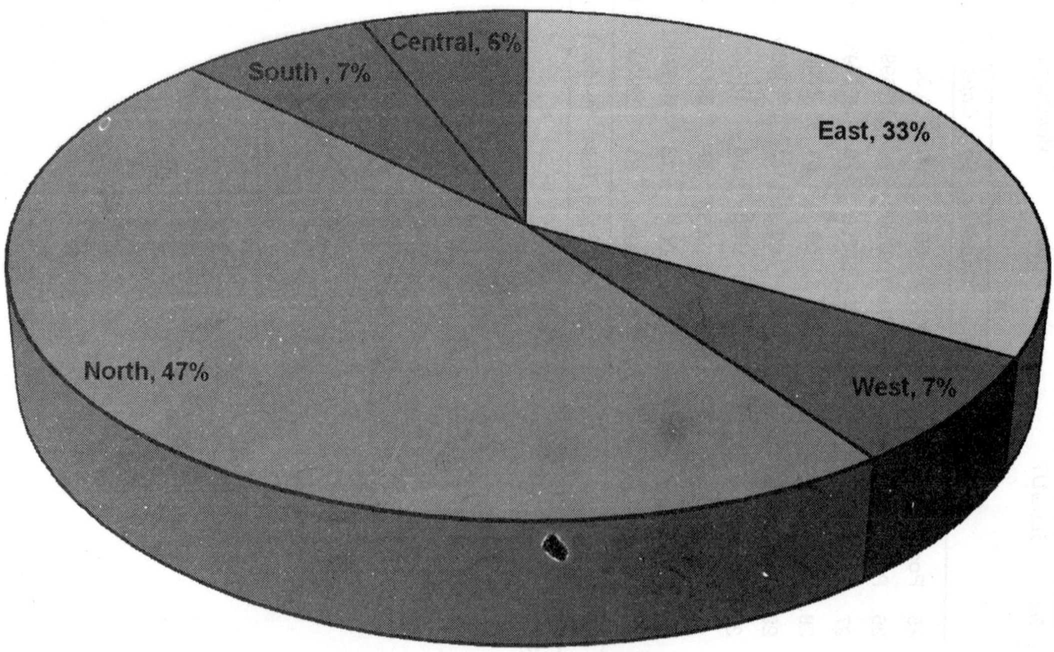

Figure 5.22: Region-wise distribution of cold stores in India (2007).

unnecessary control. Most of the States except West Bengal have either repealed their cold storage order or removed some provisions such as compulsory licensing, rent control etc., to encourage demand driven growth of cold storages. The state of West Bengal has also relaxed some provisions under State Cold Storage Order, except for the fixation of rent for storage of potatoes.

Types of Cold Stores

The cold stores can be classified as follows:

- ☆ ***Bulk Cold Stores:*** These cold stores generally store a single commodity and hence, operate on a seasonal basis. Commodities like chillies, apples etc., are stored in bulk cold stores.
- ☆ ***Multipurpose Cold Stores:*** These store a wide range of commodities like fruits, vegetables, spices, milk and milk products etc., and hence, they operate round the year. These units are mainly located near the consuming centres.
- ☆ ***Mini units/Walk-in cold stores:*** These units are seen in hotels, restaurants, malls, super markets etc., to stock items like ice creams, chocolates, cold drinks etc.
- ☆ ***Ripening chambers:*** These units are meant for ripening produce like bananas, mangoes sapotas etc.

vi. SWOT Analysis

Considering the importance of cold storage units and the existing gap between their present number and actual requirement, in view of the increasing production of perishable commodities, it is felt appropriate to go for SWOT analysis of cold store units (Figure 5.23).

Strengths

- India is the second largest producer of fruits and vegetables and leading producer of milk in the world'
- Safe storage for perishable commodities
- Benefits both the farmers and traders.
- Value addition

Weakness

- Not exploring the possibilities of storing more varieties of commodities in multi chambers
- Inadequate forward and backward linkages
- Inadequate availability of pledge finance to farmers
- Inadequate working capital support for cold storage unit owners
- High operational cost due to inefficient cold storage technology

Opportunities

- Cold storages are to be established in good production and marketing areas
- Export potential can be explored
- Agro processing industries can also be started in adjoining areas

Threats

- Mismatch between demand and supply of commodities leads to under utilization of the cold store units
- More number of cold storage units in specific pockets led to decline in rental value of commodities
- Rising trend in tariff of electricity

Figure 5.23: SWOT analysis for cold store units.

vii. Measures to Prevent loss during Storage

During storage of food grains, the following measures should be followed to prevent losses:

- ✰ ***Drying:*** At the time of harvesting, the moisture content in grains and seeds varies from 15-35 per cent. Besides, grains have a tendency to absorb moisture from the atmosphere. If the moisture content is high, seeds are prone to be infested with pests and diseases and this, in turn, will affect their quality. Hence, moisture content in grains and seeds must be reduced to below nine per cent for their safe storage. To achieve this, seeds have to be spread out and dried in the sun before storing them in the shade. Mechanical hot air driers could also be used.
- ✰ ***Hygiene:*** The storage structures should be well cleaned before stacking fresh grains. Any refuse, dirt, husk of the previous grain etc., should be swept. The premises should be well maintained and any crack or hole in the wall, floor or ceiling should be sealed. Before reusing old gunny bags, they should be turned inside out, dusted thoroughly, exposed to sun and fumigated. If earthen pots are used, they should also be cleaned well and exposed to sun before storing fresh grains.
- ✰ ***Prophylactic treatment:*** As a preventive measure, godowns should be treated with pesticides. Gunny bags should also be sprayed. Grains that have to be used for seed purpose must be mixed with pesticides.
- ✰ ***Use of modern storage structures:*** Research organizations have developed improved storage structures, which have proved comparatively safe for the storage of grains. In these structures, temperature, moisture, oxygen and carbon-di-oxide can be adjusted to protect the grains. The structures are airtight, moisture impervious, thermally insulated and rodent proof. Changes in outside temperature do not affect the grains stored in these structures. Some of these storage structures are Pusa Bin, Pusa Cubicle, Pusa Kothar and Pant Kuthla.
- ✰ ***Pest control:*** It is not safe to mix fungicides/insecticides with grains meant for human or animal consumption. If any of these are used, they should be within safety norms. Spraying of pesticides like BHC wettable power, malathion etc., can be done as a prophylactic treatment of the surface area of the store house. Fumigants like Aluminum phosphide tablets, Methyl bromide, Ethyl Dichloride Carbon Tetrachloride will also be used to control the pests. But, the recommended measures include application of plant products like neem kernel powder, crushed pepper etc., to control various pests and diseases. This treatment prevents laying of eggs, reduction in egg hatching and prevention of larval development.

viii. Warehousing

When the storage of agricultural commodities is done on a large scale and in a specified manner, it is called as 'warehousing'. The place, where goods are kept is called 'warehouse'. These warehouses are scientific storage structures constructed for the protection of quality and quantity of the stored produce. In other words, it can be called as the *'protector of the national wealth'*. The person in-charge of warehouse is called 'warehouse-keeper'. Warehousing refers to the activities involving storage of agricultural produce on a large-scale in a systematic and orderly manner and making them available to the marketing agencies conveniently when needed. In other words, warehousing means holding or preserving goods in huge quantities from the time of their purchase or production till their actual use or sale. Warehousing is one of the important auxiliaries to agricultural trade. It creates time utility by bridging the time gap between production and consumption of produce. It also refers to the holding and preservation of produce by the farmers or traders, until they are distributed to the consumers.

Warehousing scheme in India is an integrated scheme of scientific storage, rural credit, price stabilization and market intelligence and is intended to strengthen the co-operative institutions. The need for a warehouse arises due to the time gap between production and consumption of products. By bridging this gap, storage creates time utility. There is a need for storing the commodities, so as to make them available to buyers as and when required. Storage of commodities in warehouses enables the farmer to carry on production in anticipation of demand in future. Need for warehouses arises also because, some commodities are produced only in a particular season (paddy, wheat, mangoes etc), but are demanded throughout the year. Similarly, certain products are produced throughout the year (fertilizers, pesticides etc), but demanded only during a particular season. Traders and co-operatives at market centres need storage structures to hold produce, when the transport facility is inadequate. The Government also needs storage structures to maintain buffer stocks to offset the effects produced by the vagaries of nature. Hence, there is necessity to store the produce for different periods primarily for commercial reasons. The farmers, processors, transporters and warehouse men have to develop storage facilities for proper storage of food grains, oilseeds, commercial crops (chillies, vegetables and fruits etc.), seeds, fertilizers, pesticides etc., for continuing both production and marketing programmes. Considering these essentials of storage, it can be emphasized that, storage is an important aspect for both crop production and post-harvest operations. Further, storage of agricultural commodities also provides protection against moisture, weather, pests and diseases, rodents, birds and any other type of infestation and contamination. Among the above, storage of agricultural commodities is done mainly considering their moisture content, as it is key factor that determines the storage life of commodities. Due to lack of storage facilities, farmers sellout their commodities immediately after harvest at low price. With reference to agricultural marketing, warehousing implies the provision of facilities for safe storing of the produce of the farmer and raising funds for him so that, his holding power is increased and he is able to fetch good prices by selling it at the appropriate time. The role of co-operatives in this field can hardly be over-emphasized. In India, the private initiative, in this field has not been adequate and therefore, the Government has to assume the responsibility of developing a strong warehousing system.

In India, about 70 per cent of farm produce is stored by farmers for their own consumption. Farmers store grain in bulk, using different types of storage structures made from locally available materials. The pre-treatment necessary for better storage life is cleaning and drying of the grain, but storage structure design and its construction also play a vital role in reducing or increasing the losses during storage. Storage losses constitute a major share of food grain loss in post-production operations. When scientifically constructed storage structures are available, it is essential that, the grain being loaded is of good quality. Therefore, the grain is cleaned to remove impurities, fungus infestation and rotten seeds and then dried to a safe storage moisture level. The present trend is to harvest the crop at high moisture content. Therefore, grain moisture is generally twice the safe limit at the time of harvesting. Oilseeds and vegetable seeds are harvested at 3-7 times higher moisture than their

Table 5.63: Upper limit of grain moisture content for safe storage.

Commodity	*Moisture Content (Per cent wet basis)*
Paddy, rice (raw)	14
Rice (Parboiled)	15
Wheat, Kabuligram, Bengal gram	12
Sorghum, maize, barley, ragi, bajra, pulses, turmeric, wheat atta maida, besan	12.5
Coriander, chillies	10
Groundnut pods	6-7
Mustard seed	5-6

Source: Agricultural Engineering Directory, 1983.

safe storage moisture content. The safe storage moisture limits for major agricultural produce are given in Table 5.63.

Various national committees like The Agriculture Finance Committee (1945), The Rural Banking Enquiry Committee (1950) and All India Rural Credit Survey Committee (AIRCSC, 1954) were set up to investigate the feasibility of warehousing. The Agriculture Finance Committee remarked warehousing might be considered as extension of the system of transport and the planning of the location of warehouses must be done as part of the extension and improvement of transport facilities. The warehouse system should be operated by a public sector. The Rural Banking Enquiry Committee emphasized the importance of storage and warehousing in relation to rural credit and rural banking. It suggested the formation of a Warehousing Development Board with a large capital for the purpose of giving loans and subsidies to those, who were prepared to take up this line of activity on a business. AIRCSC suggested strengthening of warehouses in the country and it recommended the following:

- ✩ The establishment of National Co-operative Development and Warehousing Board (NCDWB) and All India Warehousing Corporation
- ✩ The NCDWB should be in-charge of district level of development concerned with storage and warehousing
- ✩ Plan and finance the activities of warehousing development
- ✩ Developing storage and warehousing in important centres
- ✩ At all levels, the godowns and warehouses should also distribute fertilizers and other inputs to the farmers.

As per the recommendations of AIRCSC, the Agricultural Produce (Development and Warehousing) Corporation Act was enacted and enforced from 1st August 1956. This Act will have the following impact on agri-business:

- ✩ Customers would be encouraged to store their produce in registered warehouses.
- ✩ Grading and quality standards of agriculture commodities would be encouraged.
- ✩ Better return of their produce will be assured.
- ✩ There will be demand for scientific warehouses in rural areas

This Act provided for the establishment of NCDWB, the CWC and the SWCs. NCDWB was set up in 1956 with 22 members. It was assigned the following functions:

- ✩ To advance loans and grants to State Governments for financing co-operative societies engaged in marketing, processing and storage of agricultural produce, including contribution to the share capital of these societies
- ✩ To provide funds to warehousing corporation and the State Government for providing finance to co-operative societies for the purchase of agricultural produce on behalf of the Central Government
- ✩ To plan and promote programmes, through co-operative societies, for the supply of seeds, manures, fertilizers, implements and other inputs needed for the development of agriculture
- ✩ To subscribe to the share capital of the CWC and advance loans and grants to the CWC and SWCs.
- ✩ To administer the National Warehousing Development Fund.

The Board was essentially a policy making and financing body. To enable it to carry out its functions, it receives grants from the Government of India. In 1962, the Government of India decided to bifurcate the Act of 1956 into two separate Acts *viz.*, NCDC Act (1962) and Warehousing Corporation Act (1962). Subsequently in March, 1963 the NCDWB was converted to the NCDC and was exclusively responsible for co-operative development. Under this Act, 1962 two types of corporations have been set up *viz.*, the CWC and the SWCs in each state.

a. Need for Warehousing

The following are the two approaches that highlight the importance of warehouses:

Conventional approach (Push concept): According to this approach,

- ☆ Warehouses offers protection against delays and uncertainty in transportation
- ☆ Warehouses ensure availability of products on desired time and at desired place
- ☆ Warehouses provide adjustment between the time of production and use of the product
- ☆ Serve as a reservoir of goods

Modern approach (Pull concept): According to this approach,

- ☆ Warehouse is a location, where inputs received from production are converted into outputs to the consumers, at shortest possible time
- ☆ Warehouse serves as a location of temporary storage facility and from where they are dispatched with the main objective of maintaining the flow of goods throughout the logistics system
- ☆ Warehouse adds to the cost of distribution, but the financial overall benefits outweigh the costs.

b. Basic Components of a Warehouse

There are three basic components *viz.*, Space, Equipment and People.

- ☆ ***Space:*** Space provided in the warehouse allows for the storage of goods, when there is mis-match between demand and supply of goods.
- ☆ ***Equipment:*** Warehouse equipment includes loading and unloading devices, handling devices, storage racks, information processing system etc. These equipments help in product movement, storage and tracking.
- ☆ ***People:*** People are the most critical component of a warehouse. Space and equipment mean nothing without competent people. A primary reason for establishing a warehouse is to increase customer service levels. This often requires individual attention to special customer requests like specialized packaging, grading, quality maintenance of goods, arranging shipments etc. So, these warehouses play critical role in every part of the supply chain.

c. Characteristics of ' Ideal Warehouses'

An 'ideal warehouse' must possess the following characteristics:

- ☆ Warehouse should be located at a convenient place near highways, railway stations, airports and seaports, where goods can be loaded and unloaded easily.
- ☆ Mechanical appliances should be there to load and unload the produce. For this, sufficient parking space should be provided inside the warehouses. This reduces the wastages in handling and also minimizes handling costs.

- ☆ Should have ample space to stock the goods in proper order.
- ☆ For stocking perishable goods like fruits, vegetables, milk etc., warehouses must have cold storage facilities.
- ☆ Scientific management must be provided to protect the produce both from biotic and abiotic agents.
- ☆ Tight security arrangements should be there to avoid theft of goods.
- ☆ The building should be fitted with latest fire-fighting equipments to avoid loss of goods due to fire.
- ☆ Should have sufficient storage capacity and that can be expanded in future keeping in view the increasing production trends of the commodities.

d. Types of Warehouses

In order to meet the storage needs of different types of agricultural commodities, various types of warehouses came into existence, which may be classified as follows (Figure 5.24):

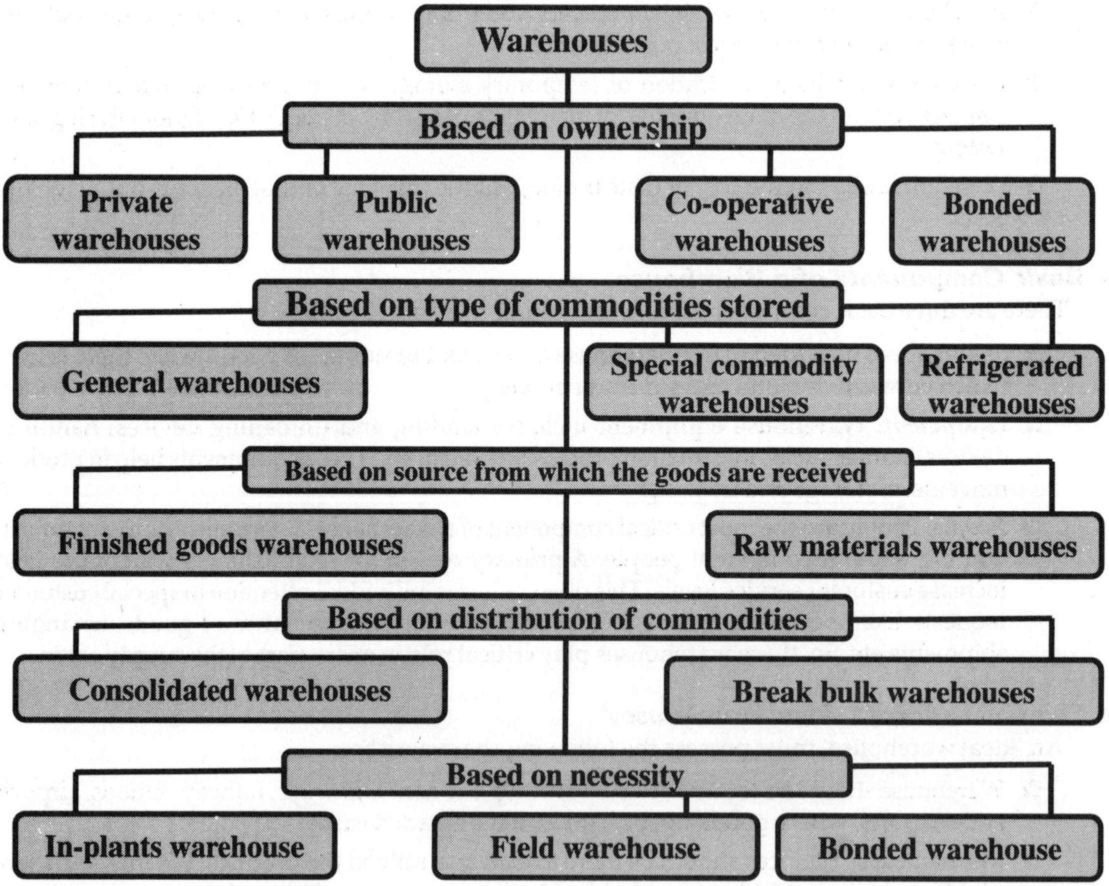

Figure 5.24: Types of warehouses.

1. *On the Basis of Ownership*

Based on this criterion, the following are the different types of warehouses:

- ☆ ***Private warehouses:*** These warehouses are also called as Proprietary warehouses. The warehouses, which are owned and managed by the large farmers or traders to store exclusively their own stock of commodities are known as private warehouses. So, these warehouses are meant to fulfill the own storage needs of large farmers and traders. Big business firms, which needs large storage capacity on a regular basis and who can afford huge money for their construction will maintain these private warehouses. A big farmer or trader may have a network of his own warehouses in different parts of the country. The private warehouses are licensed to private persons and only the goods imported by or on behalf of the licensee are stored in such warehouse. Generally, these warehouses are constructed by the farmers near their fields, by traders near the market yards. The design and the facilities provided therein are according to the nature of produce to be stored. These warehouses also store the produce of others on hire basis. Thus, in a private warehouse, manufacturing and warehousing take place under the same roof and both activities are controlled by the same firm. The warehouse is just a part of the overall firm's operations, which may be producing or wholesaling or retailing. It is very risky to use commodities in private warehouses as collateral for loans, as stringent control is required from the banks to ensure that the goods are really present.

Table 5.64: Public warehouses vis-à-vis Private warehouses.

Parameters	*Public Warehouse*	*Private Warehouse*
Ownership	By Government (rented)	By firms owning the product
Flexibility	Highly flexible	Rigid
Return on investment	Contributes a considerable amount	Contributes less
Operating cost	Low operating cost	High operating cost
Economics of scale	Yes	No
Range of services	Greater	Lesser
Scale of operations	Large	Small

- ☆ ***Public warehouses:*** These warehouses are a specialized business establishments that provide storage facilities to the general public for a certain charge. These warehouses are owned, managed and controlled by Central or State Governments (Table 5.64) or public corporations or local authorities. Both Government and private enterprises may use these warehouses to store the commodities. They are generally located near the junctions of railways, highways and waterways. They, therefore, provide excellent facilities for the easy receipt, dispatch, loading and unloading of goods. A public warehouse is also known as 'Duty paid warehouse'. Public warehouses are very useful to the small farmers and traders, as they can meet their storage needs easily and economically without heavy investment. They provide facilities for the inspection of goods by prospective buyers. They also permit packaging and grading of goods. The public warehouses receipts are good collateral securities for borrowings. There are three agencies in the public sector, which are engaged in building large scale storage/ warehousing capacity namely, FCI, CWC and SWCs. Over a period of time, sizeable scientific storage/warehousing capacity has been developed by these public sector agencies and they are implementing plans to increase it further. While the capacity available with FCI is used mainly for storage of food grains, that with CWC and SWCs is used for storage of food grains as well as certain other items. The storage capacity of FCI, CWC and SWCs as on 31st March, 2010 is shown through Table 5.65.

Table 5.65: State-wise storage capacity available with different storage agencies (public sector) in the country as on 31st March, 2010 (Lakh m tonnes)

State	*FCI*	*CWC*	*SWC*	*Total*
Andhra Pradesh and Andaman and Nicobar islands	38.18	15.04	20.32	73.54
Arunachal Pradesh	0.22	–	–	0.22
Assam	2.73	0.65	2.50	5.88
Bihar	6.72	1.26	2.57	10.55
Chhattisgarh	8.43	2.84	10.24	21.51
Delhi	3.67	1.51	–	5.18
Goa	0.15	0.41	–	0.56
Gujarat	6.80	7.65	1.53	15.98
Haryana	24.45	5.37	16.24	46.06
Himachal Pradesh	0.25	0.07	–	0.32
Jammu and Kashmir	1.31	–	–	1.31
Jharkhand	1.19	0.35	–	1.54
Karnataka	8.37	7.11	10.44	25.92
Kerala	5.37	1.23	2.25	8.85
Madhya Pradesh	9.50	5.17	20.82	35.49
Maharashtra	19.98	16.64	12.48	49.10
Manipur	0.20	–	–	0.20
Meghalaya	0.26	–	0.14	0.40
Mizoram	0.23	–	–	0.23
Nagaland	0.34	0.13	–	0.47
Odisha	6.43	3.17	4.06	13.66
Philippines		0.80	–	0.80
Pondicherry UT/Chandigarh	3.55	0.19	–	3.74
Punjab	73.73	6.95	57.00	137.68
Rajasthan	16.08	4.01	7.72	27.81
Sikkim	0.11	–	–	0.11
Tamil Nadu	9.70	6.31	6.47	22.48
Tripura	0.51	0.24	–	0.75
Uttar Pradesh	26.87	11.63	32.33	70.83
Uttarakhand	2.37	0.71	–	3.08
West Bengal	11.06	6.54	2.15	19.75
Grand total	288.36	105.98	209.26	603.60

Source: <cewacor.nic.in>

SWOT Analysis of Public Sector Warehouses vis-à-vis Private Sector Warehouses

A comparative analysis of the strengths, weaknesses, opportunities and threats (SWOT) of the public sector warehouses and private sector warehouses is shown through Table 5.66.

Table 5.66: SWOT analysis of Public sector warehouses vis-à-vis Private sector warehouses.

Public Sector Warehouses	*Private Sector Warehouses*
Strengths	**Strengths**
☆ Homogeneous storage practices	☆ Hygiene compared to public warehouses
☆ Standardized scientific storage facilities	☆ Plays crucial role in international trade
☆ Wide range of products can be stored	☆ Facilities are highly systematic and scientific compared to public warehouses
☆ Huge quantities of produce can be stored	☆ Ensures more control, flexibility, cost, and other intangible benefits
☆ Satisfactory record-keeping and financial accounting	☆ Private warehousing is usually considered less costly than public warehousing because, private facility costs do not have a profit markup
☆ Warehouse receipts are negotiable by most banks	☆ Operating cost can be 15 to 25 per cent lower, if the firm achieves sufficient utilization. This is possible, even if the firm achieves at least 75 per cent utilization and if not, it would be best to use public warehousing
☆ ISO certification for the products	☆ There is greater care in handling and storage, when the firm's own workforce operates the warehouse
☆ Good reputation with most of its private clientele	☆ When a firm distributes its products through a private warehouse, it gives the customer a sense of permanence and continuity of business operations. The customer perceives the company as a stable, dependable, and lasting supplier of products
☆ High level of capacity utilization	
☆ Public warehouses often are more efficient or may operate at lower wage scales	
☆ From a financial perspective, public warehousing may have a lower variable cost than comparable privately operated facilities	
☆ Public warehousing offers flexibility in that, it is easy to change the location, size, and number of facilities, allowing a firm to quickly respond to supplier, customer, and seasonal demands	
☆ Public warehousing can also offer significant scale economies	
☆ Permit freight to move at lower rates	
☆ Reduced risk (Low opportunity cost)	
☆ If firm's operations have seasonality, then having public warehouse allows the user to rent as much of warehouse space during peak season and less space can be taken for rent during lean marketing season	
☆ Zero capital investment in warehousing *i.e.*, the cost of public warehousing is a variable cost component	
Weaknesses	**Weaknesses**
☆ Not well integrated with private supply chains	☆ Limited professional development
☆ Biased towards consumption centres	☆ Heterogeneous practices
☆ Public ownership diminishes freedom to expand and improve services	☆ Limited space for stocking of goods
☆ Lack of proper maintenance	☆ High rental costs for storing the produce
☆ Poor stacking of cargo	☆ Private warehouses are relatively fixed and difficult to change because buildings have to be constructed or sold
☆ Excessive bureaucracy	☆ It is too costly, because of its fixed size and costs
☆ Most public warehouse facilities provide local services, which may not be useful for the big MNCs, who require more specialized services	
☆ Shortage of space can happen in some places especially during peak season, and this may affect the firm adversely	

Contd...

Table 5.66–*Contd...*

Public Sector Warehouses	*Private Sector Warehouses*
Opportunities ☆ Partnerships and alliances with the private sector ☆ Alliances with banks and other lenders to channel credit to depositor ☆ Alliances with commodity exchanges to act as delivery points ☆ Alliances with transport providers ☆ Scope to improve facilities on par with international standards ☆ Public warehousing provides economical and practical means to reach out to new markets	**Opportunities** ☆ Development of collateral management along supply chains ☆ A formal accreditation system can make services more marketable ☆ Storage role of rural entrepreneurs can be developed through training and accreditation
Threats ☆ Increased competition from private sector warehouses ☆ The setting of storage rates for farmers is to a significant extent a political process, even though politicians may not be directly involved, and rates are set at levels that often compromise profitability ☆ Low profitability ☆ Low storage rates negotiated for farmers and co-operatives pose a problem in expanding services to non-Governmental customers ☆ Users of SWCs more often store non-agricultural produce	**Threats** ☆ Increased competition from public sector warehouses in the ports ☆ Restrictions from the Government regarding the produce to be stored and in its quantity ☆ High opportunity cost (high risk) ☆ Low rate of return ☆ High start-up cost *i.e.*, firms have to generate enough capital to build or buy a warehouse

There are several reasons, which make a farmer or trader select a public warehouse over a private one. In general, small and medium farmers or traders do not have sufficient financial strength to construct own warehouses and hence, they contract a third party warehouse facility. They usually depend on public warehouse facility, as they incur less cost on rents. However, large farmers and big traders have sufficient financial strength to have their own warehouses. They have their own trucks that transport goods from one location to another. So, the reasons for using a public or private warehouse have to do with the size of the marketing agency needing the storage and its amount of revenue on hand that can be used to either lease a public facility or private facility or own a private one. These factors make all of the difference.

- ☆ ***Co-operative warehouses:*** These warehouses are owned, managed and controlled by co-operative societies. These societies provide warehousing facilities to their members at the most economical rates.
- ☆ ***Bonded warehouses:*** A warehouse licensed to keep imported goods awaiting clearance because of non-payment of custom duties is called a bonded warehouse. Thus, bonded warehouses are used to store imported goods for which import duty is yet to be paid. So, the warehouse is required to give an undertaking or 'Bond' that it will not allow the goods to be removed without the consent of the custom authorities. The goods are held in bond and cannot be withdrawn without paying the customs duty. These warehouses are very helpful to importers and exporters. If an importer is unable to pay customs duty immediately after the arrival of goods, he can store the same in a bonded warehouse. He can withdraw the goods in installments by paying the customs duty proportionately. These warehouses are

situated near the port and are owned, managed and controlled by Government or private agencies and supervised by port trust and custom authorities. Private bonded warehouses have to obtain license from the Government. Goods lying in a bonded warehouse can be packaged, graded and branded for the purpose of sale. CWC operates 75 Custom Bonded Warehouses with a total operated capacity of nearly 0.5 million m tonnes. These warehouses are generally owned by dock authorities and found near the ports. The goods stored in this warehouse are called 'bonded goods'. The goods on which duty has been paid is kept in duty-paid warehouse for which a warehouse receipt is given. The goods on which duty has not been paid is kept in the bonded warehouse for which dock warrant is issued, which serves like a warehouse receipt. Following services are rendered by bonded warehouses:

- The importer of goods is saved from the botheration of paying customs duty all at one time because, he can take delivery of the goods in parts.
- The operations necessary for the maintenance of the quality of goods *i.e.*, spraying and dusting are done regularly.
- Entrepot trade (re-export of imported goods) becomes possible.
- The importer can effect sale of goods or transfer the title of goods by endorsing the warehouse receipts. By this, it enables to save much time on part of the importer.
- The owner of the goods is allowed to brand, grade, label and pack the goods when these are in bonded warehouses. By this provision, it facilitates proper marketing of goods.
- Goods kept in bonded warehouse are used as collateral security for bank loan.
- Goods kept in bonded warehouse are safe and are free from the fear of being stolen or damaged.

Thus, bonded warehouses are used by the importer when :

- he does not has his own godown,
- he may not be able to pay custom duties immediately,
- he wants to re-export the imported goods for entrepot trade.

2. On the Basis of Type of Commodities Stored

This criterion classifies the warehouses into the following categories:

- *General warehouses:* These warehouses are meant for storage of different types of agricultural inputs and commodities like fertilizers, pesticides, food grains etc. In constructing these warehouses, no commodity-specific requirement is kept in view.
- *Special commodity warehouses:* These warehouses are commodity-specific *i.e.*, they are specially constructed for the storage of specific commodities like rice, wheat, cotton, tobacco etc. This is so because, some commodities or items are so special that, they are impracticable to combine their storage with others. They need special protection from fire say, in case of cotton bales, tobacco, oil products, woolen materials etc. So, warehousing has to be protected from fire and there should be no scope to spread fire.
- *Refrigerated warehouses:* In these warehouses, temperature is maintained as per requirements usually below 30° to 50°F or even less so that, the produce may not get spoiled by external high atmospheric temperature. These warehouses are meant for stocking highly perishable commodities like chillies, milk, fish, fruits, vegetables etc.

3. On the Basis of Source from which the Goods are Received

We know, any warehouse performs the following functions:

- ☆ *Receive* the goods from a source.
- ☆ *Store* the goods until they are required.
- ☆ *Pick* the goods from the storage area, when they are required.
- ☆ *Ship* the goods to the appropriate user.

Depending upon the nature of goods stocked, there are two types of warehouses *viz.*, finished goods warehouse and raw materials warehouse. However, in both these warehouses, the above four functions are to be performed *i.e., 'receive-store-pick-ship'* in a sequence. So, both these store rooms can be regarded as ware houses. The only way to distinguish these two warehouses is based on the source from which the goods are received and the user to whom the goods are shipped.

- ☆ ***Finished goods warehouses:*** In these warehouses, the goods are received from an inside source (say, cotton from the farmers and traders), stored, picked and shipped to an outside user (say, exported to a foreign nation).
- ☆ ***Raw materials warehouses:*** In these warehouses, the goods are received from an outside source (say, importing raw materials for manufacturing fertilizers), stored, picked and shipped to an inside user (say, domestic fertilizer plant).

4. On the Basis of Distribution of Commodities

According to this criterion, there are two types of warehouses:

- ☆ ***Consolidation warehouses:*** These warehouses pull together small shipments from a number of suppliers (say, farmers or traders) in the same geographical area and combines them into larger, more economical, shipping loads intended for the same area. That is, these warehouses make small, flexible shipments into large, economical shipments (Figure 5.25). So, the consolidation warehouse receives and consolidates goods from a number of farmers or traders destined to specific customers on a single transportation shipment. These warehouses allow both inbound movement from different farmers or traders to the warehouse and outbound movement from warehouse to the customers through a single shipment. It is advantageous, as it combines logistical flow of several small shipments to a specific market area. Further, it ensures lower distribution cost for farmers or traders, as they may join together and use for-hire consolidation service.
- ☆ ***Break bulk warehouses:*** In break bulk warehouses, the warehouses receive combined customer orders from farmers or traders and they are arranged for local customers through different transportation shipments (Figure 5.25). So, these warehouses split the single transport from farmers group or traders group and arrange for local delivery for different customers. This Break bulk warehousing is similar to Consolidation warehousing except that, no storage is performed.

5. On the Basis of Place of Necessity

Under this category, there are three types of warehouses as given here under.

- ☆ ***In-plants warehouses:*** A large number of manufactures find it convenient to distribute the products from the plants directly to the retailers or to customers. For this, they construct and maintain in-plants warehouses, as it reduces cost of handling and improves the service.

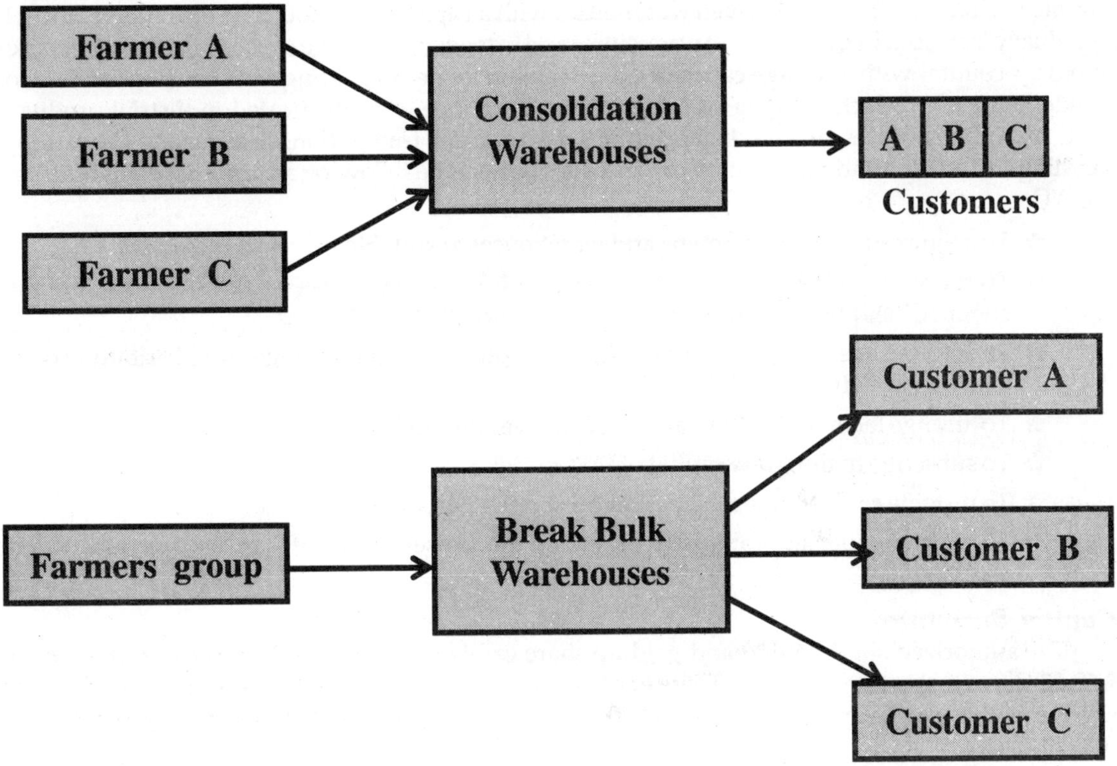

Figure 5.25: Consolidation warehouses vis-à-vis Break bulk warehouses.

Examples include fertilizer manufacturers, pesticides manufacturers etc.

- *Field warehouses:* These are centrally located warehouses from where distribution is done to wholesalers and retailers. This is necessary, where products from different plants are to be mixed together.
- *Bonded warehouses:* These warehouses are located near ports. They enable the unloading of commodities from a ship safely into a place until the owner of the goods takes delivery of them.

e. Structure of Warehouse System in India

In India, three tier system exists in storage *viz.* at the National, State/district and village levels. At the apex level *i.e.*, at the centres of All-India importance, the CWC has been providing warehousing facilities. At the state and district levels, the SWCs and the State Governments are providing the requisite storage facilities. At the Village/taluka level, CMSs are providing storage facilities.

1. CWC

This Corporation was established as a statutory body in New Delhi on 2nd March, 1957. The CWC was set up under the Agricultural Produce (Development and Warehousing) Corporations Act, 1956. The said Act was subsequently replaced by the Warehousing Corporations Act, 1962 (CWC Act, 1962). The CWC offers safe and reliable storage facilities to the farmers and traders. The area of operation of the warehouses under CWC is Centre and inter-State importance. The CWC made an humble beginning in 1957 with seven warehouses with a capacity of 7,000 m tonnes and in 2010-11, it gradually increased its operations to new centres and extended its storage capacity to 479 Warehouses across the country with a storage capacity of 9.98 million tonnes providing warehousing services for a wide range of products ranging from agricultural produce to sophisticated industrial products. CWC also offers services such as disinfestations, pest control, fumigation, clearing and forwarding, handling and transportation, procurement and distribution The following are the important functions of CWC:

- ☆ To acquire and build godowns and warehouses at suitable places in India.
- ☆ To run warehouses for the storage of agricultural produce, seeds, fertilizers and notified commodities for individuals, co-operatives and other institutions,
- ☆ To act as an agent of the Government for the purchase, sale, storage and distribution of the above commodities.
- ☆ To arrange facilities for the transport of above commodities.
- ☆ To subscribe to the share capital of SWCs
- ☆ To provide cold storage facilities for preserving perishable produce
- ☆ To provide special storage facilities for the preservation of hygroscopic and fragile commodities.

Capital Structure

The authorized share capital and paid up share capital of the Corporation is ₹ 100 crores and ₹ 68.02 crores respectively in 2011. There has been no change in the CWC's subscribed share capital and the paid up share capital. Figure 5.26 clearly depicts the growth in warehousing capacity of CWC over the years.

Physical Performance

As on the 31st March, 2011, CWC operated 479 warehouses with a total storage capacity of 102.47 lakh m tonnes (Table 5.67 and Figure 5.27). During 2010-11, the average capacity utilization was improved to 87 per cent from 74 per cent achieved during 2006-07. As on 31.3.2011, the CWC operated 66 custom bonded warehouses with a capacity of 4.34 lakh m tonnes.

Table 5.67: Physical performance of CWC.

Item	*As on 31st March, 2009*	*As on 31st March, 2010*	*As on 31st March, 2011*
Number of warehouses	499	487	479
Capacity (Lakh m tonnes)			
Constructed/Owned	67.60	68.46	69.85
Hired	23.15	21.32	17.81
Open	14.50	16.20	14.81
Total	105.25	105.98	102.47
Commodity-wise utilization (Lakh m tonnes)			
Food grains	36.50	43.88	48.72
Fertilizers	3.41	2.23	2.08
Others	41.45	45.76	39.97
Total	81.36	91.87	90.77

Source:Annual Report, 2010-11, CWC.

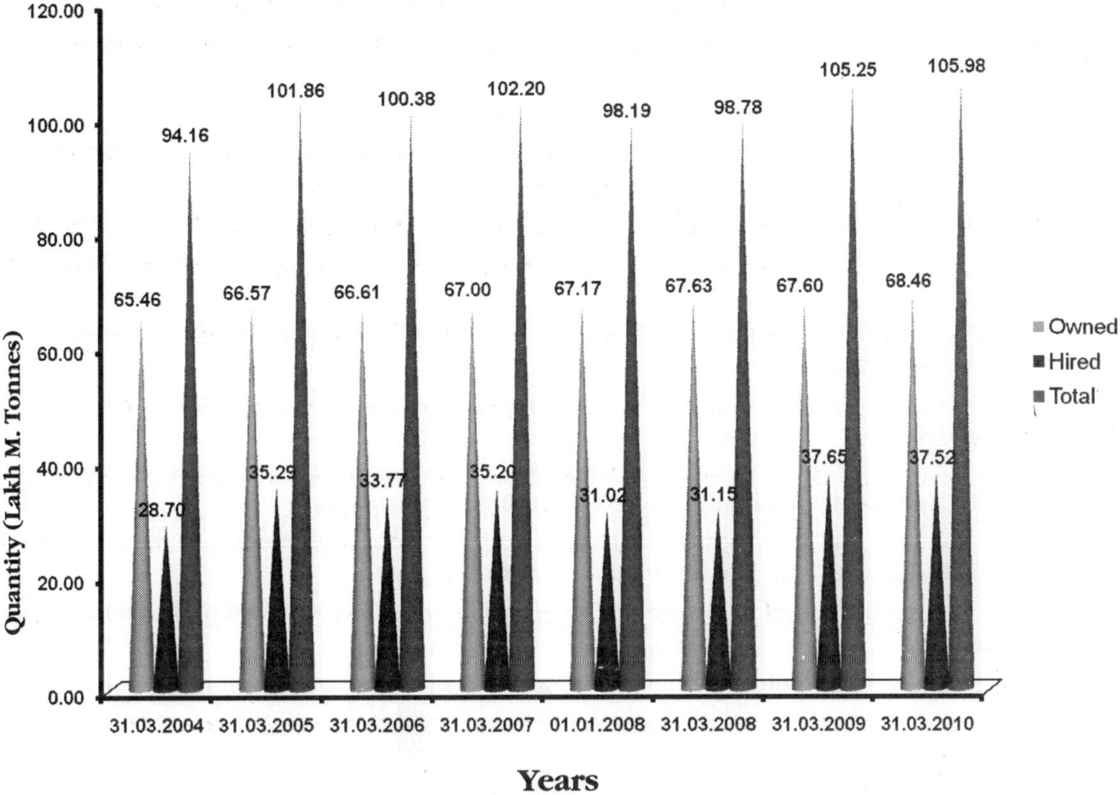

Figure 5.26: Growth in warehousing capacity of CWC.

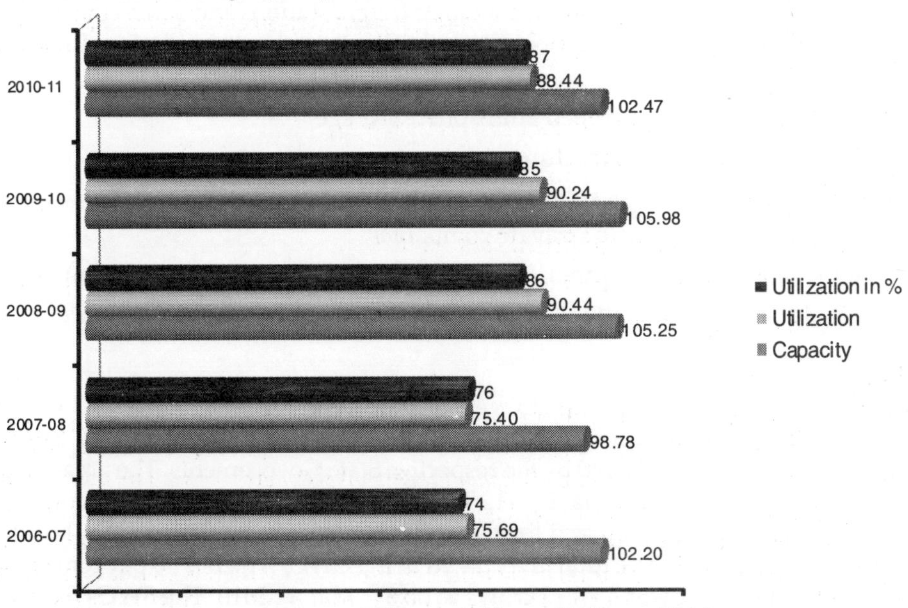

Figure 5.27: Storage capacity of CWC and its utilization as on 31st, March (Lakh m tonnes).

The CWC has significantly contributed to the economy of the country in the following ways.

- It provides storage facilities through a chain of over 479 warehouses spread all over the country.
- It helps in preservation of stocks and minimizing in storage losses.
- It provides specialized storage for highly hygroscopic and delicate commodities like jaggery, spices, pharmaceuticals, chemicals and other consumer goods.
- Handling, clearance and distribution of imported cements were also undertaken by CWC.
- Air cargo complex for import/export and un-accompanied baggage at airports was handled by CWC.
- Managing air-conditioned warehouses for exportable shellac and other delicate commodities, was within its purview.
- It supervises orderly marketing through proper receipt, handling, grading and delivery of products.
- It provides disinfectant service to the produce stored by farmers, traders, co-operatives, millers and others, and fumigation of exportable commodities.
- Renders extension and educational services to farmers, which educate them about the scientific storage and benefits of public warehousing.
- It provides training of personnel and expert advice on matters related to warehousing and storage.
- It provides an incentive in the form of a rebate of 10 per cent in storage charges to co-operative societies and individuals/farmers who avail of the corporation storage facilities.

New Activities of CWC

The CWC started diversifying its services in the 1980s through establishing custom-bonded warehouses (capacity 800000 m tonnes), container freight stations and some cold storages. Given the low profitability of its general warehousing activities, it is now considering a range of additional activities including:

- A cold chain with controlled atmosphere storage
- The provision of infrastructure in the ports
- Provision of rail freight terminals
- Custom warehousing for private companies
- Bulk handling of food grain, in conjunction with an international company
- Liquid warehousing, involving edible oils and industrial chemicals

2. SWCs

Separate warehousing corporations were also set up in different States and UTs. The first SWC was set up in Bihar in 1956 and presently, 17 SWCs are operating. These 17 SWCs work under different Warehousing Acts enacted by the respective State Governments. The areas of operation of the SWCs are centres of district importance. The total share capital of the SWCs is contributed equally by the concerned State Government and the CWC. Thus, SWCs are under the dual control of the State Government and the CWC. The total investment of the CWC, which is 50 per cent shareholder in the equity capital of the SWCs was ₹ 60.12 crores as on 31st March, 2010. The SWCs paid a total dividend

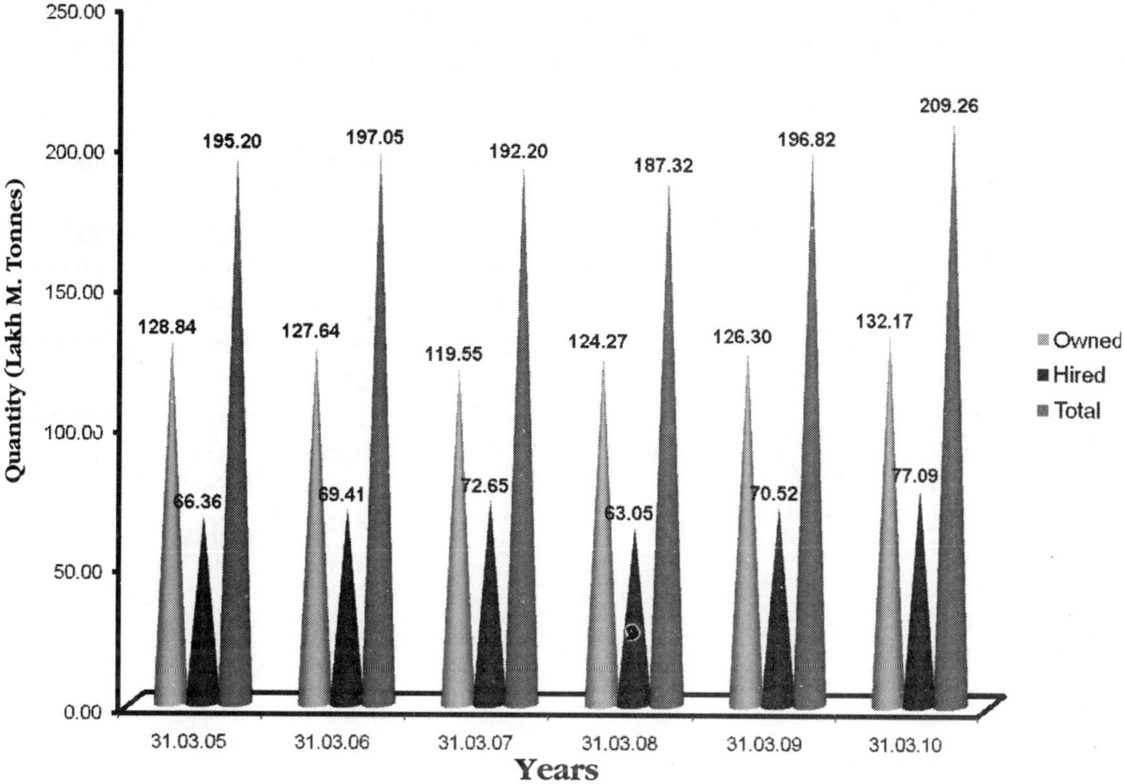

Figure 5.28: Growth in warehousing capacity of SWCs.

of ₹ 4.59 crores to the CWC during 2009-10. The following Figure 5.28 clearly depicts the growth in warehousing capacity of SWCs over the years.

Physical Performance of SWCs

CWC holds 50 per cent equity in the 17 SWCs with the respective State Governments holding the remaining 50 per cent of the equity. As on 31st March, 2011, these SWCs operated a network of 1585 warehouses with an aggregate storage capacity of 211.27 lakh m tonnes (Tables 5.68 and 5.69 and Figure 5.29).

Table 5.68: Number and storage capacity of SWCs (as on 31st, March).

Year	*Number*	*Storage Capacity (Lakh m tonnes)*			
		Constructed	*Hired*	*Plinth*	*Total*
2009-10	1595	123.97	77.09	8.20	209.26
2010-11	1585	126.34	74.89	7.04	211.27

Table 5.69: State-wise number of centres and storage capacities of SWCs as on 31st March, 2011.

Name of SWC	*No. of Centres*	*Capacity (Lakh m tonnes)*				*Occupancy*	
		Constructed	*Hired*	*Plinth*	*Total*	*Actual (Lakh m tonnes)*	*Per cent*
Andhra Pradesh	133	5.34	16.92	—	22.26	21.65	97
Assam	44	2.12	0.42	—	2.54	1.83	72
Bihar	41	1.45	1.00	—	2.45	2.05	84
Chhattisgarh	111	5.64	4.08	—	9.72	9.46	97
Gujarat	45	1.45	0.03	—	1.48	0.83	56
Haryana	107	10.13	3.01	2.06	15.20	13.39	88
Karnataka	121	6.63	3.07	—	9.70	8.02	83
Kerala	59	1.80	0.14	—	1.94	0.91	47
Madhya Pradesh	258	11.62	14.16	0.40	26.18	18.17	69
Maharashtra	167	11.79	0.43	0.13	12.35	10.38	84
Meghalaya	6	0.14	—	—	0.14	0.11	79
Odisha	59	3.95	0.16	—	4.11	3.80	92
Punjab	116	24.18	28.99	4.45	57.62	50.80	88
Rajasthan	90	7.71	0.15	—	7.86	5.23	67
Tamil Nadu	58	6.26	0.12	—	6.38	5.30	83
Uttar Pradesh	140	24.76	4.42	—	29.18	24.26	83
West Bengal	30	1.37	0.79	—	2.16	1.20	56
Total	1585	126.34	77.89	7.04	211.27	177.39	84

Source: Annual Report, 2011, CWC.

Functions of SWCs

The statutory functions of the SWCs are:

- ☆ Acquire and build godowns and warehouses at such suitable places within the State.
- ☆ Run warehouses in the State, for the storage of the agricultural produce, seeds, manures, fertilizers, agricultural implements and notified commodities.
- ☆ Arrange facilities for the transport of agricultural produce seeds, manures, fertilizers, agricultural implements and notified commodities to and from warehouses.
- ☆ Act as an agent of the CWC or the Government for the purpose of purchase, sale, storage and distribution of agricultural produce, seeds, manures, fertilizers, agricultural implements and notified commodities.
- ☆ Undertakes disinfestation extension services at the door steps of the farmers, traders, industrialists etc., on short calls as well as on yearly contract basis as per the request of the customers.
- ☆ Offers its helping hand to the public sector organizations in designing and constructing buildings and godowns making use of the technical know-how and personnel available with the Civil Engineering Wing.

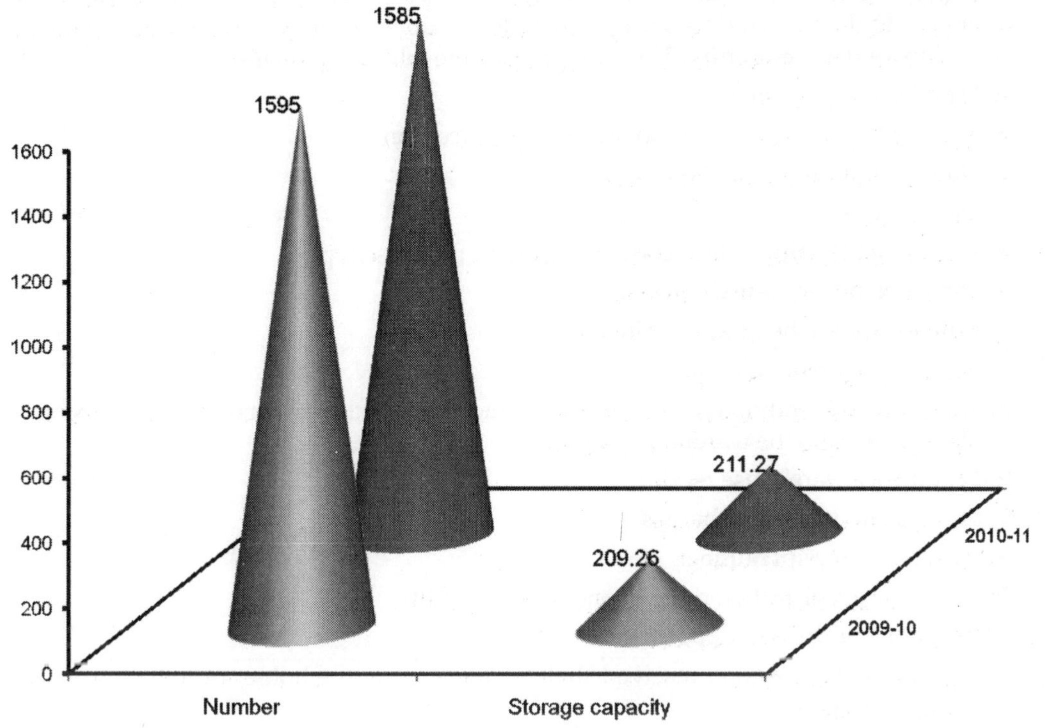

Figure 5.29: Number and storage capacity of SWCs as on 31 st, March (Lakh M tonnes).

All types of agricultural commodities and other commodities notified by the Government from time to time are accepted for storage in a warehouse. The idea behind the whole warehousing scheme is to give scientific storage to the agricultural produce and also to arrange easy credit and holding power to the agriculturists to enable them to get a better price for their produces. A Warehouse Receipt incorporating the quantity, quality, market value and particulars of insurance against fire, theft and burglary will be issued to the depositor. The Warehouse Receipt is negotiable in nature and can be pledged in any bank.

3. CMSs

Co-operatives are providing storage facilities at the primary and marketing societies level, which are located at village/taluka.

f. Working of Warehouses

The following are the features of the warehouses:

- ✰ ***Acts:*** The Warehousing Corporations Act came into operation on 18th March 1962. The Act defines the specific functions and the area of operations of CWC and SWCs. It enlarged the list of the number of commodities meant for storage.
- ✰ ***Eligibility:*** Any person may store notified commodities in a warehouse on agreeing to pay the specified charges.

☆ ***Warehouse receipt (Warrant):*** The warehouse receipt is simply a warrant issued by the warehouse to the person depositing the goods/produce in the warehouse indicating his ownership for the commodity. This receipt states the following details:

- Kind/Type of product
- Specified quantity, quality and grade of produce stored
- Name and location of warehouse
- Date of issue
- Approximate value of the produce indicating the present price
- Owner of the (deposited) product
- Whether or not the product is insured
- Name of insurance company
- The terms and conditions of the storage contract between the farmer or trader or processor (depositors) and the warehouse operator
- Feature of a warehouse receipt
- Storage and handling charges
- Warehouse receipt number
- Warehouse registration number and date of validity
- Name and address of depositor
- Statement that the good received shall be delivered to the holder or to the order of a named person
- Whether warehouse receipt is negotiable or non negotiable
- Statement of the amount of any advance made or any liability incurred for which the warehouse man claims his lien
- Date and signature of the warehouse man or his authorized agent
- Declared shelf life of goods

Warehouse receipts are issued by all approved warehouses after quality certification of the stored goods. The main advantage of warehouse receipt is that, it converts agricultural produce or other inventory to a tradable warrant, which can be sold or used to raise a loan and even used for delivery against a derivative instrument like futures contract. Hence, warehouse receipt finance is also called as 'inventory credit'. The amount of loan advanced depends on the market price, MSP and the guidelines issued by the bank. In general, the banks issue loans to the extent of 75 per cent of value of produce. The interest charged also depends on the norms of the bank for that type of commodity. With the priority sector lending norms making it mandatory for banks to advance 18 per cent of loans to agriculture, warehouse receipt financing is being considered as the next big opportunity in catering for the agricultural sector. The Planning Commission has also reinforced its commitment to double the farm sector's growth from a low two per cent per annum to four per cent a year during the 11th FYP.

Different banks have different guidelines for providing loans against the stored produce. By and large, there are two systems prevalent *viz.*, Hypothecation/Pledge of stocks (Issuing storage receipt) and Warehouse receipt system. Among these, the former is most common.

1. Hypothecation/Pledge of Stocks (Issuing Storage Receipt)

This system is prevalent mostly for the small private warehouses, where the entire warehouse is occupied with material belonging to a single entity. The material deposited in the warehouse is pledged/hypothecated as security. Once the loan is repaid after negotiations with the buyer, the borrower is free to express his control over the stocks. This scheme is also popular as 'Produce Marketing Loan'. The term 'Storage receipts' is closely associated with pledge financing, which is most common for private gowdons/private licensed warehouses. These receipts are issued by collateral managers of an agency known to the bank, who will have control over the commodity by lock-key method. In this way, the bank finances the depositor, only if the collateral manager approves the stock through the storage receipt.

2. Warehouse Receipt System

This is the most popular system, where the financing is done based on warehouse receipt issued by the warehouse. The financing against warehousing receipts is important because of the following reasons:

- For storage of agricultural produce which are excess in demand at the time of harvest / production.
- For protection to farmers against unfavorable prices
- For scientific post-harvest management
- For reducing wastage
- For trade support to Farmers

The warehouse receipt duly endorsed in favour of the Bank is to be deposited at that bank and the material is released once the bank receives the payment from the buyer. Financing against pledge of warehouse receipts is available to farmers, traders, rice millers, oil mills, floor mills, exporters, agro-industries etc. Financial Institutions extends warehouse receipt financing for CWC, SWCs, warehouses managed and controlled by Collateral Managers like NBHC/NCMSL/Origo/GeoChem etc., private warehouses under specific tie-up with the bank. As mentioned earlier, warehouse receipts can be negotiable or non-negotiable.

- ***Negotiable Warehouse Receipt (NWR):*** For the first time in the country, in 2011, Government of India has introduced a NWR system in the country. This system has been formally launched by Prof. K. V. Thomas, Hon'ble Minister of State (Independent Charge), for Consumer Affairs, Food and Public Distribution on 26.04.2011 at New Delhi. NWR is a paper or electronic document of title in which the ownership of the produce represented by the warehouse receipt is transferrable by delivery or endorsement. The Government introduced this NWR system, which will help farmers gain access to loans from banks and avoid distress sale of agricultural commodities. NWRs allow transfer of ownership of that commodity stored in a warehouse without having to deliver the physical commodity. These receipts are issued in negotiable form, making them eligible as collateral for loans. Banks will have more faith in such NWRs and farmers would be able to seek loans easily against these receipts. It will also enhance banks' lending in respect of agricultural produce deposited by farmers in the registered warehouses. Further, these NWRs will increase liquidity in the rural areas and encourage scientific warehousing of goods. The warehouse receipts are made negotiable under the Warehouse (Development and Regulation) Act, 2007 and regulated by the WDRA. The following are the important advantages of NWR:

- Avoid distress sale by farmers
- Encourage scientific warehousing of goods.
- Lower cost of financing as a result of reduced risk
- Shorter and more efficient supply chains.
- Increased trade and finance in commodities.
- Higher returns to farmers and better services to consumers

Thus, in NWR system, the warehouse owner will offer guarantee regarding quality and quantity of produce and thereby, facilitates the farmers to get the loans on collateral basis. The warehouses that issue NWR should get registered by WDRA.

- ***Non-Negotiable Warehouse Receipt (NNWR):*** NNWR is a paper or electronic document that cannot be transferred by endorsement or delivery to another holder or person. That is, under NNWR, the warehouse owner do not take ownership of produce ie., it will not offer guarantee regarding quality and quantity of produce. So, the farmer will not get the loan directly unless if the collateral manager provides 50 per cent collateral to get loan from the bank. This NNWR authorizes only one party to release the goods from the warehouse. This party may also transfer or assign the goods to another party, for example a bank. The warehouse company must be so notified by the transferor before the transfer or assignment becomes effective. It is to be noted that, if nothing is mentioned on the warehouse receipt, it is considered as NWR.

Advantages of NWR over NNWR

With the introduction of NWR system, there would be following advantages:-

- Increased liquidity in rural areas for farmers.
- Encouragement of scientific warehousing of goods.
- Lower cost of financing by banks.
- Shorter and more efficient supply chains.
- Enhanced rewards for grading and quality.
- Better price risk management by farmers.
- Higher returns to farmers and better services (quality) to the consumers.
- Avoid distress sale of agricultural produce by farmers during peak marketing season.

The NNWR in the past did not enjoy the fiduciary trust of depositors and banks, as there was fear of not being able to recover the loans in events, such as fraud, or mis- management on behalf of the warehouse or insolvency of depositor. The available legal remedies were also time consuming and inadequate. Further, the format of warehouse receipts used in the country was not uniform. Hence, there were impediments in the negotiability of warehouse receipts creating difficulties for the farmers and other depositors of goods. Therefore, it was envisaged to establish a NWR system for all commodities including agricultural commodities. On the one hand, it will make warehouse receipts a prime tool of trade and facilitate finance against it throughout the country. On the other, it will allow banks to improve the quality of their lending portfolio and enhance their interest in lending against goods deposited by farmers in registered warehouses.

Electronic Warehouse Receipts (EWR)

As per the Warehousing (Development and Regulation) Act, 2007, NWRs can be in both paper and electronic forms. The format of the NWR has been finalized in consultation with various stakeholders and Indian Banks' Association (IBA), and paper NWRs are now being issued by the registered warehouses across the country. The Act empowers WDRA to regulate and develop the electronic system of holding and transfer of credit balances of fungible goods deposited in the warehouses. The objective is to electronically connect farmers and consumers, which will go a long way in eliminating intermediaries, and help increase farm gate and lower consumer prices. The advantages of electronic warehouse receipts over the paper warehouse receipts include: reduction in manual paper handling, elimination of transportation of paper warehouse receipts, reduction in chances of forgery, and quick access of information. The importance of EWRs include:

- Farmers will have access to a large number of buyers nationwide
- Farmers will enjoy better bargaining power
- Farmers realize higher prices by selling graded produce
- Avoid distress sales of produce by getting loans from banks against EWRs
- Enable multiple transfers without physical movement of goods
- Provide alternate channels of marketing to famers and reduce the cost of intermediation for consumers
- Facilitate standardization of farm produce, grading, scientific warehousing, packing and logistics
- Enable consumers (industries, processors, wholesalers, retailers etc.) to procure graded produce at competitive prices at locations of their choice
- Promote an efficient clearing, settlement and delivery system; and
- Bring transparency in trading of agricultural produce.

Advantages of Warehouse Receipt System

The potential benefits of a well-functioning licensed warehouse receipt system include,

- Warehouse receipts can be used by farmers, smallholders and co-operatives to finance production
- Processors will receive better and more consistent quality produce and can finance inventory
- Collateralizing inventories will increase availability of credit, reduce cost and mobilize external financial resources
- Warehouse receipts contribute to the creation of cash and forward markets thus, increasing competition
- Warehouse receipts allow a transaction to take place informally or on an organized market or exchange.
- The warehouse receipt is the basis for a spot or cash market
- Warehouse receipts can be combined with price hedging instruments
- Warehouse receipts increase the confidence of participants, particularly those in the private sector, in market transactions.
- Warehouse receipts programmes can increase productivity and the quality of commodities produced by small farmers

- Warehouse receipt programs provide concentration and accumulation of products at one point, attracting more buyers. The more buyers in the market place, the keener the competition, resulting in higher prices being paid
- Warehouse receipt programme provide an opportunity to the Government to purchase stored commodities to offset shortages caused by a crop failure
- Exporters can gain pre-export financing against warehouse receipts under the terms of their documentary credits thus, assisting them to procure stock.
- The produce accepted at the warehouse is preserved scientifically and protected against rodents, insects and pests and other infestations. Periodical dusting and fumigation are done at the cost of the warehouse in order to preserve the goods.
- The warehouse receipt serves as a collateral security for the purpose of getting credit.
- The warehouse receipt has to be surrendered to the warehouse owner before the withdrawal of the goods. The holder may take delivery of a part of the total produce stored after paying the storage charges.
- Warehouse Receipts provide farmers with an instrument that allows them to extend the sales period of modestly perishable products well beyond the harvesting season
- When delivering the product to an accredited warehouse, the farmer obtains a Warehouse Receipt that can be used as collateral for short-term borrowing to obtain working capital. That way, the farmer does not need to sell the product immediately to ease cash constraints
- Warehouse Receipts contribute to the creation of cash and forward markets and thus, enhance competition
- They can form the basis for trading commodities, since they provide all the essential information needed to complete a transaction between a seller and a buyer
- A Warehouse Receipt system provides a way to reduce the need of Government agencies in procurement of agricultural commodities.
- Government intervention in agricultural markets usually has two main objectives: to support prices, by buying directly from producers, and to guarantee a measure of food security. In order to support prices, governments can accept Warehouse Receipts when prices drop below a support floor, rather than taking delivery of physical inventories.
- Since Warehouse Receipts guarantee the existence of stocks, governments can achieve their food–security objectives by merely holding these receipts.

***Limitations on the Use of Warehouse Receipts:* They Include**

- Lack of incentives for the development of a private storage industry owing to government intervention in agricultural markets
- Lack of an appropriate legal, regulatory, and institutional environment to support a system of Warehouse Receipts
- Need for splitting the Warehouse Receipt in case the depositor has an obligation to transfer only a part of the commodities
- Need to move the Warehouse Receipt from one place to another with risk of theft/mutilation, etc. if the transferor and transferee are at two different locations
- Risk of forgery

Advantages of Electronic Receipts over Paper Receipts

The advantages of electronic receipts over their paper counterparts include:

- ☆ Reduction in manual-paper handling;
- ☆ Transporting paper documents is eliminated along with the attendant risks
- ☆ Information is moved faster;
- ☆ Multiple keypunching of data is reduced;
- ☆ An audit trail of receipt activity is kept, and the electronic receipt system serves to back-up receipt data for the warehouse;
- ☆ Chances of forgery are reduced.
- ☆ Electronic Warehouse Receipts are different from paper Warehouse Receipts in that any part can be fractionalized to thousandths of the whole

g. Mechanism of Storing Agricultural Produce in Warehouses and Recovery of Market Finance

The chief advantage of storing the produce in the warehouses is to avoid distress sales of produce at the farmers' level. Even the traders also enjoy the same benefit in stocking the produce at times of low prices and thus, enable them to get higher prices by destocking the same, when prices shoot up in the market. The farmers take the produce to the nearest accredited warehouse instead of selling it in the local market at low prices during the peak marketing season (Figure 5.30). At the warehouse, the farmer can stock the produce till the prices shoot up in the market. By stocking the produce in the warehouse, the farmer gets a NWR, which serves as a collateral instrument in getting the loan from the nearest bank to meet his immediate liquidity needs. The warehouse owner meanwhile checks the quality of the produce and classifies the same into different grades. If any trader wishes to purchase the produce at a particular price and if that price is accepted by the farmer, he can simply sale the produce by transferring the NWR. The trader and the farmer approach the bank (or the trader give the money to the farmer in advance based on mutual trust) and release the NWR from the bank by repaying the loan and interest charges. Then, the farmer (and trader) approach the warehouse keeper and collect the produce after submitting the NWR and paying the storage charges. Then, the farmer will hand over the produce to the trader. Thus, banks will recover the marketing loan by linking credit with marketing through warehouses.

In view of the significance of warehouses in protecting the interests of farming community and in linking credit (market finance) with marketing for their recovery, financial assistance was provided to various States for construction of godowns at the rural level under the scheme of National Grid of Rural Godowns, which stands transferred to State Governments with effect from 1st April, 1992. The FCI constructs godowns for storage of food grains procured by it for maintenance and distribution of buffer stocks. It constructs storage capacity at certain nodal points, keeping in view its requirements. The CWC constructs/creates storage capacity for FCI as well as for the general warehousing. The Department of Agriculture and Co-operation introduced a central sector scheme, the 'GBY', in March 2001 to promote the construction of rural godowns. The main objectives of the scheme include the creation of a scientific storage capacity with allied facilities in rural areas to meet the requirements of farmers for storing farm produce and to prevent distress sale of produce. Initially, the scheme was approved for two years *i.e.*, up to 2003. The scheme was later approved for implementation up to 31st March, 2007 with some modifications. Under the revised scheme, a subsidy at the rate of 25 per cent was given to all categories of farmers, agriculture graduates, co-operatives and the CWC/SWCs. All

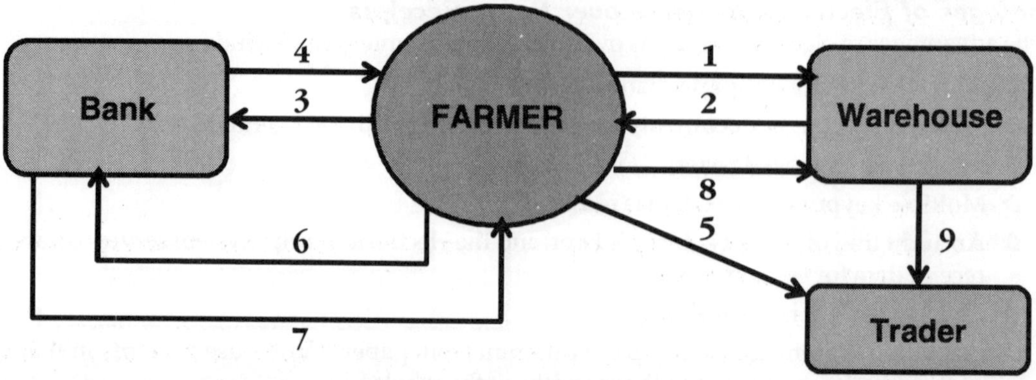

1. To overcome distress sales, farmer stock the produce in a warehouse	**5. Farmer finds a trader when price rises for the produce and negotiates with him**
2. Warehouse issues warehouse receipt to the farmer	**6. At the negotiated price, farmer approach the bank and repays the loan amount to the bank**
3. Farmer offers warehouse receipt as collateral security to the bank	**7. Bank deduct the loan amount and interest charges and issues warehouse receipt back to the farmer**
4. Bank offers pledge loan to the farmer to the extent of 75% of value of produce	**8. Farmer redeems receipt at the warehouse to collect the produce after paying the storage charges.**
	9. Trader collects produce from the farmer at warehouse

Figure 5.30: Mechanism of marketing agricultural produce by the farmers using NWR.

other categories of individuals companies and corporations are entitled to a subsidy at the rate of 15 per cent of the project cost. In North-Eastern states/hilly areas and SC/ST entrepreneurs and their co-operatives, subsidy is provided at the rate of 33.33 per cent.

In 2011-12, an exclusive Scheme for Financing Warehousing Infrastructure under Rural Infrastructure Development Fund (RIDF) for development of warehousing sector with allocation of ₹ 2000 crores has been announced by Government of India. This RIDF scheme is an important scheme for routing bank funds for financing rural infrastructure. The scheme will be open to the State Governments as also to entities owned and/or supported by the Central and/or State Governments and to other agencies engaged in setting up warehousing infrastructure. The scheme would cover the following requirements:

- ✩ Storage of agricultural produce, especially food grains, at centres of production, distribution and consumption.
- ✩ All dry, wet and cold storage needs of agricultural produce like fruits and vegetables, integrated loans to food parks, etc, as also the needs of sectors allied to agriculture, *viz.*, dairy (bulk milk coolers, etc.), poultry, meat processing, fisheries (*viz.* cold chains, cold storage), etc, and
- ✩ Storage of agricultural inputs like seeds, fertilizers, etc.

NABARD has obtained approval from Government of India to extend loans from RIDF to the private sector also for the construction of warehouses. The interest rate would vary, based on the creditworthiness of the entrepreneurs. Loans to Government entities like CWC and the SWCs backed by the Government guarantee would be eligible for 6.5 per cent interest rate provided the loan is for a maximum period of seven years as per RIDF norms. In case any institution wants longer repayment period, the interest rate would be somewhat higher.

In case of highly perishable goods like milk, meat, eggs, vegetables, fruits, ornamental flowers etc., installation of scientific cold storages is essential for preserving them to ensure a longer shelf life and by preventing them from rotting due to humidity, high temperature and micro-organisms. The DMI of the Department of Agriculture and Co-operation provides consultancy and technical services to prospective entrepreneurs for the construction, maintenance and operation of cold storages. Other activities that the DMI is involved in are preparation of a master plan for cold storage requirements at micro/macro levels, trains managers and operators of cold storages to store products at the optimum temperature, maintain machinery, hygiene and administration, conducting seminars, problem oriented studies and coordinating research in cold storage.

ix. Costs and Returns on Storage

Let us discuss the benefits derived from the storage of agricultural produce. This can be studied through analyzing both costs and returns from the storage of a commodity in a storage structure.

Costs of Storage

The various costs incurred in the storage of produce include,

1. The costs incurred towards repairs and maintenance of storage structure
2. Depreciation on capital equipment
3. Interest paid on borrowed capital
4. Interest foregone
5. Rent paid for hiring the storage structure
6. Value of the quantitative and qualitative loss during storage
7. Risk premium for a possible price fall and damage during storage
8. The cost of protective material like gunny bags, polythene covers, electricity, spraying of pesticides etc.
9. Payments made to labour for loading and unloading of produce, arranging the produce in storage structure etc.

It is essential to note that, all the above costs may not arise in a particular situation. For example, if the farmer or trader hires a storage structure, the items 1, 2 and 3 will not arise. On the other hand, if the farmer or trader owns a storage structure, item 5 will not arise and if the farmer or trader construct the storage structure with his owned capital, items 3 and 5 will not arise. On the other hand, for a professional warehouse owner, the repairs and maintenance costs of storage structure, salaries of permanent staff, depreciation of the building, taxes, record keeping etc., will arise. Further, the above costs can be conveniently divided into fixed costs and variable costs. For a professional warehouse owner, the repairs and maintenance costs of storage structure, salaries of permanent staff, depreciation of the building, taxes, record keeping etc., are fixed costs. However, for a farmer or trader, who is willing to sell the commodity after stocking the same temporarily in the hired storage structure, the storage costs fall under variable costs.

Gross Returns on Storage

Gross returns on storage may be defined as the increase in the price of the stored product *i.e.*, price difference between at the time of 'destorage' and at the time of 'storage' in the storage structure. Net returns from the storage is the difference between gross returns realized from the storage of the product and storage costs incurred. This is given by,

Gross returns on storage = Price of the commodity at the time of destocking – Price of the commodity at the time of storage

Net returns=Gross Returns – Storage costs.

From the above concepts, it is evident that, if the price of the commodity at the time of destocking of produce is less than the price at the time of stocking the produce, then it is a loss on the part of the farmer even though, he stocked the produce with the anticipation of rise in price of the commodity. Further, he has to bear the storage charges and this further reduces the net returns on the part of the farmer. So, if the farmer stocks the produce, the price at the time of destocking the produce from the storage structure must be sufficiently higher than the price prevailing at the time of stocking of produce such that, the price difference covers the storage costs, so as to ensure fair net returns to the farmer. So, we can express the same as, $GR=P_1-P_0$ and $NR = GR-S_C$, where, GR=Gross returns from storage, P_1=Price of the produce at the time of destocking from the storage structure, P_0= Price of the produce at the time of stocking in the storage structure, NR=Net returns due to storage, S_C=Storage costs incurred by the farmer/trader. So, if NR>0, it implies positive returns due to storage and if NR<0, it implies, negative returns due to storage. The percentage margin from storage (M_S) may be calculated as:

$$M_S= [(P_1-P_0-S_C)/(P_0+S_C)]*100$$

The above concept can be best explained with the help of the following example. Assume, a farmer brought cotton produce (one quintal) to the market yard and he kept the produce on the auction platform. The farmers has two options before him *viz.*, he is willing to sell the produce when its current price is ₹ 3000/quintal and above or if the price is less than ₹ 3000/quintal (say, ₹ 2500 in the current period), he will stock the produce in the warehouse godown and sell the same at a later date say, after six months. Further assume that, the farmer is aware of the following costs (Table 5.70) and based on these, he can decide at what price in the future, he will transact the produce after stocking the produce for six months.

Table 5.70: Different costs incurred by the farmer due to storage of cotton for six months.

Items	*Costs (₹/quintal)*
Charges of loading at the auction platform and unloading in the warehouse.	10.00
Transportation charges from the auction platform to the warehouse.	15.00
Warehouse charges for six months.	50.00
Loss during storage and transit (approximately 1 per cent quantity is lost)	25.00
Loss of interest on ₹ 2500 foregone for 6 months @ 10 per cent per year.	125.00
Transportation charges from warehouse to the auction platform.	15.00
Loading at the warehouse and unloading at the auction platform.	10.00
Total	250.00

From the above discussion, we can compute the price at which farmer sells the produce: This is given by,

$GR = NR + S_C = 2500 + 250 =$ ₹ 2750/quintal

Thus, if the market price of the cotton is above ₹ 2750 only, after six months, the farmer has to sell the commodity, as such high price only will cover the storage costs, thereby, NR will be positive.

The above analysis can be done across different types of storage structures and Benefit-Cost Ratios (BCRs) can be used to compare different storage methods. Assume there are three different types of storage structures (SS_1, SS_2 and SS_3) that the farmer can construct on his own and based on the above methodology, let us compare the BCRs realized by the farmer across these structures. Assume that, SS1 and SS2 are temporary storage structures and SS3 is a permanent storage structure constructed by the farmer by employing equity (owned) capital. The details pertaining to total storage costs incurred across these three storage structures, NR due to storage, percentage margin from storage and BCR were shown through Table 5.71. The findings revealed that, SS3 storage structure yields more benefits to the farmer, as indicated by the lowest total storage costs (₹ 295/quintal), highest NR due to storage (₹ 205/quintal), highest percentage margin from storage (7.33 per cent) and higher BCR (0.69).

Table 5.71: Benefits due to storage across different storage structures.

Item	*SS1*	*SS2*	*SS3*
Initial cost of storage structure	100	110	150
Initial market price of cotton (before storage)	2500	2500	2500
Treating and conditioning cost	10	10	10
Interest on initial investment at the rate of 10 percent/year	10	11	15
Loss of interest on the value of the stored grain at the rate of 10 per cent of the value/year	250	250	250
Value of the quantitative losses (@2 per cent)during storage	50	50	10*
Depreciation charges	0	0	10**
Total storage costs/quintal	320	321	295
Initial market price of cotton (before storage)+Storage charges	2820	2821	2795
Actual selling price/quintal of cotton after storage for one year	3000	3000	3000
Net return per quintal of cotton	180	179	205
Percentage margin from storage(per cent)	6.38	6.35	7.33
BCR	0.56	0.56	0.69

Notes: *: Losses are less, as SS3 is a permanent storage structure

**: Depreciation charges are incurred, as SS3 is a permanent storage structure

x. Risks in Storage

The storage of agricultural commodities involves three major types of risks. They include,

- ☆ ***Quantity risk:*** Quantity risk in storage arises due to thefts, fire accidents, floods etc. Dehydration also leads to loss in weighment of produce.
- ☆ ***Quality risk:*** Quality risk in storage arises due to several factors likes pests and diseases, excessive moisture conditions, extreme high and low temperatures etc. The loss in quality of

produce adversely affects both domestic and export competitiveness of the commodity. In the modern days of agri-business, quality of the product as per the norms of SPS is gaining more importance.

☆ ***Price risk:*** Generally, the farmers or traders will store the commodities in the godowns, when prices are low in market with the anticipation that, the prices may rise in the future. But, there is an equal chance of further fall in the prices even in the future depending upon the supply and demand for the commodity in the market. In such cases, storage cost is an added burden to the farmers or traders. Sometimes, prices do not always rise enough during the storage period to cover the storage costs. So, price risk is also an important consideration both to the farmers and traders, while stocking the produce especially at times of greater price volatility.

xi. Importance of Storage/Warehousing

We know, warehouses preserve agricultural produce on a large-scale in a systematic and orderly manner. They provide protection to the produce against both biotic and abiotic agents. This is the basic function of every storage structure. In addition to this, warehouses now a days also perform a variety of other functions as discussed here under.

☆ Warehouses store large stock of goods *i.e.,* both inputs and produce from the time of their purchase or production till their consumption or use.

☆ Warehouses provide protection to goods from both biotic and abiotic factors thereby, cuts down losses due to spoilage and wastage during storage.

☆ Once the farmers or traders stock the goods in the warehouse, the responsibility of these goods passes on to the warehouse-keeper. Thus, the risk of loss or damage to goods in storage is borne by the warehouse keeper. Since it is bound to return the goods in good condition, the warehouse becomes responsible for any loss, theft or damage etc. Thus, warehousing takes all precautions to prevent any mishap.

☆ For some commodities like paddy, wheat etc., require processing to convert them into palatable form. Sometimes warehouses also undertake these activities on behalf of the farmers or traders, who stocked the produce.

☆ On request, warehouses also perform the function of grading of produce on behalf of the marketing agencies. The warehouses also provide facilities for packaging of goods for the convenience of handling and sale.

☆ In some cases, warehouses provide transport arrangement to the bulk depositors. It collects goods from the place of production and also sends goods to the place of delivery on request of the depositors.

☆ Generally, at the time of harvesting, the prices of agricultural commodities tend to be lower because of positive supply situation and farmers often do not get remunerative price for their produce. By depositing their produce in a registered warehouse and obtaining NWR, farmers can use it as collateral security for obtaining short-term borrowing from banks for their working capital requirement for the current sowing season. Further, when the prices become favourable, the farmer can sell the same, repay the loan and get a surplus income. The introduction of NWR system in the country would not only help farmers to avail better credit facilities and avoid distress sale, but would also safeguard financial institutions by mitigating risks inherent in credit extension to farmers. The pledging/collateralization of agricultural produce with a legal backing in the form of NWR would lead to increase in flow of credit to the rural areas, reduce the cost of credit (due to certainty of recovering credit by the bank) and would spur other related activities, like standardization, grading, packaging and insurance services in the agricultural sector. With the increased requirement of quality storage, warehousing industry would also get a boost in rural areas. This would also fill gaps in the logistic chain of agri-business in the rural sector.

- ☆ Apart from individual farmers, co-operative societies and SHGs would also be able to assist their members particularly small and marginal farmers by aggregating their surplus produce and keeping the same in registered warehouses. An established NWR system would also act as an incentive for farmers to produce commodities of good quality and specifications amenable to standardization and storage in registered warehouses.
- ☆ Warehousing regulate price levels by regulating the supply of commodities in the markets. More commodities from the buffer are released when supplies are less, and less is released when supplies are more in the market. Thus, the demand levels are monitored. This tends to cushion the price fluctuations and stabilize prices, as equality between supply and demand is ensured.
- ☆ The storage of goods from the time of production to the time of consumption, ensures their continuous flow in the market.
- ☆ Storage protects the quality of perishable and semi-perishable produce from deterioration. For some commodities like mangoes, bananas etc., storage hastens ripening.
- ☆ Some of the goods (water melon), have a seasonal demand. To cope with this demand, production on a continuous basis and storage become necessary.
- ☆ Many commodities like rice, wheat etc., are produced during a particular season, but are consumed throughout the year. Warehousing ensures regular supply of such seasonal commodities throughout the year.
- ☆ Warehouses enable the farmers to carry on production continuously without bothering about the storage of raw materials. They help to provide seasonal raw materials taking up the production on sustained basis.
- ☆ Warehouses help in price stabilization of agricultural commodities by checking the tendency to making post-harvest distress sales by the farmers. Thus, warehousing gives withholding power to the farmers to tide over difficulties and helps them to secure better prices for their produce. It also gives purchasing power to traders.
- ☆ Warehouses offer the price, supply and demand information to persons, who hold their produce in them, so as to develop selling and buying strategies.
- ☆ A trader can pay customs duty on installments by storing the imported goods with the custody of a bonded warehouse.
- ☆ Warehouses are generally located at convenient places near road, rail or waterways to facilitate easy movement of goods. Further, convenient location reduces the cost of transportation of goods.
- ☆ Modern warehouses are generally provided with mechanical appliances for easy loading and unloading of goods. This also reduces cost of handling of such goods and minimizes wastage.

- ☆ The warehouses offer same benefits to all the farmers irrespective of their land holdings in storing their produce. Moreover, construction of own warehouse requires heavy capital investment, which small and marginal farmers cannot afford. In this situation, by paying a nominal amount as rent, they can preserve their raw materials as well as agricultural produce in public warehouses.
- ☆ Warehouses create employment opportunities both for skilled and unskilled workers in every part of the country. It is a source of income for the people to improve their standards of living.
- ☆ By simply transferring NWR, the ownership of produce can be easily changed from the seller to buyer.
- ☆ Warehousing forms the basic platform of delivery based trading in commodity futures. Warehouses play an important role in commodities futures, as most of trades are settled with delivery. That is, if the seller chooses to handover the commodity instead of the difference in cash, the buyer must take physical delivery of the underlying commodity. Hence, warehouses play a significant and decisive role in the proper and efficient functioning of commodity futures trading.
- ☆ Buyers and sellers have different roles in a transaction. So, an impartial collateral manager is essential to act as an interface between buyers and sellers. Warehouse keeper acts as collateral manager who is neutral, reliable and creditworthy to both the parties. He supervises all aspects of the underlying assets and sanctions the warehouse receipt, which can be used as collateral security.
- ☆ It plays a very important role in implementing the agricultural price policy of the Government.

xii. Reasons for Low Utilization of Warehouses by Farmers

Though the godowns of CWC and SWCs are mainly built for use by farmers, these are being mainly used by traders, wholesalers and for Government procurement. Since their inception warehouses have gradually become popular and are in continuous usage. In the beginning stages of 1955-60, the utilization was around 50 per cent and gradually it touched to 95 per cent. So, the purpose is served and hence, investment is proved worthy. However, the storage (warehouse) capacity of the CWC is utilized for other commodities followed by food grains and fertilizers during 2009 and 2010 (Figure 5.31). But, during 2011, food grains occupied major share (48.72 per cent) followed by other commodities and fertilizers. But, it is disheartening to note that, the food grains stocked in the warehouses mainly belong to traders and only insignificant portion was stored by the farmers or farmers' co-operatives. Thus, though significant progress was achieved, the farmers are not making best use of it and the following are the major reasons responsible for low utilization of warehouses by the farmers:

- ☆ Majority of the farmers are still ignorant of this facility.
- ☆ Illiteracy on the part of the farmers about the procedural formalities involved in stocking the goods in warehouses.
- ☆ Farmers are not accustomed to do paper work and hence, avoid Government warehouses
- ☆ The warehouses are not within the reach of the farmers, as they mostly located in urban areas
- ☆ Majority of the Indian farmers are small and marginal and they have only small quantum of marketable surplus after meeting their domestic retentions

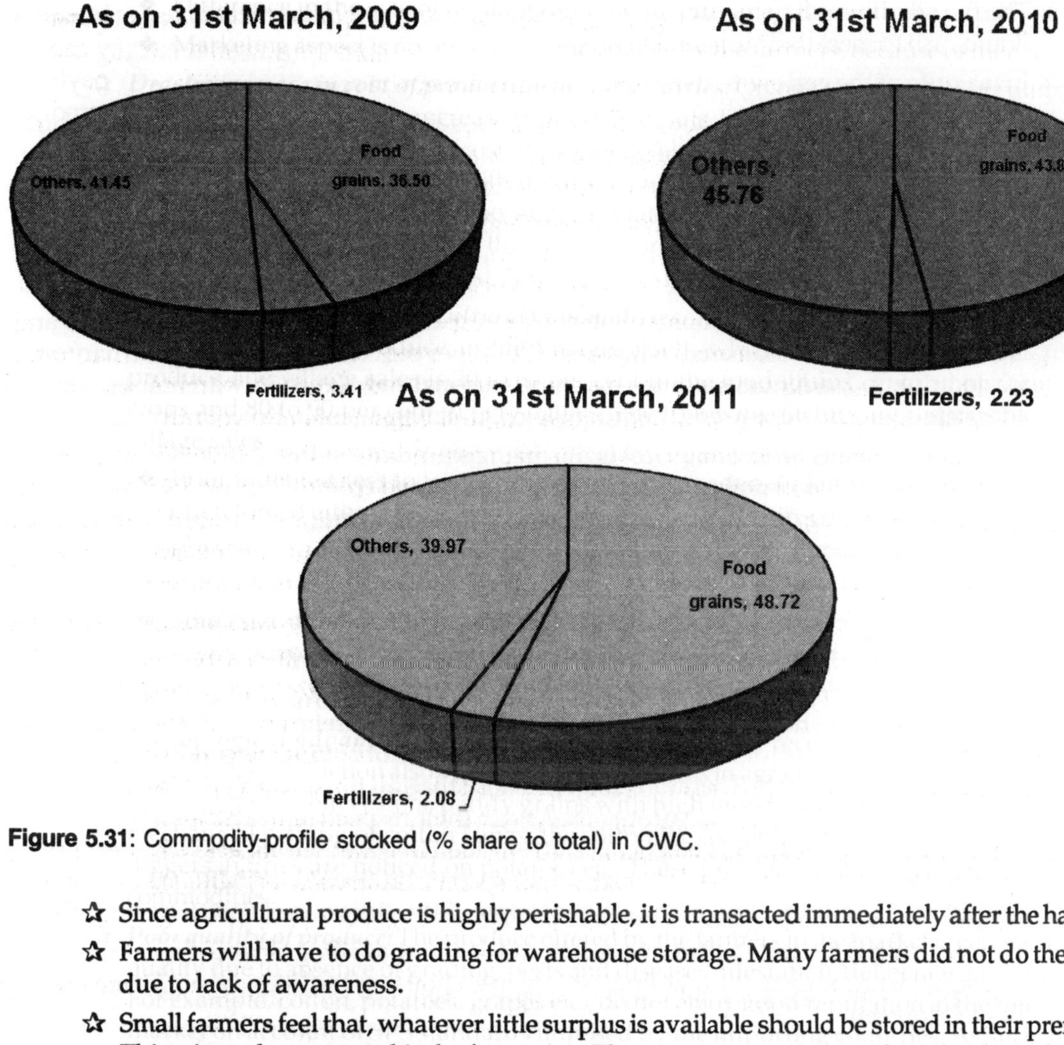

Figure 5.31: Commodity-profile stocked (% share to total) in CWC.

- ✩ Since agricultural produce is highly perishable, it is transacted immediately after the harvest.
- ✩ Farmers will have to do grading for warehouse storage. Many farmers did not do the same due to lack of awareness.
- ✩ Small farmers feel that, whatever little surplus is available should be stored in their premises. This gives them some kind of security. They are not accustomed to the formalities of Government warehouses.
- ✩ Since the warehousing storage is on rental basis, many farmers store their produce in their homes than in warehouses. They fail to realize the safety and care being taken at the warehouses.
- ✩ Since many farmers are small and marginal, they want to have facilities to sell goods whenever they want, without formality hassles.
- ✩ Farmers are not aware of differences they enjoy due to scientific storing at warehouses and at their premixes. They are yet to learn that, losses due to insects, pests, moistures and pilferage at their premises will be much more than the rent they would have paid at warehouse godown.
- ✩ Non-existence of nationalized banks in villages thereby, the problem arises for arranging of pledge finance based on the NWR.

- ✰ The farmers immediately dispose their produce, as they are in immediate requirement of liquid cash to meet their urgent obligations
- ✰ In majority of the godowns, the storage facilities are not on scientific lines
- ✰ The storage capacity available with the warehouses are primarily meant for storing central stock of food grains for the buffer stock, PDS and other Government schemes, which consequently leaves only a marginal capacity for other players to store their produce.
- ✰ The cost of warehousing per unit produce is high.

The following are the other important constraints of storage of agricultural commodities:

- ✰ Even with significant development of storage capacity sanctioned under NABARD and NCDC schemes, it is estimated that, around 20-30 per cent of the total food grain harvest is wasted due to inadequate storage capacity, regional imbalance in warehouses, lack of adequate scientific storage and inefficient logistic management in the country.
- ✰ Food grain (mainly wheat and rice) is the main commodity stored, while the other major crops storable in the godowns include oilseeds, spices and cotton. Though the Government has started focusing on building storage capacity through various schemes, the focus is still largely on the storage of wheat and rice, which leads to acute shortage for storage capacity for other agricultural commodities.
- ✰ Most of the warehouses owned by State agencies are over 15-20 years old and hence, the facilities are not on scientific lines. This calls for upgradation to contain wastage levels, but it demands huge investment from the Government to the tune of ₹ 35-40 billions.
- ✰ There is an acute shortage of trained manpower in the warehousing sector, both in the public and private sectors. Most of the warehouses are not having trained warehousemen and other technical staff. Therefore, there is a need that, all these warehousemen and other personnel should be imparted proper training in their respective fields. A comprehensive course content for different category of staff engaged in warehousing activities needs to be prepared and training arranged through some expert institutions like IGMRI Hapur, National Institute of Agricultural Marketing (NIAM) Jaipur, CWC, Institute of Food Security, FCI etc.
- ✰ Warehousing is a capital intensive sector and without availing finance from banks and financial institutions at cheaper interest rate, warehousing facilities cannot be created by the entrepreneur.
- ✰ Land requirements for constructing the warehouses are difficult to meet due to rising cost of land in urban, semi urban and rural areas. Procedures for conversion of agriculture land for industrial or commercial use are cumbersome and difficult.
- ✰ Warehousing has not been accorded the status of full-fledged infrastructure and all financial and other benefits available to infrastructure sector are not available for warehousing sector.
- ✰ Storage charges and other handling charges offered by the Government agencies to private warehouse operators are not attractive.
- ✰ Modern technology has not been introduced in the warehousing sector.

xiii. Suggestions to Improve Storage/Warehousing in India

Storage/warehousing is of great importance to the fast developing businesses of today. It helps lessen the expenses of farmers and traders in having to transport all the goods to different places. Storage warehouses serve as depots for assembling the produce, from where it can be distributed to the needy consumers. The following are the important suggestions that are to be looked into to improve storage/warehousing in India:

- ☆ It is high time that, both the Central and State Governments should think about the severity of the problem of storage losses, as it is estimated that, the country might be losing every year ₹ 5000 crores to 7000 crores worth of food grains due to defective storage. As a beginning, the farm level storage practices must be strengthened. One strategy could also be that, the FCI can keep the stocks with the farmers instead of purchasing the grain from them. As per the present categorization in vogue that, if a farmer stores the grain without infestation, then he should be given some incentive apart from normal MSP. This would encourage the farmers to store the grains properly and ultimately it would lead to zero loss level and the pressure on storage space with FCI can also be reduced. FCI can initiate this programme in a few selected villages on a pilot scale and the same can be extended, if it is successful.
- ☆ The Government should assume responsibility for minimizing the storage loss at various stages of handling and storage. An adequate organization needs to be set up at the Center and in the States to take coordinated action in this direction.
- ☆ Emphasis should be on systems approach rather than piece meal. The existing post-harvest system should be greatly revamped such that, it significantly minimizes the losses at all levels of post-harvest handling of produce.
- ☆ Establishment of National Grid of Rural Godowns as per ecological zones is a must. This will go a long way in reduction of losses in storage. For individual small farmers, metal bins under this project need to be supplied because, a good storage structure is the first line of defense in protecting and preserving the produce in sound condition. Rats will not be able to damage the grain at all.
- ☆ With a view to minimize post harvest losses in food grains at farm level, the Government of India had launched a Central Sector scheme known as Save Grain Campaign (SGC) in 1971. The scheme was implemented by Department of Food and Public Distribution, Government of India. The objectives of the scheme were:
  - ❖ To popularize the scientific techniques of storage of food grains among farmers by organizing training, demonstration and publicity programmes.
  - ❖ To popularize scientific storage structures like metal bins, Reinforced Cement Concrete (RCC) ring bins etc., among the farmers.
  - ❖ To suggest improvements in traditional storage structures used by farmers.

The SGC used to organize stipendiary training programmes of five days duration in villages for farmers including farm women on scientific storage of food grains. The training programmes used to be organized at village level or block level in association with State Governments, NGOs and Corporate Houses. These programmes were very popular among farmers. However, the scheme was discontinued in 2009 by the Government of India on the recommendations of Report of Expenditure Reform Commission (ERC), which had found the scheme obsolete. In view of its importance, SGC needs to be strengthened, extended and storage technology should not only be taken to the door steps of the farmer, but ensured that it is adopted by him. All media of mass contact should be used to spread the message of 'Save the Grain' far and wide followed by Campaign team's visit to the farmer's house.

- ☆ Wherever possible solar energy for drying the produce need be harnessed.

- Sufficient emphasis should be laid on grain handing and grain storage in the institutes, where Agricultural Engineering is taught. The subject of grain storage should also be included in the job card of extension workers. This should be a nation-wide educational program.
- Storage warehouses must be equipped with cranes, forklifts and container trucks, which are used in loading and unloading cargos. They should also be provided with automated storage systems and would no longer require manpower inside the facility. Those storage warehouses that hold perishable goods should be geared up with large refrigerators, which keep the produce safe and fresh.
- The recent development of storage warehouses presents double purpose. Storage warehouses can now be manipulated as retail stores. The traditional high ceilings with decorative shelves are replaced with sturdy industrial racks. These racks encompass ready to sell items placed just below the racks, while products which are in pallets are usually stored on top of the shelves. In this way, one building can be used as warehouse and at the same time offers retailing that can save a lot of resources compared to having two separate buildings for each purpose. Today, many exporters are using storage warehouses as a means to establish retail outlets for a particular region or country. This new system also facilitates in reducing the cost of products for the consumers and relatively increases the production sales. Warehousing is considered a great tool by many traders in directly reaching consumers without the need of middle men or agencies. This saves traders from additional fees, as well.
- Since the objective of warehousing is to prevent distress selling and to secure fair price to the farmers, the bulk of warehousing programmes should be undertaken in the areas of concentrated production. Some warehouses may be located in centre of export or heavy consumption.
- A list of principal agricultural commodities in each region, their production levels and marketable surplus etc., should be submitted by every State to the Central Government for which warehousing facilities are considered necessary.
- As the godowns are built in the interest of the primary producers, they should be helped to derive benefits of storage and warehousing to the fullest possible extent by providing concessional charges to farmers or farmers co-operative societies.
- There is a need to improve the efficiency of the cold storage units through conversion of existing single commodity cold storage units to multi-purpose storage, if feasible.
- The farmers may be encouraged to store their agricultural produce in registered warehouses. At initial stage, some rebate on storage charges, transportation charges and quality testing may be allowed to farmers, so as to make the farmers quality conscious and to induce them to store their produce in scientific warehouses which is the very spirit of the Warehousing (Development and Regulation) Act, 2007. At present CWC is providing 30 per cent rebate in storage charges to farmers. Besides, some State Governments through their SWCs are also giving rebate in storage charges. All the State Governments should be requested to allow rebate in storage charges for small and marginal farmers through their respective SWCs.
- During the peak marketing season of food grains, there is a glut in the market, as most of the farmers are keen to sell their produce to Government agencies at MSPs. This creates a lot of problem for procurement agencies as well as for farmers. If the Government announces a markup on the MSP for crops brought to the mandi, a few months after the procurement season is over, this would enhance the farmers' holding capacity, improve returns to the

farmers and avoid distress sales. This markup on MSP should be so determined, as to cover the additional costs incurred by farmers in transporting and storing the produce in scientifically managed warehouse facilities. On the other hand, the procurement agencies and Government would find this option attractive, if they save some of the money, which they would have spent on storage and handling during the interim period. The pledge finance and crop loans should be treated at par, as both are classified as direct agriculture loan and sanctioned to the same categories of loanees *i.e.*, farmers. There is no economic reason for a higher interest rate for produce pledge loan particularly, when pledged loans guaranteed by commodities are more secure than the crop loan. Pledge loans will help the farmers through better price discovery and the volume of loan to the farm sector will almost double. Pledge loans at reduced interest rate is desirable for minimizing intermediation, to free the farmers from the clutches of money lenders and also empower them to quote their own price for sale of their goods. Agriculture credit policy of the Government of India for the year 2011-12 stipulates a target of ₹ 4, 75, 000 crores for credit flow to small and marginal farmers, at a concessional rate of interest. The farm sector loan is available to farmers only as a pre-harvest loan. The post-harvest loan sanctioned against NWRs to the farmers is between 11 to 13 per cent rate of interest. At present, short term crop loans facilities @ 7 per cent interest rate per annum with interest subvention of 3 per cent for prompt repayment with maximum limit of ₹ 3.0 lakh is provided to the farmers by Government of India. Thus, the effective rate of interest for such farmers will be 4 per cent per annum. So, provision of loans against NWR at the same rate of interest as short-term crop loan would encourage farmers to deposit their food grains in warehouses and avail the benefits of warehousing in terms of better price realization. This provision of pledge loans at low rate of interest will enhance the farmers' holding capacity so that, they are not forced to sell their produce immediately after harvesting. Initial requirement of finance under NWR will be small, but will go on multiplying year and year within the overall annual target of agricultural credit. Further, the small and marginal farmers will have requisite liquidity for timely availability of inputs used in agriculture. They will also get empowered to experiment new package of practices for enhancing productivity and for diversification into other crops including cash crops. There are substantial post-harvest losses, both in food grains as well as in other agricultural commodities including perishable commodities. The NWR system will encourage scientific storage of agricultural commodities and minimize storage losses in food grains and other agricultural commodities. This, in turn, will contribute to the fulfillment of the objectives of National Food Security. Warehousing of food grains such as wheat and paddy under NWR, in the private warehouses will avoid glut at the procurement centres. It will not only ease the storage constraints of Government agencies like FCI and CWC but will also help in preserving the quality of such food grains against deterioration. It is, therefore, recommended by the Working Group on Warehousing Development and Regulation for the Twelfth Plan Period (2012-17) that, farmers may be provided post-harvest loans against NWRs at a rate of interest of 7 per cent per annum with interest subvention of 3 per cent for timely repayment of loan. It was also suggested that, interest subvention on NWR finance may be extended to private sector banks also and the limit for warehouse receipt finance by banks may be enhanced from ₹ 10 lakhs per farmer to ₹ 50 lakhs per farmer.

☆ The Government of India should provide incentives in form of subsidy for the purchase of metallic storage structures and construction of improved non-metallic and modern storage structures like RCC ring bins etc. Subsidy to the tune of 33 to 50 per cent may be provided on the purchase of metal bins ranging from 2 to 10 quintals storage capacity manufactured by State Agro Industries Corporations or other private manufacturers, as per design approved by IGMRI. Subsidy may also be considered for the purchase of appropriate scientific storage structures such as silo bags of larger capacity by the farmers for the storage of food grains and other agricultural produce. Farmers including farm women of weaker sections of the society including SC and ST farmers may be considered for 75 per cent subsidy on the purchase/construction of these scientific storage structures for food grains and other agricultural commodities.

- Short duration training programme on all aspects of scientific storage of food grains and other agricultural commodities should be organized to the farmers by the Government at village/panchayat/block level. State and Central Government Agriculture Departments, SAUs, Krishi Vigyan Kendra (KVKs), NGOs etc., may be involved for organizing such training programmes. These programmes should be stipendiary in nature and practical demonstration on various techniques of insect, pests and rodent control should be organized. Further, there is an acute shortage of trained manpower in the warehousing sector, both in the public and private sectors. Most of the private warehouses are not having trained warehousemen and other technical staff. Moreover, they are not familiar with the code of practices of scientific storage of agricultural and other commodities. Graders, weighmen and staff engaged in sampling and physical analysis of agricultural and other commodities are not properly trained. Therefore, there is a need that all these warehousemen and other personnel should be imparted proper training in their respective fields.
- Publicity programmes by distributing leaflets, booklets, and posters, by organizing film and video shows on benefits of scientific storage of food grains and other agricultural commodities should also be organized in the villages. Farm women, should be involved in training and publicity programmes.
- Small and marginal farmers are sometimes not able to bring their produce to the nearby warehouses due to lack of transportation facilities. In such cases, the services of aggregators (co-operative farmers group, SHGs etc) may be used. Aggregators at the village level may collect food grains from the farm-gate or from the houses of farmers and may transport the same to nearby warehouses. These aggregators may be considered for some transport subsidy for bringing the collected food grains to the warehouses.
- The Rural Godown Scheme is aimed at the creation of scientific storage capacity to meet the requirements of farmers. The Scheme has made a valuable contribution to the creation of storage capacity. The Rural Godown Scheme must continue during the 12th FYP as well. However, the norms with regard to subsidy and project cost were revised in the year 2008 based on the Cost Index of the year 2002. Further, godowns with maximum storage capacity of 10,000 m tonnes only are covered. The Rural Godown Scheme needs to be revised, so as to cover godowns having a capacity of up to 25,000 m tonnes and the capital cost needs to be revised keeping in view the current cost of construction of godowns. The subsidy component @25 per cent of project cost may be fixed up to 10,000 m tonnes capacity and for the warehouses more than 10,000 m tonnes to 25,000 m tonnes, the subsidy component may be reduced to half *i.e.* 12.5 per cent of project cost. Linkage of Rural Godown Scheme with NWR System will encourage farmers to store their produce in these warehouses/godowns. The farmers will be able to seek loans from the banks and scientific storage and warehousing will minimize the storage losses, thereby saving grains for national food security.

- Post-harvest storage has been covered under the Viability Gap Funding Scheme (*The Viability Gap Funding Scheme (2004-05) provides financial support in the form of grants, one time or deferred, to infrastructure projects undertaken through PPPs with a view to make them commercially viable. Government of India has established a Viability Gap Fund to aid the PPP infrastructure projects, which face the viability gap due to inherent nature of the project. The Scheme is administered by the Ministry of Finance.)* of PPPs. The Central and State Agencies like FCI, CWC, and SWCs etc., may take the lead in inviting private entrepreneurs for the construction of warehouses. This scheme has not made much progress and therefore, fresh initiatives are required to popularize this scheme.
- Government may plan to set up Agricultural Economic Zones (AEZs) on the lines of SEZs. In these AEZs, land may be acquired by the Government and may be given on lease for setting up of warehouses, agro processing units, farmers training centres and agricultural produce marketing facilities. Setting up of warehouses in AEZ may be treated as priority sector and easy loans at lower interest rates may be provided to entrepreneurs for constructing them. This would be helpful in development of modern and scientific warehousing facilities for agricultural and other commodities.
- The Central Government extends price support to paddy, coarse grains and wheat through FCI and the State Agencies. All the food grains conforming to the prescribed specifications offered for sale at specified centers are bought by these agencies. The producers have the option to sell their produce to FCI/State Agencies at the support price or in the open market as is advantageous to them. The procurement policy is open ended and no targets, as such, are fixed for the procurement of food grains. There is a huge rush among farmers to bring the produce to the market, if the MSP is attractive and/or higher than the prevailing market price. This leads to a glut in the market; delays in procurement and lifting and issues like quality, drying, cleaning of the grain takes the back seat. Government of India/FCI may explore the possibility of use of NWRs for procurement of food grains for Targeted PDS (TPDS) to avoid a glut in the market and bring about efficiency improvements in storage and handling operations. In this way, the grain would be properly dried, cleaned, graded and stored in a scientific environment, which could be accessed by the procurement agencies from the warehouse as and when required for distribution. The farmers would, of course, be paid immediately on delivery of the NWRs. It is recommended that procurement through NWR may be undertaken in Decentralized Procurement (DCP) States on a pilot basis in a few districts.
- Loans to farmers against NWRs issued by registered warehouses are treated as priority sector lending by the banks. The Boards of Public Sector Banks may prescribe targets for sanctioning of loans against NWRs.
- The farmers can realize the best price for their agricultural produce, in case they have up to date information on the current price and other relevant factors like stocks, overall availability of the commodity in question etc. The development of a robust information dissemination system would significantly increase the efficiency of distribution from farmers to the consumers, reduce the intermediation costs and maximize the returns to the farmer. Hence, it should be made mandatory for all warehouses to provide latest information on prices, stocks, etc on a real-time basis.
- Banks and financial institutions support financing of storage infrastructure by both public and private entities, with credit plus intervention to improve efficiency in implementation

and ensure effectiveness in utilization of infrastructure created. PACS would find it financially viable to create new storage capacity and upgrade existing warehouses in case, they are permitted and encouraged to take up other activities like retail sale of products of daily use, gas agencies, petrol pumps etc. To provide backbone support to farmers, member friendly multifunctional services could be taken up by PACS in their area of operation and they could link up with supply chain logistic providers. Integration of PACS with NWR System will help in reduction of storage losses and provide scientific storage facilities to farmers. The Working Group on Warehousing Development and Regulation appointed by the Government of India felt that, it would be of great use, especially for small and marginal farmers, to utilize the PACS godowns for storage of food grains under NWR System. The PACS godowns, however, need improvement/construction as per BIS specified structure to make them storage worthy and WDRA complaint. The funds for this purpose can be provided under RIDF and subsidy can be provided under Rural Godowns Scheme. CWC/SWCs/ private warehousing companies may be involved in hand holding of these PACS for technical, logistics and marketing activities. There is a need to have a special incentivized scheme with specific regulations to achieve this objective, which may be administered by the Department of Agriculture and Co-operation through NABARD and NCDC. Provisioning of logistics for scientific storage and marketing of produce needs to be an integral component of the stipulated scheme. Since the PACS godowns are essentially of small capacities which are not financially viable, there is a need to relax the requirement for registration of these godowns by the WDRA without in any way compromising with their storage worthiness and interest of the farmers depositing their agricultural produce.

- We know, an agri-logistic park is an area, wherein integrated facilities for warehousing, cold chain for perishables and non-perishable agriculture produce and transportation facilities are available. Most of these agri-logistic parks are in SEZs areas. The basic criterion for such parks is that, they should be near National Highways and in metros and close to the producing areas. Numerous benefits accrue from agri-logistic parks, which are as follows:
  - Scientific storage facilities for agri-produce, both perishable and non-perishable lead to minimization of storage losses
  - Proper grading of agri-commodities and better prices to farmers.
  - No glut of commodities in the Mandis during marketing season.
  - Reduces logistics cost.
  - Facilities for packing, re-packing and testing and distribution are available under one roof and
  - Better quality to consumers.

The Working Group was of the opinion that, Planning Commission may identify the Ministry/ Department which could formulate schemes for the promotion and regulation of agri-logistic parks.

- Only a few scientific studies have been conducted in the past for the assessment of post-harvest losses at farm level for food grains and pulses. Authentic data on post-harvest losses in various agricultural commodities at farm level are very important for formulating schemes to minimize these losses. Institutions like IASRI, New Delhi, IGMRI, Hapur, Uttar Pradesh, ICAR institutions etc. may be assigned these studies for major agricultural commodities such as cereals, pulses, spices, oil seeds, edible nuts, fruits, vegetables etc.

- ✰ The WDRA should notify regulations to include cold storages/temperature controlled warehouses in association with National Horticulture Board (NHB), in the NWR system so that farmers producing fruits and vegetables may avail the benefits of NWR issued by the registered cold storages/temperature controlled warehouses for seeking loans from the banks.
- ✰ In view of the increase in the economic activities, it is essential to promote efficiency in the warehousing sector with improvement in basic standards and operational procedures for warehouses. The receipt issued by any warehouse in respect of goods deposited by farmers or traders should enjoy the fiduciary trust of depositors, receipt holders and banks with legal remedies in case of default with willful violation. Considering the increasing importance of warehousing sector in the economy of the country and in order to enhance the trust and confidence of the stakeholders, it is desirable that the WDRA proactively encourages registration of all warehouses receiving deposit of goods of others. The WDRA may prescribe varying standards based on capacities, locations and commodities to be stored so that, the farmers can benefit from a large number of registered warehouses spread throughout the length and breadth of the country.

The various suggestions discussed above, if implemented by the Government enable the farmers to obtain better return for their produce and secure efficient post harvest management infrastructure that help reducing post harvest losses. The establishment of scientific storage structures avoid gluts, reduces transport bottlenecks during peak periods of production and marketing. It also helps the farmers to avoid distress sale and realize better prices in the post-harvest season. The consumers are assured of constant supply of quality produce at reasonable price.

xiv. Future Role of WDRA

- ✰ The WDRA should not play the role of a regulatory body only by registering warehouses interested to issue negotiable warehouse receipt. It should play a vital role in developing an orderly, robust and reliable warehousing system in the country not only for food grains and other dry commodities but also for perishable commodities like fruits and vegetables wherein post-harvest losses are reported to be about 30 per cent.
- ✰ For the storage of perishable agricultural commodities, warehouses with controlled conditions of temperature and humidity would have to be created. Such type of cold storages and cold chains are lacking in the rural areas resulting in huge wastages in such commodities. The Authority should encourage testing and grading of such commodities so that consumers may get good quality of fruits and vegetables.
- ✰ Storage space in warehouses particularly in rural areas should be reserved for agricultural commodities for which enough provisions are there in the Warehousing Act.
- ✰ To develop efficiency and expertise among the warehousemen, intensive training programmes for the officials of warehouses, both public and private, should be initiated by the Authority in collaboration of some expert institutions.
- ✰ Accreditation of warehouses for which there are provisions in the Act, should be encouraged among warehousemen.

xv. IGMRI

In order to develop facilities for applied research and apex level training in the field of food grain storage and preservation, a Grain Storage Research and Training Centre was established at Hapur in 1958, which was later on expanded into IGSI with two field stations at Ludhiana and Bapatla (later shifted to Hyderabad) with the financial assistance from UNDP in 1968. Three field stations at Jabalpur, Jorhat and Udaipur were subsequently established in 1981. In 1996 the IGSI was renamed as IGMRI. Field Stations at Jabalpur and Udaipur have however been closed in October, 2002. While the Research and Development and Training activities are under taken by IGMRI, the work of popularizing scientific methods of food grain storage developed by IGMRI among farming community is implemented through the SGC teams.

Objectives

The main objectives of the Institute are:

- ✩ To develop code of practices for proper grain storage and handling by recommending cost effective techniques for control of insects, rodent, birds and micro organisms
- ✩ To investigate the nature, extent and degree of losses due to these factors under different agro-climatic conditions
- ✩ To develop improved type of storage structures, grain dryers, grain handling, cleaning and grading equipment besides, improvement in traditional storage structures using locally available and Eco-friendly materials.
- ✩ To develop suitable publicity material and semi-technical literature on grain storage and quality control in food grains
- ✩ To train personnel from various organizations *viz.* FCI,CWC, SWCs, Civil Supplies Corporations etc. as well as trainees from developing countries sponsored by FAO, UNDP and Commonwealth Secretariat on Grain Storage Management Practices.

Activities

The main functions or activities of IGMRI are as follows:

- ✩ ***Pest Control and Storage Division:*** The important activities include,
  - ❖ Testing of new insecticides, rodenticides, fungicides and non-chemical methods for control of stored grain insects, rats, mites and fungi.
  - ❖ Assessment of food grain losses and their quantitative and qualitative estimation.
  - ❖ Evaluation of pest control methods and development of code of practices for safe storage of food grains.
  - ❖ Study on Insecticidal/pesticidal resistance in stored grain insect/pest.
  - ❖ Monitoring of pesticide residues and mycotoxin contamination in food grains.
- ✩ ***Training division:*** The IGMRI, Hapur and its field stations at Hyderabad, Jorhat and Ludhiana conduct the following training courses.
  - ❖ Long Term Training Course (two months duration) on 'Scientific Methods of Storage and Inspections of food grains' for the officials of FCI, CWC, Civil Supplies Corporation, candidates sponsored by SAUs and private candidates graduated in Agriculture/ Biology/Chemistry as well as trainees from developing countries sponsored by FAO, UNDP, Commonwealth Secretariat and Colombo Plan etc.
  - ❖ Short Term Training/refreshers courses (1-2 weeks duration) on storage pest management and fumigation, pesticide residue analysis, chemical and physical analysis of food grains, Rodent control etc.

- ❖ Specialized Training Course (two weeks duration) on Post-Harvest Technology of food grains.
- ❖ Artisan training course (ten days duration) for the rural Artisans on the fabrication of modern and traditional storage structures.

II. Exchange (Trading) Functions

The process of passing goods from sellers to buyers is called function of exchange. It includes buying and selling. So, exchange functions involve the transfer of ownership, a process implying the existence of processes for price discovery and determination of other terms of sale. These exchange functions are what is commonly thought of as marketing. They involve finding a buyer, a seller, negotiating price and transferring ownership (but not necessarily physical transfer). These functions take place at the 'market' - that is, the physical meeting point for buyers and sellers at the point of production or *via* some other means of communication. However, in the modern era of agricultural marketing, face to face contact of buyers and sellers is not necessary for transferring the produce, as the transactions are even possible through internet. So internet, here forms an arrangement and such arrangement also is included in the market. These buying and selling functions are probably the most visible functions of agri-marketing. Among these two, buying is the first step in the process of marketing. It involves careful planning and needs setting up of policies and procedures towards the purchase of the commodities or products and the following points are generally considered before buying a particular product:

- ✩ What to buy? (Nature of product)
- ✩ When to buy? (Time of purchase)
- ✩ Where to buy? (Place of purchase)
- ✩ How much to buy? (Quantity of purchase)
- ✩ From whom to buy? (Source of supply)
- ✩ On what terms and conditions the purchase is to be made? (Negotiations of purchase)
- ✩ At what price the product is to be purchased? (Price of purchase)

After buying function, the buyer (wholesaler) goes for assembling the stocks of produce purchased from different sources. So, assembling starts after the goods have already been purchased. Buying and assembling are not one and the same. Buying involves transfer of ownership of the goods, where as assembling involves creating and maintaining of stock of goods purchased from different sources. Thus, assembling is a process of collection of produce from different places (big or small town, village etc.) and putting it together at a central place or its further movement to the processing firm or consumers. The main advantage of assembling is that, it facilitates widening of market, grading, reduces per unit cost of transportation and handling, regularize the supply of seasonal products, facilitates for raising finance etc. But, some problems are encountered in carrying out this function, which includes requirement of special skills, lengthy marketing channels, unorganized and unregulated market practices, involvement of large number of middlemen, imperfect standardization etc. Thus, between buying and selling functions, the traders perform assembling of produce and assembling begins after the produce has been purchased. It refers to produce or commodities already purchased from different places at pooled at one central place. It facilitates easy transportation and storage of produce. So, assembling is a function separate from buying, as buying involves transfer of ownership of the goods, whereas assembling involves creating and maintaining of stock of produce purchased from different sources. The important problems encountered in assembling of agricultural produce include:

- ✩ Seasonal production
- ✩ Difficulties in controlling quantity and quality
- ✩ Non-availability of information about sources of supply
- ✩ Low quantity of marketable surplus.

Selling is a complementary function to buying. In fact, without selling, no buying takes place in the marketing process. Like buying, selling also involves careful planning towards the sale of the products and the following points are generally considered before selling a particular product:

- ✩ What to sell? (raw produce or processed product or by-product etc)
- ✩ When to sell?
- ✩ Where to sell?
- ✩ How much to sell?
- ✩ To whom to sell?
- ✩ On what terms and conditions the sales is to be made?
- ✩ At what price the product is to be offered for sale?

Thus, buying and selling functions are directly concerned with change in the ownership of goods thereby, they create possession utility. These two functions go hand in hand and they are complementary to each other. That is, for every sale, there is a purchase and for every purchase, there must be a sale. In fact, one cannot exist without the other. And, assembling precedes a sale and assembling follows buying. Let us discuss these two functions in-detail here under.

A. Buying

It is a function that ensures the product offerings are available in sufficient quantities to meet consumers' demands. So, this function is mainly related to the consumers of a product. The following are the essential steps involved in buying function:

- ✩ Consumers realize the need for a product
- ✩ They seek information on that product
- ✩ They search for alternative products in fulfilling the same want
- ✩ They compare the products based on their relative benefits
- ✩ Contractual function (identifying the source or supplier of desired product)
- ✩ Negotiations with the supplier for the price
- ✩ Plan and purchase the best product
- ✩ Time of purchase of product, as some products have specific demand in a specific season
- ✩ Estimation of the demand for the product (a wholesaler purchase the product from the processor or from farmer directly based on the demand for the product among the consumers in the market.

According to specialists in the field of Agri-Marketing, 'buying' is the first step in the ladder of marketing functions, as the farmers have to buy raw materials or inputs for conducting crop production programme. Even for processing firm, buying is the basic marketing function, as they have to purchase the requisite raw material for processing the raw produce intc final products. Similarly, a wholesaler has to buy finished goods for the purpose of sale to the retailer and a retailer has to buy goods for resale to the consumers. In case of large sized businesses, the firms even maintain a separate department namely 'purchasing department' for the purpose of buying. The primary purpose of buying is to assemble goods upon demand, for resale or use in production or personal consumption. Efficient buying is essential for successful selling. There are different kinds of buying such as:

- ✰ ***Hand to mouth:*** It is buying in small quantity thereby, involves less capital. The expenses in transport, storage etc., are saved.
- ✰ ***Forward buying:*** It is also known as 'Speculative Buying'. This is practiced by the traders (wholesalers), as the purchase the produce in peak marketing season (when prices are low), stock the produce and sell the produce in the lean season, when prices shoot up in the market.
- ✰ ***By inspection:*** Under this method, buyer purchases the produce after examining it for its quality and features in the seller's premises.
- ✰ ***By sample:*** A purchase by sample is made after the buyer examines the sample of produce supplied by the seller. For example, a consumer purchases rice for his family consumption only after verifying the sample of produce.
- ✰ ***By description:*** Some sellers issue catalogues containing description of commodities or products offered for sale. The intending buyer places an order specifying a particular number mentioned in the catalogue. For example, a farmer purchases tractor or machinery or irrigation equipment by this mode.
- ✰ ***By grading:*** Based on the quality specifications laid down for the produce and after classifying the commodities into different grades, the buyers purchases the requisite grade either by physical inspection or through telephone or mail.

Thus, buying function is concerned with seeking out the sources of supply of produce, activities associated with its purchase and its assembling. It can be either the assembling of raw produce from the production areas or the assembling of finished products, so as to transfer the same into the hands of other middlemen in order to meet the demands of the ultimate consumer. In case of buying of inputs, buying function implies the acquisition of the materials that are used in the production of the produce or products that are to be sold to the ultimate consumer. So, buying function deals with a wide range of activities like purchase of inputs to produce agricultural produce, purchasing the raw produce from the assembling markets, transferring the finished or processed products from one agency to another agency and finally to the consumer. Further, buying may be done in three types of situations (with reference to products) *viz.*, repeated, modified and new. The purchases of rice, wheat, tea powder, coffee powder etc., fall under repeated buying. Purchases of wheat flour mixed with coarse grains (jowar, ragi etc), masala oats, diabetic atta etc., represent buying under modified situation. Purchases of RTE and RTC foods represent buying under new situation. It is essential that, the produce must be purchased at the right time to minimise the time lag between purchase and actual use or sale as well as to ensure that, they are purchased at the most favourable price, where prices are seasonal. In case of production of processed products, purchasing of raw agricultural produce must be done in sufficient quantity to ensure continuity of production and in good quality so that, processing costs are maintained at the lowest possible level. Scientific and economic purchasing would naturally increase the ultimate profitability of the operation.

i. Buyer vis-à-vis Consumer

A 'buyer' may be defined as a person or a private business firm or a Government organization (FCI, CCI etc) that purchases the produce by spending the money to obtain them either for own use or for resale. A buyer may not necessarily be a consumer. Anybody, who uses or consumes a product is called as consumer. For example, FCI procures paddy and wheat from the farmers to maintain buffer stocks and re-distribute the same to the needy beneficiaries at issue prices through FPS. So, FCI is the buyer, but the actual consumers are the beneficiaries, who purchase the produce through FPI at issue prices and consume them. Similarly, when the wholesaler purchases mangoes from the main market by paying some money, he is a buyer, but not the consumer because, he buys mangoes for resale. Now, when a person purchases and consume them, he become a consumer.

ii. Reasons for Buying

We buy agricultural produce or products for the following reasons:

- ☆ To meet consumption requirements
- ☆ The nature of products purchased indicates the standard of living of consumers. For example, rich consumers generally prefer to consume processed products like jam, butter, RTE and RTC foods etc. Poor consumers prefer to consumer jowar, bajra, ragi etc.
- ☆ To consume nutritious products for health safety.
- ☆ To respond to changes in tastes and fashions. For example, now the consumers are preferring to have RTE and RTC foods rather than traditional way of cooking foods. Similarly, we find in recent times, people have become more health conscious and are spending money more on health products like diabetic atta, horlicks lite etc.
- ☆ To run a business (Food processing), the entrepreneur has to buy machines, raw materials, employ people to serve and perform different jobs and also pay for securing water and power supply. Traders have to buy products to be sold at a profit. They have also to pay for the services of employees to run the business.

iii. Factors Influencing Buying Decisions

There are several factors that influence buying of goods and services. If a consumer wishes to purchase mangoes for his family consumption, he may have them from local market or go to a distant market. So, the decisions like what to buy, where to buy, when to buy, from whom to buy and how to buy also influence the buying decisions. Further, assume that, a farmer wishes to go for mango plantation and he buys the mango saplings not only considering above criteria, but also the discount factor in computing discounted costs and returns for the project during its life time. The following are other important factors that guide the buying decision:

- ☆ ***Capacity to buy:*** Every buyer buys according to his capacity *i.e.*, he buys as much as he can afford. So, capacity to buy depends upon his level of money income. This also influences what to buy, where to buy, when to buy etc. Further, buying capacity of a buyer is also influenced by eligibility for availing loans, his fixed expenses, debt repayments and other monetary obligations. It is essential to note that, the buying capacity of a buyer is not the same as his earning capacity and buying decisions of buyers changes according to their capacity to buy.
- ☆ ***Advertisement:*** Advertising about products (RTE and RTC foods) and services through various media like newspapers, hoardings, radio, television, magazines, posters etc., positively influence the buying decisions.
- ☆ ***Sales promotion schemes:*** We know, the entrepreneurs of seeds, fertilizers, pesticides, drip irrigation firms etc., engage different sales persons and they motivate and stimulate the buyers to purchase the inputs. Further, food processing firms also announce special schemes

by giving discounts, Buy-One-Get-One-Free (BOGOF) schemes, offering a mix of products etc., to increase their sales and all these influences the buying decision of a buyer. For example, when a buyer goes to the shop to purchase 'Horlicks' and he came to know that, if he buy 'Boost' he will get a 'pencil box' free. This may influence the buyer to change his decision and buy 'Boost' instead of 'Horlicks'. Thus, sales promotion schemes also influence buying decisions.

- ☆ ***Social status:*** The status of the buyer in the society influences his buying decision to a great extent. For example, rich and affluent consumer generally prefers processed products to fresh produce.
- ☆ ***Intellectual status:*** The knowledge of a buyer about the products also influences his buying decisions. For example, an educated consumer is very much selective to have nutritive foods and organic foods instead of oily foods and fat rich foods.
- ☆ ***Availability of products:*** Mostly buyers choose products to buy from those which are available easily. Very rarely do people wait for a product to be available. For example, if a buyer wishes to purchase 'Kissan' fruit jam and if it is not available readily, he may prefer to go for other brand and he will not wait till 'Kissan' brand is made available in the market.
- ☆ ***Urgency:*** At times, due to urgency, the buyer compromises on the available product. For example, when a buyer wishes to purchase old stock of rice (as he is a diabetic patient) and if it is not readily available, he cannot postpone his consumption needs and thereby, he purchases rice of freshly harvested produce.

B. Selling

Selling is more than an economic activity. It does not mean, mere offering of a product to the buyers or consumers, but includes much more than that. The concept of selling mainly revolves around three S's, namely: service, satisfaction, and self-interest. Every time a product is offered for sale in the market, a service is made available for the benefit of the consumers *i.e.,* the benefit and advantage of being able to use the product. When that product fulfills the consumer's need or a want, it yields satisfaction to the consumer. In the end, the seller (farmer) is able to serve his own self-interest, *i.e.,* he enjoys more sales and more profits in the business. So, selling may be defined as the *'personal or impersonal process of assisting or persuading a prospective consumer to purchase a product or service or to act favorably upon an idea that has commercial significance to the seller'*. Looking from different angle, selling implies offering the right product by the seller (farmer), at the right place, in the right quantity, at the right price, at the right time and under the right impressions to the trader or consumer. All these righteousness is made possible by performing the sales function. Through selling function desires are created among the consumers and hence, it is also called as *'Creative function'*. Selling is otherwise referred as *'Distribution function'* because, distribution makes the products to move from the place of production to the place of consumption. This is achieved only through selling function. Thus, selling ensures transfer of title of produce to the buyer. Its main objective is to create demand for a product even against keen competition. It involves creation of demand for the product, find out potential buyers, giving them the information of the availability of products to satisfy them and assure them of reasonable prices. Thus, selling is more than merely passively passing the product to the buyers at the offered price. It involves various sub-functions like,

- ☆ Product planning and development – It is the starting point to decide the precise nature of product. For this, the physical factors (colour, design, feature etc.) and non-physical factors (utility, value, quality etc.) are to be analyzed.

- ☆ Physical arrangements of display of goods
- ☆ Advertising and other promotional devices to influence or create demand. (Demand creation can be done through advertising, personal salesmanship and non-personal selling techniques to match goods and services to customer needs). Considering this sub-function, there is another view of understanding about the concept of 'selling'. It is to be noted that, selling has such a broad connotation as to include advertising, publicity and public relation activities, sales promotion, personal as well as non-personal selling techniques in moving the product towards the consumers. All these activities attempts to create or promote demand for the product among the customers. So, by selling process, latent demand is converted into effective demand. It is an art. Personal (Direct) and non-personal (Indirect) methods are used to convert the latent demand into effective demand. Advertisement and publicity are direct methods and demonstrations, exhibitions, providing samples etc., are the indirect methods.
- ☆ Identifying potential buyers
- ☆ Negotiating with the buyers for the settlement of price
- ☆ Setting terms and conditions under which the sales are to effected
- ☆ Decisions regarding quantity of sale, proper packages, proper time and place of delivery of products
- ☆ Final disposal (transfer) of products to buyers.
- ☆ Provision of customer service is gaining more significance in the new marketing environment especially in transacting processing machinery, drip irrigation, sprinkler irrigation, tractors etc. Customer service is generally considered as 'after-sales service' to satisfy customers and establish direct relations with customers.

i. Selling vs. Marketing

There is a misconception that, both selling and marketing is one and the same. But, there exists lot of difference between these two. This is because, while selling focuses attention on the needs of the sellers and is thus, preoccupied with the seller's need to convert his product or service into cash. On the other hand, marketing is also concerned with the needs of the buyer and as such is responsible for providing him with requisite products and services. In 'marketing', the focus is on the 'consumer's needs and their satisfaction'. 'Selling', on the other hand, focuses on 'product' *i.e.,* emphasizes on selling what has been produced. In fact, selling is a small part of the wide process of marketing, wherein emphasis is initially on promotion of goods and services and eventually on increase in sales volume. But, in marketing, the consumer is treated as the king, whose needs must be satisfied. In selling, the product is supreme and the entire focus is on its sale. Marketing starts before production and continues even after the exchange of goods and services has taken place. It is so because, as mentioned earlier, provision of after sales service (customer service) is an important component of marketing process. Selling starts after the production and ends as soon as the exchange of goods and services has taken place. Further, the difference between selling and marketing also lies in the attitude towards business. The selling concept takes an inside-out perspective. It starts with the farmer, focuses on the produce or processed products produced by him and calls for heavy selling and promotional strategies in moving the same to potential consumers at remunerative prices. But, the marketing concept takes an outside-in perspective. That is, it starts with a well-defined market, focuses on identifying consumers' needs, coordinates all the marketing functions that ensure remunerative price

Table 5.72: Selling vis-à-vis marketing with reference to attitude towards business.

Selling	*Marketing*
Focuses on seller's needs	Focuses on consumer's needs.
Product enjoys supreme importance.	Customer enjoys supreme importance
Converting product into cash.	Converting customer's needs into product.
Profits through sales volume.	Profits through customer satisfaction.
Emphasis is placed on sale of products already produced.	Emphasis is given on product planning according to consumer's needs.
Stresses on needs of the seller.	Stresses on needs of the buyer.
Focus is on earning profit through maximization of sales.	Focus is on earning profit through maximization of customer's satisfaction.
Selling has short-term prospective of only increasing the sales volume.	Marketing has long term perspective of winning over consumer loyality to the product by providing him maximum satisfaction (and even offering after sales services)
The principle of caveat emptor (let the buyer beware) is followed.	The principle of caveat vendor (let the seller beware) is followed.

Starting Point	*Focus*	*Means*	*Ends*
Selling Concept			
Farm gate	Produce or processed products	Demand creation	Increasing profits through more sales.
Marketing concept			
Market	Consumer needs	Coordinating varoius marketing functions	Consumer satisfaction

to the sellers and offer the requisite goods and services to the consumers at affordable prices (Table 5.72). So, the selling concept is a supply driven, where as marketing concept is mainly demand driven (though recognizing the sellers' welfare perspective).

ii. Importance of Selling Function

We know, the ultimate aim of every farmer or business is to earn profits and for realizing this aim, selling place an important role. Nothing really happens until somebody sells something. Selling enables a farmer to satisfy the needs of consumer. It is the process through which ownership of produce is transferred from the farmer (seller) to the buyer. Sales are the source of income for the farmers, processors, wholesalers and retailers. The importance of selling has increased significantly with an increase in the number of products offered for sale by large number of firms. When production was on a small basis, the farmers had no problem to dispose off their produce. But now, with the increase in the volume of production, selling has become a problem and the farmer or firms have to induce people to sell his products. The following points further highlight the importance of selling function:

- ☆ Selling creates value for a product and recognition for firm at the point of contact with a potential buyer.
- ☆ Selling creates more employment opportunities to the people engaged in that process.

- ☆ For performing other marketing functions like processing, storage, transportation etc., selling is a pre-requisite.
- ☆ A firm lives or dies by what it sells. Sales of the product serve as the engine of sustainability and growth of the business.
- ☆ Selling function alone ensures distribution of products and services to the consumers in requisite quantity and quality at right time and at right place.

iii. Proportion of Selling Expenditure

The proportion of selling expenditure varies across different types of products (Figure 5.32) and at different stages of purchase (buying) process (Figure 5.33). So, every sales situation is in some way unique. It is evident that, across different types of products (Figure 5.32), for selling capital goods, the share of expenditure of 'personal selling' is highest followed by the shares of 'sales promotion and publicity' and 'advertising'. However, in case of selling RTE and RTC foods, the share of expenditure incurred through 'sales promotion and publicity' is highest followed by the shares of 'advertising' and 'personal selling'. 'Sales promotion and publicity' expenditure will have highest share in total selling expenditure (Figure 5.33) for creating awareness about the product and service among the consumers followed by the shares of 'advertising' and 'personal selling'. However, for convincing the consumers to purchase the product or service, the share of 'personal selling' is highest followed by the shares of 'sales promotion and publicity' and 'advertising'.

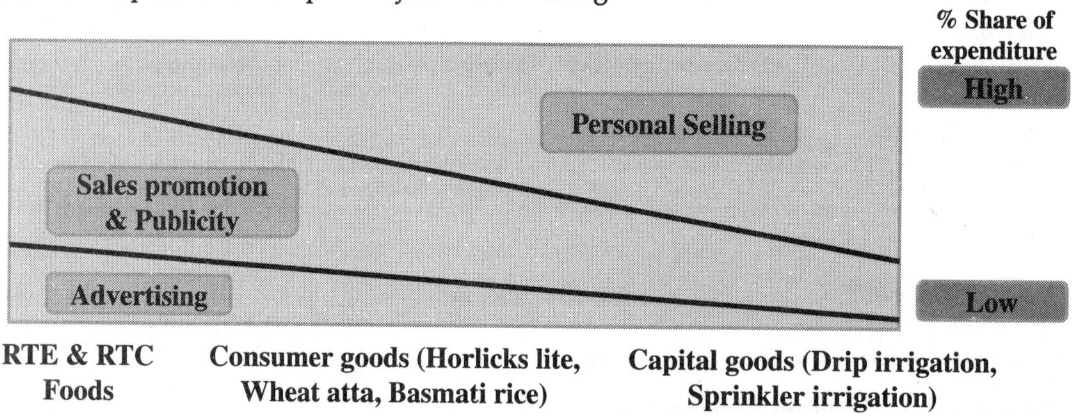

Figure 5.32: Per cent share of selling expenditure across different types of products.

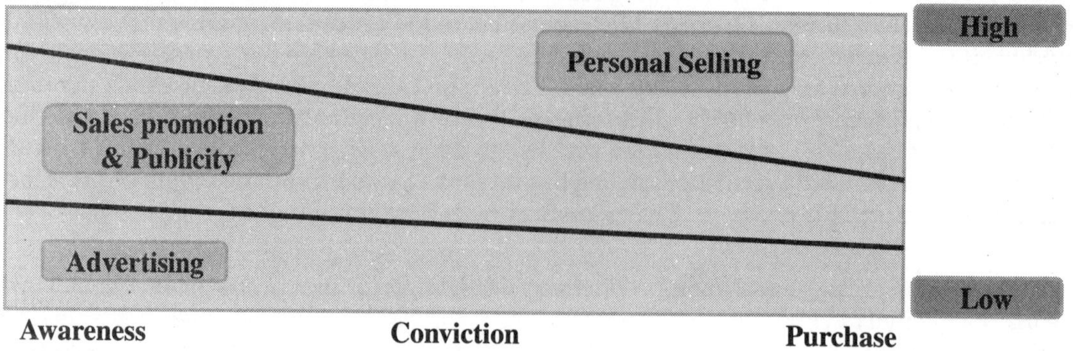

Figure 5.33: Per cent share of selling expenditure across different stages of purchase process.

iv. Performance of Selling Function by different Marketing Agencies

The mode and efficiency of performing selling function varies across the marketing agencies involved in the marketing channel. As marketing channels vary across the products transacted, the selling function performed by these agencies also varies across the products. In marketing of agricultural produce, the farmers or traders (processors or wholesalers) does not come into contact with the consumers directly. Only retailer in the marketing channel for majority of the products comes in direct contact with the consumers. However, in case of fruits and vegetables marketing, with the advent of Apni Mandi in Punjab and Haryana, Rythu Bazaars in Andhra Pradesh, Uzhavar Santhaigal in Tamil Nadu, Krushak Bazaars in Odisha, Raithara Santhegalu in Karnataka etc., there is direct contact between farmers and consumers. In case of marketing of agricultural inputs like seeds, fertilizers, pesticides etc., the firms engage sales persons and they play vital role in transacting these inputs to the buyers (farmers). Irrespective of the type of agencies involved in selling and buying the produce or inputs, the selling strategies come into play for transacting the same between a seller and a buyer and these are discussed here under.

Farmers

The selling function performed by farmers traditionally involves selling the fresh/raw produce immediately after harvest in the villages itself to the local traders. However, with the intervention of Government in the form of regulated markets and Apni Mandi in Punjab and Haryana, Rythu Bazaars in Andhra Pradesh, Uzhavar Santhaigal in Tamil Nadu, Krushak Bazaars in Odisha, Raithara Santhegalu in Karnataka etc., and with the improvement in transportation facilities (construction of link roads), the distress sales of produce in the villages is effectively checked. Since majority of the farmers are largely illiterate with reference to modern marketing techniques and lack of awareness regarding the procedural formalities involved in using modern marketing facilities at regulated markets, they usually employ 'commission agent' as their intermediary in guiding the sales transactions.

Wholesalers

Wholesalers typically hold produce acquired from farmers and sell them to retailers. They purchase the produce in large quantities from the farmers directly (say, vegetables) and transact the same in small quantities to the retailers. In some cases, they sell the produce to processors in large quantities (say, paddy, wheat, mangoes etc) for processing the produce. In some cases, processors may directly purchase the produce in large quantities from the farmers (say, sugarcane) and sell the processed product (sugar) in large quantities to wholesalers. So, irrespective of the agency from whom the produce is purchased by the wholesalers, it will be made in large quantities only. Further, they will not sell significant quantities to both retailers and consumers directly. Compared to farmers, the wholesalers are dynamic in the marketing process in the context of performing several marketing functions efficiently like transportation, scientific storage, grading and standardization, packaging and packing, access to market news, risk bearing etc. Especially, in marketing processed products, these wholesalers even employ sales personnel to meet with retailers to convince them to carry products in their stores. They even go for advertising about their products, so as to stimulate and motivate the consumers in knowing and purchasing about their products.

Retailers

Retail sales employ many sales people because, retailers maintain the final connection with the end consumer in the traditional distribution process. These retailers employ sales executives and they are mainly involved in identifying potential buyers for finding and selecting their products. These retailers handle the produce in small quantities compared to wholesalers and they sell only insignificant quantities to the final consumers. These retailers are so dynamic in accessing the market news and handle the quantum of sales accordingly. Unlike wholesalers, the retailers will not go for advertising about the products, for two reasons *viz.*, they handle a wide range of products and it is too costly on the part of them to meet heavy advertising expenditure (considering low-volume of business).

Other marketing agencies like processors, local exporters etc., handle the produce in large quantities and they perform selling function in a more sophisticated way compared to the farmers. This is because, they handle the produce in large quantities and they have to necessarily maintain the quality standards of the products as per the quality norms and suiting to the consumers' tastes and preferences. The recent increase in demand for RTE and RTC food products is due to their quality assurance. These agencies can even go for advertising on large scale, so as to make the consumers aware about their products and to gain rapid acceptance.

v. Different Methods of Sale of Agricultural Produce

There are different kinds of selling, which includes personal selling, sales through advertisement, sales by inspection, sales by samples, sales by description etc. In addition to these methods, many more methods are recognized for sale of agricultural produce. The marketing of agricultural produce in India is generally transacted in one of the following ways:

1. Hatta system or undercover sale or Sale under cover of cloth or Secret sale
2. Private negotiations
3. Quotations of samples taken by commission agents
4. Moghum method of sale or Unbhav Sale
5. Dara method of sale
6. Open auction method of sale
7. Closed tender method of sale
8. Mutual agreement
9. State trading
10. Forward sale
11. Jalap sale
12. Sale by open agreement:
13. Direct marketing
14. Co-operative marketing
15. Contract farming
16. Internet marketing (e–Marketing)

1. Hatha System or Undercover Sale or Sale Under Cover of Cloth or Secret Sale

Under this system, the price of the produce is settled between the buyer and the commission agent of the seller (Kaccha Arhatiyas) by pressing or twisting the fingers of each other under cover of a piece of cloth usually a towel or a kerchief or a dhoti or front portion of kurta or shirt. Under this system, the commission agent covers his hand with a cloth and invites offering individually from each buyer. The buyers make their offers in a secret language, which is mostly understood only by them and the commission agent. Most of this secret language is a sign language that involves pressing fingers and finger joints. Certain code symbols are associated with the twisting of the fingers and the codes include: touching 1 finger = ₹ 10, 1 tapping = Re 1, holding three fingers and then tapping the fingers two times would communicate price of ₹ 32/kg etc. If the prices are acceptable to both the parties, clapping hands signals the deal has done. The negotiations in this manner continue till a final price is settled. When all the buyers have given their offers, the name and the offer price of the highest bidder is announced to the seller (farmer) by the commission agent and if the seller agrees to sell the commodity at that price, the produce was transacted to the highest bidder. Thus, a sale gets confirmed. This system was earlier common in cattle sales and even today, it is practiced in fruit and vegetables markets. This method of sale is well justified by the wholesalers in the markets. This is because, with sharp increase in the number of retailers visiting the markets and the prices offered to the wholesale buyers are negotiated *via* talking or announcing loudly through loud speakers, even the retail buyers would demand a similar (wholesale) price, which would not be acceptable to the wholesale buyers. So, with a view of maintaining confidentiality of wholesale prices, the hatha system is often practiced in the markets. However, the sale under this system is open to several malpractices, as the farmer remains in the dark all the time and it is only the commission agent and the ultimate buyer, who really knows the negotiated price. Further, there is no guarantee that, the benefit of this higher price always goes to the farmer, as he is unaware of the price offered by the buyer. So, this system of sale has been banned by the Government and is not legally permitted to be practiced in the regulated markets because of the possibility of cheating the farmer, as he was not taken into confidence until the final bid is cleared. Since, this system operates to the disadvantage of the farmers and hence, they are shifting from this method to private negotiations.

2. Private Negotiations

This method of sale is more common in unregulated markets and this the most common method of sale. The price is settled between the buyer and commission agent engaged by the farmer (Kaccha Arhatiya) based on personal negotiations. To negotiate the price, the individual buyers approach the commission agent's shop, inspect the produce and offer the price. The commission agent conveys the same to the farmer and the produce is sold to the buyer, only if that price is accepted by the farmer. Further, the seller may invite several offers for his produce and may sell to one, who might have offered the highest price for the produce. So, this method of sale offers good price to the farmers, as one buyer does not know the price offered by other buyers. But, this method of sale is more time consuming, not suitable when either large quantities have to be sold or a large number of buyers exist in the market.

3. Quotations of Samples Taken by Commission Agents

In this method of sale, instead of buyers visiting the sellers' shops, the seller or the commission agent (Kaccha Arhatiya) will move to the buyers' shops. The produce brought by the farmer is graded by the commission agent based on laid out quality specifications. From each grade, the commission agent will take a sample of produce and he moves to the neighbouring villages to contact the buyers for selling the commodity through showing the samples of grades. Based on the negotiations between commission agent and buyers, a price will be settled. The commission agent makes a number of rounds to prospective buyers, until none is ready to bid a higher price than the one offered by a particular buyer. The commission agent will inform the highest price realized for the produce to the farmer for his final decision to dispose the produce. If the farmer accepts that price, the produce is transacted to the highest bidder. This is the most convenient method of sale, where the produce is systematically graded and saves the cost of transportation and inspection on the part of the buyers. However, utmost honesty in the dealing is to be followed. For example, during the transactions of chilies, the buyer quotes the price for the produce by looking at the samples. This sale by sample is also practiced in food grain sector. Food grains such as wheat, rice etc., are also sold by this method of sale.

4. Moghum Method of Sale or Unbhav Sale

In this method of sale, there exists a verbal understanding (written agreement) between buyers and sellers. For example, a dealer of different fruits like mangoes, approach the mango growers and enter into a verbal agreement with them by just looking at the flowers setting. This method is more common in villages, where the farmers are highly indebted to the local traders or commission agents in cultivating the crops. We know that, the farmers, in general, depend upon local traders or commission agents for non-equity capital (borrowed funds) right from the plough share point till the efficient marketing of agricultural produce. So, these traders or commission agents will put a condition to the farmer that, after harvesting the crop, the produce must be sold to him only and not in the outside open market. Regarding the price of commodity, there is no pre-settlement and the produce will be purchased at the price prevailing on that day or the price at which other sellers of the village sold the produce. The farmer is bound to deliver the produce to the buyer on a set date within a period prescribed by a verbal understanding between them. In return, the buyer is supposed to pay the price ruling on the delivery day. This method is usually adopted in interior areas, where price fluctuations get communicated within a conventional time lag. It is usually beneficial to the farmer, as he dictates the price and the buyer should pay on a future date of his choice. However, the traders pay low price to the farmers quoting the reason that, the produce is of inferior in quality. So, this method of sale will not ensure profitable or remunerative price to the farmers, as they are at the mercy of these local traders or commission agents and further, they have to depend on them for loans for the ensuing crop season. But, this method of sale is inevitable in the villages, as the farmers are in requirements of liquid cash immediately after the harvesting season, more procedural formalities involved in getting loans from the financial institutions, the financial institutions will not provide credit both in terms of adequacy and in time, lack of adequate transportation facilities to transact the produce in distant markets for getting still higher prices etc. All these factors cumulatively influence the dependency of the farmers on local traders or commission agents and thereby, entering the verbal agreement regarding this method of sale.

5. Dara Method of Sale

In this method of sale, the agricultural produce of different qualities is mixed and sold as one lot at one price. That means, no grading is done for the produce and good quality produce is mixed with poor quality produce and is offered for sale in the market at uniform or fixed price. The advantage of this method is that, a large quantity of produce is sold in a short period of time. But, the disadvantage is that, good quality produce is mixed with poor quality produce and hence, the quality of final produce is not up to the mark, as desired by the buyers and there is loss of incentive to the farmers to produce quality goods. This method is most commonly followed for crops like zeera, onions, potato.

6. Open Auction Method of Sale

In this method of sale, the seller (farmer) will bring the produce to the market yard and approach the commission agent (Kaccha Arhatiya) for making necessary arrangements for transacting the produce to the prospective buyer. The commission agent after inspecting the produce, grade the produce based on quality specifications and he will inform through mike to invite the potential buyers for his shop for conducting the auction. The potential buyers will inspect the produce and openly announce the prices. Then, the commission agent will inform the farmer about the highest price bidded by the potential buyer of the produce. The produce will be sold to that buyer at that highest price, only if the farmer satisfies with that highest price. So, the produce is given to the highest bidder after taking consent of the farmer-seller. So, in this method of sale, the commission agent does the auctioning under the supervision of APMC officials. In some markets, an auctioneer, who belongs to the staff of the APMC conducts the auctioning of the produce. The commission agent will charge commission from the farmer for the services rendered by him. This auction will be conducted for each lot of produce (of different qualities or grades) separately and hence, it gives a competitive price to the farmer-seller. In most of regulated markets, the sale of produce is permissible only by this method. In Andhra Pradesh, both in Coastal Andhra and Telangana regions, open auction method is prevailing.

Sometimes, auction on a sample basis takes place in front of the APMC office, which is centrally located. Before start of the auction, heaps of produce (say, jaggery) belonging to the same commission agent are graded and lumped side by side by APMC staff. The commission agent and the farmer select one heap, generally of the best quality for representative auctioning and jointly assign price differences to other heaps according to the difference in qualities. In other words, the price difference between the selected heap and other heaps are assigned by mutual agreement between the concerned farmer-seller and the commission agent. Then the auction is performed over the selected heap, with the buyer bidding at once for the entire bunch of heaps with their different qualities. This way, auctioning could be carried out speedily and under the given conditions, this procedure maximizes the number of transactions per unit of time. However, this approach involves an element of mutual agreement (over price differences across different grades of produce) and it is not as competitive as would be the auctioning of every lot separately. In some markets, the bidding took place at the commission agents' shops on a rotation basis *i.e.*, the auction starts every day at a different shop, so that, everyone had an equal chance of being at the start or the end of the auction.

The timing and sequence of auctioning of different products differ from market to market. Generally, large markets start auctioning earlier than smaller ones. Traders in small markets wait for information from the big markets to guide them in their price formation.

Merits

The following are the merits or advantages of open auction method of sale:

- ✰ The auction serves as a meeting place for the supply of and demand for goods thereby, there is proper interaction between demand and supply forces.
- ✰ A wide variety of goods and of different grades can be offered to the buyers.
- ✰ This method of sale reduces number of marketing middlemen in the transaction process.
- ✰ The buyers of small lots are not put to disadvantage over buyers of large lots of produce.
- ✰ This method of sale is conducted under the close supervision of APMC officials and hence, there is less scope for cheating the farmers.
- ✰ This method of sale ensures prompt payment of sales proceeds to the farmers after conducting the auction.
- ✰ The seller is able to follow the bidding easily.
- ✰ This method of sale disposes of the market supply promptly.
- ✰ The buyers and sellers know clearly the price quoted and leaves no room for being cheated thereby, it is better than any other system as sale.
- ✰ This method of sale ensures fair dealing to all the parties and secures a premium for superior quality produce due to the open competition among traders.
- ✰ This method of sale makes prices transparent, since all of the buyers and sellers at the auction hear the bids and are therefore aware of prevailing price levels.

Demerits

However, this method of sale suffers from the following limitations:

- The major drawback is that, it is time-consuming for the traders, because of the variation in the quality of the various lots. Similarly, sellers have to wait for the entire day for completing the transaction. One way of speeding up the auctioning process is to lump up several lots of similar quality into one big lot for the purpose of bidding, but to pay the farmers according to their relative shares in such lumped lot.
- In big market centres, especially in the peak marketing season, the time allotted for auction is short. As a result, sellers may not receive the competitive price.
- In open auction, buyers sometimes join hands (collude) and hence, their active participation is reduced and they purposefully quote less price to the produce.
- The auction leads to a 'buyers market', as buyers have full information about the supply of, and demand for, the product.

Different Systems of Auction

There are three systems of open auction, as discussed here under.

- ***Phar system of open auction:*** In this method, one bid is given for all the lots in a particular commission agents shop and all the lots are sold at that price. One extreme case of this method is, when one bid is given for the produce in the whole market.
- ***Random bid system of open auction:*** In this method, the commission agent invites few buyers, when the produce is brought to his shop for sale. All the prospective buyers are not informed. As a result, the competition is poor.
- ***Roster bid system of open auction:*** Bidding starts from a point in the market at a notified time about which the prospective buyers are given information in advance. The bidding party after the auction of the produce at one shop, moves to the next in a clock-wise or anti-clockwise direction till the auction of the produce at all shops is over, or the scheduled auction time expires. On the following day, the auction starts from the next point, and so on. The auction is supervised by the auction clerk or the person nominated by the APMC. This method of auction is more systematic when compared to above two systems.

7. Closed Tender Method of Sale

In this method of sale, the procedure is more or less similar to open auction method, but instead of announcing the prices openly, each buyer will write his name, date and time, price of the produce and the grade he wishes to purchase in a prescribed form ('bid slip', a blank form supplied by the APMC). Buyers inspect the produce after visiting the shop and they will place their slips in the container or box placed before the commission agent's shop. At the end of day, the commission agent will open the box and list out the prices quoted by the buyers and the same will be informed to the farmer. Based on the willingness of the farmer, the commission agent will sell the produce to the buyer, who bids the highest price to the produce. In case, if two or more bidders offer the same highest price, the one who deposited his offer first gets the lot. Regulated markets in Tamil Nadu and Rayalaseema region of Andhra Pradesh have closed tender method of sale.

Merits

This method of sale ensures the following advantages:

- It is less time consuming

- ☆ Involves minimum physical labour
- ☆ No possibility of collusion among the buyers

Demerits

This method of sale too suffers from the following drawbacks:

- ☆ This method does save traders time, but the competition is probably not as perfect as in the case of open auction because, it leaves more room for collusion.
- ☆ One buyer does not know the price offered by the other buyer and hence, the demand for the commodity may not be known exactly.

8. Mutual Agreement

Under this system, the individual buyers may come at any time during the transaction hours and make their individual offers for the produce exhibited at the commission agent's shop. These offers may or may not be acceptable to the seller. His final decision is conveyed only towards the close of the day. In this system, the scope for competition is limited.

9. State Trading

In this method of sale, Government purchases huge quantities of food grains for distribution to the beneficiaries. State trading plays a major role in purchasing agricultural produce from farmers at a fixed procurement prices. The government undertakes this to enable centralized and equal distribution of the produce among the consumers and assumes better returns to producers. This method of sale is discussed in-detail in Chapter 8.

10. Forward Sale

In this system, process of purchase and sale is done at present, but commodities and money are exchanged at some specific date. This method of sale reduces the risks by allowing the farmer and trader to decide a price today for goods to be delivered on a particular future date. Thus, in this method of sale, commodities are delivered at some time in the future, unlike cash markets that call for immediate delivery. These advance sales help both buyers and sellers in long-term planning. This method of sale is discussed in-detail under Section 'III-D' of this Chapter.

11. Jalap Sale

In this method of sale, the trader purchase the standing crop of the farmer well-in advance of the harvest at a price fixed on the date of the bargain. The price is usually quoted in lump sum for the entire crop and the seller may receive 50 per cent of the value of the sale transaction in advance. Usually, this method of sale goes against the farmer-seller, as the price is fixed not on the basis of the prevailing market price. If the crop fails to give the expected output, the buyer may force the farmer to reduce the price further. However, this method facilitates the farmer to sell the produce without undertaking any further risk and marketing expenses.

12. Sale by Open Agreement

This method of sale is more common in case of commodities that enjoy oligopsonistic market (few buyers). Since the buyers for the produce are few, the seller (farmer) deals with the individual buyer to enter into a sale/transaction with him. So, in this method of sale, there are no middlemen and it is the seller (farmer), who moves from one buyer to another in search of a remunerative price or offer. Commodities like sugarcane, tobacco, cotton, jute enjoy this method of sale. This method of sale is disadvantageous to farmers, as the buyer may exploit him through unauthorized deductions and allowances and delays in making payments. Since, buyers often join their hands, it results in depression of prices.

13. Direct Marketing

In the field of agriculture, marketing determines the value of the agricultural product in terms of money and delivers them to the final consumer. Most of the farmers sell their produce through village level markets, fairs, mandies, CMSs etc. In these markets, the middlemen exploit farmers as well as consumers. Generally, the middlemen and wholesale businessmen purchase the agricultural produce from the farmers at a lower price. They also get the commission from the farmers for the transactions made. Similarly, traders (processors, wholesalers etc) purchase fresh vegetables and fruits at lower prices from the farmers at peak marketing season and are sold out to retailers at higher prices and they in turn sell the same further at higher prices to the consumers. As a result, the farmers get only the lower price for their produce (low FSCR), whereas the consumers have to pay higher price for the same produce. Hence, in order to eliminate these middlemen between farmers and consumers, the State Governments introduced the new concept, namely 'Direct markets' in the country. Thus, under direct marketing, farmers sell their produce directly to the consumers. It was developed mainly to deal with perishable commodities like vegetables, fruits and flowers. Thus, farmers' markets facilitate the sale of farmers' produce directly to the consumers. The physical infrastructure is provided by the State Governments. Farmers bring their produce to the market yard daily and transactions take place on the price fixed on a daily basis by the APMC Officials. Such initiatives include Apni Mandi in Punjab and Haryana, Rythu Bazaars in Andhra Pradesh, Uzhavar Santhaigal in Tamil Nadu, Krushak Bazaars in Odisha, Raithara Santhegalu in Karnataka etc. The following are the important objectives and features of these 'direct markets':

- ☆ To facilitate direct contact between the farmers and public.
- ☆ To provide fresh produce (vegetables and fruits) at reasonable price daily without any interference of middlemen.
- ☆ To provide correct measurement to the consumers.
- ☆ To give full satisfaction to the farmers and public.
- ☆ To aim for providing higher price than that of wholesale price to the farmers for their vegetables and fruits.
- ☆ To provide the fresh fruits and vegetables at the lesser price than that of retail price to consumers.
- ☆ These markets also serves as a Technical Information Centre and Technical Training Centre to the farmers
- ☆ Seeds and other inputs are also provided in these markets for the benefit of farmers.
- ☆ The prices of the vegetables and fruits are daily fixed at the average of 20 per cent higher than the wholesale prices and 15 per cent less than retail prices by the Committee consisting of the representatives of farmers and officials.
- ☆ The prices of vegetables and fruits are daily displayed in front of each shop as well as exhibited in big signboards of the market and it is ensured by the Department staff that the fruits and vegetables are sold at the fixed rates.
- ☆ Weighing scales are provided at free of cost for the use of farmers and they are retrieved after the sale proceedings are over.
- ☆ Sanitation is being maintained. In some markets, vermi composting of vegetable wastes is also done.
- ☆ To sell unsold items on the next day, storage facilities are available.

It is estimated that, about 10 per cent of total produce is marketed through direct marketing. At present, these markets are being run at the expense of the State exchequer as a promotional measure, to encourage marketing by small and marginal farmers of fruit and vegetables without the help of the middlemen. Since, the produce is transacted by the farmers (licensed) to the consumers directly in these markets, the farmers enjoy 100 per cent FSCR. Further, direct marketing encourages farmers to undertake grading of farm produce at the farm-gate. Considering the vastness of the country, more and more such markets need to come up in the organized sector so that, they can be developed in tune with the strengthening of backward and forward linkages. The APMC Acts will also have to be amended to permit private and co-operative sectors to take up direct marketing of agricultural commodities from the producing areas and the farmers' fields, without the necessity of going through licensed middlemen in regulated markets. Such a reform will spur private initiative in building consumer oriented market infrastructure in the country. This method of sale is discussed in-detail under Section V of Chapter 8.

14. Co-operative Marketing

Agricultural marketing co-operatives are co-operative businesses owned by farmers, to undertake transformation, packaging, distribution, and marketing of farm products (both crop and livestock.). A marketing co-operative is a business organization owned by the farmers to collectively sell their produce, as each individual farmer cannot enjoy the benefit of transacting large quantum of marketable surplus, because of small and fragmented land holdings. So, the co-operative approach among the farmers allows them to accomplish marketing functions collectively and to enjoy greater power in the market place. In addition, marketing co-operatives can give farmers more control over their produce, as they make their way to consumers by allowing them to bypass one or more middlemen in the market channel. So, farmers capture more of the returns that would otherwise go to others. These agricultural marketing co-operatives perform many functions like assembling the produce from farmer-members, more efficient handling, competitive sales, and then grade and transport the bulk produce to market. Some marketing co-operatives perform first-stage processing such as milling, ginning, decorticating etc., while others vertically integrate by-processing products for the consumer or industrial markets. In the recent period, Agricultural bargaining co-operatives are a special type of marketing co-operative that is gaining popularity. These co-operatives negotiate with buyers, usually processors, on behalf of their farmer-members for price and other terms such as quality, grade and timing of delivery. By handling large quantum of produce, it gives the co-operatives (bargaining associations) more market power and allows them to be more effective in negotiations than individual farmers can perform. These bargaining co-operatives usually do not own the farm produce, but guide the farmers in efficient transacting of produce to potential buyer. These bargaining co-operatives are most common in processing fruits, vegetables, dairy, sugarcane, sugar beet etc. A new concept of co-operatives called New-Generation Co-operatives are coming up in a big way and they are so designed to allow farmer-members to plan for value-added activities for their produce. These new-generation co-operatives are significantly different from those of more traditionally structured marketing co-operatives. First, new-generation co-operatives typically sell stock that have delivery rights and obligations for a specific quantity of product. These delivery rights result in the co-operative's membership being open only to those, who purchase the stock. Second, farmers often must make a substantial, up-front investment to purchase stock and delivery rights in a new or existing new-generation co-operative. Over time, delivery rights may increase in value. However, producers normally have the right to sell their membership and delivery rights to other qualified producers and capture the increase in value. In many traditional co-operatives, membership is open to anyone wanting to sell product and usually requires little or no up-front investment.

15. Contract Farming

Another channel of linking farmers directly with consumers and agro-processors is through contract farming. In the recent times, contract farming is becoming very popular. Since the land holdings are getting smaller, the farmers cannot make heavy investment in agriculture nor are they ready to take the risk of diversifying to other crops. In this situation, business corporates are coming into picture. They enter into a contract with the farmers and take their piece of land on a sort of mortgage. They provide all the inputs to the farmer, including the seeds, fertilizers etc. The farmer has to provide the services of sowing the crop, watering and taking other care. After the crop is ready, the entire crop is lifted by the business corporate, at a price, which is decided in advance. So, contract farming may be defined as an agreement between business firms (processing and/or marketing firms) and farmers for production support at pre-determined prices. This stipulates a commitment on the part of the farmers to provide a specific commodity in terms of quality and quantity as determined by the purchaser and commitment on the part of company to support the farmer for production through inputs and other technical support. Several domestic and multinational firms have entered into agreement with farmers to produce agricultural commodities of desired quality with assurance of procurement at pre-announced prices. This method of sale of produce through contract farming helps farmers to reduce price risk, remove input supply bottlenecks, technology constraints and access to viable procurement options. Contract farming is also seen as medium for diversification of cropping system in the country. The APMC Model Act recognizes and encourages its expansion in the country.

Contract farming arrangements of different types have existed in various parts of the country for centuries for both subsistence and commercial crops. The commercial crops like sugarcane, cotton, tea, coffee etc., have always involved some forms of contract farming. Even in the case of some fruit crops and fisheries, contract farming arrangements, involving mainly the forward trading of commodities have been observed. However, in the wake of economic liberalization, the concept of contract farming in which national or multinational companies enter into contracts for marketing of the horticultural produce and also provide technologies and capital to contract farmers has gained importance. According to this, bipartie agreements are made between the farmer and the company and the latter contributes directly to the management of the farm through input supply as well as technical guidance and also markets the produce. The main features of this type of contract farming are that selected crops are grown by farmers under a buy back agreement with an agency engaged in trading or processing. In such cases, the centralized processing and marketing agencies supply technology and resources, including planting materials and occasional crop supervision. Under such contracts, the farmer assumes the production related risks, which the price risk is transferred to the company. In some cases, the company also bears the production risk, depending on the stage of crop growth at which the contract is made. If the contract is made at flowering or fruiting stage, the company bears the production risks also. In any case, the company bears the entire costs of transaction and marketing. It is this variant of contract farming, which is said to be one of the ways by which small farmers can participate in the production of high value crops like fruits, vegetables, flowers etc., and benefit from market led growth. Thus, contract farming offers advantages both to the farmer and to the business firm, as the farmer gets a remunerative price and the business firm gets a good quality produce at a fixed price throughout the year.

However, the contract farming also suffers from serious limitations. Theoretically, farmers stand to gain from contractual agreements that provide lower transaction costs, assured markets and shifts in both production and pricing risks. From the buyers' side, this method of sale ensures the advantages of more assured supplies and reasonable control over quality and other specifications of the produce.

However, in practice, there are practical problems that emerge in this method of sale that can result in losses to both farmers and firms (buyers). The major limitations include:

- ✰ Contracting agreements are often verbal or informal in nature, and even written contracts often do not provide the legal protection in India that may be observed in other countries.
- ✰ Lack of enforceability of contractual provisions can result in breach of contracts by either party. In India, there have been instances of farmers refusing to sell to contracting firms, when market prices exceed the contract price, and of firms refusing to purchase contracted quantities or pay contracted prices due to market conditions. Neither the contracting firm nor the farmers are keen to contest these issues in a court.

Most often, it is mutual understanding and faith that drives contractual relationships and it takes a long time to win mutual trust and confidence. Contract farming arrangements are often criticized for being biased in favor of firms or large farmers, while exploiting the poor bargaining power of small farmers (Bhavani *et al*, 2007). In such situations, a viable approach seems to be to form clusters of small farmers that can create a scale effect and also enhance the bargaining position of the farmers. Success in developing contracting models or other forms of farm-firm linkages that are effective for small holders will be a key challenge to small holder participation in the transformation of Indian agriculture. This method of sale is discussed in-detail under Section V of Chapter 8.

Private Sector Companies involved in Contract Farming in India

Contract farming is becoming popular in recent years and the following are the list (Table 5.73) of private sector companies involved in contract farming in India:

Table 5.73: Private sector companies involved in contract farming in India.

Sl.No.	*Name of the Company*	*States*	*Commodity*
1.	Appachi Cotton Company Zamin Uthukuli, Pollachi, Coimbatore	Tamil Nadu, Karnataka	Cotton
2.	AVT Natural Products Ltd 72-E, Chesney Estate, Ethiraj Salai, Chennai - 600 008	Karnataka	Marigold, Caprica, Chilly
3.	Cargill India Pvt Ltd DLF Gateway Tower 11 and 12th Floor, DLF City, Phase III, Gurgaon - 122 002	Madhya Pradesh	Wheat, Maize and Soybean
4.	Escorts Ltd Agri Machinery Group 18/4 Mathura Road, Faridabad - 121 007	Punjab	Basmati
5	The Global Green Company Pvt Ltd. (Naan) 220, Alembic Glass Complex Whitefield Hoskote Road, Whitefield, Bangalore	Karnataka, Andhra Pradesh	Gherkin, Babycorn, Paprika
6	Hindustan Lever Ltd 165/166, Backbay Reclamation, Mumbai - 400 002	Madhya Pradesh	Wheat
7.	Ion Exchange Enviro Farms Ltd Neeta TowersPune-Mumbai Highway Dapone - 411 012	Tamil Nadu, Madhya Pradesh, Gujarat, Haryana, Maharashtra	Organic Products of Banana, Pineapple, Papaya, Wheat, Basmati, Cotton
8.	ITC - IBD ITC Green Centre Ist Floor, Sector 3210, Industrial Area, Gurgaon - 122 001	Madhya Pradesh	Soybean

Contd...

Table 2.1–*Contd...*

Sl.No.	*Name of the Company*	*States*	*Commodity*
9.	Ken Agritech Pvt Ltd Factory and Field Operations #131/1A, Anchatageri Karwar Road, Hubli (Karnataka) – 580 024	Karnataka	Gherkin
10.	Marico Ltd Rang Sharda, 5th Floor, K C Marg, Bandra Reclamation, Bandra (W), Mumbai- 400 050	Maharashtra, Madhya Pradesh, Gujarat, Karnataka, Chattisgarh, Rajasthan	Safflower
11.	Mahindra Shubhlabh Services Ltd Mahindra Towers, 3rd Floor, Worli, Mumbai	Maharashtra, Punjab	Many crops
12.	Natural Remedies Pvt Ltd 364, II Floor, Ashwathlakshmi Mansion, 16th Main, 4th T Block, Jayanagar, Bangalore - 56 0041	Karnataka	Coleus
13.	Nestle India Ltd Nestle House, Jacaranda Marg, M-Block, DLF City, Phase-II, National Highway 8, Gurgaon - 122 002	Punjab	Milk
14.	Nijjer Agro Foods Pvt Ltd Maharbhan Pura, P O Jandiala Guru, Dist Amritsar - 143 115	Punjab	Tomato and Chilly
15.	Pepsi Foods Pvt Ltd 3B, DLF Corporate Park, S Block, Qutub Enclave, Phase III, Gurgaon - 122 002	Punjab, Tamil Nadu	Chillies, Groundnut, Seaweed, Tomato and Basmati Rice
16.	Rallis India Appejay House, 7th Floor, 3 Dinshaw Vaccha Road, Churchgate, Mumbai - 400 020	Punjab, Uttar Pradesh, Madhya Pradesh, Maharashtra, Karnataka, Tamil Nadu	Basmati, Wheat, Fruits, Vegetables
17.	Suguna Poultry Farm Ltd Corp Off: 5th Floor, Jaya Enclave, 1057, Avinashi Road, Coimbatore - 641 018	Tamil Nadu, Andhra Pradesh	Broiler
18.	Satnam Overseas Ltd Star Trading House, 201, VIPPS Centre, 2 Community Centre, G K II, Masjid Moth, New Delhi - 110 048	Punjab	Basmati
19.	Super Spinning Mills Ltd "ELGI Towers" PB#7113, Green Fields, 737-D, Puliakulam Road, Coimbatore - 641 045	Tamil Nadu	Cotton
20.	The Ugar Sugar Works Ltd Ugarkhurd Taathani, Distt: Belgaum, Ugarkhurd - 591 316	Karnataka	Barley
21.	Unicorn Agrotech Ltd (Zeraim) 1-7-139/3, Sarojini Devi Road, Hyderabad - 500 003	Karnataka	Gherkin
22.	United Breweries UB House1/1, Vittal Mallaya Road, Bangalore - 560 001	Punjab	Barley
23.	Venkateshwara Hatcheries Ltd 1-5, Upper Ground Floor, World Trade Centre, Babar Road, Cannuaght Place, New Delhi - 110 001	Tamil Nadu	Broiler
24.	CG Herbals S12-Greenparadise, Vishalnagar, Telibandha, Raipur - 492 006	Chattisgarh, Odisha	Patchouli, Veiver, Aromatic Crops for essential Oil
25.	Sanjeevani Orchards Pvt. Ltd.	Madhya Pradesh	Pomegranate

The contract farming needs to be further developed after identifying areas, commodities and markets for market oriented and demand driven production planning. However, while providing for this system of alternate marketing under the APMC Act, it is necessary to draft any appropriate legislation separately for ensuring definition of terms and conditions of the agreement keeping in view of the objectives.

16. Internet marketing (Email Marketing or E-marketing)

This method of sale has gained more significance in the recent period with the advent of IT enabled services. This is a low-cost method of offering the food and agricultural products and helps to reach a wide variety of consumers even across the countries. In the era of SPS standards for the commodities, they are exported and imported across the countries even without physical observation of the commodities. As computers become less expensive, greater numbers of consumers are going online. This internet marketing facilitates direct marketing between farmers and consumers. Further, consumers are increasingly viewing the internet as a tool with which to gather product information or compare prices for a wide range of products. If shipping fresh produce is a problem, farmers may wish to consider developing processed or value-added products. The consumers are also expanding their purchases in product areas such as organic foods, jams, honey, dried herbs, baked goods, RTE, RTC foods etc. Special occasions and holidays can also present opportunities for marketing via the internet. If what the consumer perceives as an 'authentic' or 'quality' product isn't available in the local market or food store, the internet provides a method of purchasing directly from a known or reliable source. The internet marketing further offers several other advantages like save time, labour, money, materials, advertising costs and other selling expenses.

The above discussion infers that, there are several methods of sale to ensure the process of effecting the availability of agricultural goods from the producers to the consumers. Many types of intermediaries have evolved in this process, who provides a variety of services to the farmers as well as to the consumers. In the times to come, newer forms of agricultural marketing such as the direct marketing, contract farming and e-marketing are likely to become popular.

III. Facilitating Functions

They are also called as Ancillary functions. Facilitating functions are activities that allow physical and exchange functions to take place smoothly. They are not mere passive functions in the process of marketing and are often recognized as the indispensable ingredients of modern agricultural marketing system. They are often considered as the grease that makes the wheels of marketing process turn and ensure smooth performance of the physical and exchange functions.

A. Standardization and Grading

Promotion of standardization and grading of agricultural commodities is an important aspect of agricultural marketing. This function seems to enjoy more importance in the modern days of agri-business keeping in view of the following factors:

- ✰ Increasing quality consciousness of the consumers across the countries
- ✰ Since, the agricultural commodities are heterogeneous, it is very essential to grade these commodities as per standards to command better price both in domestic and international markets
- ✰ Good quality produce meets the end user (consumer) specifications with respect to his consumption and safety requirements

- Establishes a rational relationship between the quality of a produce and its price
- Uniform standards enable the country to compete in world trade.
- Establishes a common trade language and avoids the need for physical checking and handling at many points

These two terms *viz.*, standardization and grading are quite often misunderstood and misused because of similar and identical meaning. But, these two terms differ in the following aspects:

- Standardization refers to the establishment of certain quality specifications based upon physical properties and intrinsic qualities of a product (Table 5.74).
- The function of sorting out heterogeneous produce into different homogeneous lots based on quality specifications (standards) laid down earlier refers to grading.
- Standardization signifies that, a produce has certain specific quality features or standards and these standards forms the base in dividing the entire produce into different grades. Hence, it is remarked, *'established standards grades'*.
- Chronologically standards are fixed first and grading follows. Thus, *'where standardization ends, grading begins'*. Thus, we can infer that, grading is the sub-function of standardization or grading follows standardization or standardization precedes grading.

Table 5.74: Grade specifications (Standards) followed across different agricultural commodities.

Crop/Produce	*Grade Specifications (Standards)*
Paddy, Wheat	Foreign matter, Admixture, Damaged grains, Immatured grains, Weeviled grains, Moisture, Size, Shape, Colour, Brokens etc.
Maize	Foreign matter, Other varieties grains, Admixture, Damaged, Immatured grains, Weeviled grains, Shrivelled grains, Moisture, Uric acid and aflatoxin contents, Rodent hair and excreta, size, shape, colour etc.
Sorghum	Moisture, Organic foreign matter, Inorganic foreign matter, Other edible grains, damaged grains, Immatured grains, Weeviled grains, Shrivelled grains, Ergot affected grains, size, shape, colour etc.
Pulses (Red gram, Gram)	Moisture, Organic foreign matter, Inorganic foreign matter, Other edible grains, damaged grains, Immatured grains, Weeviled grains, Shrivelled grains, Ergot affected grains, size, shape, colour, toxic or noxious seeds, Uric acid and aflatoxin contents, etc.
Groundnut	Foreign matter, Damaged pods, Shrivelled pods, Immature pods, Pods of other varieties, Shelling percentage, Shape of the pods, Moulds of insects, Moisture etc.
Castor seeds, Sesame seeds, Safflower seeds and Sunflower seeds	Foreign matter, Damaged seeds, Discoloured seeds, Shrivelled seeds, Immature seeds, Seeds of other varieties, Broken seeds, Moisture content, Weeviled seeds, Shape, Size, Colour etc.
Soya bean	Oil content, Acid value of oil, Moisture content, Damaged, discoloured, insect infested beans, Immature shrivelled beans, Splits, brokens, cracked beans, Inorganic foreign matter, Organic foreign matter, Size, Shape, Colour, Pesticides residues etc.
Cotton	Colour of cotton fibres, fibre length, length uniformity, fibre strength etc.
Mangoes	Freshness, Foreign matter, Bruising, Damage due to low or high temperature, Damages caused by pests, Ripeness, Residue levels of heavy metals and Pesticides, Weight etc.
Bananas	Foreign matter, Bruisings, Pest incidence, Malformation, Damages caused by low temperature, Colour of crown, Residue levels of heavy metals and Pesticides

It is essential to note that, standards have a broader significance than grades. For example, standards may be legally prescribed to all the products of a particular manufacturing industry. This is because, standardization in the case of manufactured (industrial) goods is a function of production, as in the manufacturing industries, not only the finished products are standardized, but also raw materials, processes and performances are also standardized. So, once the standard is set up, there will arise no difficulty to maintain it. However, this is not the case with agricultural production, because of lack of uniformity of standards in the variety of natural products and they vary from one farm to another and from one season to another. Further, the production programme is largely based on natural factors, which is totally uncertain and not in the control of the farmer. So, with in the same production programme, the quality of produce (say, grain) differs. Thus, the produce can never be standardized uniformly even across the same varieties of a crop. But, based on certain quality specifications (standards) like size, shape, colour, moisture content, brix-acid ratio etc., the produce can be graded into different grades such that, homogeneity exists within the grade and heterogeneity prevails across the grades.

Considering the above special features, the Government of India had recognized the need to introduce the standardization of agricultural produce, which would enable the farmers to derive the benefits of grading in terms of fair practices according to the prescribed standards and enacted the Agricultural Produce (Grading and Marking) Act in 1937. This Act empowers the Central Government to prescribe grade standards indicating the quality of produce and specify grade designation marks to represent particular grades or qualities. The grade standards prescribed under this Act are based on both physical and chemical characteristics and are formulated after analyzing representative samples of each commodity collected from different regions and different seasons. Besides these, international standards and special requirements of overseas consumers are also taken into account, while formulating these standards for the commodities, which are meant for export. The grade standards are reviewed and amended from time to time in the light of the shift of the pattern of production, trade and changes in the consumer's preferences. Before going into the details of this function, let us define the standardization and grading of agricultural produce.

1. Standardization

In the marketing process, the determination of the shape or form or quality in which a product is to be presented for sale is important. Some of the most vital marketing problems centre around this activity *i.e.*, establishing and maintaining standards, and providing conformity to them. Further, the buyers also attach certain values to standard products in terms of its quality, value and its uses. Thus, standard provides an ethical basis for marketing transactions and also forms a basis for comparison with identical products. In other words, standards convey the idea about uniformity and identity in respect of quality specifications or quantity to the buyers.

Standardization refers to the process of setting up basic measures or standards to which the products must conform and taking steps to ensure that the products actually produced adhere to these standards. Standards reflect desirable features of a product in terms of its design, weight, colour, size, shape, moisture content, texture etc. Standardization means that, products are of a specified and uniform quality. Thus, standardization is concerned with the establishment and maintenance of uniform measurements of produce quality and/or quantity. We know, the produce brought to the market is very often contaminated with dust, stones and other foreign matter added either deliberately or by accident. Sometimes, the produce is immature or not properly dried or contains shrivelled grains or damaged and rotten material. Such a produce fetches a lower price to the farmers. Further, quality differences in agricultural produce arises for several reasons like due to differences in production methods, quality of inputs used, technological innovations, managerial ability of the farmers etc. In addition, a buyer's assessment of a product's quality is often an expression of personal preference. For example, some buyers prefer golden yellow colour gur (jaggery), while others prefer dark colour. Similarly, white sugar is considered 'superior' to yellow sugar. So, care should be exercised, while assembling the produce of different farmers so that, the good material is not mixed with the inferior material brought in by some farmers and while mixing produce of different features. It is, therefore, imperative to sort the produce into different lots based on certain specifications laid down and this process is called Standardization. It also refers to the process of fixing certain norms for the product based on the which, the produce can be classified into different grades. These norms are established by customs or tradition or by a certain authority. So, standardization ensures the quality specifications of the particular grade uniform among buyers and sellers over space and time. According to National Commission on Agriculture (NCA) 'standardization' is defined as, *'the determination of basic limits or grades in the form of specification to which manufactured goods must conform and a class into which the product of agriculture and the extra active industries may be sorted is known as standardization.'* Agricultural goods are standardized on the basis of different factors, based on which they are classified into different grades. The following are some of the factors on the basis of which standards are set:

- ✰ On the basis of quantity, weight and measures.
- ✰ On the basis of size and shape.
- ✰ On the basis of colour such as apples.
- ✰ On the basis of quality such as food grains and cotton.
- ✰ On the basis of physico-chemical properties such as edible oils.

2. Types of Standards

We know, standards convey the idea of uniformity and identity in respect of quality and/or quantity of the produce. Thus, standard is *'a measure that is generally accepted as having a fixed value'*. The standards of produce can be classified into:

a. Positive Standards and Negative Standards

i. ***Positive standards***: The standards which express positively the qualities present in the product. For example, high levels of butylated hydroxyanisole (BHA) and butylated hydroxytoluene (BHT) in vegetable oils is good to help artificial antioxidants, which prevent food from spoiling or oxidizing too quickly.

ii. ***Negative standards:*** The standards which express positively the qualities absent in the product. For example, the absence of aflatoxins content in the groundnut indicates its superior grade.

b. General Standards and Private Standards

i. ***General standards:*** These standards relate to all similar commodities, no matter who produces them. For example, Nicotine content (in cigarettes) is a general standard.

ii. ***Private standards:*** These are the standards seen in the products produced by individual manufacturers of the products. For example, Crocodile T-Shirts, alcohol content in beverages etc., are private standards.

c. Mandatory Standards and Voluntary Standards

Public concern over food safety has increased dramatically in the last decade. Both mandatory and voluntary standards address food safety from alleviating problems caused by contaminated food to addressing the safety of equipment used to process food and even the expertise of people that work with food.

i. ***Mandatory Standards***: These are the standards or quality specifications laid down by the Government that must be maintained in the food, so as to eliminate unnecessary trade barriers. Even for facilitating smooth international trade, these mandatory standards are embodied in trade treaties and agreements and in National Law. In the international trading system, mandatory standards generally apply to the characteristics of the product, not the process that produced it. These mandatory standards protect the consumer by ensuring a label accurately reflects what is inside. For example, packaged food products sold in India are required to be labeled with a mandatory mark in order to be distinguished between vegetarian and non-vegetarian. The symbol is in effect since 2006 following the *Food Safety and Standards (Packaging and Labeling) Act of 2006*, and got a mandatory status after framing of the respective regulations (*Food Safety and Standards (Packaging and Labeling) Regulation* in 2011. According to the law, vegetarian food (biscuits, noodles, chocolates etc) should be identified by a green symbol and non-vegetarian food (KFC (Kentucky Fried Chicken – www.kfc.com) is a global chain of fried chicken fast food) with a brown symbol (Figure 5.35). As per the present specifications, egg-products would be considered as non-vegetarian, while milk and milk-products are vegetarian and are marked with the green symbol. But, there is a common misconception that, the brown dot denotes egg-products and red dot denotes meat-products. But, there is in fact no such provision in the approved standard. These mandatory standards are related to

   - ***Standards of identity:*** These regulations specify the type and amounts of ingredients that certain foods must contain, if they are to be called by a particular name on the food label. For example, for some foods, there is a maximum or minimum concentration of a certain component that they must contain. In peanut butter, 80 per cent of the fat is in unsaturated form. This higher proportion help lower the risk of heart disease by lowering LDL (bad) cholesterol.
   - ***Standards of quality:*** Standards of quality have been defined for certain foods (*e.g.*, canned fruits and vegetables) to set minimum requirements on the colour, tenderness, mass and freedom from defects.
   - ***Standards of Fill-of-Container:*** The fill-of-container standards define how full the container must be and how this is measured.

ii. ***Voluntary Standards:*** These standards are not legally required to export the product to a particular geographic market. Voluntary standards are considered consensus standards, since they are developed using a process that allows participation by all interested stakeholders including representatives of farmers, processors, users, consumers and Government agencies. Voluntary standards, with which compliance is not mandatory, may also pose problems in international trade, if they differ widely from country to country. These standards can apply to the characteristics of the product, its production process or both. Voluntary standards are largely organized by a number of different bodies, both domestic (DMI, Bureau of Indian Standards (BIS)) and international (ISO, Codex). These standards cover agricultural products, environmentally friendly products and organic food products. A number of foods, including meat, dairy products and eggs, are graded according to their quality *i.e.*, from standard to excellent according to their origin, tenderness, juiciness, flavor and appearance. Specification of the grade of a food product on the label is voluntary, but many food manufacturers opt to do this because, superior grade products can be sold for a higher price.

3. Need for Standardization

Standardization of food-products has gained importance in the past few years because, quality guarantee has become a necessary component of the agri-business. Hence, any firm, which is getting into the processing industry or is already into it, needs to know about the various techniques that are meant to assure the quality of products as per the laid down norms and procedures. The need for standardization of food-products assumes crucial significance in the context of the following factors:

- ☆ Consumers have become more demanding and knowledgeable. They are more concerned about the ingredients of the food products and hence, maintaining optimum quality standards in food items has become very necessary.
- ☆ Since, people are more concerned about the environment protection, employing environmental friendly methods in the food industry has become necessary.
- ☆ There is more demand for organic foods because, people have realized the ill effects of inorganic foods like damage to the environment and health. Hence, quality standards are to be maintained such that, the chemicals are within the specified limits in any food product.
- ☆ With the development in technology, various methods for food processing and other food related activities have emerged. These methods provide safer and higher quality food to the consumers. Today, all organizations related to food processing and marketing have to ensure good quality systems to compete in the market.
- ☆ Food being a critical element of life, the regulatory agencies have also put in place stringent requirements for its safety and quality.

Realizing the importance of 'standards' in ensuring food quality and safety to the consumers, October 14th every year is celebrated as 'World Standards Day'. This day honours the efforts of the thousands of experts, who develop voluntary standards within standards development organizations such as the International Electro-technical Commission (IEC), the ISO and the International Telecommunication Union (ITU). The aim of the World Standards Day is to raise awareness among regulators, industry and consumers regarding the importance of standardization to the global economy. October 14th was specifically chosen to mark the date, in 1946, when delegates from 25 countries first gathered in London and decided to create an international organization focused on facilitating standardization. Even though ISO was formed one year later, it wasn't until 1970 that the first World Standards Day was celebrated.

The various standards followed for grading the produce of different agricultural produce are shown earlier through Table 5.74. It is essential to note that, the grade specifications (standards) vary from produce to produce and these may include physical and physico-chemical properties. After fixing the standards for the produce and based on these criteria, the produce is graded into different 'Grades', highlighting the differences in quality of produce ranging from superior quality (grade) to inferior quality (grade). Thus, before grading the produce, standardization is to be done based on various quality specifications (both physical and physico-chemical properties). This infers that, grading is the sub-function of standardization and grading follows standardization.

4. Organizations Involved in Formulating Grading Standards in India

A number of organizations in our country are engaged in the formulation of grade standards and monitoring their quality. As discussed above, mandatory standards are being implemented through a number of legislative measures. Under the voluntary system, there are two main organizations *i.e.*, DMI through Agricultural Produce (Grading and Marking) Act, 1937 (popularly known as 'AGMARK') and the other is BIS through BIS Act 1986, popularly known as 'Indian Standards Institute (ISI)'. Apart from these two organizations, under the present concept of self-quality assurance certification system, a number of other agencies have been authorized to certify the quality of a product.

International standards are now being recognized as means to overcome technical barriers that arise between different countries. This led the member countries to align their grade standards in tune with the international standards as well as promulgation of General Agreement of Tariffs and Trade (GATT) standards code, which recognizes international standards as suitable code standards for harmonizing among National Rules and Regulation. The GATT standard code has recognized international standard of the ISO and International Electro-Technical Commission (IEC), as suitable means of achieving necessary harmonization amongst national standards and technical standards. The increased acceptability of international standards in the world market has prompted many countries to the GATT standards codes to participate effectively in International Standardization Process so that, potential obstacles in international trade could be eliminated. Under the present quality concept system, it is not only necessary to standardize a product, but standardization of inspection techniques, test procedures, sampling technique, packaging materials and other process are necessary for successful marketing of a product up to the satisfaction of consumer acceptability. The process of standardization has in fact to start right from the production level up to the entire processing till the product is marketed and consumed. The mechanism of fixing standards and later followed by grading is a continuous process and the same needs review and changes depending upon the technological changes and requirements. But, it has been observed that, such process through R&D activities has not been given much importance in our country and in most of the cases, standards are being adopted depending upon minimum requirements without a proper scientific study.

5. Benefits of Standardization

Today, many consumers take little time to think about where or how their food was prepared, especially if it came from halfway across the world. Food safety is one of the few concerns on consumers' minds, when they are purchasing the products. They buy food (processed and packaged) made in places they have never been to and in factories they have probably never seen. These food items are manufactured under guidelines and regulations and are supposed to be prepared according to international and local regulations and laws. These regulations and guidelines are meant to standardize the products *i.e.*, they dictate how food is to be processed and packaged to ensure that consumers get the highest quality. Thus, the term 'food standard' is the required or agreed level of quality of food, while 'standardization' is defined as the 'process of conformity to standards'. The required or agreed level of quality and safety of food are enforced and regulated world-wide by four main organizations, namely, CAC, the WTO, the WHO and the FAO. This establishment of uniform quality and grading standards for the food products can be beneficial in the following ways:

- ✰ With standardized products, there is homogeneity and consistency in quality. Hence, a consumer can buy a product without much hesitation and risk, as the quality of product will not change over a period of time, thus leading to better goodwill.
- ✰ When products are standardized and classified into various types, one can use differential pricing for different grades of the product. This assures better returns to both buyer as well as seller. Ex. Prices of apples, rice, wheat etc., differ across the standards laid down.

- As there is uniformity in size, form, quality and utility of all products of a firm, the customers rely on these products. So, sale of standardized products increases and the consumers develop loyalty towards the firm.
- The standards of products are determined keeping in mind the habits, tastes and nature of consumers thereby, the risk of selling of standard products is reduced.
- When there is large-scale production of standard goods, it results in many types of saving in production, distribution, advertising and sales promotion. So, the processor is in a position to produce best quality products at minimum cost. This further attracts a number of consumers.
- Finance can be obtained easily for standard goods, as financial institutions prefer giving loans to these enterprises, as these goods will have assured market.
- There is no need for inspecting the quality of a product, as the form, size, quality and utility of all the products in a lot are uniform in all respects and the consumers are convinced with it. Thus, trade can be ensured even without physical examination of product.
- Standardization helps to reduce inventory items and forms a base for the inventory analysis.
- It helps in evolving better means of communications about a product.
- The specifications of items can be more clearly spelt out, making quality control.
- In a developing economy like India, where the need is to promote exports insistence on standards helps in creating assured international market.
- By using National standards, it is easier to locate sources of supplies of products. It could also be used in advertising the products.
- Price quotations are more meaningful
- The sale of commodities by sample or description becomes possible
- Small lots of commodities, produced by a large number of small farmers can be assembled into economic loads, if these supplies are similar in grade or quality
- Consumers get a range of different varieties of graded produce and this enable him to choose the desired quality of product in accordance to his purchasing power
- Standardization simplifies buying and selling and reduces marketing costs
- Standards permeate all business activities and even the day-to-day life of the man. They play an important role in the agro-produce and agri-input manufacturing industries and in the sale of their products in national and international markets. In the modern business environment, agro-produce and agri-input manufacturing enterprises do not themselves produce all the parts and components they need in taking up the manufacturing activity, but buy these from ancillary industries, often situated in far-off countries. The standardization of parts and components gives enterprises a wider choice of lower-cost supplies. It also enables them to maintain inventories at comparatively low levels and gives them the flexibility to use substitute components on the assembly line. Standards, thus, help industries to cut costs and improve productive efficiency.
- Standards are indispensable for the international trading of agro-products, as they convey consistent and understandable information to the buyer. For example, if a foreign buyer wishes to purchase a product from India and if he knows the standards of the same, he is able to assess its quality. Standards thus, help reduce disputes over specifications and the quality of products (and services) exported and imported.

- Standards are also used by the Government to promote its social goals. Government agencies at the national, state and local levels lay down various regulatory standards to protect the health and ensure the safety of the population and to conserve the environment. Such regulations cover both product characteristics and the materials and processes used in producing them.
- When the standards of the produce differ considerably across the countries, they can become barriers to trade. In such case, buyers are generally reluctant to purchase (imported) goods that differ in their standards from those produced in their own country. So, the exporting countries have to adjust their production processes and produce the products with standards, as desired by the importing countries. In such case, the producing (exporting) countries have to revamp their production strategies and install new machinery and this prevents them from taking advantages of economies of scale. Further, regulatory authorities of importing countries call for stringent product testing at their ports and it will affect the reputation of the exporting countries in the international trade. So, the desired solution to these problems lies in harmonizing standards at the international level and in developing guidelines for determining conformity to standards.
- Graded produce is likely to be more equitably priced than non-standardized produce. This will bring stability not only to market prices, but also to the quality offered.
- Prices quoted against a recognized grade assist farmers and traders to assess and market their products. This will also benefit the consumers in more stable prices with assured quality.
- Greater conformity in quality through standardization will provide the processors with the consistency necessary for optimum performance.
- The sanitary hazards associated with the inter-country movement of produce can be reduced, if clear defined standards are enforced.
- Standards are 'vendor-neutral'. That is, if the buyers are aware of the standards of the product they wish to purchase, they can have the same from any vendor. Thus, fixation of standards for the produce creates an open language and this greatly checks the malpractices in the marketing.
- Advertising backed by standards is more effective. It, therefore, partly helps in the demand creation function also.
- Standardization helps in the efficient performance of the various marketing functions, particularly in buying and selling.
- Standardization determines grades of agricultural products.
- Standardization helps the farmers to set higher prices.
- Standardization assures the consumers of the specific quality of the product.
- Standardization is a necessary process for ensuring effectiveness in any food product. Without it, quality cannot be guaranteed, and as a buyer, purchasing a non-standardized food product is a gambling process. In the absence of standards, the buyers have to purchase the products of different qualities from different locations at different times and they expect a different experience with each and every purchase. However, by fixing standards for the products, the buyers enjoy guarantee benefits, as expected from purchasing the products.

- It is mentioned in the ISO official site, *'Standards make an enormous and positive contribution to most aspects of our lives'* highlighting the importance of standards. This implies that, the significance of standards may be better appreciated when considering what would happen in their absence. When a product is made according to pre-defined standards and meets customer expectations, it is often taken for granted by the buyers in purchasing process. However, in an environment without standards, people would very likely voice concerns about poor quality and unsafe products. Whether a business produces products or provides services, when it meets standards relevant to its industry, it ensures that, positive characteristics such as quality, durability, efficiency, safety and environmental friendliness are reinforced.
- Standards benefit businesses because, they can focus their resources on producing goods and services that meet the quality requirements of potential buyers. These standardized products also enjoy comparative advantage in the international market and this help the country to earn significant amount of foreign exchange. Standardization at all levels *viz.*, raw materials, production process, processing and supplying the products enable the firms of an industry to enjoy cost effectiveness and price competitiveness.
- For Governments, standards provide tools to assess and evaluate conformity, and provide a legitimate base for health and safety legislation. Having standards in place also facilitates fairer trade between regions by ensuring that, both the importer side and exporter side of foreign trade transactions know about the relevant set of standards. For example, for Governments in developing countries, ISO standards give information about expected attributes of products and services for export markets. This information gives a basis for Governments to allocate resources accordingly, and produce goods and services that will enjoy comparative advantage in the international market.
- Standards benefit end consumers by safeguarding their interests and by ensuring that the products and services they purchase are safe and reliable. The standards also help to secure environmental protection through producing the products with accepted levels of technical ingredients. This further drives the production systems on sustained basis.
- Standards guarantee that, processed foods sold in the market are prepared using modern and healthy methods and processes. They ensure that, micro-organisms, such as bacteria and foreign substances, such as dust, do not contaminate food. They also screen and prevent sub-standard food products from entering the country.
- Standardization ensures that, sellers label their products properly and inform consumers about vital information such as, the ingredients used and their nutritional value. This prevents firms from incorrectly labeling* food products and making sure that an important right of consumers is protected *i.e.*, the right to information. Also, sellers are compelled to have guidelines that dictate how food should be prepared and processed and to make sure

---

* Labeling is a means for the identification and presentation of a container or package, a means for protecting the contents against pilfering, as they form a security-seal, gives information regarding batch number or quality control reference, may either be printed directly on the packaging or on the printed label itself by the packer applying the label. Following details on each package clearly and indelibly marked:

1: The grade designation mark approved by the Agriculture Marketing Advisor (AMA) or an officer authorized by him in this behalf; 2: Name of the commodity and variety; 3: Grade designation; 4: Size code (if prescribed), Lot/batch/code number; 5: Country of origin; 6: Net weight/Number of units; 7: Name and address of the packer/exporter; 8: Best before date (where applicable); 9: Storage condition, if any; 10: Date of packing.

they meet international food standards. One also needs to consider that, poor standards are just as detrimental as having no standards at all. Poor standards encourage the production of low quality and unsafe food utilizing unhealthy food preparation practices that may cause sickness or death of the consumers.

- Standardization of food also involves the acceptable and recommended way of processing or preparing specific foods. Milk, for example, is to be pasteurized to kill pathogens that it may contain and which might be harmful, if not deadly, to consumers. Standardization of food may also recommend or ban the use of certain chemicals or methods in preparing food, as they may be harmful to the health of consumers. Some of these chemicals may include pesticides and chemicals used before, during or after the processing of food products. These chemicals may leave residues, which may be ingested, touched or inhaled by consumers and may cause sickness or death.
- Food standardization protects consumers from misinformation from firms. Currently, all processed food requires proper labeling, which includes the expiration date and the ingredients. The expiration date informs a consumer, when a product is no longer safe or gives the date when the food item should no longer be consumed. The list of ingredients dictate, what has been used to make the food item and is very important to consumers avoiding specific foods or those who may have allergies. Firms are also required to ensure that, warning labels are present and that these labels provide allergy information, whenever there is need for them.
- Standardization of food is an integral aspect of food safety. This ensures that, millions of consumers world-wide are protected from food-borne disease and dangerous substances. In some countries, food standards are not effective, as they are either ignored or inefficient thus, causing cases of food-borne diseases and sometimes death. It is important that, the various non-governmental and governmental agencies world-wide, especially in India, enforce standardization of food and ensure that, their citizens are protected, whenever they consume either packaged or processed foods.

6. Grading

Due to world-wide commercialization of agricultural products, uniformity in grading and standardization is considered as one of the main tools for marketing of commodities in the internationally accepted markets. As discussed above, the objective of standardization is to provide uniform quality and is very important in case of agricultural produce (because of heterogeneity of produce). Once the standards and specifications are set, physical dividing the produce into different classes is termed as grading. Thus, *'grading is the process of sorting of heterogeneous lot of produce into different homogeneous lots based on various specified standards laid down'*. So, the quality standards for commodities are laid down first and then, the commodities are sorted out into different grades according to accepted standards. So, each homogeneous lot has substantially the same characteristics as far as quality of produce is concerned. Grading is mostly done in case of raw agricultural produce. Since, agricultural produce is bulky and heterogeneous and further assembled from various farmers at a common market place, it is essential to grade the produce into different classes or grades in conformity with certain accepted quality standards *viz.*, shape, size, form, weight, and other physical and technical characteristics, so as to benefit both the sellers and buyers in view of the fact that, the farmer gets price commensurate with the quality of produce produced by him and consumer gets a right quality product at his affordable prices. There is no need to grade industrial products, as they are produced according to pre-determined standards. The grade standards are helpful for orderly marketing of agricultural commodities and effective application of quality control laws. Unless the commodities are graded according to different standards, the buyers will have little or no assurance about their quality. Especially, traders in foreign markets cannot buy with confidence, if there are no standards by which they can specify the kind and quality of the goods to be shipped. Therefore, grade standards play a very useful role in facilitating international trade and provide a common language between the buyer and the seller across the markets. Further, grading also helps the farmer to get reliable prices for their products because, different prices may be fixed for different lots.

The parameters fixed for grading of a commodity depends upon the nature of the commodity and purpose for which it is to be used. Grading of a commodity can be done at farmer's level, processor's level, at wholesaler's level and at retailer's level depending upon its final use. Graded produce leads to not only fetching a better price, but also open a venue to the consumer to select a product of his choice. As the quality of agricultural produce is subject to considerable variation on account of differences in soil, climate, type of seeds, cultivation, production technology, processing technique, packaging, storage etc., the importance of adopting suitable criteria for grading of a commodity is, therefore, obvious. For assessing the quality of a produce, grading is done and for the purpose of grading, various physical or chemical parameters (standards) are laid down depending upon the commodity. The earlier concept of grading mostly relied upon the physical parameters depending upon the size, colour, foreign matter, maturity, moisture content, pest infestation etc., but under the present quality concept, importance is laid down on the safety parameters such as presence of pesticide residues, metallic contaminants, aflatoxins, mycotoxins etc. From this, we can infer that, standardization precede grading, standardization and grading are interdependent activities.

7. Types of Grading

In order to gain the confidence and establish a rational relationship between the quality of a produce and its price, it is necessary to devote some attention towards grading of produce based on certain accepted quality standards. Depending upon various criteria (Figure 5.34), grading can be broadly classified into different categories.

a. Based on Variability of Grade Specifications

According to this criterion, there are two types of grading *viz.*, Fixed grading and Variable grading.

i. Fixed Grading

It is also called as Mandatory grading. In this grading,

- ☆ The goods are sorted out according to the size, quality and other characteristics, which are of fixed standards. That is, these standards do not vary over time and space.
- ☆ It is mandatory for a person/grader to follow these grade standards, if he intends to sell the graded products.
- ☆ The AMA of DMI will fix these grade standards for a number of agricultural commodities and it is compulsory to grade the produce according to these grade specifications.
- ☆ The use of mandatory standards is compulsory for export of the agricultural commodities to various countries.
- ☆ Farmers and traders are not free to use their own standards to grade the products, especially for the products intended for export purpose. That is, the grade standards fixed by the AMA of DMI are considered mandatory for exporters.

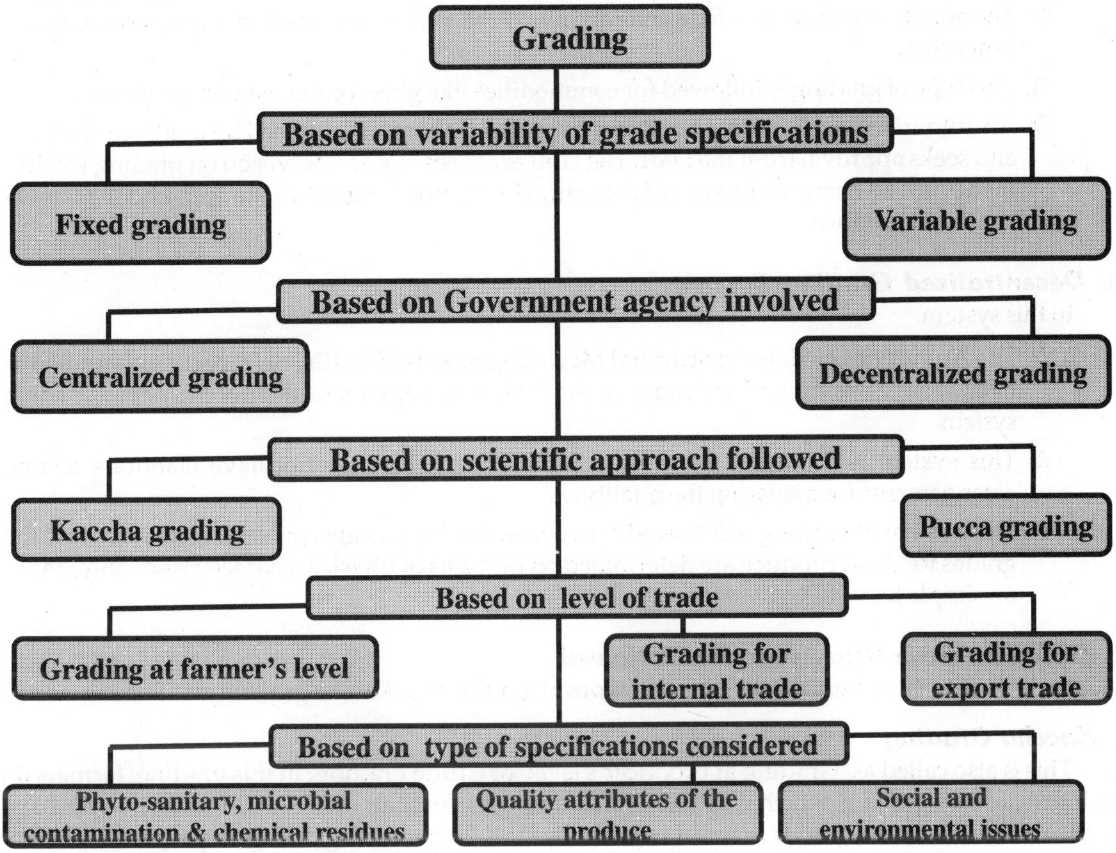

Figure 5.34: Different types of grading.

ii. Variable Grading

This grading is also known as Permissive grading. In this grading,

- ✩ The goods are graded according to the standards, which vary over time and space.
- ✩ The grading specifications vary according to the quality of the produce produced in that particular year.

In India, grading by this method is not permissible.

b. Based on Government Agency Involved

According to this criterion, there are two types of grading *viz.*, Centralized grading and Decentralized grading.

i. Centralized Grading System

In this system,

- ✩ The DMI will fix the grade specifications (standards) followed for each commodity.
- ✩ The grade specifications are fixed after conducting elaborate testing procedures meant for analyzing both the quality and purity of the commodity.

- ☆ The quality specifications for a commodity will be laid down based on its physico-chemical properties.
- ☆ This type of grading is followed for commodities like ghee, butter, vegetable oils etc.
- ☆ An authorized packer can set up his own grading laboratory manned by qualified chemists and seeks approval from the DMI. The DMI exercises close supervision on grading work of the approved chemists by way of periodical inspection* of grading stations and the quality of graded produce.

ii. Decentralized Grading System

In this system,

- ☆ The Authorities of State Agricultural Marketing Boards (SAMBs) or Departments under the overall supervision and guidance of the DMI will implement the decentralized grading system.
- ☆ This system is followed for those commodities, which do not have elaborate testing arrangements for assessing the quality.
- ☆ This system of grading is followed for vegetables, fruits, eggs, pulses, cereals etc., and the grades for these produce are determined on the basis of physical characteristics only, based on simple tests.

c. Based on Scientific Approach Followed

According to this criterion, there are two types of grading *viz.*, Kaccha grading and Pucca grading.

i. Kaccha Grading

This is also called as 'Grading at producer's level' or Crude grading. In this grading, farmers on their own also do grading, which is actually categorizing as per their own thinking. They do not have an idea of standards and grades as recommended by DMI. Farmers do grading only from the point of view of getting fair prices for their produce. They may avail the grading services offered in the markets on free basis. Marketing agents like commission agents and brokers also help the farmers in grading their agricultural produce and their opinion is given importance in price fixing and negotiations. But, this grading is not at scientific lines.

---

* Inspection in relation to a commodity means 'the process of determining whether a commodity in a particular lot complies with the standard specifications applicable to it or any other specification stipulated in the trade contract'. Either whole lot or selected samples or sample, which represents the whole lot may be inspected in order to observe whether the lots offered for grading are in sound and in merchandisable condition, free from moulds, weevils, obnoxious smell and excessive pesticidal residues. In case any part of the consignment is found sub-standard, it should be segregated and sampled separately and if any lot(s) contain pesticidal residues beyond the prescribed limits or is polished or coloured with prohibited dyes, such lot should also be sampled separately and the representative sample sent to the appropriate laboratory for chemical examination. The grading of such lots should not be carried out until the results are received. The commercial grading is subjective by nature and done generally by inspection method. As such, it not only requires technical expertise but adequate experience also to judge the commercial quality and to allot the grade on the spot. Therefore, before taking up the grading work, the standard samples must be prepared in respect of the important commodities of that market area, which will help in comparing the quality of the produce offered for grading. The grade and other details may be recorded in a register before examining next lot. On the basis of comprehensive inspection, the grades may be announced/allotted, as early as possible and a small plate indicating the grade or any appropriate indicator may be used to denote the grade.

ii. Pucca Grading

This is also called as 'Grading at trader's level'. In this grading, the traders in the market grade their produce and wish to secure a profitable price commensurate with the quality of produce. They traders may avail the grading services offered in the market on payment basis. Several regulated markets and warehouses are manned with grading personnel with the central assistance and the traders may seek their help to grade their produce, as they are more aware about the benefits of grading compared to farmers. So, pucca grading is somewhat more scientific than kaccha grading.

d. Based on Level of Trade

According to this criterion, there are three types of grading *viz.*, Grading at farmers' level, Grading for internal trade and Grading for export trade.

i. Grading at Farmers' Level

For securing adequate return to the farmers, a scheme of Grading at the Producers' Level was introduced by establishing grading units in Regulated markets and Marketing Co-operatives. The main objective of the scheme is to subject the produce to simple tests and assign a grade before it is offered for sale. Grading helps the farmer to get price commensurate with the quality of his produce. Cereals, pulses, oilseeds, spices, fruits and vegetables, fibres, arecanut, coconut and tobacco are some of the important commodities graded under this programme. The scheme is being implemented by all the States and UTs.

ii. Grading for Internal Trade

Grading of agricultural commodities for internal trade is voluntary in nature and is carried out under the provisions of Agricultural Produce (Grading and Marking) Act, 1937. However, as per provision in PFA Rules, 1955 certification of blended edible vegetable oils and fat spread is compulsory under Agmark even for internal trade. The commodities graded under 'AGMARK' label for internal trade are classified into two groups; *viz.* decentralized commodities and centralized commodities. As discussed earlier, Decentralized commodities include those items, which do not require elaborate testing facilities. Assessment of their purity and determination of grade is mostly done on the basis of physical factors or relatively simple tests. Decentralized commodities include wheat, rice and other cereals, oilseeds, edible nuts, fibre crops, fruits and vegetables etc. This programme is being implemented through the Marketing Departments of the concerned State Governments under the guidance of the DMI. Certain powers such as, renewal of Certificate of Authorization, drawl of check samples etc., under the provisions of General Grading and Marking* Rules, 1988 have also been delegated to Officers of States and UTs for grading of decentralized commodities. The commodities, which require elaborate testing arrangements for assessment of quality and determination of grades, are classified as Centralized commodities. These include ghee, butter, vegetable oils, oil cakes, powdered spices, honey, wheat atta etc. These require elaborate testing for chemical parameters. The commodities are tested by the authorized packers in primary grading laboratories for various prescribed parameters and grades are assigned as per the analytical results. Packers/manufacturers, who cannot establish their own laboratories because of cost involved and recurring expenditure, get themselves attached with State Grading Laboratories (SGLs)/co-operative/commercial laboratories.

---

* Marking is the action and result of stamping, inscribing, printing, labeling etc., marks, symbols, numeral letters etc. upon a product itself or on its package, for the purpose of identifying the product and its origin and giving its basic characteristic its intended use etc.

iii. Grading for Export Trade

The Export (Quality Control and Inspection) Act, 1963 is administered by the Ministry of Commerce. This Act empowers the Government to

- ☆ Notify commodities, which will be subject to quality control and/or inspection prior to export
- ☆ Establish standards of quality for such notified commodities, and
- ☆ Specify the type of quality control and/or inspection to be applied to such commodities.

Fish and fishery products, honey, egg products, milk and milk products, black pepper for USA, Basmati rice for EU have been notified for compulsory pre-shipment inspection and certification. DMI has attended to certification of grapes from different centers in Maharashtra, Andhra Pradesh and Karnataka. All the exported agricultural produce from the country must bear a label called 'AGMARK', which signifies both purity and quality of the produce.

e. Based on Type of Specifications Considered

According to this criterion, grades and standards can be grouped into three categories *viz.*, Grades and standards that relate to phyto-sanitary, microbial contamination and chemical residues, grades and standards that relate to quality attributes of the produce and grades and standards that relate to social and environmental issues.

i. ***Grades and standards that relate to phyto-sanitary, microbial contamination, and chemical residues:*** These standards are mandatory for all exporters, set by regulatory agencies in importing countries.

ii. ***Grades and standards that relate to quality attributes of the produce:*** The traders set grades and standards that relate to quality attributes of the produce such as colour, size, taste, appearance, texture, uniformity and packaging, as they significantly influence the consumers. These grades and standards are obligatory for exporters, but vary from place to place and from time to time depending on consumer preferences and trading norms.

iii. ***Grades and standards that relate to social and environmental issues:*** Grades and standards that relate to social and environmental issues are newly emerging market requirements that are more flexible and still partly voluntary. The strictness on conformity to these standards varies from market to market. These standards are usually embodied in codes of practice, which set the rules, regulations and practices that producers and exporters must meet. The main objective of these standards is to ensure humane treatment of employees, insure safety (of the workers and consumers), reduced use of agrochemicals and the conservation of the environment. Current trends indicate that, the importance of these standards will increase and eventually play a very significant role in the future. Already UK and other European retailers are asking their suppliers to comply with these standards.

8. Importance of Grading

The following are the notable advantages of grading function:

- ☆ This function simplifies buying and selling activities and thereby, reduces marketing costs by enabling buyers to specify precisely what they want and suppliers to communicate what they are able and willing to supply with respect to both quantity and quality of produce.
- ☆ In the absence of grades and standards, trade either becomes more expensive to conduct or impossible altogether. This is because, when good quality produce gets mixed with inferior

quality, it adversely influences the reputation of the firm in the market. So, the firm has to leave the market, as its product is no longer transacted in the market at profitable prices.

- Grading is required at many stages of marketing of both raw materials and processed products. So, it also clarifies the actual state of technology that is essential for the production and efficient marketing of products. Further, research goals can be set more rationally keeping in view the dynamic preferences and tastes of the consumers and this call for demand driven research in producing and marketing the products in tune with the quality standards of products preferred by the consumers.
- Grading and standardization of products in a country and their harmonization with corresponding internationally accepted technical standards essentially serve three important advantages *viz.*, increase in production efficiency, increase in standard of living and more protection to consumers through offering good quality and safe products.
- Grade standards ensure orderly marketing of commodities and effective application of quality control laws. Unless the commodities are produced to meet certain standards, foreign buyers will have little or no assurance that the commodities purchased would be of given identity and quality. Traders in foreign markets cannot buy with confidence, if there are no standards by which they can specify the kind and quality of the commodities to be shipped. Therefore, grade standards play a very useful role in facilitating international trade and provide a common language between the buyers and the sellers across the countries.
- Expansion of trade by a country not only depends upon the demand for its products, but is also based upon the confidence that the importer places on the quality of its supplies. This confidence of the importer can be promoted through an assurance provided by an effective export quality control system, which guarantees that, the product complies with the standards prescribed by the importing authority. If the shipments fail to meet the legislative and grade standards requirements of the importing country or with the quality requirements in the sale contract, it can cause heavy losses to exporter, affect the foreign exchange earnings of the nation and damage its prestige and economy. Rejection by one country can adversely affect trade with the other nations.
- Standardization and Grading facilitate buying and selling of products by sample or description. When products are of standardized quality, consumers do not insist on detailed inspection.
- Standardization and Grading enable the farmers and traders to direct the products of different qualities towards the market best suited to the consumers. The task of middlemen becomes easy because, they can communicate well the characteristics of standardized products to consumers.
- Transportation, storage and advertising expenses can be reduced by handling different grades or lots.
- Standardized goods enjoy a wider market.
- Standardization and Grading facilitate trading of goods on the commodity exchange. Hedging, speculation and price comparisons become easy.
- Standardization and Grading helps in raising finance because, standardized products enjoy a ready market and they are readily accepted as a collateral security for granting loans.
- Standardized products can be easily valued and their prices fluctuate less widely compared to raw produce.

- Grading of the produce before sale enables farmers to get better price for their produce.
- Grading helps the consumers to get standard quality produce at fair price.
- After grading, it is easier for the consumer to compare the prices of different qualities of a produce in the market.
- Disputes in the market can be solved amicably by grading and standardization of produce.
- Arbitrary fixation of price by middlemen is eliminated.
- Grading and standardization also helps in implementation of contract farming.
- Grading protects consumers from unfair trade practices by traders and also enables farmers to get price as per quality and grade.
- Grading works as a basis for fixing price and premium price as per quality.
- Consumers do not bother to enquire about the products graded, as they are very easy to be sold. Therefore, marketing of graded products becomes easy.
- Grading simplifies marketing, as due to standardisation and appropriate grading orders can be placed on the telephone or through letters or through internet and only the price remains to be determined. Thus, grading widens market, as buyers from distant places can send orders on the basis of grades. That is, the sale by description or sample is possible.
- Grading reduces marketing costs by avoiding the necessity for inspection of products and the avoidable cost of returning unsuitable products.
- Grading helps to protect products in transit from damage and reduces the cost of marketing by achieving economies in freight handling.
- Grading reduces the bulkiness of produce and thereby, lowers storage charges.
- The comparison of values of different qualities of a product in a single market and of differences in prices of the same grade in different markets can be made. Such comparisons help in keeping prices in different markets in line with each other.
- Grading helps to remove the element of speculation.
- Standardization and grading enable the firms to use the brand names effectively.
- Grading facilitates marketing based on the size, colour, qualities and other specifications of the produce well known to both the parties, and there is no need on the part of the seller to give any assurance about the quality of the produce.
- Grading makes it possible for the farmer:
  - to get a higher price for their produce
  - to get easy finance, when commodities are stored
  - to get the claims settled by the railways and insurance companies
  - to get storage place for the produce
  - to get market information
  - to improve the 'keeping' quality of the stored products by removing the inferior goods from the good lot
  - to facilitate futures trading in a commodity.
- Grading contributes to market competition and pricing efficiency. The product homogeneity resulting from grading can bring the market closer to perfect competition, encourages price competition among sellers, and reduces extraordinary profits.

- ☆ Storage of graded produce will have the following advantages:
  - ❖ easy to assess the value and provide pledge loan and loan against the warehousing receipt
  - ❖ quality disputes can be resolved
  - ❖ storage with special case is possible for different grades as per their requirements; and
  - ❖ helps in planning the plant protection measures.
- ☆ Procurement of graded produce will have the following advantages:
  - ❖ easy to assess the quality and decide the price
  - ❖ easy to isolate the lots
  - ❖ easy to plan its storage
  - ❖ easy to plan distribution
  - ❖ price disputes can be resolved easily
  - ❖ farmers get proper price and exploitation can be checked
  - ❖ farmers will get encouraged to produce the best
  - ❖ easy to match the demand and supply
- ☆ Distribution of graded produce ensure the following advantages:
  - ❖ easy supply to different consuming segments
  - ❖ consumers will be able to select their choice
  - ❖ farmers will be able to plan the production matching with the demand
  - ❖ e-trading is possible for well-defined quality of the produce
  - ❖ global distribution is largely based on the grade of the produce
  - ❖ consumer loyalty may be maintained.

9. Norms for Grading

They include,

- ☆ Grading should be on the basis of standards, which are largely used and easily understood
- ☆ Grading should be done on measurable quality factors.
- ☆ Grade standards should cover maximum percentage of the produce.
- ☆ Cost of grading operations should be less.
- ☆ Unless the standards are economically meaningful to the trade, they are not accepted.

10. Criticisms of Standardization and Grading

They include,

- ☆ We know, standards are mainly established for grading agricultural produce, so as to make the farmers more competitive both in domestic and international markets and to ensure protection for the consumers in terms of food quality and safety. But, these requirements of the two groups are not necessarily compatible.
- ☆ In improving the quality standards, the general experience is that, huge fixed costs will be incurred for a little improvement in quality.

- The techniques of standardization and grading have not been developed fully in India and hence, there is lack of standards and grades for many products on par with international standards.
- Lack of public consciousness about standardization and grading due to their low standard of living, illiteracy and dominance of rural population.
- Lack of scientific testing facilities in the development of standardization and grading of agricultural produce.
- In the case of agricultural produce, the quality cannot be maintained always. Thus, grading becomes meaningless.
- It is very often criticized that, standardization destroys the individual skill and ingenuity. In the case of processed products, the workers are treated only as a part of machinery and they cannot exercise any personal skill in the final product.
- They limit markets, as some of the buyers and consumers do not prefer to purchase low grade products.
- Some of the agricultural produce are highly perishable and even graded products thereby, by the time they reach the consumers, they get deteriorated in quality resulting in low prices.
- Grading of agriculture produce is more difficult, since the produce is bulky.
- The farmer does not get higher prices for graded products, since the consumers are more concerned about getting the products at lower prices and not very quality conscious in the Indian market.
- Grades often make a negative impression with consumers (especially for lower grades), although the difference may be only marginal.
- Consumers' preferences vary based on number of factors and are therefore, it is very difficult to meet their individual requirements exclusively based on standardization and grading of products.
- Farmers, especially small and marginal farmers are in immediate requirements of cash after harvesting the produce and they can't delay the produce sale for grading work.
- It is felt that, in case of grains, grading is not required, as the buyer of grains goes by feel and rates. Perhaps grading is required only for exports.

Prima facie, these disadvantages and criticisms may seem to be true, but are not deep rooted. These arguments are baseless and are due to wrong conception of the idea. The earlier discussion on the benefits of standardization and grading of agricultural produce indicated that, standardized and graded products actually enjoy a wide reputation and are capable of capturing new markets. The problem is only to maintain the standards. Taking into consideration the various advantages offered by this function, one cannot simply ignore it. But, it is important to note that, standardization is effective only, when it is legalized. This is because, where customs play an important part in the production of a commodity, standardization has no effect. Ignorance on the part of the consumer is another obstacle. In India, the agricultural products have been and more are being brought under standardization with the establishment of DMI. An important legislation *i.e.*, The Agricultural Produce (Grading and Marking) Act was passed in 1937, to promote standardization and grading of agricultural produce. This Act prescribes the qualities of certain agricultural produce and to market them with a standard name, 'AGMARK' (In the word 'AGMARK', 'AG' denoted agriculture and 'MARK' stands amended in 1943 to include more products. 'AGMARK' is a standardized grade name for agricultural products and is now attached to 205 notified commodities (Table 5.75). The other enactments in this area include *viz.*, Food Products Order, 1946, Vegetable Oil Products Order, 1947, ISI (Certification Mark) Act 1952, PFA Act, 1954, The Fruit Products Order 1956, Solvent Extracted Oil, De-oiled Meal and Edible oil (Control) Order 1967, The Meat Food Products Order, 1973, Vegetable Oil Products (Standard of quality) Order 1975 etc.

Table 5.75: Group-wise list of the commodities for which AGMARK grade standards have been formulated under the Agricultural Produce (Grading and Marking) Act, 1937 (as on 31st July, 2011).

Name of the Group	*No. of Commodities Notified*
Food grain and allied products	30
Fruits and Vegetables	44
Spices and Condiments	26
Edible Nuts	8
Oil Seeds	15
Vegetable Oils and Fats	19
Oil cakes	8
Essential oils	8
Fibre crops	5
Livestock, Dairy and Poultry products	10
Other products	32
Total	205

11. Suggestions for Popularizing Standardization and Grading

Following are the important recommendations of NCA for popularizing standardization and grading function:

- ✰ Standardization and grading have to cover all types of agricultural commodities and be made compulsory for trading in agricultural commodities at all levels *i.e.*, for local inter-state and export trade.
- ✰ Samples of graded products should be exhibited in all markets for the benefits of traders and consumers.
- ✰ Awareness should be created about the benefits of using AGMARK products among consumers.
- ✰ Highlight the benefits of selling graded products to the producers.
- ✰ There should be adequate training for graders and people with agricultural knowledge should be encouraged to become qualified graders. Only this way, there is a possibility of making the grading system more efficient.
- ✰ There should be only one agency to formulate grades and standards. This will create uniformity.
- ✰ The grade standards should have common terminology and understood in the same manner at the All-India level. Grade standards set must be easy to understand, explain and measure whatever they are measurable. The standards should have consistency so that, frequent changes are not required.

12. Organizations Involved in Standardization of Agricultural Products

There are two levels for assigning standards to agricultural products *viz.*, Domestic organizations and International organizations and they are discussed in-detail here under.

a. Domestic Organizations Involved in Standardization of Agricultural Products

There are two organizations that deal with voluntary standardization and certification systems in the domestic food sector *viz.*, DMI and BIS. The DMI looks after the standardization of raw agricultural produce and BIS looks after the standardization of processed food products.

i. DMI

The DMI is an attached Office of the Ministry of Agriculture. It was set up in the year 1935 to implement the agricultural marketing policies and programmes of the Central Government. Since its very inception, the Directorate continues to be responsible for bringing about an integrated development of marketing of agricultural and allied produce in the country with a view to safeguard the interests of farmers as well as the consumers. It maintains a close liaison between the Central and the State Governments in the implementation of agricultural marketing policies in the country. The Directorate is headed by the AMA to the Government of India. The Directorate has its Head Office at Faridabad (Haryana), Branch Head Office at Nagpur (Maharashtra) and 11 Regional Offices (headed by Deputy Agricultural Marketing Advisers) at Delhi, Mumbai, Chennai, Kolkata, Hyderabad, Chandigarh, Jaipur, Lucknow, Bhopal, Kochi and Guwahati and the Central Agmark Laboratory at Nagpur. Besides, there are 34 Sub-Offices, 23 Regional Agmark Laboratories (RALs) spread all over the country (Table 5.76).

Table 5.76: DMI and its offices.

Regional Office	*Sub-Office under Regional office*	*RALs under Regional Office*
1. Delhi	1. Karnal	1. Okhla, 2. Ghaziabad
2. Kolkata	1. Patna 2. Bhubneshwar	1. Kolkata 2. Patna 3. Bhubaneshwar
3. Mumbai	1. Nasik 2. Ahmedabad 3. Rajkot 4. Jamnagar 5. Surat 6. Panaji 7. Pune 8. Sangli 9. Margaon 10. Unjha	1. Mumbai 2. Rajkot 3. Jamnagar
4. Bhopal	1. Khandwa	1. Bhopal
5. Chennai	1. Bangalore. 2. Mangalore 3. Tuticorin 4. Raichur 5. Virudhunagar 6. Tirupur, 7. Bodinayaknur 8. Nagapattanam	1. Chennai 2. Bangalore. 3. Mangalore 4. Tuticorin 5. Virudhunagar
6. Cochin	1. Alleppey 2. Calicut 3. Trivandrum	1. Kochi 2. Alleppey 3. Calicut
7. Guntur	1. Hyderabad 2. Vishakhapattanam	1. Guntur
8. Guwahati	1. Shillong	1. Guwahati 2. Shillong
9. Lucknow	1. Kanpur 2. Varanasi	1. Kanpur
10. Jaipur	Nil	1. Jaipur
11. Chandigarh	1. Jammu 2. Amritsar 3. Abohar 4. Shimla	1. Amritsar

Central Agmark Laboratory has the following specialized commodity divisions for carrying out research on standardization work more efficiently:

- ☆ Agricultural Products (Food grains)
- ☆ Spices and Essential Oils
- ☆ Oils and Fats
- ☆ General Chemistry
- ☆ Livestock Products including microbiology
- ☆ Toxicology

The main functions of Central Agmark Laboratory are:

- ☆ To work as apex laboratory for challenged sample under Agricultural Produce (Grading and Marking) Act, 1937

- To evolve/standardize methods of analysis/tests of agricultural and allied commodities and meat products.
- To advise on technical matters to various quality control agencies and SGLs, in relation to grading of various agricultural commodities, food under Agmark.
- Formulation of specifications for new commodities for bringing under the purview of Agmark.
- Revision of specifications of various agricultural, allied products, meat products etc., periodically.
- Training to the personnel engaged in the analysis of various commodities under Agmark.
- To create awareness with regard to grading, standardization and quality of various agricultural and food products.

The RALs are engaged in analysis of agricultural and food commodities for evaluating the quality of the product. The main activities of RALs are as follows:

- Analysis of commodities covered under Agmark
- Technical advice to approved grading laboratories
- Training of Grading chemists of the private approved laboratories, SGLs (Table 5.77) and other similar organizations.
- Associate with Central Agmark Laboratory, in collaborative studies/research/standardization work of various agricultural, food and livestock products.
- To organize awareness programmes in grading, standardization and quality control.

We know, the grading activities are directed specifically for the benefit of farmers and consumers, while they also help the traders in efficient movement of the produce from the farmers to the consumers. In India, promotion of standardization and grading of agricultural and allied produce is one of the important activities of the DMI. The Directorate promotes standardization and grading and implements Scheme of Certification of Agricultural and Allied Products for orderly marketing under the provisions of Agricultural Produce (Grading and Marking) Act, 1937 as amended in 1986. This Act, 1937 empowers the Government to fix quality standards, known as 'AGMARK' standards and to prescribe terms and conditions for using the seal of 'AGMARK'. So far, grade standards have been notified for 205 agricultural and allied commodities. The purity standards under the provisions of the PFA Act, 1954 and BIS Act, 1986 are invariably taken into consideration, while framing the grade standards. International standards framed by Codex/ISO are also considered so that, Indian products can compete in international market. This Certification mark under the Agricultural Produce (Grading and Marking) Act, 1937 *i.e.*, 'AGMARK' is issued by the AMA of DMI and the grades are designated as the 'AGMARK' grades (Table 5.78). A Central Agmark Laboratory at Nagpur along with RALs are assisting to provide adequate laboratory facilities for fixing grade standards for new commodities, for revising old grade standards and for routine quality control work. Grading is carried out in accordance with the standards notified, following meticulous procedure of sampling, testing, packaging, marking and sealing as per the instructions issued under the provisions of the Act and Rules. It serves as a means of describing the quality of commodities to be purchased or sold by the buyers or sellers all over the country and abroad. This establishes a common trade language and avoids the need for physical checking and handling at many points. The following are the other important activities of DMI:

- Promotion of standardization and grading of agricultural and allied produce under the Agricultural Produce (Grading and Marking) Act, 1937 as amended in 1986 with the main

Table 5.77: State-wise grading facilities in India (2010-11).

States	*No. of Grading Labs*	*No. of Grading Labs Per '000*	
		Sq.Km	*m tonnes*
Andhra Pradesh	44	0.17	1.12
Arunachal Pradesh	–	0.00	0.00
Assam	–	0.00	0.00
Bihar	12	0.13	0.39
Chhattisgarh	2	0.01	0.20
Goa	–	0.00	0.00
Gujarat	43	0.22	1.75
Haryana	82	1.85	3.93
Himachal Pradesh	10	0.18	3.28
Jammu & Kashmir	46	0.21	9.82
Jharkhand	1	0.01	0.12
Karnataka	44	0.23	1.62
Kerala	52	1.34	7.80
Madhya Pradesh	46	0.15	1.62
Maharashtra	68	0.22	2.17
Manipur	–	0.00	0.00
Meghalaya	–	0.00	0.00
Mizoram	–	0.00	0.00
Nagaland	–	0.00	0.00
Odisha	6	0.04	0.33
Punjab	62	1.23	1.92
Rajasthan	209	0.61	8.67
Sikkim	–	0.00	0.00
Tamil Nadu	47	0.36	1.86
Tripura	–	0.00	0.00
Uttar Pradesh	153	0.64	2.09
Uttarakhand	5	0.09	1.39
West Bengal	24	0.27	0.58
All India	956	0.29	2.03

Source: <agmarknet.nic.in>

objective of making these products to compete in world trade. These standards are being harmonized with international standards keeping in view the WTO requirements. Certification of agricultural commodities is carried out for the benefit of farmer and consumer. Certification of adulteration prone commodities *viz.*, butter, ghee, vegetable oils, ground-spices, honey, wheat atta etc., is very popular. Blended edible vegetable oils and fat spread are compulsorily required to be certified under AGMARK.

Table 5.78: State-wise Agmark Nodes in India

States	*Agmark Nodes*		
	Number (Aug 2010)	*Per 1000 sq.km*	*Per 1000 MT*
Andhra Pradesh	334	1.28	8.50
Arunachal Pradesh	15	0.18	29.74
Assam	23	0.29	1.93
Bihar	58	0.62	1.91
Chhattisgarh	73	0.54	7.18
Goa	10	2.70	31.37
Gujarat	319	1.63	13.00
Haryana	150	3.39	7.20
Himachal Pradesh	39	0.70	12.79
Jammu & Kashmir	41	0.18	8.75
Jharkhand	26	0.33	3.03
Karnataka	171	0.89	6.29
Kerala	92	2.37	13.80
Madhya Pradesh	267	0.87	9.40
Maharashtra	346	1.12	11.06
Manipur	5	0.22	5.37
Meghalaya	11	0.49	11.53
Mizoram	9	0.43	24.34
Nagaland	14	0.84	17.17
Odisha	91	0.58	4.99
Punjab	199	3.95	6.17
Rajasthan	166	0.49	6.89
Sikkim	7	0.99	27.82
Tamil Nadu	190	1.46	7.54
Tripura	21	2.00	14.90
Uttar Pradesh	257	1.08	3.51
Uttarakhand	21	0.39	5.84
West Bengal	56	0.63	1.34
All India	3011	0.92	6.40

Source: <agmarknet.nic.in>

- ☆ Render advice on Statutory regulation, development and management of agricultural produce markets of the States/UTs
- ☆ Market research, surveys and planning
- ☆ Agricultural marketing reforms
- ☆ Agricultural marketing information network

- ✰ Promotion of cold storage
- ✰ Construction of rural godowns
- ✰ Development of marketing infrastructure, grading and standardization
- ✰ Training in agricultural marketing
- ✰ Marketing extension

AGMARK' Label

'AGMARK' is the official symbol given by the Government of India to agricultural and livestock commodities, which confirm to various grade standards laid down under the Agricultural Produce (Grading and Marking) Act, 1937. 'AGMARK' is the symbol of purity, quality, unadulterated and Government guaranteed food products. It is a quality certification mark under the Act, 1937. As mentioned earlier, the DMI under the Department of Rural Development in the Ministry of Agriculture is vested with the responsibility to enforce and implement the Agricultural Produce (Grading and Marking) Act, 1937 (as amended in 1986). This Act, 1937 empowers the Central Government to fix quality standards, known as 'AGMARK' standards and to prescribe terms and conditions for using the seal of AGMARK. So far, grade standards have been notified for 205 agricultural and allied commodities. Grading under the provisions of this Act, 1937 is voluntary. The main objective of giving AGMARK label was to ensure supply of quality food products to consumers and this was effective since 1986. With the demand for products and competition in the market growing rapidly, a few manufacturers were adulterating agricultural products like edible oil, dal, masala powders etc. So, the Government had come up with the Agmark grading to make sure that, consumers were supplied with quality food products. The Agmark laboratories were grading the notified food products such as vegetable oils, ghee, honey, butter, wheat flour, gram flour, sago, groundnut, spices, rava, maida desiccated coconut powder etc., by testing scientifically and physically and affixing the 'AGMARK' labels on products depending on their quality. The consumers were asked to buy only those products, which were graded by Agmark laboratories with a seal 'AGMARK' on the label. The AGMARK label shows the information pertaining to name and address of the packer, date of packing in month and year, place of packing, year of harvest, net weight and price.

For export trading, the AGMARK label is compulsory. The AGMARK label will be attached for the container of the produce such that, the container cannot be opened without tampering the label. This is mainly done to prevent adulteration of the produce. This AGMARK label indicates the purity of product and if the product purchased by the consumer found defective, the product will be replaced by the seller or consumer will be get back his money. This 'AGMARK' label was written both horizontally and vertically on India map coinciding at the letter 'M' (Figure 5.35). On right side of label there are letters, '*Bharathiya utpatthi*' written in Hindi, Tamil and Telugu languages and on the left side of label, there is a rising sun with words 'Produce of India'. This 'AGMARK' label is also given for commodities of internal consumption like ghee, vegetable oil butter, atta, besan, honey, ground masala like pepper, dhania products etc., but not for mixed spices like garam masala.

ii. BIS

It is formerly called as the ISI, which came into existence on January 6th, 1947 and in June, 1947 Dr. Lal C. Verman took over as its first Director. Considerable progress has since been made by the ISI with its multifarious activities like standards formulation, certification, testing of products international co-operation and standards promotion. However, a need was felt to provide recognition and status to the organization to enable it to discharge its functions effectively. Thus, the Government enacted the

Figure 5.35: Grade labels for trade of agricultural commodities.

BIS Act, 1986 which led to the establishment of the BIS as a statutory body on April 1st, 1987. Thus, ISI is renamed as BIS on April 1st, 1987 and it serves as the National Standards Body of India working under the aegis of Ministry of Consumer Affairs, Food and Public Distribution, Government of India. Its headquarters is at New Delhi, with regional offices in Kolkata, Chennai, Mumbai, Chandigarh and Delhi, and 20 branch offices. It also works as WTO-Technical Barriers to Trade (TBT) enquiry point for India.

ISI mark (Figure 5.35) is a certification mark employed on industrial products in India. The mark placed on a consumer product certifies that, it conforms to a set of standards laid by the BIS. The ISI mark is by far the most recognized certification mark in the Indian sub-continent. The ISI mark is mandatory for certain products to be sold in India, like most of the electrical appliances *viz.*, switches, electric motors, wiring cables, heaters etc., and other products like LPG valves, LPG cylinders, automotive tyres etc. But, in the case of most other products, it serves only as a consumer advisory status. The domestic consumers are familiar with the ISI mark on a product, which is an assurance that the product conforms to the requirements as laid down in the specification. Conformity to the standard is ensured through regular surveillance of the manufacturing process, surprise inspections and testing of samples drawn from the factory as well as from the market. As mentioned earlier, fraudulent and unauthorized use of the ISI mark is a violation of the law punishable under the BIS Act.

It is very common in India to find products with fake ISI marks, that is, affixing ISI marks on the product without actually getting certified. This is a punishable offense by the law, but the practice is common. The ISI mark in the prescribed format would contain, along with the ISI logo, a code number of the standard specification preceded with the letters IS. For example, the code IS:14543 denotes a standard specification for PET containers. The letters IS stands for 'Indian Standard'.

In the BIS, similar to standards bodies the world over, standards are formulated through technical committees, which have 25 representatives drawn from the Central or State Governments, industry, scientific and research institutions and consumer organizations. The standards are documents of consensus, which are finalized after taking the views of all those who may have an interest in it. It is engaged in the preparation and implementation of standards, operation of certification schemes both for products and systems, organisation and management of testing laboratories, creating consumer awareness and maintaining close liaison with international standards bodies. It has identified certain items like food colours/additives, vanaspati and containers for packing, milk powder and condensed milk for compulsory certification.

BIS Product Certification Scheme is basically voluntary in nature. However, keeping in view the health and safety of the consumer, it has been made mandatory for 68 items by the Government through various statutory measures such as PFA Act, Coal Mines Regulations and Indian Gas Cylinders Rules besides BIS Act. Some of the items brought under mandatory certification on consideration of health and safety are milk powder, packaged drinking water, LPG cylinders, oil pressure stoves, clinical thermometers etc. As on 31st March 2008, 20025 certification marks licences are in operation under the Scheme, covering about 1000 different items ranging from food products to electronics.

All foreign manufacturers of products, who intend to export to India are required to obtain a BIS product certification licence. Towards this, BIS launched its Product Certification Scheme for overseas manufacturers in the year 1999. Under the provisions of this scheme, foreign manufacturers can seek certification from BIS for marking their product(s) with BIS Standard Mark (Figure 5.35). Under this scheme, about 101 licences of foreign manufacturers in about 15 countries are in operation as on 31st March 2008. Under the scheme for Indian importers, Certification Marks Licence can be granted to Indian importers for the product imported into the country and are not covered under Regulatory Requirements. Indian importers can apply for BIS licence on such products and they are treated as Indian manufacturers.

As a founder member of ISO, BIS continues to actively participate in international standardization. As a member of the ISO Council, it participates in its policy-making bodies like the Committee on Developing Country matters (DEVCO), Committee on Conformity Assessment (CASCO), Committee on Information (INFCO) and Committee on Consumer Policy (COPOLCO). Besides, it also holds secretarial responsibilities of various Technical Committees and Sub-committees and maintains participation status and observer status on most of the Technical Committees. BIS is also actively involved in IEC and has participation status or observer on all the important Technical Committees.

iii. Other Important Quality Marks

There are other foodstuffs too, which are sold in a packed form and bear a quality mark. Spices, biscuits, jams, tea are some of the examples. These quality marks also help the consumers in the right selection of food items. They include:

- ☆ ***Fruit Products Order (FPO) mark:*** The FPO mark is an inverted 'V' shaped certification mark (Figure 5.35) mandatory on all processed fruit products sold in India such as packaged fruit beverages, fruit-jams, crushes and squashes, pickles, dehydrated fruit products and fruit extracts, following the Food Safety and Standards Act of 2006. The FPO mark guarantees that, the product was manufactured in a hygienic 'food-safe' environment thus, ensuring that, the product is fit for consumption. The FPO standards have been in force since 1955 by the law of Fruit Products Order, after which the mark is named, but the mark itself got a mandatory status only after passing the Food Safety and Standards Act, 2006. An FPO license is, in fact, necessary to start a fruit processing industry in India. The agency that develops standards for this purpose and that which issues this certification mark is the MOFPI of the Government of India.

☆ ***ECOMARK:*** The issue of environmental protection has brought the consumers, the industry, and the Government to a common platform, where each has to play its own role. The Government and legislatures are using their influence to reduce environmental and health hazards due to industrialization and to stimulate the development of technologies that ensure production of quality products. However, the environment is under tremendous stress from rapid industrialization, unplanned urbanization and changing consumption patterns in the race to achieve better living standards. It is clear that, regulatory actions by pollution control agencies alone cannot restore the environment to its pristine state. Pro-active and promotional roles should also be geared up in harmony with the overall environmental protection strategy. The time has come for consumers to take the lead in prompting manufacturers to adopt clean and eco-friendly technologies and environmentally-safe disposal of used products, along with preventive and mitigative approaches. To increase consumer awareness, The Ministry of Environment and Forests, Government of India launched the eco-labeling scheme known as 'ECOMARK' on 21st February, 1991 for easy identification of environment-friendly products. Any product which is made, used or disposed of in a way that significantly reduces the harm, it would otherwise cause the environment could be considered as Environment-Friendly Product. The scheme is operating on a national basis and provides accreditation and labelling for household and other consumer products, which meet certain environmental criteria along with quality requirements of the Indian Standards for that product. This scheme is voluntary and invites participation from common citizens and concerned industrial sectors in the larger interest of environment. The specific objectives of the ECOMARK scheme are as follows:

- To provide an incentive for manufacturers and importers to reduce environmental impact of products.
- To reward genuine initiatives by companies to reduce adverse environmental impact of their products.
- To assist consumers to become environmentally responsible in their daily lives by providing information to take account of environmental factors in their purchase decisions.
- To ensure citizens to purchase products which have less harmful environmental impacts.
- Ultimately to improve the quality of the environment and to encourage the sustainable management of resources.

The Ministry of Environment and Forests have so for identified the following 16 products categories to be covered under the scheme of ECOMARK. They include,

Soaps and Detergents	Paints	Paper
Food items	Food additives and preservatives	Wood Substances
Textiles	Batteries	Lubricating oil
Packaging	Plastics	Aerosols
Cosmetics	Pesticides and drugs	Electronic goods
Vegetable oils		

An 'earthern pot' has been chosen as the logo for the ECOMARK scheme (Figure 5.35) in India. The familiar earthern pot uses a renewable resource like earth, does not produce hazardous waste and consumes little energy in making. Its solid and graceful form represents both strength and fragility, which also characterizes the eco-system. As a symbol, it puts across its environmental message. Its image has the ability to reach people and can help to promote a greater awareness of the need to be kind to the environment. The logo for the ECOMARK Scheme, signifies that, the product which carries it does the least damage to the environment.

- ☆ ***Vegetarian and Non-vegetarian food marks:*** As discussed earlier, packaged food products sold in India are required to be labeled with a mandatory mark in order to be distinguished between vegetarian and non-vegetarian. The symbol is in effect since 2006 following the *Food Safety and Standards (Packaging and Labeling) Act of 2006*, and got a mandatory status after the framing of the respective regulations (*Food Safety and Standards (Packaging and Labeling) Regulation* in 2011. According to the law, vegetarian food should be identified by a green symbol and non-vegetarian food with a brown symbol (Figure 5.35).
- ☆ ***Date marks:*** Certain foods like breads, fish, chilled meat, milk products and other foods which have a shelf-life of less than 40 days are required by law to carry a Date mark. They are:
  - ❖ ***Best before:*** This is the date until which the food should still retain the qualities, the manufacturer claims it will have, if it has been kept under the conditions specified on the label.
  - ❖ ***Packed on:*** This is the date on which the food was placed in the package in which it will be sold to the consumer.
  - ❖ ***Use by:*** This is the date after which the food, provided it has been stored under the correct conditions, may not retain the qualities the consumer would normally expect.

b. International Organizations Involved in Standardization of Agricultural Products

Standards for agricultural products make an enormous contribution to most aspects of human lives, although very often, that contribution is invisible. It is, when, in their absence only, their importance is intensely felt by the people. Say for example, a consumer purchased a product, but it is found out to be of poor quality, as it is does not meet the expectations in terms of its usage efficiency or not compatible with the equipment the consumer have or not safe for consumption etc. This makes the consumers to insist upon standardized goods, as they ensure good quality, safety, reliability, efficiency etc., at an economical cost. At international level, there are two important organizations involved in the standardization of agricultural products *viz.*, ISO and CAC. When the large majority of products or services in a particular business or industry sector conform to the International Standards prescribed by these two International organizations (ISO and CAC), a state of industry-wide standardization can be said to exist. This is achieved through consensus agreements between national delegations representing all the economic stakeholders concerned – producers, processors, suppliers, users,

Government regulators and other interest groups, such as consumers. They agree on specifications and criteria to be applied consistently in the classification of materials, in the manufacture and supply of products, in testing and analysis, in terminology and in the provision of services. In this way, international standards provide a reference framework or a common technological language, between producers and their consumers, which facilitates trade and the transfer of technology across the countries.

i. ISO

ISO is the world's largest developer of standards. ISO is a network of the national standards institutes of 163 countries, one member per country, with a Central Secretariat (Head quarters) in Geneva, Switzerland, that coordinates the system. ISO is a non-governmental organization (came into existence on 23rd February,1947) that forms a bridge between the public and private sectors. On the one hand, many of its member institutes are part of the Governmental structure of their countries or are mandated by their Governments. On the other hand, other members have their roots uniquely in the private sector, having been set up by national partnerships of industry associations. Therefore, ISO enables a consensus to be reached on solutions that meet both the requirements of business and the broader needs of society. ISO has three membership categories:

- Member bodies are national bodies that are considered to be the most representative standards body in each country. These are the only members of ISO that have voting rights.
- Correspondent members are countries that do not have their own standards organization. These members are informed about ISO's work, but do not participate in standards promulgation.
- Subscriber members are countries with small economies. They pay reduced membership fees, but can follow the development of standards.

Participating members are called 'P' members as opposed to observing members, which are called 'O' members. Because 'ISO' would have different acronyms in different languages ('IOS' in English, 'OIN' in French for *Organization internationale de normalisation*), its founders decided to give it also a short, all-purpose name. They chose 'ISO', derived from the Greek *'isos'*, meaning 'equal'. Whatever the country, whatever the language, the short form of the organization's name is always ISO. The following points highlight the importance of ISO standards:

- ISO standards make the development, manufacturing and supply of products and services more efficient, safer and cleaner
- They facilitate trade between countries and make it fairer
- They provide Governments with a technical base for health, safety and environmental legislation, and conformity assessment
- They share technological advances and good management practice
- They disseminate innovation
- They safeguard consumers, and users in general, of products and services
- They make life simpler by providing solutions to common problems.

Features of ISO Standards

They include,

- ***Democratic:*** Every full member of ISO has the right to take part in the development of any standard which it judges to be important to its country's economy. No matter, what the size or strength of that economy, each participating member in ISO has one vote. Each country is on an equal footing to influence the direction of ISO's work at the strategic level as well as the technical content of its individual standards.

- ☆ ***Voluntary:*** ISO standards are voluntary. As a non-governmental organization, ISO has no legal authority to enforce the implementation of its standards. ISO does not regulate or legislate. However, countries may decide to adopt ISO standards - mainly those concerned with health, safety or the environment - as regulations or refer to them in legislation, for which they provide the technical basis. In addition, although ISO standards are voluntary, they may become a market requirement, as has happened in the case of ISO 9001 quality management systems or of dimensions of freight containers and bank cards.
- ☆ ***Market-driven:*** ISO only develops standards for which there is a market requirement. The work is mainly carried out by experts from the industrial, technical and business sectors, which have asked for the standards and which subsequently put them to use.
- ☆ ***Consensus:*** ISO standards are based on international consensus among the experts in the field. ISO standards require a periodic review at least once in every five years to decide whether they should be maintained, updated or withdrawn. In this way, ISO standards retain their position as the state of the art.
- ☆ ***Globally relevant:*** ISO standards are technical agreements, which provide the framework for compatible technology world-wide. They are designed to be globally relevant and useful everywhere in the world.

We know, food products regularly cross national boundaries at every stage of the supply chain. ISO International Standards create confidence for the consumers in consuming the food products. These ISO standards are developed through common understanding and co-operation with all stakeholders like agricultural producers, food manufacturers, laboratories, regulators, consumers etc., across the countries. Out of more than 19000 ISO International Standards, some 1000 are specifically dedicated to food, and deal with subjects as diverse as agricultural machinery, logistics, transportation, manufacturing, labeling, packaging and storage. ISO 22000 is a standard developed by the ISO dealing with food safety. This is a general derivative of ISO 9000.

ISO 22000 Standard

This International Standard specifies the requirements for a food safety management system that combines the following generally recognized key elements to ensure food safety along the food chain, up to the point of final consumption.

- ☆ interactive communication
- ☆ system management
- ☆ prerequisite programs
- ☆ HACCP principles

ISO 22000 has been aligned with ISO 9001 in order to enhance the compatibility of the two standards. It can be applied independently of other management system standards or integrated with existing management system requirements. It integrates the principles of the HACCP system and application steps developed by the CAC. ISO is developing additional standards that are related to ISO 22000. These standards will be known as the ISO 22000 family of standards. At the present time, the following standards will make up the ISO 22000 family of standards:

- ☆ ISO 22000 - Food safety management systems - Requirements for any organization in the food chain.

- ISO 22001 - Guidelines on the application of ISO 9001:2000 for the food and drink industry (replaces: ISO 15161:2001).
- ISO/TS 22002- Prerequisite programmes on food safety–Part 1: Food manufacturing
- ISO/TS 22003 - Food safety management systems for bodies providing audit and certification of food safety management systems.
- ISO/TS 22004 - Food safety management systems - Guidance on the application of ISO 22000:2005.
- ISO/22005 - Traceability in the feed and food chain - General principles and basic requirements for system design and implementation.
- ISO/22006 - Quality management systems - Guidance on the application of ISO 9002:2000 for crop production.
- ISO/22000 is also used in the Food Safety Systems Certification Scheme FS22000. FS22000 is a Global Food Safety Initiative approved scheme.

Beneficiaries of ISO Standards

ISO has developed over 19000 International Standards on a variety of subjects and more than 1000 new ISO standards are published every year. ISO standards provide technological, economic and societal benefits to different stakeholders as listed below.

- For businesses, the wide spread adoption of International Standards means that, suppliers can develop and offer products and services meeting specifications that have wide international acceptance in their sectors. Therefore, businesses using International Standards can compete on many more markets around the world.
- For innovators of new technologies, International Standards on aspects like terminology, compatibility and safety, speed up the dissemination of innovations and their development into manufacturable and marketable products.
- For customers, the world-wide compatibility of technology, which is achieved when products and services are based on International Standards gives them a broad choice of offers. They also benefit from the effects of competition among suppliers.
- For Governments, International Standards provide the technological and scientific bases underpinning health, safety and environmental legislation.
- For trade officials, International Standards create 'a level playing field' for all competitors on those markets. The existence of divergent national or regional standards can create technical barriers to trade. International Standards are the technical means by which political trade agreements can be put into practice.
- For developing countries, International Standards that represent an international consensus on the state of the art are an important source of technological know-how. By defining the characteristics that products and services will be expected to meet on export markets, International Standards give developing countries a basis for making the right decisions, when investing their scarce resources.
- For consumers, conformity of products and services to International Standards provides assurance about their quality, safety and reliability.
- For everyone, International Standards contribute to the quality of life, in general, by ensuring that, the transport, machinery, products and tools we use are safe.

☆ For the planet we inhabit, International Standards on air, water and soil quality, on emissions of gases and radiation and environmental aspects of products can contribute to efforts to preserve the environment.

Thus, international standards, which ISO develop are very useful to industrial and business organizations of all types, to Governments and other regulatory bodies, to trade officials, to conformity assessment professionals, to suppliers and customers of products and services in both public and private sectors, and, ultimately, to people in general in their roles as consumers and end users. ISO standards contribute to making the development, manufacturing and supply of products and services that are more efficient, safer and cleaner. They make trade between countries easier and fairer. They provide Governments with a technical base for health safety and environmental legislation. They aid in transferring technology to developing countries. ISO standards also serve to safeguard consumers in the use of products and services and to make their lives simpler. The importance of 'ISO' standards can be highlighted from the following statement: *'when things go well, for example, when systems, machinery and devices work well and safely, then it is because, they conform to standards and the organization responsible for many thousands of the standards, which benefit society world-wide is ISO'*.

Criticism of ISO Standards

With the exception of a small number of isolated standards, ISO standards are normally not available free of charge, but for a purchase fee, which has been seen by some as too expensive.

ii. CAC

The CAC was established in 1963 in Rome, Italy as an intergovernmental agency of the United Nations under the FAO and the WHO to develop food standards, guidelines and related texts such as codes of practice under the Joint FAO/WHO Food Standards Programme. The main purposes of this CAC are protecting health of the consumers, ensuring fair trade practices in the food trade and promoting coordination of all food standards work undertaken by international governmental and non-governmental organizations. We know, food standards are becoming more important, as international trade in food opens up and consumers are more concerned about safety and quality. Standards must provide a high level of consumer protection and not unnecessarily restrict trade. The CAC is recognized by the WTO, as an international reference point for the resolution of disputes concerning food safety and consumer protection. The CAC international food standards contribute to the safety, quality and fairness of international food trade. Consumers can trust the safety and quality of the food products they buy and importers can trust that the food they ordered will be in accordance with their specifications. In the event of a trade dispute, Codex standards would become accepted reference documents for its settlement. This highlights the increasing importance of Codex in international law. The CAC established food standards are termed as '*Codex Alimentarius*'. The *Codex Alimentarius* (meaning 'Book of Food', 'Food Code' or 'Food Law' in Latin) is a collection of internationally recognized food standards, codes of practice, guidelines and other recommendations relating to foods, food production and food safety in a uniform way. Codex standards, guidelines and other recommendations ensure that, food products are not harmful to the consumer and can be traded safely between countries. The detailed work of CAC on drafting standards, codes of practice and other guidance is handled by about thirty Committees, which fall into three general categories *viz.*, Commodity Committees (covering a particular commodity; *e.g.*, Milk and Milk Products), Horizontal Committees (dealing with issues applicable across a range of commodities *e.g.*, labeling, food additives and contaminants) and Regional Committees (one for each of the seven Codex regions *e.g.*, Europe). The *Codex Alimentarius* officially covers all foods, whether processed, semi-processed or raw, but far more attention has been given to foods that are marketed directly to consumers. In addition to standards for specific foods, the *Codex Alimentarius* contains general standards covering matters such as food labeling, food hygiene, food additives and pesticide residues, and procedures for assessing the safety of foods derived from modern biotechnology. It also contains guidelines for the management of official (*i.e.*, governmental) import and export inspection and certification systems for foods.

A principal concern of national governments is that, food imported from other counties should be safe and not jeopardize the health of consumers or pose a threat to the health and safety of their animal and plant populations. Consequently, Governments of importing countries have introduced mandatory laws and regulations to eliminate or minimize such threats. In the area of food, animal and plant control, these measures could be conducive to the creation of barriers to inter-country food trade. To overcome this, Codex standards are recognized as uniform and harmonized standards for ensuring fair trade across the countries. Presently the Codex provides 28 volumes of guidelines and principles with 237 food commodity standards, 41 hygienic and technological practice codes and other recommendations relating to food quality composition and safety. These Codex standards were considered as a vital component in promoting food control systems designed to protect consumer health, including issues related to international trade and the agreements on the Application of SPS and on TBT of the WTO. International Codex standards also provide a basis for standard setting by smaller and LDCs. Capacity building activities of FAO, WHO and Codex were found to be continuing to make a substantial contribution internationally and to individual countries both in protecting their own citizens and in benefiting from an increasingly globalized market in food. The Codex standards (*Codex Alimentarius*) has resulted in evaluations of the safety of over 760 food additives and contaminants and the setting of more than 2 500 maximum limits for pesticide residues and more than 150 for veterinary drug residues. In addition, CAC has established guideline levels for a number of environmental and industrial contaminants (including radionuclides) in foods. The *Codex Alimentarius*, has become the global reference point for consumers, food producers and processors, national food control agencies and the international food trade. The code has had an enormous impact on the thinking of food producers and processors as well as on the awareness of the end users, the consumers. Its influence extends to every continent, and its contribution to the protection of public health and fair practices in the food trade is immeasurable, as the volume of world food trade is enormous and is valued at between US$300 billion and $400 billion.

Food hygiene has been a major area of activity of CAC since the Commission's establishment. The Codex Committee on Food Hygiene, hosted by the Government of the United States, was founded in 1963. Because food hygiene is best regulated at the production and processing stages in the exporting country, the Committee's main outputs have been codes of hygienic practice rather than end-product microbiological standards. Taking this philosophy a step further, CAC has also developed *Guidelines for the application of the HACCP system* and its implementation in the area of food safety through its Committee on Food Hygiene. In doing this, it has recognized HACCP as a tool to assess hazards and establish control systems that focus on preventive measures instead of relying primarily on end-product testing. So, Codex standards, guidelines and other recommendations take on unprecedented importance with respect to consumer protection and international food trade. As a result, the works of the CAC including the *Guidelines for the application of the HACCP system* - has become the reference for international food safety requirements. In view of this, it is imperative that, the Codex guidelines for the application of HACCP be unequivocal in their guidance, otherwise conflicts on food safety grounds may arise.

CAC has been actively revising much of its work in recent years to stress the so-called horizontal aspects of food regulation, including food hygiene. New considerations such as risk analysis and the determination of equivalence in different food control systems have an impact on the new approach to international food hygiene regulations.

13. AGMARK and Codex Quality (Grade) Standards of Wheat

Different quality standards as prescribed under AGMARK and Codex for wheat are given here under.

AGMARK Standards for Wheat

The following are the maximum limits of tolerance across different grade standards of wheat (Table 5.79):

Table 5.79: AGMARK standards for wheat

Grade Designation	*Foreign Matter (Per cent by wt.)*	*Other Food Grains (Per cent by wt.)*	*Other Wheats (Per cent by wt.)*	*Damaged Grains (Per cent by wt.)*	*Slightly Damaged Grains (Per cent by wt.)*	*Immature Shrivelled and Broken Grains (Per cent by wt.)*	*Weevilled Grains (Per cent by wt.)*
I	1.0	1.6	5.0	1.0	2.0	2.0	1.0
II	1.0	3.0	15.0	2.0	4.0	4.0	3.0
III	1.0	6.0	20.0	4.0	6.0	10.0	6.0
IV	1.0	8.0	20.0	5.0	10.0	10.0	10.0

Note: In foreign matter, not more than 0.25 per cent by wt. shall be mineral matter and not more than 0.10 per cent by wt. shall be impurities of animal origin. Grades I and II should be free from living insect infestation.

General Characteristics

(a) Wheat shall be the dried grains of *Triticum vulgare* and *Triticum durum* Desf,

(b) Have uniform size, shape and colour;

(c) Shall be sweet, hard, clean, wholesome and free from moulds, weevils, obnoxious smell, discolouration, admixture of deleterious substances and all other impurities except to the extent as indicated in the Schedule;

(d) Be in sound merchantable condition;

(e) Not have moisture exceeding 12 per cent.

Definitions

- ☆ ***Foreign Matter:*** It includes dust, stones, lumps of earth, chaff, stem or straw and any other impurity including non edible seeds.
- ☆ ***Other Food Grains*** : Edible food grains other than wheat.
- ☆ ***Other Wheat*** : For this purpose wheat would be divided into two classes (1) Durum or Macaroni wheat and (2) vulgare or common wheat. Durum again would be sub divided into two colour groups (1) amber and (2) red and vulgare would be sub divided into three colour groups - (1) white (2) amber and (3) red.
- ☆ ***Damaged Grains*** : Grains that are internally damaged or discoloured, damaged and discolouration materially affecting the quality.
- ☆ ***Slightly Damaged Grains*** : Grains that are superficially damaged or discoloured, damaged and discolouration not materially affecting the quality.

☆ ***Immature, Shriveled and Broken Grains*** : Immature and shrivelled grains are those that are not properly developed, Broken grains are pieces of whole grains.

☆ ***Weevilled Grains*** : Grains that are partially or wholly bored or eaten by weevils or other grain insects.

Note: Maximum tolerance limits for various food safety parameters such as poisonous metals, pesticide residues, aflatoxin, uric acid, etc. are as prescribed in Prevention of Food Adulteration Rules, 1955.

Codex Standards (International Trade Standards) for Wheat and Durum Wheat: We know, CAC implements joint Food and Agriculture Organization (FAO)/WHO Food Standards Programme. The purpose of the CAC programme is to protect the health of consumers and ensure fair practices in the food trade. The CAC is a collection of internationally adopted food standards presented in a uniform manner. SPS and TBT Agreements of WTO recognizes standards framed by CAC with respect to safety and quality aspects of wheat. Thus, for international trade, standards framed by CAC are recognized (Table 5.80) for wheat.

Table 5.80: Codex standards (Quality and Safety factors) for wheat and durum wheat.

Parameter	*Wheat*	*Durum Wheat*
Moisture Maximum	14.5 per cent m/m	14.5 per cent m/m
Organic extraneous matter, Maximum	1.5 per cent m/m	1.5 per cent m/m
Inorganic extraneous matter, Maximum	0.5 per cent m/m	0.5 per cent m/m
Test weight (wt. of a hundred litre volume expressed in kg/hectolitre), Minimum	68	70
Shrunken and broken kernels, Maximum	5.0 per cent m/m	6.0 per cent m/m
Edible grains other than wheat and durum wheat, Maximum	2.0 per cent m/m	3.0 per cent m/m
Damaged kernels, Maximum	6.0 per cent m/m	4.0 per cent m/m
Insect bored kernels, Maximum	1.5 per cent m/m	2.5 per cent m/m
Filth (impurities of animal origin, including dead insects), Maximum	0.1 per cent m/m	0.1 per cent m/m
Ergot (Sclerotium of the fungus Claviceps purpurea), Maximum	0.05 per cent m/m	0.05 per cent m/m

☆ ***Toxic or Noxious Seeds:*** The wheat and Durum wheat shall be free from the following toxic or noxious seeds in amounts which may represent a hazard to human health. Crotolaria (*Crotalaria* spp.), Corn cockle (*Agrostemma githago* L.) Castor bean (*Ricinus communis L.*), Jimson weed (*Datura* spp.) and other seeds that are commonly recognized as harmful to health.

☆ ***Heavy Metals:*** Maximum levels for Lead is 0.2 mg/kg.

☆ ***Pesticide residues:*** Wheat and Durum wheat shall comply with following maximum pesticide residue limits (Table 5.81).

☆ ***Hygiene:*** Wheat and Durum wheat, after cleaning and sorting, and before further processing:

- ❖ Shall be free from microorganisms in amounts which may represent a hazard to health.
- ❖ Shall be free from parasites which may represent a hazard to health.
- ❖ Shall not contain any substance originating from microorganisms, including fungi, in amounts which may represent a hazard to health.

Table 5.81: Pesticidal residue limits of wheat and Durum wheat (Codex standards).

Sl.No.	*Pesticide*	*Maximum Residue Limits (MRL) or (Extraneous Maximum Residue Limits) EMRL*	*(mg/kg)*
1	2,4-D	MRL	0.5
2	Aldicarb	MRL	0.02
3.	Anilazine	MRL	0.1
4	Bentazone	MRL	0.1
5.	Bifenthrin	MRL	0.5
6.	Bioresmethrin	MRL	1
7.	Bitertanol	MRL	0.1
8.	Carbaryl	MRL	5
9.	Carbofuran	MRL	0.1
10.	Chlordane	EMRL	0.02
11.	Chlormequat	MRL	5
12.	Chlorothalonil	MRL	0.1
13.	Chlorpyrifos-methyl	MRL	10
14.	Cypermethrin	MRL	0.2
15.	Dichlofluanid	MRL	0.1
16.	Diquat	MRL	2
17.	Dithiocarbamates	MRL	1
18.	Ethephon	MRL	1
19.	Fenbuconazole	MRL	0.1
20.	Flucythrinate	MRL	0.2
21.	Flusilazole	MRL	0.1
22.	Glyphosate	MRL	5
23.	Hexaconazole	MRL	0.1
24.	Imazalil	MRL	0.01
25.	Methomyl	MRL	0.5
26.	Monocrotophos	MRL	0.02
27.	Phorate	MRL	0.05
28.	Piperonyl butoxide	MRL	10
29.	Pirimicarb	MRL	0.05
30.	Prochloraz	MRL	0.5
31.	Propiconazole	MRL	0.05
32.	Pyrazophos	MRL	0.05
33.	Tebuconazole	MRL	0.05
34.	Terbufos	MRL	0.01
35.	Triadimefon	MRL	0.1
36.	Triadimenol	MRL	0.2

International food irradiation symbol: The label of a food, which has been treated with ionizing radiation shall carry a written statement indicating that, treatment in close proximity to the name of the food. The use of the international food irradiation symbol (Figure 5.35) is optional, but when it is used, it shall be in close proximity to the name of the food. When an irradiated product is used as an ingredient in another food, this shall be so declared in the list of ingredients. When a single ingredient product is prepared from a raw material which has been irradiated, the label of the product shall contain a statement indicating the treatment.

14. Role of Standards in the Context of WTO Agreement

Almost 16 years have elapsed since various trade agreements were signed under the auspices of WTO, 1995. One agreement considered most effective in reforming food and agricultural sector was the AoA. The essence of AoA liberalization was that, markets should be distortion-free, a standard thinking in neoclassical economics. AoA translated this thinking by aiming for improving market access and export competition and reduction in domestic support. This, in-turn, was to be achieved through tariffication of Quantitative Restrictions (QRs) and time-bound reduction in existing tariffs, export subsidies and domestic support. An important assumption in the neoclassical thinking is that, there is complete information in the markets and elimination of tariffs and subsidies will lead to free trade among nations. However, markets are not characterized by complete information thereby, could not result in a smooth and distortion-free trade. This aspect is extremely important in the global trade in food products. We know that, food and agricultural markets/products are considered as examples of perfectly competitive markets with homogeneous products. However, this is not true in the practical sense. This is because, individual food products are not homogeneous across countries, different countries and firms adopt different performance standards and safety and quality norms and moreover, buyers cannot ascertain quality of food products merely by physical inspection. As a result, AoA by itself cannot guarantee removal of all barriers to trade. No doubt, international standards for food quality (ISO and Codex) have played an important role in both protection of health and safety of consumers and facilitation of international trade. However, with the establishment of the WTO and the signing of the non-tariff agreements (say, quality standards of food), the international scenario has rapidly changed and opportunities are available to all countries to benefit from greater access to world markets. In the given scenario, the role of standards and conformity assessment has become very important in ensuring that, the product is safe and of the desired quality. However, while imposing standards and conformity assessment procedures by member countries, it is necessary that, certain rules and disciplines be followed so that, the standards/regulations do not act as unreasonable barriers to trade. This aspect has been taken care of through the non-tariff agreements, which basically lay down the rules and disciplines with regard to standards and conformity assessment procedures for international trade. The aspects relating to standardization of quality of food products are specified in the two agreements *viz.*, SPS Measures and Agreement on TBT to ensure the trade flow as freely as possible. These two agreements serve to prevent standards from being misused as non-tariff trade barriers by the member countries. Both SPS and TBT Agreements stipulate countries to participate in International Standardization work. The international body, which has been referenced for the purpose of human health under the SPS Agreement is the CAC. With regard to TBT Agreement, the works of ISO and IEC are important. These two Agreements are discussed in-detail here under.

a. The Agreement on the Application of SPS Measures

The SPS Agreement confirms the right of WTO member countries to apply measures necessary to protect human, animal and plant life and health. With regard to food safety measures, WTO members base their national measures on international standards, guidelines and other recommendations adopted by the FAO/WHO's CAC, where they exist. Thus, SPS agreement endorses guidelines of CAC. The SPS Agreement covers all food hygiene measures and food safety measures such as the control of residues of veterinary drugs, pesticides and other chemicals used in food production. In addition, it covers animal and plant quarantine measures. A Phyto-Sanitary Certificate (PSC) will be issued, if the product conforms to the quality standards prescribed by the CAC. The PSC is an official declaration stating that, plants and plant material that move across international boundaries are free from pests and diseases, so as to prevent introduction and spread of any pests in the newer areas. PSC paves the way for rapid screening and expeditious release of imported plants/plant material at the port of arrival. Under WTO Agreement, its adoption has assumed greater significance. The SPS Agreement states that, any measures taken that conform to International Codex standards, guidelines or other recommendations are deemed to be appropriate, necessary and non-discriminatory. Furthermore, the SPS Agreement calls for a programme of harmonization of national requirements based on international standards. This work is guided by the WTO Committee on SPS Measures, to which representatives of CAC, the International Office of Epizootics or International Animal Health Organization (OIE) and the IPPC are invited. The following are the objectives of SPS Agreement:

- To set out the basic rules for food safety, animal and plant life/health standards
- To agree to the use of harmonized SPS measures between members on the basis of international standards developed by the CAC, the OIE and the IPPC
- To allow members the appropriate level of protection of human, animal or plant life or health for their country or region.

The SPS Agreement covers the following aspects:

- Human or animal health from food-borne risks *i.e.*, from additives, contaminants, toxins, pesticides, veterinary drugs, disease-causing organisms in the food, beverages or feedstuffs
- Human health from animal or plant carried diseases
- Animals or plants health from pests and disease causing organisms
- The health of fish and wild fauna
- The health of forests and wild flora
- Certification: food safety, animal or plant health
- Processing methods with implications for food safety
- Labeling requirements directly related to food safety
- Plant/animal quarantine
- Declaring areas free from pests or disease
- Preventing disease or pests spreading to or in a country
- Other sanitary requirements for imports (*e.g.* imported pallets used to transport animals).

b. The Agreement on TBT

The Agreement on TBT (TBT Agreement) is a revision of the Agreement of the same name first developed under the Tokyo Round of GATT negotiations in the 1970s. The objective of this Agreement is to prevent the use of national or regional technical requirements or standards in general, as unjustified technical barriers to trade. The agreement covers standards relating to all types of products including industrial and agricultural products, unlike SPS agreement, which relates to food standards alone. In general, differences arise between countries in preparation and application of Technical Regulations and Standards and Conformity Assessment Procedures with reference to packaging regulations, marking and labelling requirements, testing and calibration procedures, certification procedures, inspection procedures and other requirements for a particular market. In view of this, the main objective of TBT agreement is to reduce obstacles in international trade by ensuring technical regulations, standards, and conformity assessment procedures are prepared, adopted and applied in a 'trade friendly' manner and do not result in or unjustifiable discrimination against imported goods. This TBT agreement deals with the following aspects:

- Labeling of composition or quality of food, drink and drugs
- Quality requirements for fresh food
- Volume, shape and appearance of packaging
- Packaging and labeling for dangerous chemicals and toxic substances, pesticides and fertilizer
- Regulations for electrical appliances
- Regulations for cordless phones, radio equipment etc.
- Textiles and garments labeling
- Testing vehicles and accessories
- Regulations for ships and ship equipment
- Safety regulations for toys

c. *Differences between SPS and TBT Agreements*

Under the auspices of WTO, SPS and TBT Agreements were signed along with many other agreements including AoA. However, SPS and TBT agreements have not received the kind of attention they should have from industry and researchers alike. There is a lot of confusion regarding understanding the difference between these two Agreements. The distinction between the two is as follows: The SPS articles refer to food and agricultural sector alone, while TBT measures refer to all products including food products. SPS agreement aims to protect human, animal and plant life or health from pest and diseases arising out of imports of food and agricultural products. On the other hand, TBT agreement deals with product specifications which include size, shape, weight and packaging material requirements including labeling and handling safety. Tables 5.82 and 5.83 and Figure 5.36 clearly shows the distinctions between SPS and TBT Agreements:

15. Food Quality Management Systems

It is quite difficult to define the term 'quality' of food, unless it makes reference to particular criteria or standards, as the general term 'quality' is subjective. In the most generic sense, quality refers to the combination of characteristics that establish a product's acceptability. In the food industry, this is usually an integrated measure of purity, flavor, texture, color, appearance and workmanship. In a highly competitive market, another criteria of quality can be 'value' or perception of the product's worth. There are two distinct aspects of food quality management for agro-enterprises, yet they are both are inter-related. The first approaches 'quality' in terms of conforming to certain market requirements such as a perceptible superiority of desirable traits or characteristics like size, coloration etc. The second approaches 'quality' as a synonym for food safety, which can be also used as a marketing tool to move product across different countries. Various quality management systems are

Table 5.82: Differences between SPS and TBT agreements.

SPS	*TBT*
SPS refers to food and agricultural sector alone that too with reference of quarantine measures only. *i.e.,* SPS refers to food risk aspects only	TBT deals with wide range of products compared to SPS that too of all the sectors *i.e.,* agriculture/food sector, industry and service sectors, other than quarantine aspects/food safety aspects
SPS deals with the setting of the basic rules for food safety, animal and plant life/health standards	TBT deals with product specifications which include size, shape, weight, technical requirements, certification procedures, testing procedures, labeling and handling safety, packaging material requirements of products, so as to avoid unjust obstacles in trade
SPS measures can only be introduced to the extent necessary to protect life or health on the basis of scientific justification	TBT Agreement technical regulations can be introduced for a variety of reasons such as, national security, the prevention of deceptive practices, the protection of human health or safety or the environment.
Ex: Presence of aflatoxin level in groundnut	Ex: Measures regarding pharmaceutical restrictions or the labelling of cigarettes are considered TBT measures, despite their objectives to protect human health, because they do not involve food-borne diseases or diseases carried by plants or animals.

Table 5.83: Differences between SPS and TBT Agreements with reference to different measures.

Controlling Agreement	*SPS*	*TBT*
Regulation on pesticides	If the measure relates to residues in food or in animal feed, and the objective is to protect human or animal health.	If the measure relates to the quality or efficacy of the product, or to a health risk to handlers.
Establishment of labelling requirements for foods	If the measure is related to food safety.	If the regulation concerns such matters as the size of the typeface used on the label, the presentation of information on contents, grade, etc.
Regulation on containers for the shipment of grains	If the regulation is on fumigation or treatment of containers, *i.e.* disinfection of containers in order to prevent the spread of disease.	If the regulation relates to the size or structure of the containers.
Food labelling	Regulation on permitted food safety: health warnings, use, dosage	Regulation on label's position, lettering, nutrient content, quality
Containers for shipping grain	Regulation on fumigation, disinfectant, to prevent disease spreading	Regulation on size, construction/structure, safe handling
Import of fruits	Regulation on treatment of imported fruit to prevent pests and diseases spreading	Regulation on quality, grading and labelling of imported fruits
Bottled water specifications for the bottles	☆ Materials that can be used because safe for human health ☆ No residues of disinfectant in water	☆ Permitted sizes to ensure standard volumes ☆ Permitted shapes to allow stacking and displaying shapes

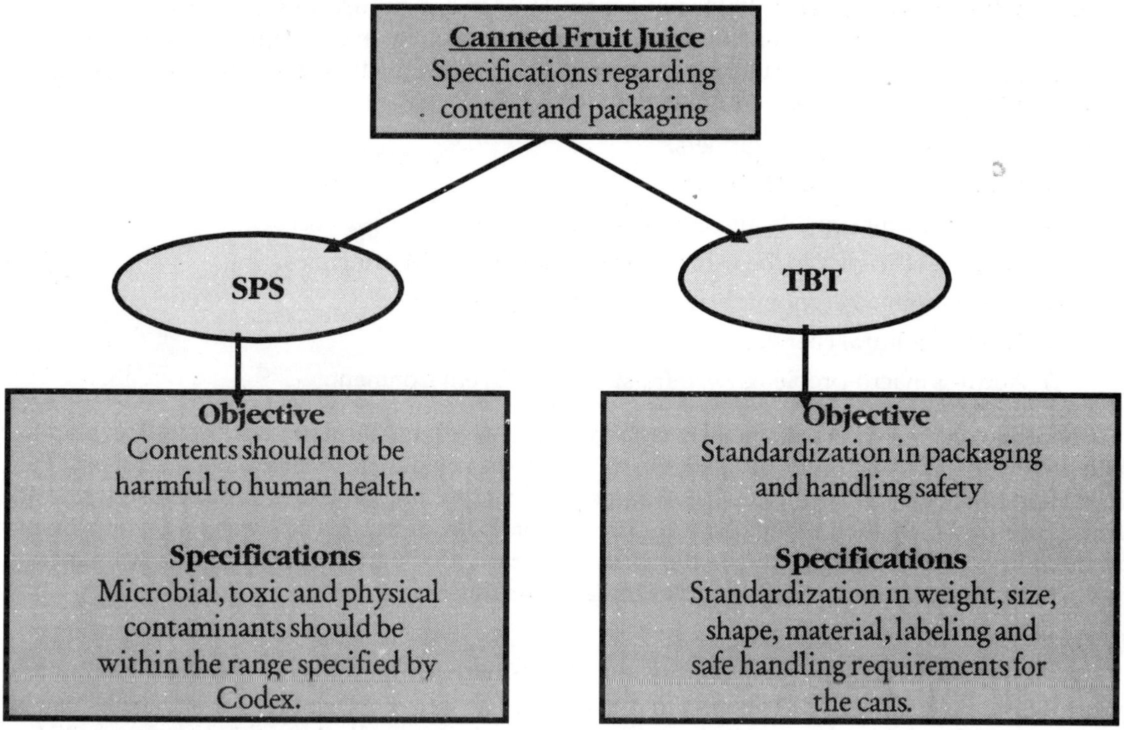

Figure 5.36: SPS vs TBT measures.

being applied to help ensure food safety and they include Total Quality Management (TQM), Good Agricultural Practices (GAP), Good Handling Practices (GHP), GMP and HACCP. These are discussed in-detail here under.

a. TQM

India enjoys significant position in the world with respect to arable land, different agro-climatic zones, production of major commodities, adequate rainfall, irrigation potential etc. By application of modern management practices and scientific techniques and converging these advantages, we can boost farmers' income and enhance competitiveness of agricultural scenario both in domestic and international markets. Despite these distinct advantages, the Indian agriculture is plagued by bottlenecks of low productivity, vagaries of monsoons and poor price realization of agricultural produce due to seasonal glut or poor production. With about 60 per cent area under rainfed cultivation, the Indian agriculture still remained gamble with monsoon, as 90 per cent of rainfall in the country is restricted to four months of the year. In many ways, onset of monsoon and rainfall decides the fate of farmers and the state of country economy. The agricultural productivity in India has been doubled in the last half century, but this jump did not succeeded in providing continuous increase in the income of farmers. Gone are the days, when farmer used to get better production and in turn better price for the produce through application of costly agricultural inputs. The Government on various forums and at different levels emphasized on the needs of management of farm produce in order to stop the huge wastage, which is currently estimated to be about 28 per cent of total agriculture produce in the country. In this context, it is essential to implement TQM in agriculture, as this is a step in the right direction to enable the farmers to enhance quality management at every stage of crop production. Through application of TQM, the farmers ensure cost-effective production and thereby, realizes higher profits. The following aspects are considered under TQM:

- ☆ Take quality thought to farmers.
- ☆ Train farmers in 'quality' skills.
- ☆ Ensure product quality and consumer satisfaction.
- ☆ Improve efficiency in the value chain by applying TQM principles.
- ☆ Bring sustained economic benefit by reducing wastage.
- ☆ Bring in cultural change.
- ☆ Address macro-problems *e.g.* infra-structure and environment.

So, at its core, TQM is a management approach to achieve long–term success through revamping both production and marketing strategies in tune with the consumers' needs and expectations. In a TQM effort, all members of an organization participate in improving processes, products, services and the culture in which they work. This application of TQM in agriculture results in improving productivity, cost reduction and better price realization of agricultural produce, enhancing competitiveness etc., and thereby, launches Indian agriculture on global map.

b. GAP

GAP are basic food safety principles associated with minimizing biological, chemical and physical hazards from field through distribution to consumers. The term GAP can refer to any collection of specific methods, which when applied to agriculture, produces results that are in harmony with the values of the proponents of those practices. To ascertain that a practice is 'good', it all depends upon the standards with which it is applied in the agriculture. FAO defined GAP as *'the collection of principles to apply for on-farm production and post-production processes, resulting in safe and healthy food and non-food agricultural products, while taking into account economical, social and environmental sustainability'*. GAPs may be applied to a wide range of farming systems and at different scales. They are applied through sustainable agricultural methods such as Integrated Pest Management (IPM), Integrated Nutrient Management (INM) and conservation agriculture. The objectives of GAPs include,

- ☆ ensuring safety and quality of the produce in the food chain
- ☆ capturing new market advantages by modifying supply chain governance
- ☆ improving natural resources use, workers health and working conditions
- ☆ creating new market opportunities for farmers and exporters in developing countries.

Thus, GAPs rely on the following four principles:

- ☆ Economic and efficient production of sufficient (food security), safe (food safety) and nutritious food (Food quality)
- ☆ Sustain and enhance natural resources (Environmental sustainability)
- ☆ Maintain viable farming enterprises and contribute to sustainable livelihoods (Economic viability)
- ☆ Meet cultural and social demands of society (Social acceptability).

These four 'pillars' of GAP (food security, safety and quality; environmental sustainability; economic viability and social acceptability) are included in most private and public sector standards, but the scope which they actually cover varies widely. These principles of GAPs may serve as a reference tool for deciding on practices and/or outcomes that are environmentally sustainable and socially acceptable at each step in the production process. The implementation of GAPs should, therefore, contribute to Sustainable Agriculture and Rural Development (SARD). However, the concept of GAPs has changed in recent years because of a rapidly changing agriculture, globalization of world trade, food crisis, appearance of pesticide resistance, soil erosion etc. GAPs applications are being developed by Governments, NGOs and private sector to meet farmers and transformers needs and specific requirements. The GAPs provide the opportunity to assess and decide on which farming practices to follow at each step in the production process.

Produce may become contaminated with pathogens anywhere in the farm-to-table chain. If produce becomes contaminated, there is no process other than thorough cooking to ensure elimination of the pathogens. Since cooking is not appropriate for produce bound for fresh markets, prevention of contamination is imperative to assure a safe product. The use of GAPs during growing, harvesting, sorting, packaging, and storage operations for fresh fruits and vegetables is key to preventing pathogen contamination. Key areas of concern when implementing a GAP programme are prior land use, adjacent land use, water quality and use practices, soil fertility management, wildlife, pest, and vermin control, worker hygiene and sanitary facilities, and harvesting and cooling practices.

GAPs Related to Soil

They include:

- ☆ Reducing erosion by wind and water through hedging and ditching
- ☆ Application of fertilizers at appropriate time and in adequate doses
- ☆ Maintain or restore soil organic content, by manure application, crop rotation etc.
- ☆ Reduce soil compaction issues (by avoiding using heavy mechanical devices)
- ☆ Maintain soil structure by limiting heavy tillage practices
- ☆ *In situ* green manuring by growing crops like cowpea, horse gram, sunhemp etc.

GAPs Related to Water Management

They include,

- ☆ Practice schedule irrigation with monitoring of plant needs and soil water reserve status to avoid water loss by drainage
- ☆ Soil test based nutrient application to overcome soil salinity/alkalinity problems
- ☆ Avoid crops with high water requirements in a low availability region
- ☆ Avoid fertilizer run-off
- ☆ Maintain permanent soil covering in particular, in winter to avoid nitrogen run-off
- ☆ Manage the water table level
- ☆ Restore or maintain wetlands
- ☆ Provide good water points for livestock
- ☆ *In situ* water harvesting by digging catch pits, crescent bunds across slope

Potential Benefits of GAPs

They include,

- ☆ Appropriate adoption and monitoring of GAPs helps improve the safety and quality of food and other agricultural products.
- ☆ It may help reduce the risk of non-compliance with national and international regulations, standards and guidelines (in particular of the CAC, OIE and the IPPC regarding permitted pesticides, maximum levels of contaminants (including pesticides, veterinary drugs, radio-nuclide and mycotoxins) in food and non-food agricultural products, as well as other chemical, microbiological and physical contamination hazards.
- ☆ Adoption of GAPs helps promotes sustainable agriculture and contributes to meeting national and international environment and social development objectives.

c. Good (Food) Handling Practices (GHP)

These are general food safety procedures that should be followed to help reduce the risk of contamination and mishandling at all levels of a food establishment. From the time the food is delivered to the minute it is served to the consumer, food safety should be on the top of the list. Following are the basic procedures that help keep food safe and prevent food borne illnesses:

i. Purchasing and Receiving

All foods must come from an approved source. So, while purchasing the food, plan with the supplier(s) to ensure the foods are transported by the safety standards. When receiving a delivery, time and temperature is the biggest concerns. Foods must be received and stored as soon as possible. All refrigerated foods should be put away quickly to prevent time and temperature abuse. Frozen foods should not have large ice crystals, be discolored or dried-out. Canned goods should have labels, no swelling and flawed seams, rust or dents. Never accept home-canned foods because of the risk of botulism.

ii. Proper Food Storage

It is important to store foods properly to prevent cross-contamination, food spoilage and pest control issues.

- ☆ ***Dry goods storage:*** All canned foods and dry ingredients must be stored in a designated area. This area should be well ventilated and pest free. Dry storage areas can become a food source for rodent and insects. Keeping containers closed, in good condition and off the floor help to keep the storage area pest free. Stock rotation is a good management practice. Foods should never be stored in areas such as rest rooms, furnace rooms, stairwells or hallways. Never store food products on the floor or near chemicals. Chemicals should be stored below and away from foods to prevent chemical contamination.
- ☆ ***Walk-in refrigerators:*** The walk-in refrigerator is the major cold storage area in a food service establishment. The temperature of a walk-in refrigerator must be sufficient to adequately hold the food temperature at 41°F or below. The temperature of a walk-in refrigerator is usually colder than 41°F to compensate for the opening and closing of doors and demands of adding additional foods for storage and cooling.
- ☆ ***Prevent cross-contamination:*** Foods need to be stored in a certain way in order to prevent cross contamination. Cross-contamination occurs when bacteria are transferred from one source to another. Most commonly, it is because of the way food is stored, transported, and

the ways the food preparation areas are cleaned. All cooked foods and foods that will receive no further cooking should be stored above other foods. Store foods in the following order (from top shelf to bottom shelf):

- RTE foods (fruits, vegetables, cooked items, etc)
- Raw seafood (fish, shrimp, shellfish, etc)
- Raw whole pork and beef (roasts, ribs, steaks, briskets, etc)
- Raw ground meats (beef, pork, turkey, etc) and eggs (pooled or shelled)
- Raw poultry and stuffed meats

☆ ***Temperature control:*** Temperature is the key factor that controls the growth of bacteria in food. Many types of pathogens and spoilage bacteria are prevented from multiplying to significant levels that cause food borne illness with proper holding practices.

- ***Room Temperature Storage:*** All foods need to be stored cold 41°F and below or hot 135°F or above. When food is held at room temperature, there are certain control measures that can be taken. Control measure to ensure foods are out of temperature for very limited times include:
  - ❑ Refrigerate foods before preparation.
  - ❑ Prepare foods not further in advance than necessary.
  - ❑ Prepare small batches of food and return them to the refrigerator before cooking and serving.
- ***Cooking Temperatures:*** It is important to cook foods to the appropriate temperatures. The time needed to cook raw protein, specifically, is based on the pathogens and bacteria found in the food. The amount of heat required to kill the various bacteria depend on the species of microorganisms. To kill all pathogens in foods, cooking must bring all parts of the food to required temperature for a correct length of time. It is important to remember that, it does not matter what temperature the grill, stove, or oven is set to, the temperature that matters is the internal temperature of the food being cooked. You should insert a food thermometer into the middle (or thickest part) of the food in order to get an accurate internal temperature. The following are the requirements for different foods:
  - ❑ Hot holding of cooked food products - 135°F
  - ❑ Beef, pork, fish, seafood, eggs - 145°F
  - ❑ Ground or fabricated meat - 155°F
  - ❑ Poultry, stuffed meats, casseroles, or re-heated foods - 165°F
- ***Cold holding of food:*** Cold holding is storing food under refrigeration at 41°F or below. Refrigeration prevents food from becoming a hazard by slowing the growth of most bacteria. Keeping foods cold does not stop the growth of bacteria, but it will slow it down. If you notice food that appears to have mould growth or smells off, discard it. A good motto to adopt in this instance is, '*When in doubt, throw it out.*'
- ***Hot Holding of Foods:*** Once a food is heated or cooked, the food must be maintained at a temperature to limit the growth of bacteria. The correct hot holding temperature is 135°F. If the food falls below this temperature, it must be corrected within 2 hours or the food must be thrown out.

☆ ***Thawing, Reheating and Cooling of Foods***

- ***Thawing foods:*** Thawing foods may take several hours or days depending on the size of the food item being thawed. Thawing must be done so that, the risk of cross-contamination is reduced, and the time that potentially hazardous food is in the temperature danger zone (41°F to 140°F) is kept to minimum.
- ***Reheating foods:*** Proper reheating can eliminate pathogens. Proper reheating temperature is 165°F within two hours. The more a food is cooled and the reheated, the greater the risks. There is an increased risk from contamination caused by personnel, equipment, procedures and the fact that food will go through the danger zone several times which also increases the risk of bacteria growth causing food borne illnesses.
- ***Cooling of foods:*** When cooling foods, it is important to remember that, time is of the essence. When food is cooled gradually or over an extended period of time, it allows bacteria the opportunity to grow. The ideal temperature for severe pathogen growth is between 80 - 120°F, but it is important to limit the time food is under 135°F and above 40°F. The ideal way to cool foods quickly is to place the food in an ice water bath. Place the food in an uncovered container into a basin. Fill the basin around the food container and cover the ice with water. This allows a greater surface area for the ice water to cool. Once it reaches 41°F or below, you can cover the food and store for a longer period of time.

iii. Serving Procedures

Develop good serving procedures to protect food and consumer's health. They include,

- ☆ Employees should wash their hands after cleaning tables and after touching any item that can contaminate their hands
- ☆ Avoid touching RTE foods with bare hands. Use a utensil such as a tong, deli tissue or gloves.
- ☆ Do not re-serve unwrapped bread, rolls, crackers, salad dressings, or relish trays.
- ☆ Avoid touching the food-contact surfaces of glasses, cups, plates or tableware.
- ☆ ***Proper glove use:*** Using gloves does not replace the need for hand washing. The server must wash his hands properly before and after wearing gloves. When using gloves, it is important to note that, gloves must be changed according to the task. Discard gloves when they become soiled or damaged.
- ☆ ***Proper Handling of RTE Foods:*** It is also important to remember that food handlers are required to use utensils or wear gloves when handling RTE foods. There should never be any bare hand contact of RTE foods.

d. Good Manufacturing Practices (GMPs)

GMPs are guidelines that a company uses to evaluate the design and construction of food processing plants and equipment. These standards require that, all stainless steel and plastics used during the processing steps must meet food-grade specifications. The GMPs also direct all persons working in direct contact with food, surfaces that food might contact and food packaging materials, to conform to sanitation and hygiene practices to the extent necessary to protect against contamination of food from direct or indirect sources. The procedures like wearing white uniforms, hair-nets, disposable gloves, face masks and other protective gear fall under GMPs. Similarly, standards for cleaning and sanitizing practices in food processing plants and equipment are also outlined in the GMPs. The treatment of water to make it of drinkable quality, the filtering of air, and the treatment of food processing wastes are also addressed in the GMPs. The management of unavoidable pests in food processing plants and warehouses is also done to ensure that GMPs are used. In effect, the GMPs cover every aspect of the processing of food.

e. HACCP System

In the 1960s, NASA wanted to guarantee that, the food for astronauts on space flights was totally safe. The task of producing 'Zero Defect' food was contracted to the Pillsbury Corporation, who developed the first system of HACCP (pronounced Hassep). HACCP is a risk management tool that provides a more structured and arguably more effective approach to the control of processing or manufacturing hazards than that achievable by traditional inspection and quality control procedures. HACCP is the main platform for international legislation and good manufacturing practices for all sectors of the food industry. Rather than testing the end product for failure, it functions to prevent failure by systematically controlling the process. It requires systematic analysis for potential risks and then identification of appropriate control and monitoring systems, particularly those deemed critical to the safety of the product. HACCP significantly reduces the risk of food contamination in two ways: First, it anticipates potential problems or failures and does not depend only on a final inspection. Second, because it identifies problems during the process rather than at the end of the process or once the product moves into the supply or marketing chain there is a greater likelihood of resolving the problem at hand as opposed to pursuing a product recall. Not surprisingly, HACCP can thereby, also yield potential cost savings in product wastage, reprocessing, or recalls should problems occur. An important CAC guideline for food processing companies is to follow HACCP food quality management system. The importance of HACCP guidelines can be elicited from the following points:

- ✰ United States (US) and European Community (EC) have already made this system mandatory for food processing firms.
- ✰ EC put a ban on imports of fish from companies in Gujarat, which did not adopt HACCP system.
- ✰ Herbal product exports from India were severely affected, as US planned to impose ban on imports of these products if they did not confirm to HACCP.
- ✰ Indian seafood processors, in their bid to remain competitive in the US market, are taking help from foreign consultants at exorbitant cost to implement HACCP in their production units.
- ✰ Multinational companies like Nestle-India have already planned to implement HACCP for coffee processing.

Thus, the use of HACCP guidelines in export manufacturing processes conveys a level of professionalism and safety that often facilitates foreign trade transactions. Since HACCP has the potential to identify areas of concern, where failure has not yet been experienced, it is particularly useful for new operations. Many people, particularly those from industry, feel that the use of the HACCP system will significantly reduce the need for traditional Government inspection. However, the ability of the HACCP system to deal with emerging pathogens has still to be determined. In fact, HACCP is far from perfect, one weakness is related to the pre-existing presence of microorganisms, chemical and physical hazards in foods and the receiving of such contaminated foods into the processing plant. Once contaminated raw materials have been allowed to come into a plant, an incredible amount of remedial work has to be carried out in order to ensure that, these microorganisms are killed during processing and that cross-contamination does not take place. Since many products exposed to cross-contamination do not have a definitive killing-step for microorganisms in their processing, the end result can be contaminated products. Although it is a significant improvement in overall process control, HACCP is definitely not a panacea for the problems of microbial contamination in foods. There has been discussion of introducing HACCP for all processed food, but it is not yet legally required. However, it has become a de facto requirement for most supermarkets.

Thus, food products are expected to meet the limits prescribed under AGMARK/BIS/ISO/Codex or other standards etc. The parameters to be satisfied include moisture, foreign matter, damaged or insect infested materials, uric acid, aflatoxins, poisonous metals, insecticides, rodent hair, excreta etc. The International Food Trade is enormous and is growing fast with the ever-more rapid means of communication and transport as well as the new technologies of food preservations. Now in the trade, the provision for health, safety and requirements for various ingredients are to be primarily and essentially ensured. In this context, observation of standards and their harmonized versions on product specifications, terminology, sampling, methods of tests and even certain codes of practice are unquestionably important. Under these situations, international trade for food products particularly from developing countries like India is likely to face enormous trade hardship, if one does not take note of these- aspects right now. Much attention is now required to be paid at the primary production level of the agricultural produce. TQM, GAPs, GHPs, GMPs and HACCP relating to food management must be strictly followed to ensure a healthy trade environment.

B. Market Financing

It refers to the provision of capital (credit) to carry out the various activities in the marketing process. Considering the gestation period in the crop production programme and inevitable time lag between the harvest and efficient marketing of farm produce, the farmers are in essential requirement of finance at various stages. In the context of agricultural marketing, market finance is essential to the farmers to overcome distress sales of produce immediately after harvest and to sustain them in the agri-business. The Indian farming community consists mostly of small and marginal farmers. Micro level studies indicate that, small farm holdings contribute about 54 per cent of total marketable surplus and distress sale by these small farmers account for about 50 per cent of the total marketable surplus. The farmers often sell their produce to clear off their debts soon after harvesting. The solution for this problem lies in providing to them access to safe and scientific storage and easy marketing credit. There are different categories of market financing and they are discussed here under.

1. Pledge Loan Scheme

In order to avoid distress sales by the small and marginal farmers in the peak marketing season, Pledge Loan Scheme has been introduced in all the APMCs since 1994-95. It is to be noted that, in the past 15 years, loan programmes for major crops have moved from price support programmes to marketing loan programmes. So, the strategy should be promotion of pledge financing through a network of rural godowns and NWR system. Under this Scheme, regulated markets are issuing pledge loan to the farmers against pledge of their produce. The following are the important features of this scheme:

- ☆ There is no budgetary restriction for giving advances under this scheme. 25 per cent of market fee revenue is earmarked for this purpose.
- ☆ All farmers, who sell their agricultural produce in the regulated markets are eligible under this scheme.

- Every farmer shall be eligible for the Pledge Loan subject to a maximum of ₹ 1.00 lakh or 75 per cent of the value of the produce pledged, whichever is less.
- Under this scheme, the farmers can store their agricultural produce in the godowns of the regulated markets for a maximum period of six months.
- Interest will not be charged on the loans sanctioned under this scheme for the first 15 days. Interest at the rate of five per cent will be charged beyond 15 days.
- Pledge loan is also sanctioned to the farmers, who store their produce in the godowns of SWCs or CWC, where APMCs do not have their own godowns or due to non availability of space in their godowns. The warehousing corporations concerned shall have to give an undertaking to the APMC that, they will not allow the farmers to remove or sell the stocks without the specific and prior approval of the concerned APMC.
- Pledge loan facilities are extended to traders also, with the rate of interest specified from time to time. Pledge loan of 50 per cent of the purchase value up to the maximum ceiling of ₹ 1.00 lakh is given to the licensed traders. Traders can stock their purchased agricultural produce in the regulated market itself for a maximum period of three months for which the interest rate is nine per cent.
- The time limit for sanction/rejection for pledging the produce both to the farmers and traders is one week.

Objectives of Pledge Loan Scheme

They include:

- Increases the retention capacity of the small and marginal farmers to avoid distress sales of agricultural produce
- Minimizes the farmers' dependence on the commission agents and private money lenders, as the pledge finance provides financial support to them immediately after harvest period
- Participation of the farmers, irrespective of their land holdings, increases the arrivals in market yards
- Gives a sense of security to the farmers, even if their produce is not sold out in the market yard immediately
- Provide liquidity to the farmers
- Enable farmers to hold on their produce during the period of bumper crop harvest, when market prices are ruling at the lowest
- Facilitate farmers to realize remunerative price by selling their farm produce during off-season when market prices are ruling high
- Promotes cleaning, drying and grading of farm produce
- Promotes proper storage facilities
- Facilitates better price realization by farmers
- Avoids glut conditions in market.

In some States, DCCBs are directly financing individual farmers on the basis of pledge. In States like Andhra Pradesh, Tamil Nadu, Bihar, Uttar Pradesh, Karnataka, Maharashtra, Rajasthan, Haryana and Punjab the scheme of pledge finance is being operated by the APMCs (Table 5.84).

Table 5.84: Pledge finance of agricultural produce – Grant of advance by the APMCs in different states.

States	*Details of Pledge Finance Advanced to Farmers*
Andhra Pradesh	A Scheme under Andhra Pradesh (Agricultural Produce and Live Stock) Markets Act, 1966 provides advances against pledge to the producers. The advance is given to the extent of 75 per cent of the value of the produce pledged with the Market Committee subject to maximum limit of ₹ 50,000/-. The pledge stocks may be sold within 90 days. The advance is free of interest for the first 90 days. Interest @ 6 per cent per annum is charged from the 91st day till the date of disposal
Tamil Nadu	In order to avoid distress sales by the small and marginal farmers in the peak season, Regulated Markets are issuing pledge loan to farmers. Under this scheme, the farmers can store their agricultural produce in the godowns of Regulated Markets for a maximum period of 6 months and take pledge loan of 75 per cent of the total value of the produce up to a maximum of ₹ 1,00,000. Likewise Pledge Loan facilities are extended to traders also with the rate of interest specified from time to time. Interest at the rate 9 per cent from traders is charged for pledge loan facilities.
Uttar Pradesh	A scheme of pledge-finance is being operated through Market Committees. According to the scheme, advance is given to the extent of 75 per cent of the value of the produce pledged with the market committee subject to maximum limit of ₹ 5000/- and ₹ 2500/- to small and marginal farmers respectively. The advance is free of interest for the first 30 days. Interest six per cent per annum is charged from 31st day till the date of disposal. The Market Committee does not charge any godown rent for the first 7 days. From the 8th day, the charges are 10 paise per bag per month or part thereof. The pledged stocks may be sold in 90 days.
Karnataka	The state of Karnataka has implements the pledge loan scheme through which the market committees are advancing loans against the pledge of agricultural commodities to the farmers. In fact, the market committees statutorily earmark 10 per cent of their annual income towards this scheme. This scheme was initiated on 01-04-1994 and it operates only in 132 markets of the state. During the time of fall in price, agriculturist can pledge their produce in the APMC godowns or warehouses and avail short-term loan to a maximum of ₹ 50, 000/- or 60 per cent of the value of the agricultural produce whichever is lower for a period of 90 days. No interest is charged for the first 30 days, interest at the rate of 8 per cent and 12.5 per cent is charged for the next two successive 30 days period, respectively. The market committee is empowered to dispose off the produce after 90 days in case the farmer fails to repay the loan.
Maharashtra	Since 1990, the Maharashtra State Agriculture Marketing Board (MSAMB) has implemented a scheme of pledge finance for the benefit of farmers of the State. The scheme of pledge finance is already popular among the farmers growing moong, tur, urad, soya bean, paddy, sunflower and chana. From 2008-09 MSAMB has Expanded this scheme for jowar, bajra and maize. This has safeguarded the interests of the farmers, who otherwise had no option but to sell at whatever low price, that is prevalent during peak harvest season. Under the scheme, the farmer is given a loan to an extent of 75 per cent for the value of the produce prevailing in the market (for jowar, bajra and maize 50 per cent or ₹ 300/- which is less), at an interest rate of 6 per cent (well below the market rate). The farmer is allowed to avail this facility upto a period of 180 days. There are 36 APMCs who are actively participating in this scheme. So far in the year 2004-05 MSAMB has advanced ₹ 15.30 crores as pledge finance under the scheme. This scheme has evoked good response from farmers of the Vidarbha and Maratwada region.
Rajasthan	The advance to the extent of 60 per cent of the value of the produce pledged with the Krishi Upaj Mandi Samities is paid subject to a maximum limit of ₹ 15000/-. The advance attracts concessional rate of interest of 9 per cent for the first 60 days and 12 per cent for the next 90 days. Maximum duration of storage allowed under pledge scheme is of 150 days (5 months). The pledger is liable to pay the warehousing charges for the period his goods are stored in the godowns. The Krishi Upaj Mandi Samities have the right to sell the produce after five months from the date of storage of the stocks and thereafter through open auction. The Government of Rajasthan also operate schemes for 'Pay-back', fee collection to small and marginal farmers and for free transport facility to small and marginal farmers for agricultural produce.

Contd...

Table 5.84–*Contd...*

States	*Details of Pledge Finance Advanced to Farmers*
Haryana	The Scheme regarding pledge of agriculture produce is being implemented by Haryana Government through Central Bank of India and Punjab National Bank. The Government is also thinking to associate the Co-operative Banks for granting advances to the farmers against warehouse receipts of Haryana Warehousing Corporation. The Haryana State Agriculture Marketing Board is not directly participating in the scheme.
Punjab	As regards schemes for granting loan to the farmers against their produce, the matter has been linked up with the mechanization of mandi operations. The scheme for partial mechanization of mandi operation is in vogue in eight mandies on pilot project basis. Its extension along with provision of storage facility and grant of loans against such storage depends upon successful operation of mechanization programme.

As discussed through Figure 5.30, by showing the warehouse receipt, the banks will offer pledge loans to the needy farmers. This system ensures large benefits such as, increased liquidity in rural areas, lower costs of financing, shorter and more efficient supply chains, enhanced rewards for grading and quality, development of other productivity enhancing agricultural services and better price-risk management. All these developments will result in higher returns to farmers, better service to consumers (due to lower prices, better quality and greater variety) and macro-economic benefits through a more healthy trade balance in agricultural commodities. Introduction of this system for agricultural commodities will also enhance competitiveness of Indian agriculture in the domestic and international markets. The aim is to greatly expand the availability of warehousing services, while making NWR a prime tool of trade and trade financing throughout the country. It will also enable the banks to improve the quality of their lending portfolio to the agricultural sector. However, the banking institutions are, at present, hesitant in making advances against the CWC warehouse receipt, when the holder thereof, is not a person in whose favour the receipt was originally issued. Transferability of the warehouse receipt by endorsement is presently further limited by the fact that, the original holder of the warehouse receipt cannot transfer it to another person without clearing the bank loans. This inhibits the negotiability of the warehouse receipt and reduces its usefulness to the depositor, who cannot sell his produce before settling his loan with the banks. The State Warehouses Acts provide that, a receipt issued by a warehouseman is transferable by endorsement and shall entitle its lawful holder to receive the produce specified in it on the same terms and conditions on which the person who originally deposited the produce would have been entitled to receive them. Further, the present warehouse receipt is a document of title to the produce as per the Sale of Goods Act, 1930. Nevertheless, because of the imperfections in the present structure of warehouse receipts, the usage of the present warehouse receipts remained restricted to be accepted by the commercial banks as a collateral security for grant of loans against the produce stored in the warehouses and the present warehouse receipts has not yet gained its acceptability as a negotiable instrument that could be freely transferred from one persons to another. This can be accomplished by creating a secure system, where warehouse operators are accredited by the banks and where investors can build warehouses in the knowledge that they can gain accreditation provided they meet prescribed standards. A system of quality certification and grading of commodities will have to be established, with a view to minimizing disputes and permitting cost savings through the combining of stocks of different owners. The status of warehouse receipt should to be enhanced through legal changes for creating an effective system of regulation in claiming pledge loans.

Despite the implementation of pledge loan scheme in several states, still the bottlenecks in the efficient marketing of agricultural produce at times of peak marketing season cannot be removed. The major constraint arises from the availability of limited capital (pledge financing) from the formal banking channels. The quantum of financing done both by commercial banks and co-operative banks for pledge financing is very little as compared to the crop production loans. According to the RBI guidelines, bank advances upto ₹ 1 lakh can be given to farmers against pledge/hypothecation of agricultural produce (based on warehouse receipt) for a period not exceeding six months subject to the condition that, farmers have availed crop loan from the same bank. Such advances are included as direct finance to farmers for agricultural purposes under priority sector lending. There is no bar on banks extending pledge loans for periods up to 12 months. However, this would not be an automatic extension, but will depend on the nature of crops stored in the godown and the appropriate time to sell the produce and would be left to the financing banks' commercial judgment. To promote pledge loans in respect of high value crops, RBI should enhance the ceiling of advances from existing ₹ 1 lakh to up to ₹ 5 lakhs to farmers against pledge/hypothecation of agricultural produce (based on warehouse receipt) and the repayment of these loans may also be extended from the existing six months to up to one year depending on the nature of produce stored in the godowns and the appropriate time to sell the produce. Crops which are subjected to wide fluctuations in prices need to be identified and marketing credit policies should be specifically designed for them. Banks should be encouraged to augment the resources of State Marketing Co-operatives, which provide pledge financing facilities to farmers. Regional Rural Banks (RRBs) have an extensive reach through their wide spread branches all over the country. At present, NABARD does not support RRBs through its refinance for pledge financing loans. So, NABARD should provide 100 per cent refinance to RRBs, on similar lines as that of co-operative banks. Arrangements should be developed so that, the warehousing receipts/godown receipts of private sector are acceptable to the bankers for providing credit to farmers. Since pledge financing is considered to be crucial to farmers to enhance their holding capacity and to obtain remunerative price for their produce, it is recommended that, RBI should monitor pledge financing to farmers within the overall target of 18 per cent of Net Bank Credit (NBC) to agriculture, fixed for commercial banks.

2. Minimum Floor Price Scheme

To safe guard the farmers from distress sales, Minimum Floor Price Scheme has been formulated. We know, Government will announce MSPs for the produce and at times of distress sale, it arranges for procurement of the same. For this purpose, Revolving Fund has been set up with the APMCs in each State, by contributing 0.5 per cent of the market fee collected every month. This Scheme was first implemented in Karnataka in 1999 and many States are evincing keen interest to adopt the same for the benefit of farming community.

Besides above two schemes, the APMCs also implement Raitha Sanjeevini Scheme (If a farmer or his family members in the age group of 15 to 80 dies, while performing the agricultural operations or selling agricultural and horticultural commodities in the market yards, a compensation of ₹ 50,000 is sanctioned.), Hamals Insurance Scheme (If a hamali dies naturally he gets a compensation of ₹ 30,000. If the death is due to accident then the compensation to be paid is ₹ 75,000). Though these two schemes does not directly check the distress sale of produce by the farmers at times of peak marketing season, but they certainly contribute to higher marketing efficiency in transacting the produce.

C. Market Intelligence

It refers to the collection, interpretation and dissemination of market information. It is an important tool for decision making at all the stages right from farm production to ultimate consumption for all the participants in the marketing channel. It is essential for farmers in planning production and market led production. It is equally important for other market participants for trading and also for consumers. Noticing its importance, the Government of India has launched Agricultural Marketing Information Network (AGMARKNET) Scheme through DMI to bring out improvement in the present market information scenario by linking all Agricultural Produce Wholesale Markets in the States and UTs in a phased manner. The decisions regarding what to buy and sell, when to store and how much, where to transport etc., depends upon market intelligence. Agriculture sector needs 'well-functioning market' to drive growth, employment and economic prosperity. For achieving a 'well-functioning market' in the economy, systematic Agricultural Market Intelligence System (AMIS) is indispensable. This helps the farmers and traders in marketing their agricultural produce to fair price and to ensure remunerative returns to them. This further enable the marketing agencies in devising and implementing new marketing technologies to reduce post harvest losses through appropriate post-harvest operations including grading, value addition, packaging, processing and transportation, so as to derive remunerative prices in transacting the produce. We know, agricultural production has increased manifold since Green Revolution. But, the increase in production has not reflected in the levels of living of farmers. This is due to the problems faced by farmers in marketing the agricultural produce. The major problem is non-availability of reliable market information at the right time. Reliable and timely market information and intelligence to the farmers will definitely increase their net income and thereby, improve their living standards. This calls for strengthening the AMIS, as it ensures better performance in agriculture-business as farming itself. (Agri-Business is a process, which starts with a decision to produce a saleable farm commodity). The strengthening of AMIS involves:

- Providing price forecast before sowing and during harvesting of selected farm commodities.
- Providing information on high price markets.
- Improving the dissemination of market intelligence.
- Increasing the capacity to absorb and use such market intelligence by all stakeholders.
- Developing commodity market outlook, so as to ascertain both domestic and export competitiveness.

The main objective of this AMIS is to advise the farmers on the prices that could be realized during harvest time as well as in advance of sowing the crop. Again during harvest, farmers are advised on whether to store the produce or sell the same immediately. While recommending storing the produce for realizing maximum prices, farmers are advised on the method of storage, storage cost, storage losses etc. In this way, the AMIS is more useful to the farmers in taking right sowing and marketing decisions. For this benefit, the Government has started strengthening the AMIS in all the regulated markets and price forecasts are disseminated through the media for the benefit of farmers and other stakeholders. Further, the experts in the APMCs also collect price information of different agricultural commodities and they are econometrically analyzed. The results are compared with prices of futures markets for the commodity in question. These price behaviours are also informed to the traders, so as to ensure competitive environment in transacting the produce in the markets. Similarly, to assess the export competitiveness, price information is also collected for the commodities in major countries, where India enjoys comparative advantage. The impacts of EXIM policy of Government of India on the price of the commodities are also taken into account. The price forecasts are also analyzed and they are released for the benefit of farmers, traders, exporters etc. It is interesting that, in all the major market centres in India, the forecasted prices match with the actual market price with an accuracy rate of 90-

98 per cent thereby, creating confidence among the farmers and traders. Thus, the AMIS is effective and it guides the farmers in taking sowing and marketing decisions. Besides price information, the AMIS established in all the regulated markets in the country also provides useful information on market arrivals, marketable surplus and marketed surplus of different agricultural commodities, grading facilities, different grades of agricultural commodities, market charges for different functions like storage, weighment of produce etc., so as to benefit different stakeholders of marketing process.

1. Types of Market Intelligence

There are two types of market intelligence *viz.*, Market information and Market news. Though these two are used interchangeably, there exist few important differences as discussed here under.

- ☆ ***Market information:*** It relates to information pertaining to market arrivals, prices, storage charges, grading charges, processing charges of different agricultural commodities etc., over a period of time in a market. That is, market information provides historical data on the above aspects. An analysis of these aspects over a period of time helps to formulate the trends for the future so that, farmers and traders can plan their marketing strategies. Further, the analysis pertaining to market prices of the commodities also guide the farmers to plan their crop production programmes. For example, if the prices trend favour paddy in the market, then all the farmers in that locality plan for paddy based cropping system and the area under substitute crop (say, maize) will decline in the current year. Similarly, price trends also guide the farmers, traders and exporters in planning for exports into the international market based on the principle of comparative advantage. Thus, market information refers to longer term market information (Figure 5.37) and it includes:
  - quarterly or annual price reports from market information services
  - product and trader fact sheetso contacts of companies providing services (*e.g.* transport, storage) and inputs (*e.g.* seeds, fertilizers and packaging)
  - descriptions of the marketing chain and how it operates.

  This kind of market information helps farmers decide which crops to grow, how they might market their produce and to whom the produce is to be sold. Some of the information may be available from directories, catalogues and trade magazines. This specialist information needs to be drawn together from different sources and is not likely to be easily available.

- ☆ ***Market news:*** It relates to information pertaining to market arrivals, prices, storage charges, grading charges, processing charges of different agricultural commodities etc., in the current period of time in a given market (Figure 5.37). That is, market news provides latest or current data on the above aspects. The market news guides the farmers and traders regarding when and where to sell the produce. Further, it makes the farmer to decide on marketed surplus and proportion of total marketable surplus to be stocked in the godowns. It is essential to note that, the agency which first receives the market news will be more benefitted over the other. In general, traders receive the market news at a quick pace compared to farmers, as they have strong contacts with the marketing personnel. Considering its importance, the APMCs display the day-to-day price information of all notified agricultural commodities of their respective markets and nearby big markets in the notice boards for the benefit of farmers and traders. This market news forms the base for developing market information about notified agricultural commodities in the long run.

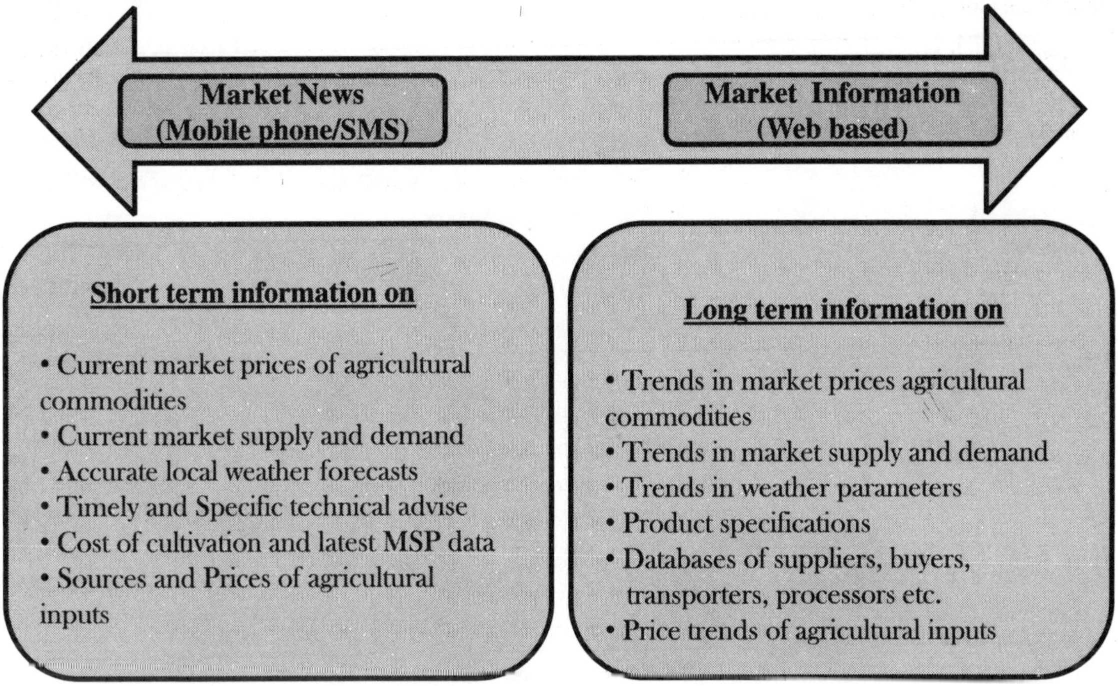

Figure 5.37: Differences between market news and market information.

Thus, market news refers to short-term information, which helps farmers and traders to make instant marketing decisions on selling their products. That is, market news includes up-to-date price information, up-to-date information on supply and demand etc. Government's market news services have been set up in all the States and UTs, but they have experienced problems. They are costly to run. The news provided by several agencies is not always accurate. The time taken to process market news is more thereby, it is generally out of date by the time it reaches the farmer. These days, the most important source of market news in many markets is often the telephone. Buyers and sellers contact each other and provide instant feedback on prices, supply and demand. Access to telephones, particularly mobile telephones, is increasingly important for market-oriented farmers.

2. Examples of Agricultural Market Information

We know, all information about the buying and selling of products and services can be called as market information. Market information is much more than just information about prices and quantities. This definition is broad enough to encompass information about both input and output markets. The Table 5.85 presents several examples of agricultural market information:

3. Importance of Market Information

Market information is critical to the social and economic activities that comprise the agricultural development process. Indian agriculture has witnessed green, white, yellow and blue revolutions and the country even experienced industrial and information technology revolutions. Good communication system and information system reinforce commitments to sustainable agricultural productivity. In this context, market information assumes greater significance, as it serves as an important facilitating function in the agriculture marketing system. It facilitates marketing decisions, regulates the competitive

Table 5.85: Categories and examples of market information.

Categories of Market Information	*Examples of Market Information*
Inputs	☆ Location and contact of input suppliers ☆ Type and quality of inputs available ☆ Prices of different inputs
Demand	☆ Size of local, regional and national markets ☆ Growth or trends in demand ☆ Seasonality of demand
Buyers	☆ Location and contact ☆ Volume requirements ☆ Quality specifications ☆ Packaging requirements ☆ Seasonal demand ☆ Prices offered ☆ Payment conditions ☆ Support services provided (inputs, credit etc)
Prices	☆ Prices in different markets ☆ Price differences according to grade or quality ☆ Price seasonality ☆ Price volatility within seasons ☆ Price trends
Competition	☆ Main supplying areas ☆ Quality of supply from different areas ☆ Seasonality of supply from different areas ☆ Imports
Marketing costs	☆ Transportation costs ☆ Market fees ☆ Unofficial charges, if any

market process and simplifies marketing mechanisms. Market information is a means of increasing the efficiency of marketing system and promoting improved price formation. It is crucial to the farmers to take decisions about what to grow, when to harvest, to which market the produce should be sent and whether or not to store it. Improved information should enable traders to move produce profitably from a surplus to a deficit market and to make decisions about the viability of carrying out storage where technically possible. Market intelligence is defined as *'a service usually operated by the public sector, which involves the collection on a regular basis of information on prices and, in some cases quantities of widely traded agricultural products from rural assembly markets, wholesale and retail markets, as appropriate and dissemination of this information on a timely and regular basis through various media to farmers, traders, government officials, policy makers and others, including consumers.'* Thus, market information is important in all the stages of marketing right from farm to ultimate consumption and simultaneously, for all the participants in marketing. The importance of market information can be studied from the following points:

- It provides the up-to-date information on the arrivals and prices of agricultural commodities of different markets thereby, help the farmers to take right decision, when, where and how much to market their produce (Figure 5.38).
- It educates the farmers and traders about the post-harvest management *i.e.*, techniques to minimize losses during post-harvest period, value addition to the produce by proper cleaning, processing, packaging, storage and transportation etc.
- It orients the farmers, traders and consumers about price trends, demand and supply situation etc.
- It orients the farmer regarding the importance of grading, proper storage, co-operative/ group marketing, direct marketing, contract farming, future trading etc.
- It provides the information about the sources of credit availability (pledge loan), various Government schemes, policies, rules and regulations etc.
- Up-to-date information on prices and other market factors enables farmers to negotiate with the traders and also facilitates spatial distribution of products from rural to urban areas and between markets.
- Market information is also an important component of early warning systems for food security, as it can assist in identifying areas of possible shortage and can highlight whether prices are above or below normal seasonal trends.

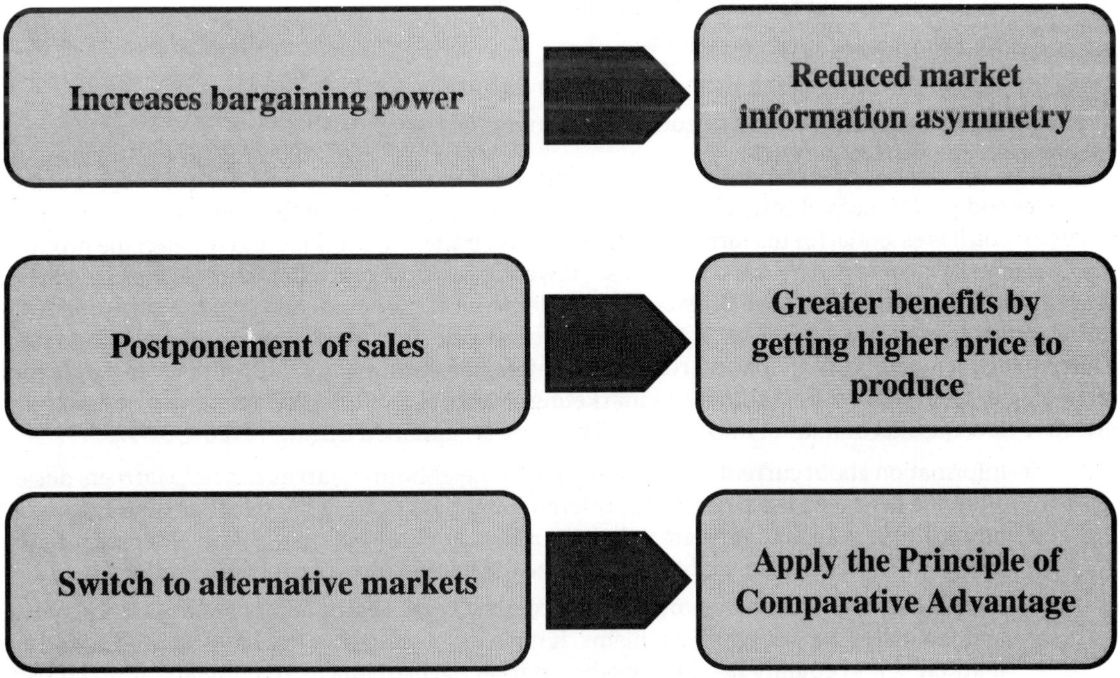

Figure 5.38: Usage of market information by farmers to improve prices and profits.

- Farmers need market information to aid them in planning their operations right from the time they plant the seeds until the produce passes the hands in the market.
- Market information helps the farmers in comparing the prices offered by different firms in different markets and also in the selection of alternative outlets available.
- The market information reduces business risks of farmer-sellers and traders.
- If the market information on commodity prices prevailing in various markets is made available, the farmers would be able to get better price to their produce by moving their produce to the market, which pays higher.
- It educates the farmers/traders about the post-harvest management *i.e.,*
  - harvesting care
  - techniques to minimize losses during post-harvest period.
  - value addition to the produce by proper cleaning, processing, packaging, storage and transportation.
- The market information is important for the farmers in answering the following questions:
  - What and how much should farmers produce?
  - Should farmers grow off-season crops?
  - Which varieties should be planted?
  - Which post-harvest operations should farmers undertake?
  - Should farmers store their crops?
  - Where should farmers sell their products?
  - Who should farmers sell to?
  - Should farmers sell individually or as a group?
  - How should farmers negotiate with buyers?

We know, markets are constantly changing, and so are the answers to all these questions. Further, the tastes and preferences of consumers are also dynamic. So, in view of the demand driven market environment, it is essential for the farmers, processors and traders to produce and transact the products that are actually demanded by the consumers. However, the changes in demand besides generating new opportunities, may also pose threats. So, in order to meet the actual needs of the consumers, the farmers and other stakeholders must have regular access to the right market and technical information. Thus, changing market conditions and consumer preferences require rapid adjustments in production technologies, post-harvest operations and marketing strategies and this all depends upon timely and effective transmission of market information. The same is explained through Figure 5.39.

- Information about current prices locally and in neighbouring areas can help farmers decide, whether to accept the price being offered, negotiate or look for another buyer. It is also important to note that, farmers will enjoy a stronger bargaining position when negotiating as a group and this also facilitates to utilize the market information efficiently.
- Market information facilitates marketing decisions, regulates the competitive market processes and facilitates marketing mechanisms. It is needed by farmers in planning production and marketing and equally needed by other market participants in arriving at optimal trading decisions. The existence and dissemination of complete and accurate marketing information is the key to achieving both operational and pricing efficiency in marketing system.

1

Which crops farmer should grow and how much area he should allocate under each crop?
- Price trends of different crops
- Potential profits associated with different product options
- Demand for the produce in the local market
- Market competition from co-farmers

2

Which off-season crops the farmer should cultivate?
- Price variations across different seasons
- Whether off-season prices are sufficiently high to justify the additional investment and costs associated with off -season cultivation?
- Selection of appropriate seed
- Sources and cost of seed of off -season crops

3

Which varieties should the farmer plant?
- Cost of seed of different crop varieties
- Varietal requirements of different buyers?
- Competition that farmer will face from other farmers and areas for different varieties

4

Where the farmer purchase inputs from?
- Identify the input suppliers
- Quality of the inputs to be purchased
- Prices at which the quality inputs should be procured
- Scope for getting credit to purchase inputs and conditions attached to get credit

5

Which post-harvest operations should farmers undertake?
- Quality specifications of buyers
- Whether buyers prefer fresh or processed, type of grade, type of packaging etc.

6

Whether the farmer should store the produce?
- Future prices of produce
- Whether the price difference is high enough to justify the costs and risks of storage?
- Duration of storage of produce

7

Where should the farmer sell his produce?
- Quality parameters of produce
- Prices of produce at farm -gate and in different markets
- Marketing costs incurred at the farm - gate vis-à-vis at other locations?
- Risks associated while supplying produce to different locations?

8

Identification of buyers
- In the local market or distant markets
- Buyers' requirements about quality of produce
- Prices and payment conditions offered
- Supply costs to buyers and net benefits

9

Whether the farmer has to sell the produce individually or in a group?
- Price in the local market
- Price offered by distant buyers
- Marketing costs incurred in transacting produce in local or distant markets.
- Reduction in marketing costs in transacting on collective basis.

10

How to negotiate with buyers?
- Are the prices offered by buyers in line with current market prices for similar quality?
- Negotiations with buyers of distant markets through internet or SMS.

Figure 5.39: Importance of market information to farmers in taking appropriate production and marketing decisions.

- ☆ Introducing and promoting the production of commodities that enjoy comparative advantage in the international market is difficult in the absence of adequate, accurate, relevant and timely market information (Figure 5.38).
- ☆ The small and marginal farmers' role in agricultural marketing can be enhanced through the Government support especially in the provision of necessary infrastructure for strengthening relevant market information.
- ☆ The availability of prompt and reliable information about what is happening in the market, what quantities are arriving and what prices are quoted for different commodities considerably improve the decision making capability of the farmers and strengthens their bargaining power.
- ☆ In the modern era, with the popularization of internet as a medium for providing market information (AGMARKNET), it offers distinct advantages like, instant access, remote access, simultaneous access and high speed search over volume of available knowledge. A large amount of data with more accuracy can be stored in computer, analyzed and retrieved within shortest possible time. AGMARKNET will ensure dissemination of data through network to any distance with the help of communication devices for the benefit of farmers, traders, consumers, etc. The improved communication system will enable the farmers to know about probable markets, where they can dispose of their produce more profitably. The traders and consumers can also derive maximum advantage out of their purchases at low communication cost. The modernization of market information system will lead to the efficiency in the markets and increased participation of the farmers.
- ☆ Farmers can use market information in two ways. Current or immediate information can be used to negotiate with traders, to decide whether or not to go to market and, in some cases, to decide which market to visit or supply. Historical information, such as a time series of prices over several years, can be used to make decisions regarding product diversification or the production of out-of-season crops. It may even be used to help basically subsistence farmers to identify opportunities for a cash income.
- ☆ The availability of market information can enable farmers to check on the prices they actually receive, vis-à-vis the announced MSPs. If farmers receive prices lower than those announced MSPs, they may conclude that, they should sell to other traders in future or negotiate more forcefully or try to improve the quality and presentation of their produce. The announced MSPs are also used as a starting point in negotiations with traders the following day and the availability of the AMIS does enable farmers to negotiate from a position of relative strength.
- ☆ There will be significant reductions in price differences between distant markets after introduction of market information services. In the past, the available information about more distant markets had been very limited because, both roads and telecommunications were bad, but as the result of an AMIS being introduced, there was a noticeable reduction in price differences.

4. Sources of Market Information

Though many sources were provided to access to market information service to farmers and traders (Table 5.86 and Figure 5.40), these have tended to experience problems of sustainability. Moreover, even when they function, the service provided is often insufficient to allow commercial decisions to be made because of time lags between data collection and dissemination. Modern

Table 5.86: Different sources of market information and their activities.

Source/Institution	*Activities for Market Information*
National Informatics Centre (NIC)	✰ Procurement, Installation and maintenance of Hardware and Software at the AGMARKET nodes, State Agricultural Marketing Directorates/Boards and DMI ✰ Providing internet connectivity to the nodes ✰ Networking of AGMARKNET nodes, State Marketing Directorates/Boards and DMI (Faridabad) ✰ Imparting training to the officials at all the three levels to ensure effective use of the computing facilities ✰ Development and implementation of a customized application software in the identified Agmarknet nodes and State Agricultural Marketing Boards/Directorates and DMI to up-link data on daily prices (Minimum, Maximum, Modal) [commodity-wise, variety-wise, origin-wise, grade wise] and arrivals data [commodity-wise], to the respective State Marketing Boards/Directorates and DMI Headquarters for upgrading in the AGMARKNET server for Internet/Intranet Access ✰ Providing necessary support to the users for implementation of the system ✰ Designating Nodal Officers in states for coordinating with the users at AGMARKNET nodes ✰ Design and Develop Agricultural Marketing Portal for DMI (Faridabad) as well as at State Levels ✰ Support services to AGMARKNET Nodes for two years after the expiry of one year warranty period ✰ Restoration of Computer system at AGMARKNET Nodes due to virus infection, mishandling, and voltage fluctuation. It will also harmonize/integrate software package developed by the State with Agmarknet to bring uniformity in data base and information ✰ Will manage national/state level portals developed under the scheme.
DMI NH-IV, CGO Complex, Faridabad. Web: www.agmarknet.nic.in	✰ Provides information through nationwide Marketing Information Network ('AGMARKNET' portal) ✰ Marketing extension through training to consumers, producers, graders, etc. ✰ Marketing research and surveys ✰ Publication of reports, pamphlets, leaflets, 'Agricultural Marketing' journal, Agmark standards etc.
Directorate of Economics and Statistics, Shastri Bhavan, New Delhi Wen: www.agricoop.nic.in	✰ Compilation of agricultural data on area, production and yield for development and planning ✰ Dissemination of market intelligence through publications and Internet
Directorate General of Commercial Intelligence and Statistics (DGCIS), 1, Council House Street, Kolkata-1 Web: www.dgciskol.nic.in	Collection, compilation and dissemination of marketing related data *i.e.* export-import data, inter-state movement of food grains etc.
CWC 4/1 Siri Institutional Area, Opp. Siri fort, New Delhi-16 Web: www.fieo.com/cwc/	✰ Farmers Extension Service Scheme was launched by CWC in the year 1978-79 with the following objectives: o To educate farmers about the benefit of scientific storage and use of public warehouses

Contd...

Table 5.86–*Contd...*

Source/Institution	*Activities for Market Information*
	o To impart training to the farmers on the techniques of scientific storage and preservation of food grains. o To assist farmers in getting loans from the banks against pledge of warehouse receipt. o Demonstration of spraying and fumigation methods to control insects.
APMCs	☆ Provide market information on arrivals, prevailing prices etc. ☆ Provide market information of adjoining/othermarket committees P Arranges training, tours, exhibitions etc.
SAMBs at different State capitals	☆ Provide marketing related information to coordinate all the market committees in the state. ☆ Arrange seminars, workshops and exhibitions on subjects related to agricultural marketing. ☆ Provide training facilities to producers, traders and employees of the Boards.
Federation of Indian Export Organisations (FIEO), PHQ House (3rd Floor) Opp. Asian Games, New Delhi-110016	☆ Provide information to its members about latest developments of export and import. ☆ Organize seminars, workshops, presentations, tours, buyer-seller meets, sponsoring participation in international trade fairs, exhibitions and providing advisory services with specialized divisions. ☆ Provide information about market development assistance schemes. ☆ Provide useful information on India's export andimport with diverse database.
Kisan Call Centers (New Delhi, Mumbai, Chennai, Kolkata, Hyderabad, Banglore, Chandigarh and Lucknow) Mass Media Support to Agriculture Extension	☆ Provides expert advice to the farmers. ☆ These centers operate through toll free telecom lines throughout the country. ☆ A country-wide common four digit number 1551 has been allocated to these centers. ☆ Mass media support to agriculture extension has been augmented with three new initiatives. o The first component establishes a cable satellite channel for national broadcast using the existing facilities available with Indira Gandhi National Open University (IGNOU). o The second component is use of low and high power transmitters of Doordarshan for providing area specific telecast. Initially, 12 locations chosen to launch broadcasting are Jalpaiguri (West Bengal), Indore (Madhya Pradesh), Sambhalpur (Odisha), Shillong (Meghalaya), Hissar (Haryana), Muzzafarpur (Bihar), Dibrugarh (Assam), Varanasi (Uttar Pradesh), Vijayawada (Andhra Pradesh), Gulbarga (Karnataka), Rajkot (Gujarat), Daltonganj (Jharkhand). ☆ The third component is use of FM transmitter network of All India Radio (AIR) to provide area specific broadcasting through 96 FM stations.
Agriculture-Clinics and Agri-Business byAgriculture Graduates	☆ A central sector scheme "Establishment ofAgriculture-Clinics and Agri-business Managed by Agriculture Graduates" is being implemented since 2001-02. ☆ The aim is to provide opportunity to all eligible agriculture graduates to support agriculture development through economically viable ventures.

Contd...

Table 5.86–*Contd...*

Source/Institution	*Activities for Market Information*
	✰ The scheme is being jointly implemented byNABARD, National Institute of Agricultural Extension Management (MANAGE) and Small Farmers' Agri-business Consortium (SFAC) in association with about 66 reputed training institutes in the country.
Different websites on Agricultural Marketing Information	✰ www.agmarknet.nic.in ✰ www.agricoop.nic.in ✰ www.ncdc.nic.in ✰ www.ikisan.com ✰ www.fmc.gov.in ✰ www.icar.org.in ✰ www.fao.org ✰ www.agriculturalinformation.com ✰ www.agriwatch.com ✰ www.kisan.net ✰ www.agnic.org ✰ www.nafed-india.com ✰ www.indiaagronet.com ✰ www.nic.in/eximpol ✰ www.agrisurf.com ✰ www.commodityindia.com

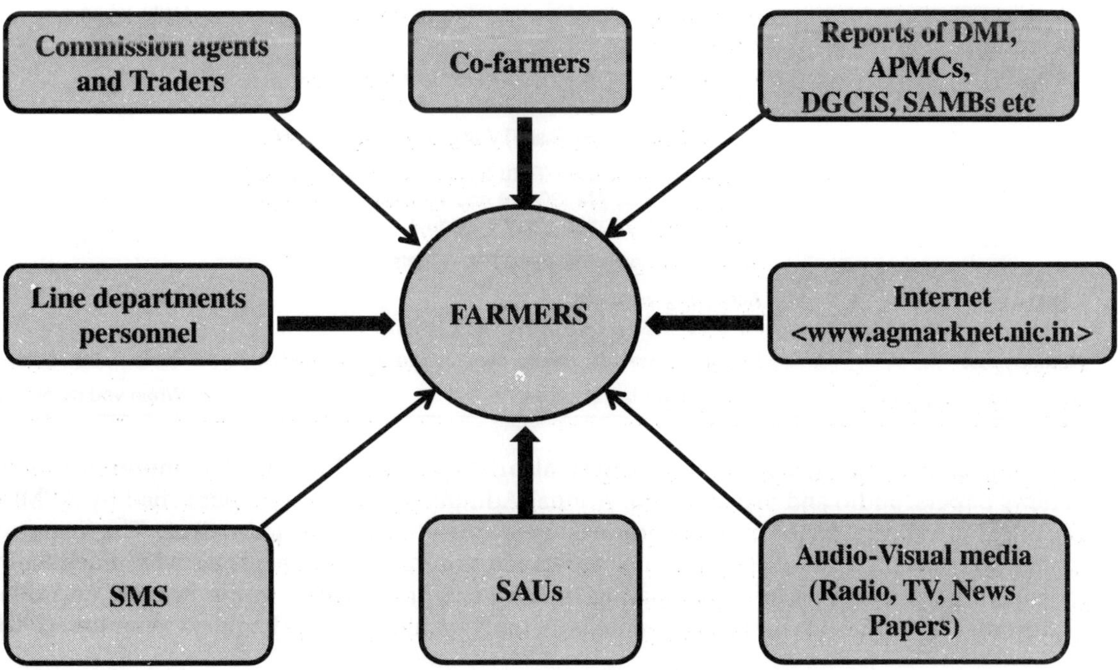

Figure 5.40: Different means of dissemination of market information to farmers.

communication technologies open up the possibility for rapid market information delivery through SMS on cell phones and the rapid growth of FM radio broadcasts in many areas offers the possibility of more localized information services. Presently, the internet became an effective way of delivering information to farmers and traders. However, problems associated with the cost and accuracy of data collection still remain to be addressed. Even when they have access to market information, farmers often require assistance in interpreting that information. For example, the market price quoted on the radio may refer to a wholesale selling price and farmers may have difficulty in translating this into a realistic price at their local assembly market. Various attempts have been made to introduce commercial market information services, but these have largely been targeted at traders, commercial farmers or exporters only, and not small and marginal farmers. Thus, small and marginal farmers have limited access to commercial market information services and hence, they are unable to get the reliable market information. This inability to access to reliable market information has denied most of the small and marginal farmers, the opportunity to effectively plan and market their produce. Furthermore, lack of access to market information is one of the contributing factors to the slow development of market opportunities.

Table 5.87: Frequency and method of obtaining of market information across different sources.

Source of Market Information	*Method of Obtaining*	*Frequency*
Radio broadcasts	Tuning in to the radio at the time the programme is being aired	Daily
Phone calls (Mobile telephones)	Toll free number in Andhra Pradesh - 18004253536	Any time
Kisan Call Centers (New Delhi, Mumbai, Chennai, Kolkata, Hyderabad, Banglore, Chandigarh and Lucknow)	A country-wide common four digit number 1551 has been allocated to these centers.	Daily
Television	Watching agricultural programmes on different channels like ☆ 'Annadata' on E-TV at 6.30 AM to 7.00AM' and 'Jai Kisan' on E-TV2 at 6.30 PM to 7.00 PM ☆ 'Palle Talli' on Saakshi TV at 6.30 AM to 7.00 AM etc. ☆ Doordarshan (DD) 'Rytu Nestam' 6.25 AM (35mts), Kisan, 6.30 AM (30mts) and Vyvasaaya Vigyanam, 6.25 PM (25 mts, (Only Tuesday) ☆ TV5 'Annapoorna' (5.30 PM, 30 mts)	Daily, when it is telecasted
Internet	Accessing websites www.agmarknet.nic.in, www.agricoop.nic.in	Any time
Market yards	Display on boards, asking commission agents and traders	When in the market
Neighbours, Co-farmers	Asking them latest prices	When you meet them

Besides above sources, market information is also disseminated through other important sources like news papers, audio and visual media, Annual Administration Reports published by APMCs, DMI, price bulletins published by Agricultural Situation in India (monthly journal) etc. The frequency and method of obtaining market information varies across different sources (Table 5.87). Each source has its own pros and cons in disseminating the requisite information to the farmers and other stakeholders (Table 5.88). On the recommendation of the Agricultural Prices Enquiry Committee (1954),

Table 5.88: Pros and cons of media involved in disseminating market information.

Media	*Pros*	*Cons*
Radio	☆ Almost every household owns a radio. ☆ Inexpensive for users ☆ Widely used	☆ Information is not 'on record' ☆ Broadcasting schedule may conflict with other activities
Television	☆ Many households own a TV ☆ Inexpensive for users ☆ Widely used	☆ Information is not 'on record' ☆ Broadcasting schedule may conflict with other activities
Printed media – Newspapers, Notice boards	☆ Information is 'on record' and can be referred to ☆ Useful for the dissemination of specific information (*e.g.* production planning, post-harvest quality issues) ☆ Can be easily set up in markets ☆ Farmers and traders in the market have easy access to the information	☆ There may be time lag between submission of market report to the newspaper office for publishing and its appearance in the paper. So, information may become out of date by the time it is printed in the news paper ☆ Farmers and traders who don't visit the market don't have access to the information (Notice boards)
Mobile phones (SMS text messages)	☆ More and more people have a mobile phone; traders definitely, but also more farmers. ☆ Relatively inexpensive to send and receive text messages ☆ Dissemination is fast	☆ Farmers may not be aware of the service ☆ There is only a limited amount of information that can be disseminated by mobile phone text messages
Internet	☆ Dissemination of information is fast ☆ Information is 'on record' and can also be printed out, if required ☆ Accessing websites or receiving e-mails with trade information can be cheaper than subscribing to trade magazines	☆ Not everybody has access to a computer or internet (can be too expensive) ☆ AMIS needs someone with skills in web site design; users need to be computer literate

the Directorate of Economics and Statistics, Ministry of Agriculture (DESMOA) set up 14 Market Intelligence Units (MIU) in the capitals of Andhra Pradesh, Assam, Bihar, Delhi, Gujarat, Karnataka, Kerala, Madhya Pradesh, Maharashtra, Odisha, Rajasthan, Tamil Nadu, Uttar Pradesh, and West Bengal. The MIUs are intended to help the DESMOA in the formulation, implementation and review of the agricultural price policy relating to procurement, marketing, storage, transportation, import, export etc. The units furnish regular reports on market arrivals, off-takes, stocks, crop prospects, and outlook of market prices. They are also required to give their appraisal of production of various kharif and rabi crops at regular intervals to help preparation of crop forecasts. Though the data to be supplied by the MIUs are of great utility, the units have ceased to be effective in discharging their functions mainly due to a lack of proper direction and control of their activities. Over the years, the staff strength of the units has been considerably reduced resulting in even worse performance

5. Type of Market Information per Type of Market Participant

Different types of market participants require different types of market information (Table 5.89). Some deal with a diversified range of commodities, others specialize in one or a very small number of commodities. Some trade within very localized areas, others buy products from different regions and supply to regional as well as more distant urban markets. Some traders and processors even import from, or export to, other countries. So, the requirements of market information differ greatly across different types of market participants in the marketing channels.

Table 5.89: Type of market information per type of market participant.

Market Participant	*Type of Market Information*
Input manufacturers	☆ Current wholesale selling prices for different inputs ☆ Trends in input wholesale prices ☆ Advantages and disadvantages of different agro-chemicals, processing equipments etc.
Local input suppliers	☆ Local availability of different inputs ☆ Local input retail prices (current and trends) ☆ Sale conditions
Local buyers of agricultural produce (Itinerant traders and small processors)	☆ Exchange arrangements with local farmers ☆ Own product requirements ☆ Product requirements of their buyers ☆ Local price trends for different agricultural products
Wholesale buyers of agricultural produce (Medium and large processors from distant areas)	☆ Production trends in different supplying regions ☆ Competitive position of different supplying regions ☆ Demand patterns (volume, quality preferences, seasonality, trends) in regional, national and sometimes international markets ☆ Wholesale purchasing prices ☆ Price seasonality and trends ☆ Marketing opportunities
Local (rural) retailers	☆ Preferences of local consumers ☆ Seasonality of local demand ☆ Local demand trends ☆ Retail prices in local markets (current, trends and seasonality)
Urban retailers (from distant areas)	☆ Preferences of urban consumers ☆ Seasonality of demand in urban market ☆ Demand trends in urban market ☆ Retail prices in urban market (current, trends and seasonality)
Transporters	☆ Direction and magnitude of agricultural product flows from the production areas ☆ Seasonality of product flows ☆ Location and addresses of important traders and processors ☆ Transport costs

6. 4 A's of Market Information

In the context of making the market information relevant to farmers and other stakeholders, it should be viewed in terms of 4 A's *viz.*, Accuracy, Availability, Applicability and Analysis.

- ☆ ***Accuracy:*** The information provided by the AMIS must be accurate and farmers and traders must understand to which product, quality etc., the prices refer. There are different issues to be looked into to provide accurate market information. As far as possible, AMIS should provide information variety-wise across different products so that, it benefits large scale of farmers and traders. However, it is quite difficult for AMIS to cover all the varieties even for a single product, but it does need to make sure that, when it reports on prices, the farmers must know, which varieties of product are being referred to. Then, there is the question of quality. Does the price information refers to Grade I, Grade II, Grade III, Grade IV and so on or whether the price information relates to something in between the two extreme grades. So, AMIS must provide this price information grade-wise throughout the country. Otherwise, the farmers have to face a peculiar situation that, say, a farmer who have poor quality produce (Grade IV) to sell would hear that, the price was high in distant wholesale market, and would go to that market, only to find that the price quoted by the AMIS was for the superior quality (Grade I). So, FAO recommends that, the concept of 'Fair Average Quality (FAQ)' should be used as the basis for price reporting. Further, price information provided the AMIS must be clear transactions-wise in the market. We know, in a given market, there are different types of transactions *viz.*, farmer selling to processor, farmer selling to wholesaler, wholesaler selling to retailer, processor selling to wholesaler, processor selling to retailer, retailer selling to consumer etc. So, there should be clarity from the AMIS regarding the price information across different types of market transactions. But, it is quite difficult to get such price information across different types of market transactions. So, two important criteria are generally followed *viz.*, which price is most useful to the farmer and, balancing that, which price is easiest to collect. This is because, the main objective of regulations offered by the Government is to safeguard the farming community. Further, traders purchase price is not the final price, as it is based on the negotiations between him and the seller and the price the trader quotes as a seller vary depending upon the profit margin he quotes, while transacting the produce to other agency in the marketing channel.
- ☆ ***Availability:*** Even if prices are completely accurate, they are of much use, if they are only available to farmers and traders in right time. The market news must be available in time to the farmers and traders on the same day on which it is collected. For this, the local media must be willing to co-operate. So, an efficient AMIS implies, price collection early in the day, compilation, analysis and speedy distribution to the relevant media. For crops that change prices rapidly, such as vegetables, prices should ideally be collected and disseminated on a daily basis. Besides media, the rapid dissemination of prices can be done through internet or SMS. Even today, small and marginal farmers in remote areas still listen to the price broadcasts on the radio and plan their marketing strategies. Presently, television is becoming increasingly available and resently, heir marketing streategies. in remote areas and traders ded ming community. even the poorest farmers have access. Recent advances through which the market information is made available to farmers and traders include the use of the internet and e-mail. Technology is developing rapidly and the possibility for people in remote areas to gain broadband access to the internet from satellite transmission is developing rapidly. These days, the availability of a suitable electricity supply is often a greater constraint than the availability of internet technology. Mobile telephone transmission has also been gaining significance, as it may be a useful way for traders to obtain price information, but small and marginal farmers will not be able to afford a phone. So, it is high time now, the Government should offer loans to the farmers to purchase mobile phones, which they let

other villagers use for a fee. Farmers can use the toll free numbers to get price information or the AMIS provides SMS messages free of charge on a daily basis (however, there is a limitation on the amount of data that can be supplied). By getting price information, farmers and traders can plan for quick commercial decisions.

- ✰ ***Applicability:*** After getting the price information from AMIS, the farmers need to be able to convert them into a realistic price at their local market or farm gate. That is, AMIS must provide price information to the farmers from markets that are most applicable to them. Usually, AMIS will provide price information that is prevailed in the large wholesale markets in the country. Then farmers have to be helped to understand how to translate those prices into something applicable to them. That is, they need assistance in understanding the prices in terms of produce variety, quality, type of transaction etc. They also need to understand the costs incurred between their farm gate and the wholesale market. Such information guides them to convert the prices of produce in wholesale markets into the prices that may prevail at their farm gate.
- ✰ ***Analysis:*** It is essential to process long term prices data, so as to estimate the price trends of major agricultural commodities. This will guide the farmers to make decisions about when to plant and harvest, what new crops to diversify into, plan off-season production, decide whether to store the produce in the expectation of higher prices later in the year etc. It also enable them to ascertain both domestic and export competitiveness of the commodities under consideration.

Besides above 4As, the other important requisites of good market information include, trustworthiness, easy accessibility, timeliness etc., so as to benefit the farmers and traders in using the same in the marketing process. Further, the AMIS must provide information on a consistent basis. Market information is not something that can be broadcasted for a week and then forgotten about for a month. It has to be broadcasted every day, every other day or every week without fail. Government should not initiate AMIS, unless it is sure that, resources will be available to maintain the service. In India, market information is supplied free of cost both to the farmers and traders. This is because, most small and marginal farmers would have trouble in paying and, anyway, it is difficult to see how they could be charged, even if they are provided with information on payment basis. For example, it is not possible to charge from the farmers and traders for supplying market information through radio, television etc.

7. Internet Based Market Information System (AGMARKNET): A New Dimension

In a country like India, with 65 per cent of its population living in the villages and depending on agriculture as their main occupation, accurate and timely information about the market prices of the agricultural commodities is of extreme significance. Agricultural Marketing in India is undergoing a significant metamorphosis because of economic liberalization and globalization. Advancement in Information and Communication Technology (ICT) has made the world a smaller place and a larger market at one go. The availability of prompt and reliable market information about what is happening in the market *i.e.,* what quantities are arriving and what prices are quoted for different commodities, considerably improve the decision making capability of the farmers and strengthens their bargaining power. Conventionally, the markets maintain information in the form of registers. The daily prices are compiled manually and written on the blackboards for the use of farmers visiting the markets. The information is even disseminated through various media like radio, television, newspapers, blackboard display and public address system at market yards. The information provided by these methods is stale and does not help the farmers sufficiently in taking decisions in marketing their produce. The farmers are also not able to know about the prices prevailing in other markets, as the APMCs are able to disseminate information in respect of their own markets only. The farmers are therefore, left with no alternative, but to dispose off their produce in the nearest market, even at uneconomic prices. Introduction of economic reforms, ushering in liberalization policies in the country has brought new challenges to the marketing system including information requirements. To fully utilize the new emerging trade opportunities for the benefit of farming community, improved agricultural market information service is considered *sine-quo-non*. Thus, realizing the deficiency of traditional market information and with the advent of ICT (internet), the Government developed the system to link the market data on agricultural commodities at the national level through AGMARKNET. ICT (Internet) is now considered as fifth factor of production besides other four factors of production *viz.*, land, labour, capital and management. The internet provides a completely new dimension to information utilization. Despite a growing awareness and attempt to commercialize the services and products within the internet, the market information has been regarded as a public good in India because of presence of large number of small and marginal farmers, who are unable to pay for information. Exploring on the internet for the farmers would provide wider marketing opportunities and social acceptance for this new media, a derivation from earlier, needs promotion. Basic functions of transfer of knowledge, changing attitude, behaviour and skills to the farmers of earlier media needs to be incorporated in the new media (Internet) so that, it takes over traditional media in the interest of the farmers.

We know, almost all the States and UTs are providing market information in one form or the other for the benefits of market users like producers, traders and consumers. However, the information is collected and disseminated by use of conventional methods, which cause inordinate delay in communicating the information to different target groups and thus, adversely affecting their economic interest. Therefore, there is an urgent need to bring improvement in the present *market information system* by linking all Agricultural Produce Wholesale markets in the States and UTs and the SAMBs and State Directorates of Agricultural Marketing, with the DMI of the Ministry of Agriculture, for effective and efficient information exchange. In view of this importance, the Department of Agriculture and Co-operation, in the Union Ministry of Agriculture has sanctioned a Central Sector Scheme 'National Informatics Centre Network (NICNET) based AGMARK-NET' in March, 2000, to link 810 nodes (735 agricultural markets and 75 SAMBs/DMI offices) to begin with, through its attached office, DMI. The scheme aims at progressively linking important agricultural produce markets spread all over India and the SAMBs/Directorates and the DMI for effective exchange of market information. The market information network, AGMARKNET (agmarknet.nic.in), is being implemented jointly by DMI and NIC, using NICNET facilities available throughout the country. This AGMARKNET website (www.agmarknet.nic.in) is a G2C e-governance portal that caters to the needs of various stakeholders such as farmers, industry, policy makers and academic institutions by providing agricultural marketing related information from a single window (Figure 5.41). It facilitates dissemination over web about daily arrivals and prices of commodities in the agricultural produce markets spread across the country. The data transmitted from all the markets is made available on the AGMARKNET portal in multilingual mode. It displays commodity-wise, variety-wise daily prices and arrivals information from all wholesale markets. Various types of reports can be viewed including Trend reports for Prices and Arrivals for important commodities. The objective of the scheme is to facilitate collection and dissemination of information for better price realization. The portal covers market prices, infrastructure and promotion-related information for efficient marketing. The Scheme is being implemented during the XI Plan period by enhancing its scope to cover more number of additional markets including remaining principal regulated markets if any, farmer/consumer markets, private markets, urban markets and special commodity markets. The following are the objectives of AGMARKNET:

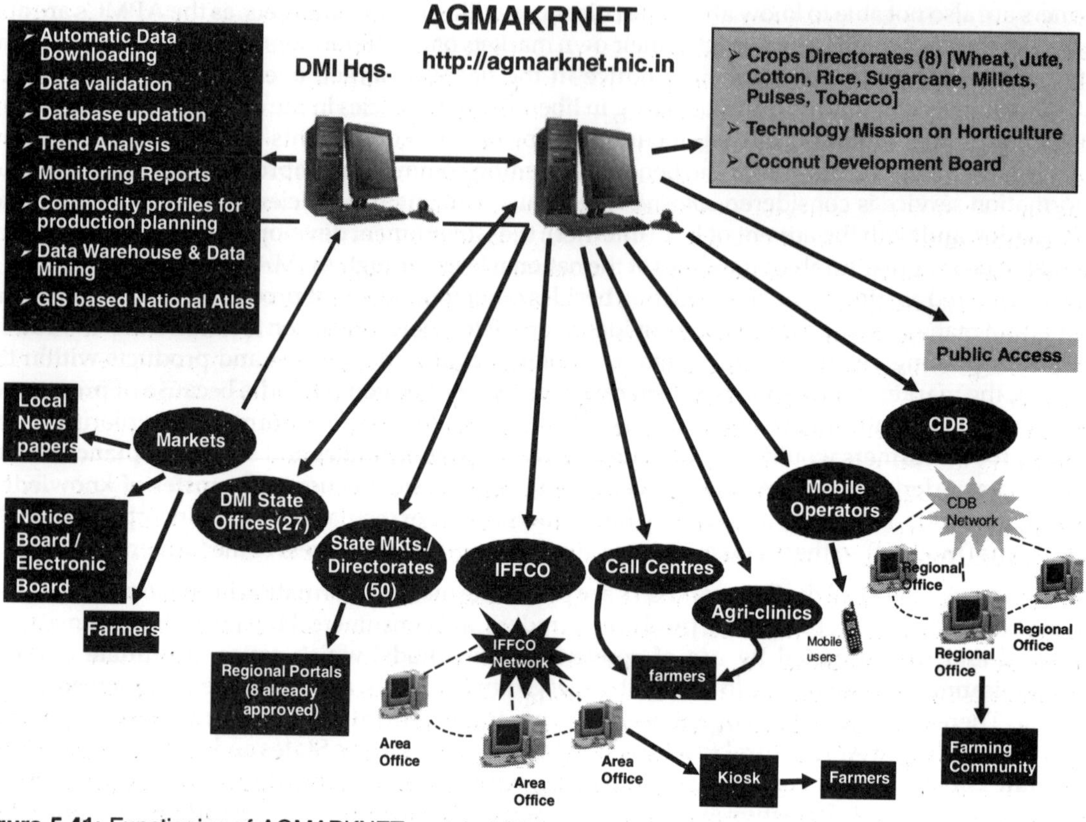

Figure 5.41: Functioning of AGMARKNET portal of DMI.

- ☆ To establish a nation-wide information network for speedy collection and dissemination of market information for its efficient utilization.
- ☆ To computerize data on market fee, market charges, total arrivals, arrivals by agencies, prices (variety-wise/quality-wise), storage, dispatches with destination, mode of transportation, costs, sold and unsold stocks, sources of supply with destination, method of sale, payment, weighment, grading facilities, quantities graded, market personnel (trained/untrained), market functionaries, market finance, development programmes, infrastructure facilities, constitution/composition of APMCs, income and expenditure and other activities of the APMCs, SAMBs and Marketing Departments.
- ☆ To ensure flow of regular and reliable data to producers, traders and consumers to derive maximum benefit of their sales and purchases.
- ☆ To increase the efficiency in marketing by effecting improvement in the existing market information system.
- ☆ To facilitate collection and dissemination of information related to better price realization by the farmers. This would cover:
  - ❖ Market related information such as market fee, market charges, costs, method of sale, payment, weighment, handling, market functionaries, development programmes, market

laws, dispute settlement mechanism, composition of market committees, income and expenditure, etc.

- Price-related information such as minimum, maximum and model prices of varieties and quantities transacted, total arrivals and dispatches with destination, marketing costs and margins, etc.
- Infrastructure related information comprising facilities and services available to the farmers with regard to storage and warehousing, cold storage, direct markets, grading, rehandling and repacking etc.; and
- Promotion related information covering accepted standards and grades, labeling, SPS requirements, pledge finance, marketing credit and new opportunities available in respect of better marketing;

- To sensitize and orient farmers to respond to new challenges in agricultural marketing by using ICT as a vehicle of extension.
- To improve efficiency in agricultural marketing through regular training and extension for reaching region-specific farmers in their own language.
- To provide assistance for marketing research to generate marketing information for its dissemination to farmers and other marketing functionaries at grass-root level to create an ambience of good marketing practices in the country.

Features of AGMARKNET Portal

To strengthen interface with farmers and other beneficiaries, AGMARKNET portal has been evolved. Over 2000 markets regularly report price-related data being disseminated through the portal. The portal also serves as a single window for accessing Web sites of various organizations. It also provides weekly trend analysis, linkage to online commodity exchange of India, FAO, Indian Farmers Fertilizer Co-operative Limited (IFFCO) Web site, and so forth. The development of the AGMARKNET portal and the state-level portal, and undertaking market-led extension activities are important components of this scheme. AGMARKNET ensures dissemination of data through the network to any distance for the benefit of citizens, farmers, traders, and consumers. The improved communication system will enable farmers to learn about probable markets in which their produce can be disposed more profitably. Also the modernization of the market information system will lead to efficiency in markets and increased participation of the farmers. Various DMI offices spread across the country prepare the state-level trend analysis monthly bulletins using the central database. The price trend reports are published on the AGMARKNET portal. AGMARKNET is a classic example of a collaborative framework between Government organizations *viz.* DMI, NIC and SAMBs/Directorates which has leveraged the potential of ICT for evolving a National Database on market-wise daily commodity arrivals and prices. Thus, the AGMARKNET portal is constantly enriched with agricultural marketing–related information. Efforts are on to reach out to the farmers in their regional languages. As of now, the portal is disseminating information in two regional languages: Hindi and Punjabi. The portal also caters to the diversified needs of these stakeholders by providing the following agricultural marketing–related information as a single window Web service over the Internet:

- ***Price and Arrivals:*** The portal provides access to commodity-wise, variety-wise daily prices and arrivals, information of various wholesale markets. Future prices from national multi-commodity exchanges are reflected online. Also linkages are established with Web sites of the FAO and Asia and Pacific Coconut Community (APCC) to access international commodity price trends.

- ☆ ***Commodities and Varieties:*** A commodity base, comprising of more than 300 commodities and about 2,000 varieties has been evolved. The commodities are being categorized into various groups: cereals, pulses, oil seeds, fruits, vegetables, spices, fiber crops, beverages, forest products, drugs and narcotics, dry fruits, flowers, forest products, livestock/poultry, and so forth to facilitate easy retrieval of market information.
- ☆ ***Grading and Standardization:*** To promote the importance of quality among the farming community, the portal emphasizes on standardization and grading aspects of the agricultural products. The information is provided in the form of documents/specifications prescribed by the act/rules of DMI and other agencies. It also links to the Codex International Food Standards, guidelines, and related texts such as the codes of practice under the joint FAO/WHO Food Standards Programme.
- ☆ ***Market Atlas:*** The market profiles and various Geographical Information System (GIS) based thematic maps are under development.
- ☆ ***E-Directory of Markets:*** With the networking of agricultural markets, timely and effective exchange of information electronically is now possible. E-mail addresses of the agencies involved are progressively published for public access, so as to facilitate direct interaction.
- ☆ ***Weather information:*** All India weather conditions, weather forecast, climate, rainfall, temperature, and soil moisture are made available through the Indian Meteorological Department Web site.

Advantages

As a step towards globalization of Indian agriculture, the DMI has embarked on this ICT project *i.e.*, NICNET based AGMARKNET. The advantages of AGMARKNET database accrue to the farmers, as they have choices to sell their produce in the markets at remunerative prices. The following are the important advantages offered by AGMARKNET:

- ☆ AGMARKNET plays a catalytic role for ushering in 'market-led agricultural extension' in India, highly scalable, planned through bottom-up process, and implemented through active involvement and collaboration of APMCs.
- ☆ AGMARKNET has strengthened the interfaces among government organizations, farmers, industry, policy makers, and other beneficiaries.
- ☆ AGMARKNET aims at empowering the farming community with market information. For maximizing the benefits, it needs to be integrated with other ICT initiatives targeting the upliftment of rural India. The project is part of National e-Government action plan of Government of India.
- ☆ AGMARKNET has helped establish a nationwide information network for speedy collection and dissemination of market data for efficient and timely utilization.
- ☆ The AGMARKNET has already emerged as the sun-shine website to bargain better prices for their produce, and marching ahead towards becoming an e-Commerce and e-Business Portal in India.
- ☆ The speedier collection and dissemination of information on commodities enables farmers to realize better prices. This also leads to a transparency in the pricing process across the country.
- ☆ Progressive sensitization and orientation of farmers helps them respond to new challenges in agricultural marketing by using ICT as a vehicle of extension.

☆ AGMARKNET project also facilitates:

- ❖ nation-wide market information for wholesale prices of agricultural produce
- ❖ information dissemination progressively through local languages
- ❖ strengthening of computer facilities at the marketso study trends in market prices at various markets
- ❖ bypass middlemen
- ❖ serves as single window for accessing web-sites of various organizations concerned with agricultural marketing.

The next steps in the pipeline are mobile based applications for automatic agricultural inflow system in order to e-bridge the farmers and other stake holders, to promote ICT under small area farm business and implementation of Agriculture Resource Information System (AgRIS) and DISNIC Plan Programme (http://disnic.gov.in) to establish grassroots level informatics of sustainable agriculture development through information modeling.

8. Constraints in Providing Market Intelligence

The following are the important constraints in the present market intelligence system in India:– The system of market information has continued to be far from satisfactory. While the traders and processors use their own informal sources, farmers depend both on formal and informal sources. Though, both market news and market information are equally important, but farmers are more interested in market news.

☆ Market news of interest to farmers is collected or compiled by APMCs, SAMBs, State Departments of Agricultural Marketing, field staff of Directorate of Economics and Statistics and Department of Food. The information is disseminated through display boards in market yards, by announcements during open auction, by newspapers, radio broadcasts and TV channels. Farmers also gather information through personal contacts with other farmers and traders. However, the market news is able to provide only a broad overview to the farmers due to several defects in the system. The price quotations are not backed by grades and the information is available with considerable time lag. This information is not linked to local grade standards. Quite often, a range of prices is made available, which is of little use to farmers. There is also a serious AMIS conception about the buying and selling price, which are distinctly different.

☆ Electronic medium has been used for transmission of market information in various industries. However, their use in agricultural markets is relatively low. Markets of some States are linked with NICNET to provide the speedy and timely dissemination of market information to the farmers. Under the scheme, about 3011 Agmarknet nodes (Table 5.78) have been promoted in the country so far. Out of these, 92 per cent have been promoted in the agricultural markets, where as remaining are used for monitoring and follow up. The availability of Agmarknet nodes per 000' Sq. Km of area is not even one. However, their availability per 000' tones of produce is six. Concerted efforts are required to expand the Agmarknet nodes in the states of Assam, Bihar, Jharkhand, Manipur, West Bengal, Odisha, Punjab, Uttar Pradesh and Uttarakhand.

☆ The reports on market intelligence of APMCs, SAMBs etc., are prepared in a stereo-typed manner. No updating is done in computing the trends in market arrivals, prices, grades of produce transacted, quantum of produce stocked etc. More bias is shown towards providing price information compared to marketable surplus, efficient marketing channels, price spread studies etc.

- Most of the officials functioning in the APMCs, SAMBs etc., are non-professionals in the field of agricultural marketing. Hence, they themselves were not aware about different technical aspects in the concepts of agricultural marketing. So, they could not provide sufficient, reliable and trustworthy market information to the farmers and traders.
- Poor data processing and data management practices, lack of standardized formats for data collection, lack of well documented instructions necessary for data collection and insufficient co-operation among data collectors and processors (at different levels) are hampering the provision of quality market information to the beneficiaries.
- Most farmers receive no or less information about changes in the prices of their products, whereas traders have reasonably rapid access to such trade-based information. Such access gaps to basic market information have been cited as a major impediment to empowering farmers to negotiate for better prices.
- Not only the dissemination of market information service, but research on this subject both at macro and micro levels was inadequate.
- Prices for a particular product in one market may differ significantly from prices for the same product in another market. In theory, the provision of market information can permit farmers and traders to take advantage of such price differences. In reality, however, this is not done as often as might be expected. In part, this is because, both farmers and traders have strong linkages with particular market say, credit obligations. Further, individual farmers may not be able to achieve the economies of scale necessary to visit more distant markets.–Lack of information is an entry barrier to both production and trade. Where farmers have had no access to information, shifts in cropping patterns towards higher value produce will be hampered. Further, in the area of trade, farmers and traders find it difficult to begin trading without information and this adversely affects their competitiveness in the markets.
- In the day-to-day marketing of their products, farmers are mainly at a considerable disadvantage. Their market information will come from a neighbouring farmer, who may have visited a market on the previous day. A trader's core skill is to read the market, assess supply and demand, and compute how these factors might affect price. Increasingly traders will triangulate their information with information from others. Given the opportunity, traders will exploit farmers' relative ignorance to buy low and, ideally, sell high. Thus, the present AMIS is not fully safeguarding the interests of the farmers in transacting the produce.

9. Suggestions to Strengthen AMIS in India

Even though encouraging growth is being observed in the agricultural sector since past few decades, solving the centuries old problems of agriculture and building necessary supporting services still requires a lot of activities. According to DMI Reports, introducing and promoting the production of marketable commodities is difficult in the absence of adequate, accurate, relevant and timely information. Very few farmers participate in commercial markets due to lack of access to market information and understanding as to how the market operates. The inability to access agricultural marketing information has denied most of the farmers, the opportunity to effectively plan and market their produce. Furthermore, lack of access to market information is one of the contributing factors to the slow development of market opportunities. In this context, it is essential to consider the following suggestion to improve the AMIS in tune with the latest developments in agri-business:

- There is a need for creation of a AGMARKNET Cell at the Department of Agricultural Economics in all the Colleges of Agriculture with some core staff and infrastructure with necessary budget provisions.
- Specific projects can be initiated with all Agricultural Colleges, where the subject on Agricultural Marketing is being taught, for development of a sound database in agricultural marketing, in a phased manner.
- It is essential to establish State Institutes of Agricultural Marketing on the pattern of NIAM, Jaipur.
- A programme for production of a series of Multi-Media Scripts and plays for broadcast/ telecast over mass-media/Internet, may be taken up on the topics of Market Information and Pre-Harvest/Post-harvest handling.
- Integrate Websites for all agencies of both State and Central Government involved in Agricultural marketing services like APEDA, APMCs, CWC, SWCs, Commission for Agricultural Costs and Prices (CACP), CCI, DMI, FCI, Jute Corporation of India Limited (JCI), KVKs, MPEDA, NAFED, The Tribal Co-operative Marketing Development Federation of India Limited (TRIFED), NCDC, National Dairy Development Board (NDDB), NHB, SAMBs, and State Trading Corporation (STC).
- Integrating AGMARKNET with State Wide Area network (SWAN) and NICNET.
- Establishment of AGMARKNET Nodes at KVKs and Panchayats with IT infrastructure along with Internet accessibility.
- Dissemination of market information through electronic media, ICT media, telecommunication media and print media.
- Linking all co-operative marketing organizations through provision of computerization and internet facility and putting them on common or inter-linked websites.
- E-networking of quality testing laboratories in the country.
- E-linking of rural business hubs with exporters, supermarkets and retailers.
- As discussed earlier, market information is key to achieve both operational and pricing efficiency in the marketing system. This accurate market information needs to reach to the farmers before they sell their produces in the early hours of the day. Hence, the data has to be collected and uploaded to the website in the early hours of the day for the benefit of the farmers. By doing this, the farmers can access the price of a commodity before transaction. Again the minimum price, average price and the quantity arrived also need to be collected and uploaded to the Agmarknet in subsequent time of the market day. Hence, the data has to be collected and uploaded more than a time in a day. The provision for communicating the price of a commodity of a particular market through SMS is also expected to be operationalized through NIC.
- Touch screen Kiosks should be established in all the wholesale markets functioning under the APMCs and they should provide latest information on market prices, arrivals etc., of all agricultural commodities in major markets.
- Farmers' dependence on informal sources such as fellow farmers, friends and relatives for market information indicates the inability of formal sources in disseminating the required information to the farmers. This calls for revitalizing the AMIS especially on dissemination aspects in public institutions like SAMBs, State Marketing Department, State Agricultural Universities (SAUs) etc. with modern communication technology.

☆ The existing marketing information is restricted to market arrivals and prices. There is a need to cover other components of market information such as quality standards, post harvest management, storage, transportation, imports/exports apart from production scenario.

D. Risk Bearing

Risk is defined as uncertainty about cost, loss or damage. It is inherent in all marketing transactions. Farmers deal with a significant amount of risk and uncertainties every day. They have little control over the market forces that drive commodity prices. Production levels and market supply and demand changes can cause large and unforeseen swings in prices. Furthermore, increasing global interaction in commodity markets and Governmental influences add to the uncertainty surrounding market prices. Changes in consumers' incomes, stability of the Government, trade policies and exchange rates all affect demand for commodities and, by extension, commodity prices. These and other unpredictable factors make price forecasting a difficult and volatile practice. Since, input prices translate to costs for farmers and output prices translate to revenues for farmer, unfavourable prices on either side can be devastating to an agricultural operation. It is, therefore, imperative for farmers to manage marketing risk both on the input and the output side in order to ensure long-term profitability in the agri-business.

i. Different Types of Risks

Risks in marketing of agricultural produce can be conveniently categorized into the following types:

1. ***Input price risk:*** In crop agriculture, input price risk has been given considerably less attention in the literature, and has been considered less substantial than output price risk and yield risk. It normally does not translate into return variability of the same magnitude. Moreover, the time window of input price risk is shorter: fertilizer and other input costs are usually incurred within a few months of the onset of production, whereas the uncertainty around output price and yield usually remains considerably for a longer period. However, the magnitude of input price fluctuations can be significant and there is no system in place to hedge against input prices (although there is the possibility of storing some items on the farm).
2. ***Output price risk:*** Regarding output price risk, it arises due to the biological lag inherent in agricultural production. Obviously, farmers must make production decisions months before they have a product to sell, before the actual crop prices are known. During this period, output prices may change dramatically in response to shocks in supply and demand. This may put farmers in a difficult situation, if commodity prices decrease drastically during the production and marketing cycle. Macroeconomic factors like variability in inflation rates, exchange rates and interest rates also impact the volatility of agricultural prices. The effect of shocks on the agriculture and food system is compounded by low elasticities of both supply and demand. Indeed, since food is an essential product with no substitute (apart from alternative foods), demand responses to price increases and decreases are typically low. Supply responses are also very low short-term, until production decisions can be made for the next season or more land brought into or taken out of production.
3. ***Quantity risk:*** Quantity risk relates to reduction or loss in quantity of produce during the marketing process. For example, when produce is stored in godowns, quantity risk may

arise due to fire accidents, thefts, rodent damage etc. Sometimes, the packing material of the produce may not be good and it may lead to physical damage. These together accounts for a large part of the loss of the produce at the individual as well as at the macro level. Such losses are loss to the society also and must be prevented to the extent possible.

4. ***Quality risk:*** Quality risk relates to loss in quality of produce during marketing process, especially during storage and transportation. The produce may get spoiled with the infestation due to pests and diseases, extreme low and high temperature etc.
5. ***Price risk:*** Regarding price risk, we know farmers generally store the produce in godowns with the anticipation that, the prices will certainly rise in the future. But, there is no guarantee, as there is equal chance of decline in the prices in future also. As a result, the farmers' income will be affected. In such a case, storage cost will become an additional burden to the farmers. Similar is the case for transportation function, that is, by the time the produce reach the distant market, the prices may fall. The prices of agricultural products fluctuate not only from year to year, but during the year from month to month, day-to-day and even on the same day. The changes in prices may be upward or downward. Price variation cannot be ruled out, as the factors affecting demand for and the supply of agricultural products change continuously. A price fall may cause a loss to the trader or farmer, who stocks the produce. Sometimes, the risks are so great that it may result in a total failure of the business, and the person who owns it may become bankrupt.
6. ***Institutional risk:*** Institutional risk relates to non-availability of market finance at right time, imposition of tariffs by the Government, large scale imports of the produce by the Government, change of Government, changes in Government policies, statutory price controls, imposition of levies etc., which may affect the price of the commodity in the domestic market.

From the above discussion, we can also list out the following different types of risks involved in the marketing process:

- ***Time risk:*** Farmers aim at selling the produce at a profit out of anticipated raises in prices in future. During the time lag, conditions might change and the prices may fall. Thus, time risk is involved in marketing.
- ***Place risk:*** Place risk arises, when prices of the same produce is different in different places. The traders may purchase produce in a market, where prices are low with view of to sell it at other places, where the prices are high. But, prices in other market also may come down causing losses.
- ***Competition risk:*** Firms have to face risk arising from forces of competition. The competing firm may introduce modern methods of production due to which quality may be improved or cost of production may be reduced. Under such circumstances, a firm may be forced to sell at a loss which is called risk of competition.
- ***Risk of change in demand:*** The firm (say, food processor) produce products on large-scale in anticipation of increase in demand in future. But, sometimes the demand of the product may not come to expectations, resulting in losses.
- ***Risk arising from natural calamities:*** Risk from natural causes is beyond human control. These include rains, earthquake, floods, heat and cold. These risks cause heavy loss.
- ***Human risk:*** This risk arises due to adverse behaviour of human beings like theft, strikes, lockouts, heavy debts etc.

ii. Measures to Overcome Risks

The agencies engaged in marketing activities worry about the risks associated at every stage and they continually try to minimize the effects of these risks. A risk cannot be eliminated because, it also carries profit. The risk can be minimized by adoption of some of the measures given below:

- ☆ ***Strategic buying and selling:*** Strategic buying of inputs and selling of output can reduce the effects of marketing risk on farm profitability. Purchasing inputs for future use, when prices are relatively low and expected to increase reduces the risk of being incurring huge costs in the future. Similarly, selling output, when prices are favourable and expected to decline reduces the risk of losses in the future. So, strategic buying may necessitate storing excess inputs planned for future use that were purchased when prices were low, and strategic selling may also necessitate storing produced outputs for future sale, if current prices are unfavorable and expected to rise. So, scientific storage of both inputs and output is essential to minimize the risk. If storage of commodities is not on scientific lines, it can also introduce other types of risk (theft, spoilage etc.). So, the use of this tool will depend on individual farm capacity for storage and ability to manage inventory.
- ☆ ***Contracting:*** A contract is a legal agreement between buyer and seller and can be used to reduce marketing risk with inputs and output. A farmer can enjoy forward contract with an input supplier for future delivery of the input at a specific price. Such a contract would protect the farmer from higher costs, if input prices increase in the future. However, if input prices fall, he is still required to pay the contracted price.

  Similarly, a farmer can contract with a buyer of his commodity to sell a given amount for a specific price at some future time. By doing so, he removes the price uncertainty, if commodity prices decline, but he also forfeits the opportunity to gain, if prices rise in the future. In addition, contracts require the farmer to deliver the agreed upon quantity of the commodity at the agreed upon time to the buyer in the future. Thus, it is also important to manage production risks to ensure the ability to meet all the terms of the marketing contract.
- ☆ ***Efficient performance of marketing functions:*** The physical loss of a produce may be reduced by the following ways:
  - ❖ Use of fireproof materials in the storage structures to prevent accidents due to fire;
  - ❖ Use of improved storage structures and giving necessary pre-storage treatment to the produce to prevent losses in quality arising out of excessive moisture, temperature, attacks by insects and pests, fungus and rodents;
  - ❖ Use of better and quicker transportation methods and proper handling of produce during transit; and
  - ❖ Use of proper packaging material.
- ☆ ***Insurance:*** Not all risks can be controlled or protected by traditional insurance. Price risk is one such risk. As discussed earlier, price changes of both inputs and output can have a dramatic impact on agri-business. However, insurance serves as one of the best compensation measure to overcome production risk.
- ☆ ***Minimization of price risks:*** This can be achieved through the following:
  - ❖ Fixation of minimum and maximum prices for commodities by the Government and allowing movement of prices only within the defined range.
  - ❖ Making arrangements for the dissemination of accurate and scientific price information to all sections of society over space and time.

- ❖ ***Operation of speculation and hedging:*** The prices associated with commodities for which the facility of forward trading is available may be transferred to professional speculators through the operation of hedging.

  Futures trading: Futures trading serves as a best measure to manage input and output price risks. To reduce input price risk, a farmer would purchase an input futures contract. Assume a farmer entered into a futures contact with an input supplier on 1-2-2012 to purchase the input (fertilizer) on 1-6-2012. By doing so, he is obligated to purchase the contract for the requisite amount of the input at the expiration date of the contract (1-6-2012). Before the contract expires and purchase is required, the farmer would resell the contract and purchase the input in the cash market (say, 1-4-2012). If input price had risen over the course of the contract (*i.e.,* between 1-2-2012 to 1-6-2012) it implies, the farmer paid a higher price for the input in the cash market (1-4-2012), but would also receive a higher price on reselling his futures contract (1-6-2012) that would largely offset his increased costs for the inputs in the cash market (1-4-2012), as increase in cash market price also tends to increase the futures price (1-6-2012).

  To reduce output price risk, a farmer would sell a commodity futures contract thus, agreeing to sell the contracted amount of the commodity at the expiration of the contract. Any gains or losses incurred in the cash market would be offset by a corresponding loss or gain in the futures market thereby, reducing price variability and market risk for the farmer. This concept is discussed in-detail in the ensuing pages.

iii. Managing Price Risk - The Futures Trading Strategy

We know, commodity market is a market, where commodities are bought and sold. The various commodities traded under futures market include bullion (gold, silver), non-ferrous (base) metals (copper, zinc, nickel, lead, aluminium, tin etc), energy (crude oil, natural gas, etc.), agricultural commodities (soya oil, palm oil, coffee, pepper, cashew etc). The existence of a vibrant, active and a transparent commodity market is normally considered as a healthy sign of country's economic development. It is, therefore, important to have active commodity markets functioning in a country.

As discussed earlier, markets have existed for centuries world-wide for selling and buying of goods and services. The concept of market started with agricultural production and hence, it is as old as the agricultural products or the business of farming itself. Traditionally, farmers used to bring their products to a central market place (called mandi/bazaar) in a town/village, where traders would also come to buy the products and transport, distribute and sell them to other markets. Thus, in a traditional market, agricultural produce would be brought and kept in the market and the potential buyers would come and negotiate with the farmers directly on the price, quantity and quality of the produce. Final transaction of produce will take place, once mutual agreement was reached between the seller and the buyer. But, in a traditional market, shortage of a commodity in a given season would lead to increase in price for the commodity. On the other hand, over supply of a commodity could result in decline in price, sometimes even below the cost of production. Neither farmers nor traders were happy with this situation, since they could not predict, what prices would prevail on a given day or in a given season. As a result, farmers could not transact the produce, since they failed to fetch their expected price and there were no adequate storage facilities available at the farm-level and at market place. It was in this context that, farmers and traders start negotiating for future supplies of commodities in exchange of cash at a mutually agreeable price. This type of agreement was acceptable to both parties, since the farmer would know how much he would be paid for his commodity and the trader would know his cost of procurement in advance. This led to the evolution of a forward contracts trading system, which subsequently led to futures market.

Being an agro-based country, the opportunity to flourish for futures trading in India is very high. It is strongly apprehended by the market experts that, the turnover of Indian commodity futures market will overtake the capital market turn over in the near future. There is enormous potential in this sector, not only in terms of trading, but also in terms of opportunities for developing value added services in terms of quality warehousing, grading and certification services, modern marketing practices, dissemination of prices and trade information etc.

iv. Types of Commodity Trading

In general, commodity trading involves basically three types of markets *viz.*, Cash market, Forward market and Futures market.

1. Cash Market

It also called as Physical or Spot market. These are traditional markets. Cash market is a place, where buyers and sellers meet face to face to buy and sell the actual commodities and payment and delivery of commodities is immediate. Price prevails in the spot market is Spot/Cash price. Example of a cash market is a mandi, where food grains are sold in bulk. Farmers would bring their produce to this market and traders would immediately purchase the produce, and they settle the deal in cash and collect the produce immediately.

2. Forward Market and Futures Markets

Forward and Futures contracts are both contracts to meet a certain obligation usually, to deliver a specified quality and quantity of a commodity for a specified price, in specified manner on a specified date in the future. Both the parties *i.e.*, seller (farmer) and buyer (trader) are required to perform their obligations in a future date known as the maturity date of the contract. These markets reduce the risks by allowing the farmer and trader to decide a price today for produce to be delivered on a particular future date. Thus, these markets allow delivery at some time in the future, unlike cash markets that call for immediate delivery. These advance sales help both buyers and sellers in long-term planning. Forward contracts laid the groundwork for futures contracts. The main difference between these two contracts is the way in which they are negotiated.

2.1. Differences between Spot and Forward Transactions

We know, there are three processes involved in any market transaction *viz.*, trading, clearing and settlement. In the trading process, the buyer interacts with the seller for transacting the produce. During the course of interactions or negotiations, they arrive at an agreement regarding the quantity, quality and price of the produce to be transacted. In the clearing process, the buyer and the seller decide on the quantum of produce and the amount of cash that would be exchanged between them. The actual exchange happens in the settlement process. In case of spot transactions, trading, clearing and settlement happen simultaneously and instantaneously *i.e.*, 'on the spot'. For example, a farmer wishes to transact jaggery in a regulated market on 1-1-2012. After the auction, he identifies the highest bidder and he transacts the commodity (say, 10 quintals) on the same day at a price of ₹ 1000/quintal and immediately receives the sales proceeds from the buyer (₹ 10000). This is a spot transaction. However, if the trader makes an agreement with the farmer on 1-1-2012 in buying jaggery at ₹ 1000/quintal to be delivered after two months *i.e.*, on 1-3-2012, it is the case of forward contract. So, in a forward transaction, cash does not change hands on the date of entering into the contract. While the trading happens on the current day (1-1-2012), clearing and settlement happen at the end of the specified period (1-3-2012). Hence, in a forward contract, the trading, clearing and settlement do not happen simultaneously as in a spot contract.

2.2. Forward Market or Contract

Forward contract is a non-standardized legally binding contract in which a seller agrees to deliver a specific commodity to a buyer on a specified date in future at a price negotiated/agreed on the date of contract. Here, payment and delivery of goods is made only after a period of 11 days. These contracts are considered to be as the fore-runners of futures contracts. In these forward markets, both buyer and seller meet face to face to negotiate on price, delivery date, quantity and quality of the produce, but delivery of the product will be done at future date (*i.e.*, after 11 days). Each contract is thus a unique and not standardized, since it takes into account the needs of a particular seller and a particular buyer only. Forward contracts (other than futures) are also called as Customized contracts, as the terms of contracts are individually accepted between the two counterparties. These forward contracts are also called as Over The Counter (OTC) contracts.

Thus, in forward contract, two parties are involved *viz.*, buyer and seller. The seller (say, farmer) offers to the buyer a certain commodity deliverable in the future, let's say, six months from now because, he presumes price may fall in the future. The buyer, who is interested in having the commodity, will purchase the commodity with the presumption that, the price may rise in the future. Both parties then make up a contract, wherein a specified quantity of the commodity may be deliverable at a particular time in the future that is agreed upon. To understand this concept, go through the following example. On 1-1-2012, Mr. Ravi enters into an agreement to buy 50 kgs of rice on 1-6-2012 at ₹ 1500/- from Mr. Sri Ram, a wholesaler. It is a case of a forward contract, where Mr. Ravi has to pay ₹ 1500/- on 1-6-2012 to Mr. Sri Ram and Mr. Sri Ram has to supply 50 kgs of rice. So, in a forward contract, a user (holder), who promises to *buy* the specified commodity at an agreed price at a fixed future date is said to be in the *'Long position'*. On the other hand, the user (holder), who promises to *sell* the commodity at an agreed price at a future date, is said to be in *'Short position'*. Thus, 'long position' and 'short position', take the form of 'buy' and 'sell' respectively in a forward contract. But, the essence of forward contract lies in the understanding between the seller and the buyer of the commodity. Sometimes, during the course of time between the agreement made and the time of delivery of commodity, the contract may change hands, as the buyer may look for other lucrative opportunities (say, getting the same commodity at still lower price from other sellers). So, there lies counter party risk (Counter party risk is the risk of non performance by the other party of the contract).

This example implies, a forward contract is a bilateral agreement in which a buyer and seller agree upon the delivery of a specified quality and quantity of a commodity on a specified future date at a pre-determined price. One of the parties to the forward contract assumes a *'long position'* and agrees to *buy* the underlying commodity on a certain specified future date for a specified price. The other party assumes a *'short position'* and agrees to *sell* the commodity on the same date at the same specified price. In simple terms, 'long' and 'short' positions indicate, whether the party to the contract has a buy position (long) or sell position (short). Other contract details like delivery date, delivery price, quantity, quality etc., are discussed in person between the two parties. As mentioned earlier, forward contracts are customized contracts normally traded outside the exchanges. The following are the important features of forward contracts:

- ✫ Forward contracts are bilateral contracts and hence, are exposed to counter–party risk.
- ✫ Each contract is custom designed and hence, unique in terms of contract size, expiration date and the commodity type and quality.

- Generally, only parties of the contract know the contract price.
- On the expiration date, the contract has to be settled by delivery of the commodity. If the party wishes to reverse the contract, it has to compulsorily go to the same counter party, which often results in high prices being charged.
- The commodity and its characteristics are known to the buyer and the seller, but may not be known to other potential buyers and sellers
- The date on which the commodity will be bought and sold is determined in advance by the buyer and seller
- The price to be paid by the buyer to the seller at the future date is determined on the day the contract is entered into between the buyer and the seller, but is not made known to other potential buyers and sellers.
- Forward contract may well be regarded as direct and alternative marketing facilitator.
- Forward contracts can work as a tool to handle the situation arising out of good harvest as well as poor harvest for stabilization of supplies along with prices.

Limitations of Forward Contracts

They include:

- The contracts are private and negotiated bilaterally between two parties. Therefore, there are no exchange guarantees.
- The prices are not transparent, as there is no reporting requirement.
- There are no regulations for establishing market stability and protection of market players.
- Lack of standardization leads to illiquidity in the absence of a secondary market.
- The profit or loss is realized only on the maturity date.
- Settlement is only through actual delivery or offsetting by cash delivery.

2.3. Futures Market or Contract

Compared to the forward market for commodities, which is highly customized, the futures market is highly standardized. So, the limitations discussed above under forward market are addressed under futures market. Thus, futures markets have evolved out of forward contracts. These futures contracts are exchange-traded versions of forward contracts *i.e.*, the commodities are traded through an 'exchange'. These contracts are an agreement to buy or sell a specified quantity of a commodity during a designated date in the future, at a price agreed upon by the buyer and seller at the time of entering into the contract. Future contracts contain standard specifications of the underlying commodity, its quality and quantity and the date and time of expiry of the contract. *A futures contract need not be settled through physical delivery of the commodity. It can be closed by entering into an equal and opposite contract.* Thus, it can be said that, futures contract is specie of forward contract. It (Futures contract) is nothing but a forward contract entered through the exchange and it is a standardized legally binding agreement made on the trading floor of a futures exchange between two parties to buy or sell a product of specified quality and quantity on a specified future date at a specified place at an agreed price. It is similar to forward contract, but differs from cash market in the sense that, trading happens on the current day of the agreement, but clearing and settlement happens on the date of maturity of the contract (Table 5.90), whereas in cash market, all these three *i.e.*, trading, clearing and settlement happen on the spot. So, in futures contract, though agreement takes place at the current time, it will be executed in a future date, like forward contract. In futures contract, all terms *viz.*, quantity, quality, and delivery time are standardized unlike forward contract (Table 5.91). Hence, futures contracts are also called as Standardized contracts because, the parties involved in the futures contracts do not decide the terms of futures contracts; but they merely accept terms of contracts standardized by the exchange. So, these futures contract matures at a future date and a reference is made for identifying the date when it is going to mature. The transaction price is discovered through

Table 5.90: Differences between cash market and futures market.

Cash Market	*Futures Market*
Actual commodities are traded	Contracts are traded
No need to pay margin money	Margin money is required to pay
Full payment is required	Full payment is not required
Actual delivery of commodities	No actual delivery of commodities in most of the cases.
Trading, clearing and settlement happens instantaneously	Trading happens today, but the clearing and settlement happens at the end of specified (maturity) period

Table 5.91: Forward versus futures contracts.

Contract Element	*Forward Contracts*	*Futures Contracts*
Price	Negotiated privately	Open outcry
Quantity	Negotiated privately	Standardized
Quality	Negotiated privately	Standardized
Delivery Period	Negotiated privately	Standardized
Delivery Location	Negotiated privately	Standardized
Transfer of Physical Product	Expected	Rarely occurs
Nature of contract	It is a customized contract	It is a standardized contract
Number of parties	Non-standardized contract between two parties	Standardized contract between two parties or can be sold even to a third party
Trading place	Trade off the 'Exchange'	Trade on the 'Exchange'
Margin money	Margin money is not required	Margin money is required
Counterparty risk	There is high counter party risk due to customized nature of transactions and parties are the only persons involved in the transactions	Counter party risk is greatly minimized by the presence of clearance house, as it provides the trade guarantee.
Transparency of price	Not transparent	Fully transparent, as the prices are displayed on the trading screen of the exchange.
Liquidity	A forward contract is less liquid due to its customized nature.	Liquidity is the measure of frequency of trades that occur in a particular commodity futures contract. A future contract is more liquid as it is traded on the exchange.
Regulations	A forward contract is not regulated by any exchange.	An exchange regulate a futures contract
Settlement	A forward contract is generally settled by physical delivery.	Futures contract is usually cash-settled, but the option of physical settlement is available.

the interaction of supply and demand in a centralized market place called 'Exchange'. In a forward contract, the buyer and seller bi-laterally negotiate all aspects of the contract. The exchange of the physical commodity between the buyer and the seller is anticipated and expected at the negotiated location. However, futures contracts are standardized and predefine the quality, quantity, delivery period and delivery location of each contract. The buyer and seller are left to negotiate, by open outcry method or electronic trading method, the price and the quantity of commodity to trade. It is to be noted that, trading of futures contracts rarely results in the actual exchange of the physical commodity.

Thus, the essential distinction between a futures contract and forward contract is that, while forward contract is one to one mutual agreement between the two parties, futures contract is a much more standardized agreement between the parties, which can be bought or sold to a third party (or between the parties) at any point of time, so as to alter or even reverse the rights and obligations arising under the original contract. This capability to altering positions, makes a futures contract an instrument with a higher level of liquidity than the forward contract. Moreover, in a forward contract, as the parties are the only persons involved in the transactions, there is a high level of counter-party risks *i.e.*, the risk of non-performance from one of the parties. But, this counter party risk is eliminated in a futures contract by the presence of clearance house, who always take an opposing position in every alternate transaction and ensure the equality of long and short positions.

Features of Futures Market

The following are the important features of futures market:

- ☆ Futures market trade futures contracts that are legally binding agreements that predefine the quality, quantity, location and period for delivery of a specified commodity.
- ☆ Open outcry method and Electronic trading method negotiates price and is determined, when there is agreement between the buyer and the seller to make a transaction.
- ☆ For every seller there must be a buyer and for every buyer there must be a seller in order for a futures trade to occur.
- ☆ The *seller* of a futures contract enters a commitment to *make* delivery of a specified commodity. This is called a *short* position.
- ☆ The *buyer* of a futures contract enters a commitment to *take* delivery of a specified commodity. This is called a *long* position.
- ☆ Note that, it is only a commitment to make or take delivery.
- ☆ There is a clearing house in the futures exchange that is responsible for settling trading accounts, clearing trades, collecting and maintaining margin moneys, regulating delivery and reporting trading data. Clearing houses act as third parties to all futures and options contracts–acting as a buyer to every clearing member seller and a seller to every clearing member buyer. The process by which a clearing house maintains records of all trades and settles margin flow on a daily mark-to-market basis for its clearing member is called Clear. Clearing house is responsible for the financial commitments of customers that clear through their exchange.
- ☆ The seller's obligation to *make* delivery of the commodity is with the clearing house in the futures exchange.
- ☆ The buyer's obligation is to *take* delivery of the commodity and pay the price to the clearing house. The clearing house pays the seller.
- ☆ Because selling a futures contract (short position) is a promise to *make* delivery of a specified quantity and quality of a commodity and buying a futures contract (long position) is a

promise to *take* delivery of the same commodity, the two transactions are essentially offsetting positions. Offsetting refers to liquidating a futures position by entering an equivalent, but opposite, transaction, which eliminates the delivery obligation. *If a trader enters the futures contract as a seller/buyer, he has to come out of the futures contract as a buyer/seller in the cash market.* Thus, offsetting refers to taking a second position opposite to the initial or opening position. Selling (or purchasing) futures contracts of the same commodity purchased (or sold) during the earlier transaction in the cash market refers to offsetting the futures contract. This offsetting feature is what makes future markets unique and attractive to both hedgers and speculators. The vast majority of futures positions are fulfilled by offsetting.

☆ The leverage of futures trading stems from the fact that, only a relatively small amount of money (known as initial margin) is required to buy or sell a futures contract. On a particular day, a margin deposit of only $1,000 might enable you to buy or sell a futures contract $25,000 worth of soybeans. The smaller the margin, in relation to the value of the futures contract, the greater is the leverage.

In the futures market, leverage refers to having control over large value of commodities with comparatively small levels of capital (margin). In other words, with a relatively small amount of cash/ margin, a trader can enter into a futures contract that is worth much more than the trader's initially pay as margin in his account. Futures positions are highly leveraged because, the initial margins that are set by the exchange are relatively small compared to the cash value of the contracts in question. The smaller the initial margin in relation to the cash value of the futures contract, the higher the leverage.

If a trader speculate in futures contracts and if the price moves in the direction the trader anticipated, high leverage can yield large profits in relation to the initial margin deposit. But, if prices move in the opposite direction, high leverage can produce large losses in relation to the initial margin deposit. This is one of the reasons, why the futures market is useful, but also very risky). So, in the futures market, price changes are highly leveraged *i.e.*, for a small change in the futures price, it will lead to huge gain or loss to the trader. So, leverage is considered as a two-edged sword. This is exemplified below:

Example 1

Assuming that, in anticipation of rising prices, a trader buys 'commodity X' in the month of January, when price index is trading at 1200 points. The initial margin requirement is ₹ 15,000. The trader purchased ₹ 3 lakhs worth of sales contract *i.e.*, the value of the futures contract is ₹ 250 times the index. This implies, each one point change in the index represents a ₹ 250 gain or loss. An increase of five per cent in the index, from 1200 to 1260, would produce a ₹ 15,000 profit (60 X ₹ 250). Conversely, a 60 point decline would produce a ₹ 15,000 loss. In either case, an increase or decrease of only five per cent in the index would, in this example, result in a gain or loss equal to 100 per cent of the ₹ 15,000 initial margin deposit. That's the arithmetic of leverage.

Example 2

A trader entered a futures contract in the month of January to purchase one tonne of wheat in the month of June at ₹ 900000/- with the expectation that price will rise in the future. Assume that, the futures trading in wheat calls for only four per cent initial margin, amounts to ₹ 36000/-. As expected in the month of June, if the price of wheat increased by two per cent, the traders gets a profit of ₹ 18000 on the deposit of ₹ 36000/- as initial margin. That means, the trader gets a profit of 50 per cent of initial margin by just two per cent increase in price of the wheat. Conversely, if the price of the wheat falls by two per cent, the trader loses ₹ 18000/- *i.e.*, 50 per cent of the margin.

The above examples clearly indicate that, transactions through futures market can be extremely risky. This is evident through the arithmetic of leverage. Highly leveraged investments can produce two results: greater profits or greater losses. As a result of leverage, if the price of the futures contract moves up even slightly, the profit gain will be large in comparison to the initial margin. However, if the price just inches downwards, that same high leverage will yield huge losses in comparison to the initial margin deposit. So, futures trading requires not only the necessary financial resources, but also the necessary financial and emotional temperament.

Thus, it is clear that, one does not need to have the physical commodity or own a contract for the commodity to enter into a sale contract in futures market. It is simply agreeing to sell/going 'short' the physical commodity at a later date. It is possible to repurchase the contract before the maturity thereby, dispensing with physical delivery of commodities. That is, the delivery of commodities is also not mandatory in futures trading. It all depends upon the agreement between seller and buyer. As futures are standardized contracts traded through an exchange, they can be used to minimize price risk by means of hedging technique. Since the exchange standardizes the quality and quantity parameters and offers complete transparency in trade, the counterparty default risk has been greatly minimized. This lowering of risk on the part of traders, in turn, lowers their cost of doing business. This savings is passed on to the consumers, in the form of lower prices for the commodities.

In India, the online commodity futures trading is regulated under the Forward Contracts (Regulation) Act (FCRA), 1952 and are under the jurisdiction of the Department of Consumer Affairs, the Ministry of Civil Supplies and Consumer Affairs, Government of India. These futures markets can only be organized through Commodity Exchanges, which are approved by the Forward Markets Commission (FMC). The FMC performs the functions of advisory, monitoring, supervision and regulation in futures and forward trading. All the exchange houses that provide the facility of online commodity futures trading have to abide by the set rules and principles of FMC otherwise, the commission is empowered to cancel their licenses. For example, in India, futures trading take place at the Multi-commodity Exchange of India Limited (MCX), Mumbai. Thus, the commodity futures trading is done at the centralized exchanges and the traders deal the trade more electronically rather than physical handing of the commodities. These exchanges operate independently under the guidelines of their byelaws approved by the FMC. The number of commodities dealt under futures trading varies from one exchange to another and there are more than 300 commodities world-wide in which a person can easily trade.

2.4. FMC

It is a regulatory body for commodity futures trade in India. This was set up under the FCRA of 1952. It is responsible for regulating and promoting futures trade in commodities. Its head quarter is located at Mumbai and regional office is located at Kolkata. Uniquely the FMC falls under the Ministry of Consumer Affairs, Food and Public Distribution and not the Finance Ministry as in most countries. This is because futures, traded in India, are traditionally on food commodities. However, this has been changing and there have been calls for change in the industry and in regulation. One proposal is the merging the commodities derivatives and securities regulation by including the FMC within the Securities and Exchange Board of India (SEBI), the primary securities regulator in India. However as of 2003 there is no clear consensus for this move. The vision of FMC include:

- ✰ To ensure that the futures trading in the commodities is conducted in a way so as to meet the twin economic functions of price discovery and risk management in an efficient and transparent way.

- The commodity exchanges in the country come up as professional organizations so as to effectively meet the aspirations of various intermediaries in the commodity's market including the farmers.
- The basket of commodities in which futures trading is permitted should progressively grow so that the participants of these commodity markets can also benefit from the futures market.
- To ensure smooth and orderly development of the commodity futures markets to enable it to perform the economic functions of the Price discovery and price risk management efficiently, through effective Regulation of the market.

Functions of FMC

The following are the important functions of FMC:

- It advises Central Government in respect of grant of recognition or withdrawal of recognition of any association.
- It keeps forward markets under observation and takes such action in relation to them as it may consider necessary, in exercise of powers assigned to it.
- It collects and publishes information relating to trading conditions in respect of commodity including information relating to demand, supply and prices and submit the same to the Government.
- It makes recommendations for improving the organization and working of forward markets.
- It undertakes inspection of books of accounts and other documents of recognized/registered associations.

Thus, FMC provides regulatory measures in order to ensure financial integrity (*i.e.*, to prevent systematic risk of default by one major operator or group of operators), market integrity (*i.e.*, to ensure that futures prices are truly aligned with the prospective demand and supply conditions) and to protect and promote interest of customers. It prescribes important regulatory measures like prevent excessive speculation, imposition of margins to prevent defaults by members/clients, imposition of penalty for defaults etc. The need for regulation is important to create competitive conditions. In the absence of regulation, unscrupulous participants could use these leveraged contracts for manipulating prices. This could have undesirable influence on the spot prices thereby, affecting interests of society at large. Regulation is also needed to ensure that, the market has appropriate risk management system. In the absence of such a system, a major default could create a chain reaction. The resultant financial crisis in a futures market could create systemic risk. Regulation is also needed to ensure fairness and transparency in trading, clearing, settlement and management of the exchange, so as to protect and promote the interest of various stakeholders, particularly non-member users of the market.

2.5. Attributes of a Futures Market

They include:

- ***Price discovery:*** In futures market, the buyers and sellers have useful information about demand and supply changes of commodities and when they are in agreement, a trade is made and a market price is reported. This price, as discovered through open outcry method or electronic trading method is a visible example of the forces of supply and demand at work. These trading methods serve as platforms to both buyers and sellers to enable direct participation in trade all over the country. This facilitates perfect interaction between demand and supply forces and thereby, establishes a long-term benchmark prices for the commodities. This, in turn, reduces the number of intermediaries involved in the supply chain. Even small farmers can take advantage of the commodity exchanges through the Aggregator Model (The Internet has generated a new sales and distribution model called the Aggregator Model). Agencies like ITC e-Choupal, The Haryana State Co-operative Marketing Federation (HAFED) etc. collect the produce from many small farmers and participate in the commodity exchanges. This price discovery also enables the farmers and traders to judge, what price the market will pay now for a future delivery. In other words, the futures price predicts the future supply and demand as estimated by the market. Prices of spot and futures market move in tandem and the national price signal links all spot market prices. Therefore, price discovery and long term price signals lead to integrated price structure throughout the country. Even in the case of physical delivery of agricultural commodities, use of objective system of grades and standards ensures that the trade is efficient and fast. It also makes the exchange the benchmark point for pricing.

- ***Price risk management:*** The offsetting ability of a futures contract allows hedgers to take equal, but offsetting futures position from that which they hold in the cash market. This allows the hedger to transfer price risk until such time as the cash position has been reduced or liquidated. This is the main economic justification for futures markets and a primary attraction to hedgers.
- ***Liquidity:*** This is a key to any successful futures market. The futures market ensure smooth and orderly price movements that allow ease of entry to and exit of participants from the market place. This shows less impact on price movements and thereby, leads to a competitive price. The greater the volume of trade without a price impact, the more the market is said to be 'liquid'.
- ***Efficiency:*** The futures market enables all participants to trade large volumes at a relatively low transaction cost. This efficiency is attractive to both buyers and sellers.
- ***Zero Sum Gain:*** Futures trading is frequently referred to as a zero sum gain, as any profit by one market participant is a result of loss by other participant in the market. For hedgers, this is not a concern, as their futures gains or losses are offset in the cash market.

2.6. Functions of an Exchange

To manage a variety of risks in agriculture led to the evolution and establishment of commodity exchanges across a large number of countries. We know, 'agriculture' involves various risks at different stages like production, storage, processing, trading, safeguarding against counterparty risks etc. Hence, a platform is needed to overcome such risks and to lower the transaction costs to the trade participants and this is done by futures exchanges. Additionally, exchanges serve as an inexpensive, efficient and transparent mechanism for price discovery and dissemination of price information. Participation in these exchanges is not limited to those, who are directly involved in the physical market. In fact, other participants like speculators impart the much-needed liquidity to these markets. So, the major functions of an exchange include:

- Price discovery
- Price dissemination
- Risk management
- Market surveillance
- Clearing and settlement

In futures market, prices are determined from the interaction of bids and offers emanating from all over the country, which converge on the trading floor of the futures exchange. The bid and offer prices are based on the expectations of prices on the future/maturity date. Two methods will be generally used by the Professionals for predicting futures prices *viz.*, fundamental analysis and technical analysis. The fundamental analysis deals with the information about basic supply and demand conditions such as, weather conditions, carryover supplies, policies and programmes of the Government and agricultural reports. Technical analysis deals with the analysis of movement of prices in the past. Many participants use fundamental analysis to determine the direction of the market, and technical analysis is used to time participants' entry and exit.

Quite often, stock exchange is confused with a futures exchange in the functional aspects. But, at stock exchange, the stocks bought and sold represent partial ownership in the company, which originally issued the stock. At a futures exchange, contracts are bought and sold. These contracts are standardized in terms of quality, quantity, delivery time and location. The only variable is price, which is discovered while trading on the exchange floor.

2.8. History of Futures Markets

What we know as the futures market of today came from some humble beginnings. Trading in futures originated first in Japan at the Dojima Rice Exchange in the 18th century and was primarily meant for the trading of rice and silk. In 1848, the USA started Chicago Board of Trade (CBOT) for using futures markets to buy and sell corn. The CBOT initiated trade in grain futures in the year 1865. The following are some of the important commodity exchanges in the world:

- New York Mercantile Exchange (NYMEX)
- London Metal Exchange (LME)
- Chicago Board of Trade (CBOT)
- New York Board of Trade (NYBOT)
- Kansas Board of Trade
- Winnipeg Commodity Exchange, Manitoba
- Dalian Commodity Exchange, China
- Bursa Malaysia Derivatives exchange
- Singapore Commodity Exchange (SICOM)
- Chicago Mercantile Exchange (CME), US
- London Metal Exchange
- Tokyo Commodity Exchange (TOCOM)
- Shanghai Futures Exchange
- Sydney Futures Exchange
- London International Financial Futures and Options Exchange (LIFFE)
- Dubai Gold and Commodity Exchange (DGCX)
- Dubai Mercantile Exchange (DME), (joint venture between Dubai and the New York Mercantile Exchange (NYMEX))

There are strong grounds to believe that Commodity futures could have existed in India for thousands of years. References to the existence of market operations similar to the modern day Futures market are found in Kautilya's 'Arthasastra". Other factors which support such a belief is the existence of words like "teji', 'mandi'. 'gali', 'phatak' etc., for centuries. Be that as it may, the Futures markets in its organized form appeared only in the late 19th Century, with the advent of the British. In India, the concept of futures trading remained underdeveloped due to Government intervention in many commodity markets to control prices. The production, supply and distribution of many agricultural commodities are still governed by the State. Both forward and futures trading are selectively introduced with stringent controls. While free trade in many commodity items is restricted under the Essential Commodities Act (ECA) 1955, forward and futures contracts are limited to certain commodity items under the FCRA, 1952.

Pre-Independence Period

The first commodity exchange was set up in India by Bombay Cotton Trade Association and formal organized futures trading was started in cotton in 1875. Subsequently, many exchanges came up in different parts of the country for futures trade in various commodities. The Gujarati Vyapari Mandali came into existence in 1900, which has undertaken futures trade in oilseeds for the first time in the country. The Calcutta Hessian Exchange and East India Jute Association were set up in 1919 and 1927 respectively for futures trade in raw jute. In 1921, futures in cotton were organized in Mumbai under the auspices of East India Cotton Association (EICA). Many exchanges were set up in major agricultural centres in North India before world war broke out and they were mostly engaged in wheat futures, until it was prohibited. The existing exchanges in Hapur, Muzaffarnagar, Meerut, Bhatinda, etc, were established during this period.

Post-Independence Period

The Government established Shroff Committee (1950) to revise the draft 'Futures Market Regulation Bill' and finally, in December 1952, the FCRA was enacted. The Forward Contracts (Regulation) rules were notified by the Central Government in 1954. The futures trade in spices was first organized by India Pepper and Spices Trade Association (IPSTA) in Cochin in 1957. Futures in gold and silver began in Mumbai in 1920 and continued, until it was prohibited by the Government by mid-1950s. Though options are permitted in stock market, they are not allowed in commodities. The commodity options were traded during the pre-independence period. Options on cotton were traded, until they along with futures were banned in 1939. However, the Government withdrew the ban on futures with passage of FCRA in 1952. To protect against unhealthy speculation, forward trading in agricultural commodities was regulated under the FCRA, 1952. This Act was enacted with a view to regulate forward contracts prohibiting options in commodities and dealing with certain other related matters. This Act has provided for the establishment of FMC in September, 1953 for the said purpose of exercising the regulatory powers on the exchanges. The Government established Dantwalla Committee, (1966) to review the functioning of the FMC amidst changing economic conditions in the country. Later, futures trade was altogether banned by the Government in 1966 in order to have control on the movement of prices of many agricultural and essential commodities. After the ban of futures trade, all the exchanges went out of business and many traders started resorting to unofficial and informal trade in futures. In 1980, the Khusro Committee recommended the re-introduction of futures trading in major commodities. As a result, the Government initiated trading in potato in the latter half of 1980 in Punjab and Uttar Pradesh. Later, futures trading was reintroduced in other commodities like cotton and jute.

Post-Economic Liberalization

To consider the implications of liberalization in nineties, Kabra Committee was set up in 1993-94 and it recommended many new commodities to be included in the futures trading. Based on the recommendations of Kabra Committee, the Government allowed futures trading in 17 commodity groups, FMC was strengthened, certain amendments were made in FCRA and permitted futures trading in all recommended commodities. The Forward Markets Review Committee, set up by the Government of India under the chairmanship of Prof. M.L. Dantwala, recognized the need for futures trading even in conditions of short supply, and upheld the view that, speculations in futures markets should be recognized as a necessary factor for their proper working. In order to give more thrust on agricultural sector, the National Agricultural Policy, 2000 has envisaged external and domestic market reforms and dismantling of all controls and regulations in agricultural commodity markets. It has also proposed to enlarge the coverage of futures markets, so as to minimize the wide fluctuations in commodity prices and for hedging the risk arising from price fluctuations. In line with this proposal, many more agricultural commodities are being brought under futures trading. As a result of this, five National Commodity Exchanges became operational from 2003. They are MCX, National Commodity and Derivative Exchange Limited, Mumbai (NCDEX), National Multi-Commodity Exchange of India Limited (NMCE), Ahmedabad and Indian Commodity Exchange (ICEX), Gurgaon (Haryana) and Ace Derivatives and Commodity Exchange, Ahmedabad. Due to mistaken apprehensions that futures trading contributes to inflation, futures trading in rice, wheat, urad and tur was suspended in early 2007.

Table 5.92: List of exchanges conducting futures trading in India.

S.No.	*Name and Place*
I.	**National Exchanges**
1.	MCX
2.	NCDEX
3.	NMCE
4.	ICEX
5.	Ace Derivatives and Commodity Exchange (ACE)
II.	**Regional Exchanges**
6.	NBOT, Indore
7.	Vijay Beopar Chamber Ltd., Muzaffarnagar
8.	Rajdhani Oil and Oilseed Exchange Ltd, Delhi
9.	Bhatinda Om and Oil Exchange Ltd, Bhatinda
10.	The Chamber of Commerce, Hapur
11.	The Meerut Agro Commodity Exchange Ltd., Meerut
12.	The Bombay Commodity Exchange Ltd., Bombay
13.	Rajkot Seeds, Oil and Bullion Merchants Association, Rajkot
14.	The Ahmedabad Commodity Exchange, Ahmedabad
15.	The East India Jute and Hussian Exchange Ltd., Calcutta
16.	The East India Cotton Association Ltd., Calcutta
17.	The Spices and Oilseeds Exchange Ltd, Sangli
18.	Kanpur Commodity Exchange Ltd., Kanpur
19.	India Pepper and Spice Trade Association, Kochi (IPSTA)
20.	The First Commodities Exchange of India Ltd., Kochi
21.	Central India Commerce Exchange Ltd., Gwalior
22.	E-Sugar India Ltd., Mumbai
23.	National Multi-Commodity Exchange of India Ltd., Ahmedabad
24.	Coffee Futures Exchange India Ltd., Bangalore
25.	Surendranagar Cotton Oil and Oilseeds, Surendranagar
26.	E-Commodities Ltd., New Delhi
27.	Bullion Merchants Association, Bikaner

The Present Status

At present, there are 27 recognized exchanges in India. The country's commodity futures exchanges can be divided into two types *viz.*, National exchanges and Regional exchanges. Out of these 27 exchanges, five are National Exchanges and 22 are Regional Exchanges (Table 5.92 and Figure 5.42). FMC is a regulatory body for all these commodity exchanges. MCX (Mumbai), NCDEX (Mumbai), NMCE (Ahmedabad), ICEX (Gurgaon) and Ace Derivatives and Commodity Exchange (ACE, located at Ahmedabad) are the National commodity exchanges. (The term, 'National' used for these exchanges does not mean that other exchanges are restricted from having nationwide operations). The leading regional exchange is the National Board of Trade (NBOT), located at Indore. The national level exchanges follow the best practices and offer better and more transparent services in comparison to the regional exchanges (Table 5.93).

Table 5.93: Comparison between national and regional commodity exchanges.

National Exchanges	*Regional Exchanges*
Compulsory online trading	Online trading not compulsory
Transparent trading	Opaque trading
Exchanges to be de-mutualised	De-mutualisation not mandatory
Exchange recognized on permanent basis	Recognition given for fixed period
Multi commodity exchanges	Generally, these are single commodity exchanges.
Professionally run	Driven by interest groups
Large expanding volumes	Low volumes in niche markets

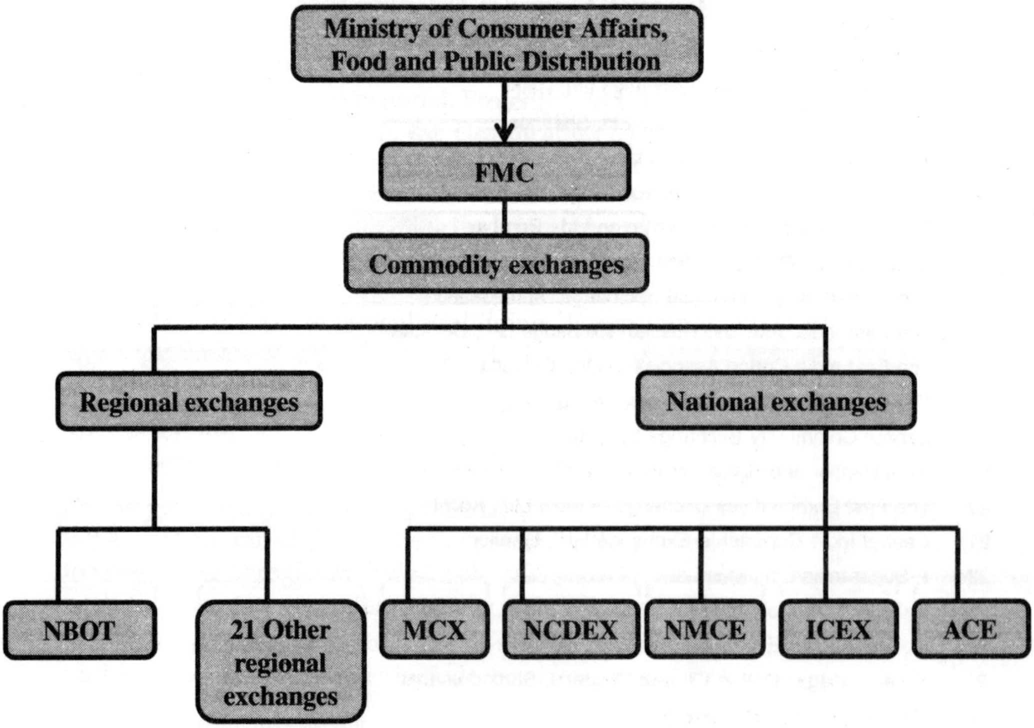

Figure 5.42: Structure of futures exchanges in India.

MCX

MCX is headquartered at Mumbai. It is promoted by State Bank of India (SBI) group and many other financial institutions including the Financial Technologies (India) Ltd. It is an independent and de-mutualized multi commodity exchange, which has permanent recognition from Government of India for facilitating online trading, clearing and settlement operations for commodity futures markets across the country. The exchange started operations in November 2003 and has risen to top most position among all commodity exchanges in India. It is the sixth largest commodity futures exchange globally in terms of the number of contracts traded on the exchange. In terms of number of futures contracts traded, it occupies first rank in silver trade and second rank in gold, copper and natural gas trade at global level. The MCX has relatively more edge in non-agricultural commodities. This commodity exchange deals above 40 commodities on its platform and has a market share of around 80 per cent in the Indian commodity market. Key shareholders of MCX are Financial Technologies (India) Ltd., SBI, Union Bank of India, Corporation Bank, Bank of India and Canara Bank. This commodity exchange facilitates online trading, clearing and settlement operations for commodity futures market across the country.

NCDEX

It is located in Mumbai and commenced its operations since December 15, 2003. This is the only commodity exchange in the country promoted by national institutions such as NABARD, National Stock Exchange of India Limited (NSE), Life Insurance Corporation (LIC) and Industrial Credit and Investment Corporation of India (ICICI) Bank Ltd. etc. It is a professionally managed online multi-commodity exchange, with an edge in agricultural commodities. It is a technology driven de-mutualised on-line commodity exchange with an independent Board of Directors and professional management.

NMCE

NMCE started its operation in November 26, 2002. It is the only exchange in India to have investment and technical support from commodity relevant institutions like CWC Ltd., Gujarat State Agricultural Marketing Board, Neptune Overseas Limited, NAFED, Gujarat Agro-Industries Corporation Ltd. (GAICL), Gujarat State Agricultural Marketing Board (GSAMB) and the NIAM.

ICEX

ICEX is latest commodity exchange of India started functioning from 27th November, 2009. ICEX is a screen based on-line derivatives exchange for commodities and has established a reliable, time tested, and a transparent trading platform. It is also in the process of putting in place robust assaying and warehousing facilities in order to facilitate deliveries. It has Reliance Exchangenext Ltd. as anchor investor and has MMTC Ltd., Indiabulls Financial Services Ltd., Indian Potash Ltd., KRIBHCO and IDFC among others, as its partners.

ACE

ACE was established in 1952. It has its registered office in Ahmedabad and the exchange is recognized under Forward Contracts Act, 1952 for futures trading since 1952, regulated by FMC. The exchange has the reputation of being first in the country to have the highest volume in Castor seed trading. This exchange is earlier known as Ahmedabad Commodity Exchange and has been in existence for more than six decades in commodity business, bringing in the best and transparent business practices in the Indian commodity space. Previously, cottonseed and groundnut oil were traded. Trading was carried on an outcry system in the trade ring. The exchanges is now anchored by the Kotak Group and is the country's fifth pan-India commodity exchange. Other key shareholders in the exchange include Haryana State Cooperative Supply and Marketing Federation Ltd. (HAFED), Bank of Baroda, Corporation Bank and Union Bank. ACE is now screen-based online derivatives exchange for commodities in India. The exchange currently trades with six agricultural commodities *viz.*, castor, chana, mustard seed, soybean, refined soyoil and sugar. ACE commenced operations as a national level commodity bourse in October 2010.

Unique Features of National Level Commodity Exchanges

The unique features of national level commodity exchanges are:

- ☆ They are demutualized, meaning thereby that they are run professionally and there is separation of management from ownership. The independent management does not have any trading interest in the commodities dealt with on the exchange.
- ☆ They provide online platforms or screen based trading as distinct from the open outcry systems (ring trading) seen on conventional exchanges. This ensures transparency in operations as everyone has access to the same information.
- ☆ They allow trading in a number of commodities and are hence multi-commodity exchanges.
- ☆ They are national level exchanges which facilitate trading from anywhere in the country. This corollary of being an online exchange.

As compared to other markets in the last ten years, commodity market has performed relatively better than other markets like bonds, equity or currency. However, the participation in future trading in Indian commodity market is very low as compared to other countries, as there is lack of knowledge about this market to the investors and traders. It is not for mere trading purpose; commodity trading is also used for hedge against inflation, price discovery of the commodity and also as a sound investment.

Organizational Set-up

In the Indian context, the scope of commodity exchanges has been limited to futures trading. The exchanges provide all organizational support for carrying out futures trading in a formal environment. These exchanges are managed by a Board of Directors composed primarily of the members of the association. There are also representatives of the Government and members nominated by FMC. The majority of members of the board have been chosen from among the members of the association, who have trading and business interest in the exchange. The board appoints a Chief Executive Officer, who with his team assists in day-to-day administration of the exchange.

2.9. Basic Concepts in Commodity Futures

1. Market

A market is conventionally defined as a place, where buyers and sellers meet to exchange goods or services. But, in the IT enabled environment, buyers and sellers from distant locations can even transact the commodities through an electronic market place. Hence, the physical market place is not necessary for the exchange of goods or services. Electronic trading and settlement of transactions has created a revolution in global commodity markets.

2. Commodity

A commodity is a product that has commercial value, which can be produced, bought, sold and consumed. Commodities are basically the products of the primary sector of an economy. The primary sector of an economy is concerned with agriculture and extraction of raw materials such as metals, energy (crude oil, natural gas), etc., which serve as basic inputs for the secondary sector of the economy.

Futures trading can be conducted in any commodity, but it should be subjected to the approval/ recognition as per the Section 15 of the FCRA. At present, 105 commodities are in the regulated list. The commodities other than those listed under Section 15 are conventionally referred to as 'Free' commodities. All the commodities are not suitable for futures trading. For a commodity to be suitable/ qualify for futures trading (Tables 5.94 and 5.95), it must possess the following characteristics:

- Commodities should be in plentiful supply. This is because, if a commodity is short in supply, a few traders may corner the whole supply and charge high price from the buyers.
- The commodity should have a large demand from a number of independent buyers so that, a single buyer may not be in a position to impose his terms for his purchase.
- The commodity must have a minimum degree of perishability *i.e.*, it must be storable for futures delivery.
- The commodity should be homogeneous and capable of being graded so that, its future deliveries may be made without problems regarding quality.
- There should be free flow of the commodity to and from the market without any external interference/control.
- The supply of the commodity should not be controlled by a few large firms (sellers). It should be available with a large number of suppliers
- The price of the commodity should be liable to fluctuations over a wide range so that, both buyers and sellers participate in transactions with presumptions about future price fluctuations.
- The commodity must not have gone through any complicated processing activity. In other words, the commodity must be in a basic, raw and unprocessed state. There are, of course, some exceptions to this rule. For example, metals, which are refined from metal ores, and sugar, which is processed from sugarcane.
- A major consideration, while buying the commodity is its price. Fundamental forces of market demand and supply determine the commodity prices.

Table 5.94: List of commodities for which futures trading is being conducted presently.

Commodity-group	*Commodities*
Metal	Aluminium, Copper, Lead, Nickel, Sponge Iron, Steel Long (Bhavnagar), Steel Long (Govindgarh), Steel Flat, Tin, Zinc
Bullion	Gold, Gold HNI, Gold M, i-gold, Silver, Silver HNI, Silver M
Fiber	Cotton L Staple, Cotton M Staple, Cotton S Staple, Cotton Yarn, Kapas
Energy	Brent Crude Oil, Crude Oil, Furnace Oil, Natural Gas, M. E. Sour Crude Oil
Spices	Cardamom, Jeera, Pepper, Red Chilli, Turmeric
Plantations	Arecanut, Cashew Kernel, Coffee (Robusta), Rubber
Pulses	Chana, Masur, Yellow Peas
Petrochemicals	HDPE, Polypropylene(PP), PVC
Oils and oil seeds	Castor Oil, Castor Seeds, Coconut Cake, Coconut Oil, Cotton Seed, Crude Palm Oil, Groundnut Oil, Kapasia Khalli, Mustard Oil, Mustard Seed (Jaipur), Mustard Seed (Sirsa), RBD Palmolein, Refined Soy Oil, Refined Sunflower Oil, Rice Bran DOC, Rice Bran Refined Oil, Sesame Seed, Soymeal, Soy Bean, Soy Seeds
Cereals	Maize
Others	Guargum, Guar Seed, Gurchaku, Mentha Oil, Potato (Agra), Potato (Tarkeshwar), Sugar M-30, Sugar S-30

Source: <www.fmc.gov.in>

We know, guar gum plays a crucial role in the fracking process. Its dispersibility in water and mineral oil assists in pushing fracking fluid through solid rock to reach petrochemical deposits in the earth's crust. While guar gum is also used for food products, Indian observers found that oil companies consistently outbid agricultural and comestible producers for guar bean and guar gum products, disproportionately contributing to food inflation. India, the main producer of guar gum, has faced crippling food inflation over the past two years. the FMC released a report accusing 4,490 different business entities of accumulating over ₹ 1,291 crore (approximately $235 million) from manipulating the commodities markets. While guar gum is also used for food products, Indian observers found that oil companies consistently outbid agricultural and comestible producers for guar bean and guar gum products, disproportionately contributing to food inflation. In March 2012, as the jurisdictional regulator of the guar gum market, the Forward Markets Commission (FMC) of India decided to close out all physical contracts of guar gum for the season. They also disallowed any new positions in the guar gum trading market effective March 21, 2012 and hiked margins on participants going long on the guar gum market. Their stated reason was that prices had quadrupled in the period between December 1, 2011 and March 27, 2012. Commodity prices fell sharply after the FMC announcement, with guar seed prices decreasing over 7 per cent and guar gum prices nearly decreasing by 6 per cent.

Staggered Delivery Mechanism

NCDEX has launched the 'Staggered Delivery Mechanism' as per the approval of the Forward Markets Commission in all 'Compulsory Delivery' contracts from June 2012 onwards. The tender period in the Staggered Delivery Mechanism shall start on the 5th of every month in which the contract is due to expire, excluding Saturdays. Seller shall have an option of marking an intention of delivery on any day during tender period up to expiry of contract and the corresponding buyer matched by the allocation process put in place by the Exchange will have to take the delivery on T + 2 day from the delivery centre where the seller has delivered the commodity.

3. Clearing House

As mentioned earlier, clearing house in the futures exchange handles all the operations including matching up of each buy and sell transactions, execution, clearing and reporting of all transactions, settlement of all transactions on maturity by paying the price difference or by arranging physical delivery etc., and assumes all counterparty risk on behalf of buyer and seller. There is no clearing house in a forward market due to which buyers and sellers face counterparty risk. In a futures exchange all transactions are routed through and guaranteed by the clearing house, which automatically becomes a counterpart to each transaction. It assumes the position of counterpart to both sides of the transaction by selling contract to the buyer and buying the identical contract from the seller. Therefore, traders obtain a position *vis-a-vis* the clearing house. That is, between seller and buyer, there is a clearing house and its main objective is to check the counter-party risk. Thus, clearing house ensures default risk-free transactions in the futures exchange.

The organizational structure and membership requirements of clearing houses vary from one exchange to the other. The Bombay Commodity Exchange (BCE) and IPSTA have set up separate independent corporations (namely Prime Commodities Clearing Corporation of India and First Commodities Clearing Corporation of India respectively) for handling clearing and guarantee of all futures transactions in the respective exchanges. While Coffee Futures Exchange India (COFEI), Bangalore has a clearing house as a separate division of the exchange and many other exchanges like The Chamber of Commerce, Hapur; Kanpur Commodity Exchange (KCE) and EICA, Mumbai run in-house clearing houses as part of the respective exchanges.

Table 5.95: Exchange-wise commodities dealt in India.

Sl.No.	Exchange	Commodity
I.	**National exchanges**	
1.	Multi Commodity Exchange (MCX), Mumbai	Metals and Agricultural Commodities
2.	National Commodity and Derivation Exchange (NCDEX), Mumbai	Metals and Agricultural Commodities
3.	National Multi Commodity Exchange (NMCE)	Metals and Agricultural Commodities
4.	Indian Commodity Exchange (ICEX)	Metals and Agricultural Commodities
5.	Ace Derivatives and Commodity Exchange (ACE)	Six agricultural commodities *viz.*, castor, chana, mustard seed, soybean, refined soyoil and sugar
II.	**Regional exchanges**	
6.	India Pepper and Spice Trade Association, Kochi(IPSTA)	Pepper (both domestic and international contracts)
7.	Vijay Beopar Chamber Ltd., Muzaffarnagar	Gur
8.	Rajdhani Oil and Oilseed Exchange Ltd, Delhi	Gur, Mustard seed its oil and oilcake
9.	Bhatinda Om and Oil Exchange Ltd, Bhantada	Gur
10.	The Chamber of Commerce, Hapur	Gur, potatoes and mustard seed
11.	The Meerut Agro Commodity Exchange Ltd., Meerut	Gur
12.	The Bombay Commodity Exchange Ltd., Bombay	Oilseed complex
13.	Rajkot Seeds, Oil and Bullion Merchants Association, Rajkot	Castrol seed, Ground nut, its oil and cake, cottonseed its oil &cake, cotton and RBD Palmolein
14.	The Ahmedabad Commodity Exchange, Ahmedabad	Castrol seed, cottonseed, its oil and oilcake
15.	The East India Jute and Hussian Exchange Ltd., Calcutta	Hessian and sacking
16.	The East India Cotton Association Ltd., Calcutta	Cotton
17.	The Spices and Oilseeds Exchange Ltd, Sangli	Turmeric
18.	Kanpur Commodity Exchange Ltd., Kanpur	Rapeseed/Mustard seed, its oil and cake
19.	National Board of Trade, Indore	Soya seed, Soya oil and Soya meals. Rapeseed/Mustard seed its oil and oilcake and RBD Palmolien
20.	The First Commodities Exchange of India Ltd., Kochi	Copra/Coconut, its oil& oilcake
21.	Central India Commerce Exchange Ltd., Gwalior	Gur and Mustard Seed
22.	E-Sugar India Ltd., Mumbai	Sugar
23.	National Multi –Commodity Exchange of India Ltd., Ahmedabad	Oilseed complex and Rubber, Sugar, Aluminum, nickel, Zinc, Copper, Lead, Tin, Pepper, Gram and Sacking
24.	Coffee Futures Exchange India Ltd., Bangalore	Coffee
25.	Surendranagar Cotton Oil and Oilseeds, Surendranagar	Cotton, Cotton seed, Kapas
26.	E-Commodities Ltd., New Delhi	Sugar
27.	Bullion Merchants Association, Bikaner	Mustard seed its oil and oilcake

Membership of clearing house requires capital contribution in the form of equity, security deposit, admission fee, registration fee, guarantee fund contribution in addition to net worth requirement depending on its organizational structure. For example, in the BCE, the minimum capital requirement for membership in its clearing house as applicable to trading-cum-clearing members is ₹ 50,000 each towards equity and security deposit, ₹ 500 as annual subscription, and additionally, members are required to have net worth of ₹ 3 lakh. Similarly, COFEI prescribes ₹ 5 lakh each towards equity and guarantee fund contribution and ₹ 40,000 towards admission fee for a trading-cum-clearing member. However, in exchanges, where clearing house is a part of the exchange, the payment requirements are lower. For example, KCE prescribes payment of only ₹ 25000, ₹ 1000 and ₹ 500 towards security deposit, registration fee and annual fee respectively for a clearing-cum-trading member.

4. Margins

The traders in the futures markets need merely a small proportion of the money to be deposited as margin in proportion to the value of the transaction made with the clearing house, so as to ensure the trade mechanism. This enables the traders to enter into the deals/contracts, which are worth many times more than the margin money they deposit in the exchange. In general, speculators, who generally purchase the product at low price, say in cash market and plan to sell the contract at high price in future date with a view of making profits. But, if the prices fall in the future date, a number of traders may face bankruptcy and even the presence of clearance houses may not be able to eliminate counter-party risks. So, to minimize or check the counter-party risk, clearing house will maintain margin money collected from the speculators (traders).

In the futures market, margin money has a definition distinct from its definition in the stock market. In stock market, margin money is the use of borrowed money to purchase securities or it is the cash down payment or money borrowed from a broker to purchase stocks. But, in the futures market, margin money represents a deposit of 'good faith money' that should be deposited with the exchange (clearing house) in order to enter into a futures contract. So, margins (also called clearing margins) are good-faith deposits of trading members kept with a clearing house usually in the form of cash. Rather than providing a down payment to have a futures contract, simply a margin amount is sufficient as the deposit in the exchange to conduct the business in the futures market. It is much like money held in an escrow account. When the contract is liquidated, margin money will be refunded back to the trader plus or minus any gains or losses that occur in the futures contract. So, the amount in margin account changes daily, as the price fluctuations yields gain or loss to the traders in the futures contract.

Types of Margins

There are two types of margins to be maintained by the traders with the clearing house *viz.*, Initial margin and Maintenance margin. They have important bearing on the success of futures. As margins are non-interest bearing deposits payable to the clearing house, up-front working capital of the trading parties gets blocked to that extent. While a higher margin requirement prevents traders from participating in trading, a lower margin makes the clearing house financially weak and hence, more vulnerable to default. Internationally, many developed exchanges maintain a low margin on positions due to their better financial strength, along with massive volume of trade resulting in large income accruing to them. However, this has not been the case with many exchanges in India. For example, the initial margin liability for transacting the minimum lot size in pepper is ₹ 30000 for domestic contracts and US$ 312.50 for international contracts. Similarly, the volume of transactions these clearing houses deal in many exchanges in India is abysmally low thereby, making their existence financially unviable.

a. Initial Margin

It is also called as Original margin. When a trader opens a futures contract, the futures exchange will fix a minimum amount of money that the trader must deposit into his account. This original deposit of money made by the trader for entering into futures trade is called the Initial margin. It is, otherwise, defined as the sum of money that the trader must deposit with the clearing house of the exchange for each futures contract to be bought or sold. When the trader's contract is liquidated/ settled, the trader will be refunded back the initial margin plus or minus any gains or losses that occur in the futures contract. That means, on any day, if profits accrue to the trader, the same will be added to the trader's margin account and if losses arises, the same will be deducted from the trader's margin account.

This pre-determined initial margin amount will be continuously under review. Exchanges continuously monitor market conditions and risks and, as necessary, raise or reduce the initial margin requirements. The margin requirements for a particular futures contract at a particular time are set by the exchange on which the contract is traded. The margin requirements are about five to ten per cent of the current value of the futures contract. At times of high market volatility, initial margin will be raised and *vice versa.* It is important to note that, the exchanges' minimum margin requirements are exactly that minimums that exchange member brokerage firms must charge. But, in general, individual brokerage firms charge higher margin amounts from their traders than the exchange-set minimums (to have more financial strength).

b. Maintenance Margin

It is also called as Variation margin. If and when the funds available in the trader's initial margin account are reduced due to losses and fall to a certain low level below the Maintenance margin requirement, the trader will be informed by the exchange/clearing house and request him to make an additional deposit into his margin account, so as to bring the margin back up to the initial margin amount. This additional amount required to bring back the existing low level margin account back to the initial margin account is called Maintenance margin. This request made by the exchange/clearing house to the trader for updating the account up to initial margin is called Marginal cail. So, a call from a clearing house to a clearing member, or from a brokerage firm to a trader, to bring margin deposits up to initial margin is called Margin call.

c. Maintenance Margin Requirement

It is the amount of deposit in the trader's account, which the balance must not fall below this level. If the deposit of the trader falls below the maintenance margin requirement, then the trader will get the marginal call from the exchange/clearing house to bring the deposit level up to initial margin by depositing the amount (Maintenance margin).

So, initial margin is the minimum amount required to enter into a new futures contract, but the maintenance margin is the amount needed to upgrade the margin up to initial margin to conduct the futures contract. If the trader's margin account drops to a certain level (below Maintenance margin requirement) because of a series of daily losses, exchange or clearing house will make a margin call and request the trader to make an additional deposit (Maintenance margin) into his account to bring the margin back up to the initial margin (Figure 5.43).

For example, a trader deposit an initial margin of ₹ 10000 on a contract and the maintenance margin requirement is ₹ 5000. A series of losses dropped the amount in the traders' account to ₹ 4000. Then the exchange or clearing house will make a margin call to the trader, requesting the trader to deposit atleast an additional ₹ 6000 (Maintenance margin) to bring his account back up to the initial

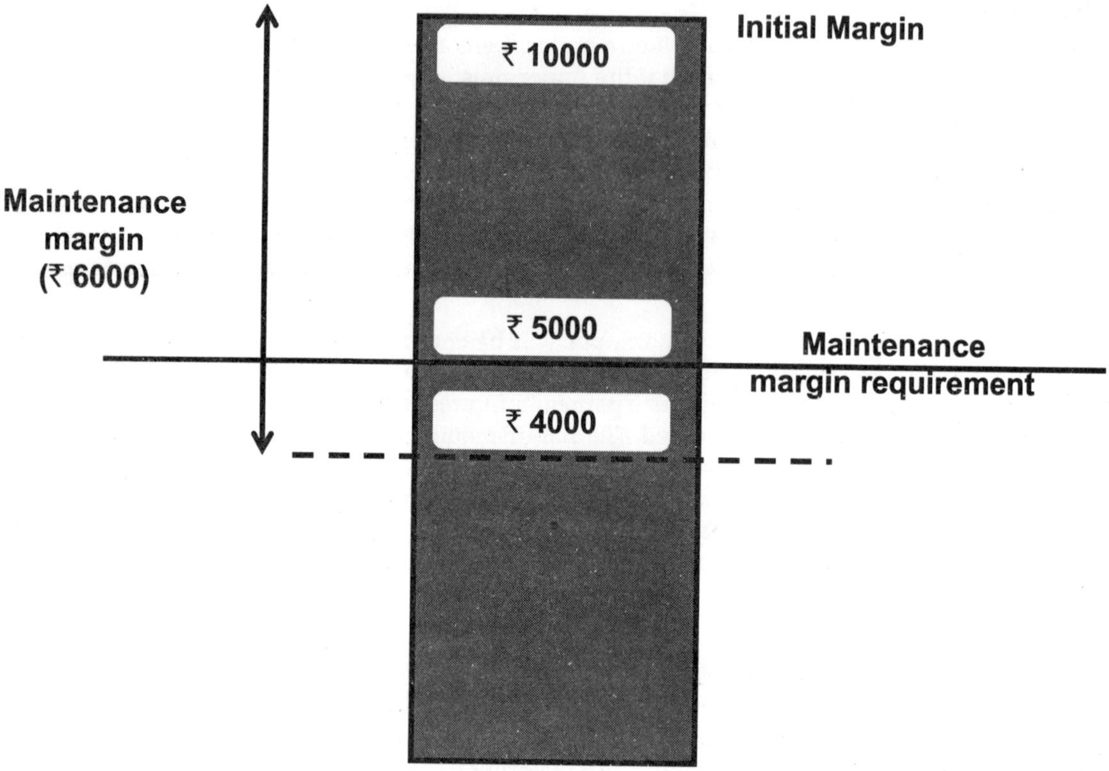

Figure 5.43: 'Margin' to be maintained in futures exchange.

margin level of ₹ 10000. When a margin call is made, the funds usually have to be delivered immediately. If they are not, the exchange can protect itself by liquidating the trader's open positions completely in order to make up for any losses it may have incurred on the traders' behalf.

The main aim of collecting margin money is to minimize the risk of default by either counter party. The amount of initial margin is so fixed such that, the possible losses which arises due to worst possible price fluctuations, can be met by the ordinary/initial margin. The exchanges fix the rates of ordinary/initial margin keeping in view to balance the high security of contract and low cost of entering into contract.

For ensuring financial integrity of the exchange and for overcoming counterparty risky trade, limits have been imposed on clearing members apart from mandatory margins. These limits, which are stringent in some exchanges and are liberal in others. These limits are normally linked to the members' contribution towards equity capital or security deposit or a combination of both and settlement guarantee fund. In the BCE, exposure limit of a clearing member is the sum of 50 times the face value of contribution to equity capital of the clearing house and 30 times the security deposit the member has maintained with the clearing house. While COFEI prescribes the limit of 80 times the sum of member's equity investment and the contribution to the guarantee fund. The IPSTA has fixed a net exposure limit of 60 units (equivalent to 1,500 quintals) for domestic contracts and 90 units (equivalent to 2,250 quintals) for international contracts. Moreover, the settlement guarantee fund helps clearing house to honour all payments in case any trader defaults. The KCE maintains a trade guarantee fund with a corpus of ₹ 100 lakh, while COFEI in addition to a guarantee fund has substituted itself as party to clear all transactions.

5. Players in Futures Market

The players in the futures market broadly fall into three categories *viz.*, hedgers, speculators and arbitragers depending on their attitude towards risk.

5.1. Hedgers

Farmers, processors, importers and exporters can all be hedgers in the futures market. Hedger is a user of the futures market, who enters into futures contract to protect himself from the risk of adverse price fluctuations in respect of his existing or future product. A hedger buys/sells in the futures market to secure the future price of a commodity intended to sell/buy at a later date in the cash market. This helps the hedger to get protection from price risks. They require some people, who are prepared to accept the risk called, speculators. This process of purchase/sale of commodity in the futures market accompanied by sale/purchase of the commodity in the cash market is called Hedging. Hedging is mainly based on two assumptions:

- ✰ The purchase/sale of the commodity should be accompanied by sale/purchase of the same commodity (simultaneous transactions) and in same quantity, but in different markets (Future and Cash markets).
- ✰ Prices in both the markets (*i.e.*, Future and Cash markets) move exactly parallel *i.e.*, move up and down together so that, the loss arise in one market is offset by gain in the other market.

Thus, in hedging, the activities of buying and selling are always in opposite direction and more over, the nature and quantity of product purchased in both the markets are equal (Panels A and B of Figure 5.44). So, hedging is the practice for offsetting the price risk inherent in any cash/future market by taking an equal but opposite position in the futures/cash market. By this mechanism, the traders transfer their price risk. As the two prices both in cash and futures markets move in tandem/together, it enables the traders in the futures/cash markets to cover their price risk by taking opposite position in the cash/futures market. So, hedging is the process of protecting or safeguarding one self against the price risk in the future *i.e.*, protection against low prices for seller in the future and protection against high prices for buyer in the future. This hedging mechanism is explained through Figures 5.44 and 5.45.

Example 1

Assume that, on 1st June, 2012, a farmer (Ravi) wants to sell his commodity X (100 quintals) to a wholesaler (Sri Ram) at ₹ 7000/qtl (Figure 5.44) after considering its average total cost of production. But, he knows that, there is no price guarantee for him, because of frequent price fluctuations in the market economy. To overcome price risk, the farmer goes for hedging mechanism in the futures market and locked the futures price with reference to 10th June, 2012 at ₹ 7000/qtl.

Case 1: Case of Price Decrease

Say, for example, in the cash market, on 5th June, 2012 the price of the commodity X is declined to ₹ 5000/qtl. So, the farmer feels that, if he waits further *i.e.*, beyond 5th June, 2012, the price may still fall and in view of this, he decides to sell the commodity in the cash market at ₹ 5000/qtl. So, he incurs a loss in the cash market to the tune of ₹ 2000/qtl (price willing to sell (₹ 7000/qtl) – price at which commodity actually sold (₹ 5000/qtl)) and a total loss of ₹ 200000.

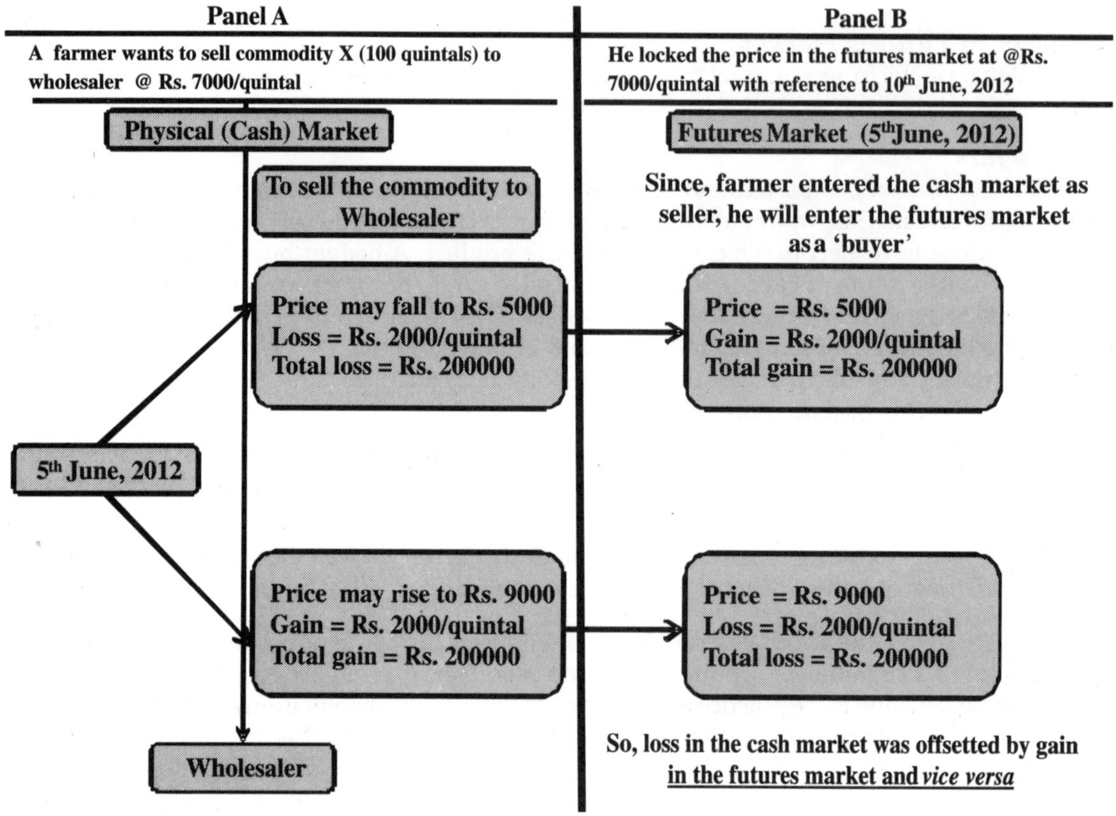

Figure 5.44: A simple example for Hedging mechanism.

However, in the futures market, the farmer gains by ₹ 2000/qtl *i.e.*, a total gain of ₹ 200000. This is because, since the farmer entered the cash market as seller, he has to enter the futures market as a buyer and so, as a buyer, he will purchase the same commodity X and in same quantity at ₹ 5000/qtl in the futures market. So, the gain is ₹ 2000/qtl (futures price locked (₹ 7000/qtl) - price at which commodity purchased in futures market (₹ 5000/qtl)) *i.e.*, a total gain of ₹ 200000. It is essential to note that, prices of a commodity both in cash and futures markets move in tandem *i.e.*, in the same direction with strong positive correlation. So, the loss in cash market (- ₹ 2000/qtl) is offsetted by the gain in the futures market (+ ₹ 2000/qtl) and hence, the farmer has successfully mitigated the price risk. It is also to be noted that, the entry into the cash market (as seller) on 5th June, 2012 is simultaneous with the entry in the futures market (as buyer *i.e.*, on 5th June, 2012). Further, this hedging mechanism is done only to overcome the price risk and not with the intention of gaining profits. Hence, the position of farmer on 5th June, 2012 is same as his position on 1st June, 2012.

Case 2: Case of Price Increase

Say, for example, in the cash market, on 5th June, 2012 the price of the commodity X is increased to ₹ 9000/qtl. So, it is felt appropriate by the farmer to sell away the commodity in the cash market, as he presume that, it is the maximum possible price for that commodity and beyond that date, the prices may fall. So, he decides to sell the commodity in the cash market on 5th June, 2012 in the cash market at ₹ 9000/qtl and gain ₹ 2000/qtl (price at which commodity actually sold (₹ 9000/qtl) - price willing to sell (₹ 7000/qtl)) and a total gain of ₹ 200000.

However, in the futures market, the farmer experiences loss by ₹ 2000/qtl *i.e.,* a total loss of ₹ 200000. This is because, since the farmer entered the cash market as seller, he has to enter the futures market as a buyer and so, as a buyer, he will purchase the same commodity X and in same quantity at ₹ 9000/qtl in the futures market. So, the loss is ₹ 2000/qtl (price at which commodity purchased in futures market (₹ 9000/qtl) - futures price locked (₹ 7000/qtl)) *i.e.,* a total loss of ₹ 200000. So, the gain in the cash market is offsetted by the loss in the futures market and this is nothing but, hedging mechanism.

Example 2

Assume that, a wholesale trader (Sri Ram) purchased commodity X (100 quintals) from farmers and he wishes to sell the same to a processor (Ravi) at ₹ 30500/tonnel on 1st January, 2012 (Figure 5.45). But, the wholesaler knows that, there is no price guarantee for him, because of frequent price fluctuations in the market economy. To overcome price risk, the wholesaler goes for hedging mechanism

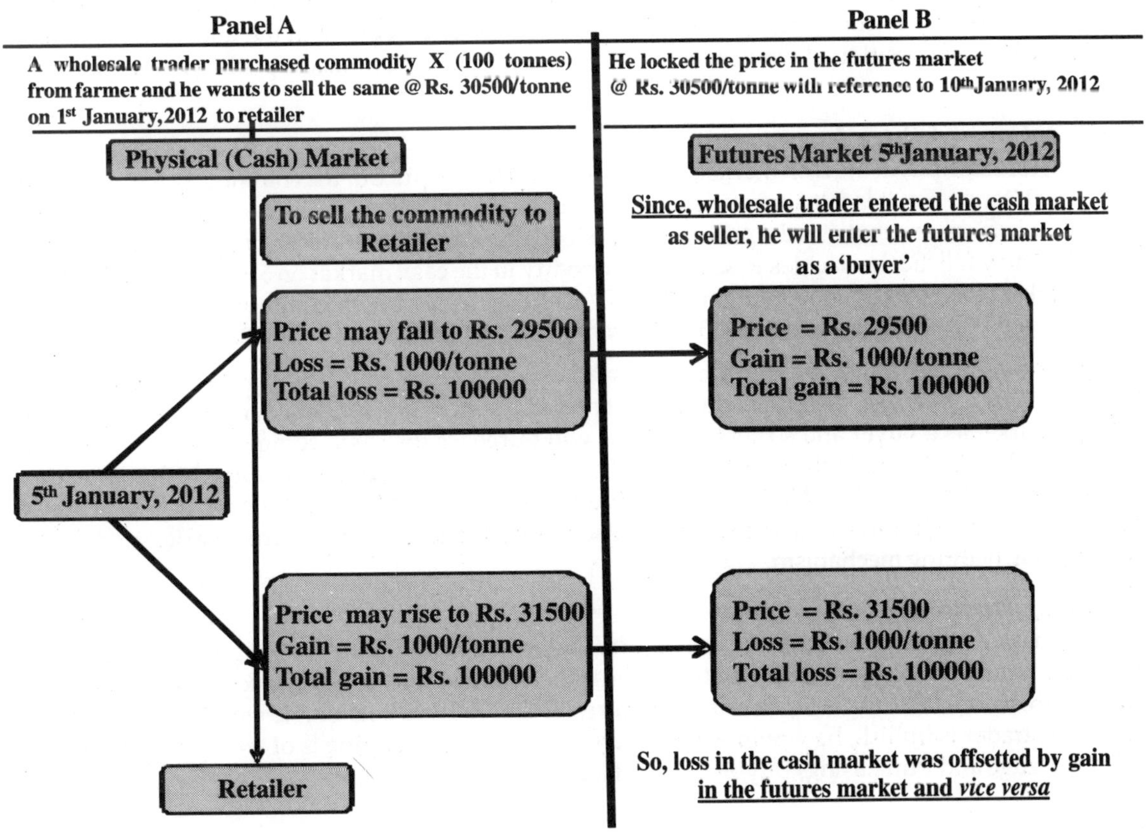

Figure 5.45: A simple example for Hedging mechanism.

in the futures market and locked the futures price with reference to 10th January, 2012 at ₹ 30500/tonne.

Case 1: Case of Price Decrease

Say, for example, in the cash market, on 5th January, 2012 the price of the commodity X is declined to ₹ 29500/tonne. So, the wholesaler feels that, if he waits further *i.e.,* beyond 5th January, 2012, the prices may still fall and in view of this, he decides to sell the commodity in the cash market at ₹ 29500/tonne. So, he incurs a loss in the cash market to the tune of ₹ 1000/tonne (price willing to sell (₹ 30500/tonne) – price at which commodity actually sold (₹ 29500/tonne) and a total loss of ₹ 100000.

However, in the futures market, the wholesaler gains by ₹ 1000/tonne *i.e.,* a total gain of ₹ 100000. This is because, since the wholesaler entered the cash market as seller, he has to enter the futures market as a buyer and so, as a buyer, he will purchase the same commodity X and in same quantity at ₹ 29500/tonne in the futures market. So, the gain is ₹ 1000/tonne (futures price locked (₹ 30500/tonne) - price at which commodity purchased in futures market (₹ 29500/tonne)) *i.e.,* a total gain of ₹ 100000. Since prices of a commodity both in cash and futures markets move in tandem *i.e.,* in the same direction with strong positive correlation, the loss in cash market (- ₹ 1000/tonne) is offsetted by the gain in the futures market (+ ₹ 1000/tonne) and hence, the wholesaler has successfully mitigated the price risk. It is also to be noted that, the entry into the cash market (as seller) on 5th January, 2012 is simultaneous with the entry in the futures market (as buyer *i.e.,* on 5th January, 2012). Further, this hedging mechanism is done only to overcome the price risk and not to with the intention of gaining profits. Hence, the position of wholesaler on 5th January, 2012 is same as his position on 1st January, 2012.

Case 2: Case of Price Increase

Say, for example, in the cash market, on 5th January, 2012 the price of the commodity is X increased to ₹ 31500/tonne. So, it is felt appropriate by the wholesaler to sell away the commodity in the cash market, as he presume that, it is the maximum possible price for that commodity and beyond that date, the prices may fall. So, he decides to sell the commodity in the cash market on 5th January, 2012 in the cash market at ₹ 30500/tonne and gain ₹ 1000/tonne (price at which commodity actually sold ₹ 31500/tonne - price willing to sell (₹ 305000/tonne)) and a total gain of ₹ 100000.

However, in the futures market, the wholesaler experience loss by ₹ 1000/tonne *i.e.,* a total loss of ₹ 100000. This is because, since the wholesaler entered the cash market as seller, he has to enter the futures market as a buyer and so, as a buyer, he will purchase the same commodity X and in same quantity at ₹ 31500/tonne in the futures market. So, the loss is ₹ 1000/tonne (price at which commodity purchased in futures market (₹ 31500/tonne) - futures price locked (₹ 30500/tonne)) *i.e.,* a total loss of ₹ 100000. So, the gain in the cash market is offsetted by the loss in the futures market and this is nothing but, hedging mechanism.

Types of Hedging

To make a profit in the business, we require that, something should be bought and sold. When trading a futures contract, it doesn't matter, if the trader initially sells or buys, as long as the trader has to do both, before the contract comes due/matures. If the trader is bearish, he would sell or go short and if the trader is bullish, he would want to buy or go long. So, hedging is of two types *viz.,* Selling (Short) Hedge and Purchasing/Buying (Long) Hedge.

a. Going Short/Short Hedge/Selling Hedge

It refers to selling a futures contract. That means, in 'Going short', the farmer/trader makes an agreement with the buyer in the current month (June), to sell the commodity in the futures market (November), with the expectations that, prices will still fall in the future (November). So, in Going short, the farmer/trader will sell the commodity in the futures market and seek protection against price fall in the future. The short holders of the futures contract (*i.e.*, the sellers of the commodity) want to secure high price in the future and hence, they will sell the product now (June) in the futures market (November). This futures contract will get closed only by purchasing equal number/quantity and same type of contract in the cash market (say, in the month of November or before the contracting period). The following are the interesting example for Short hedge:

Example 1

For example, a farmer wants to sell paddy to a miller in the month of November, 2012. But, since the prices of paddy tend to fall in November, 2012 due to glut in the market, the farmer hedges with the miller in the month of June, 2012 (even when the crop is in the field) to sell the produce at the future price (November, 2012). So, the farmer will enter the futures market with the miller in the month of June, 2012 and sell a paddy contract for settlement in next five months time (November, 2012) at a price of 1500/- per quintal. But, at the end of the five months, the price of paddy in the cash market (November, 2012) is actually 1400/- per quintal. So, the farmer is benefitted from the futures contract and escapes the declining price. Had the price of paddy rise in the cash market above ₹ 1500 (November, 2012), the farmer will incur loss or would have been better off without the futures contract. But, since the paddy market is highly volatile, this futures market will provide protection against price risk to the farmer. This case is referred as Selling (Short) Hedge.

Example 2

Assume a farmer, Ravi enters into futures contract (as a seller) because, he wants to obtain protection against declining prices in the future. He entered into futures contract in the month of January to sell pepper of 1000 kg at a price of ₹ 140/kg to be transacted to a trader in the month of July. So, this contract will expire in July. The total valuation of contract is ₹ 1.40 lakhs. Suppose, the initial margin is five per cent of total valuation of contract and Ravi will pay ₹ 9,000 as an initial margin. This contract is accepted by one buyer, Sri Ram. Then, three possible situations may arise on the trading floor *viz.*, the futures price may go lower or higher or do not change and they are discussed here under.

Case a. If the Futures Price goes Lower

As mentioned earlier, in the month of January, farmer, Ravi made an agreement with the buyer (Sri Ram) to sell a futures contract on pepper *i.e.*, 1000 kgs at ₹ 140 per kg by July, at a total valuation of ₹ 1,40,000. Here, farmer entered into futures contract as a seller to obtain protection from an expected price decrease in the future. As he expected, the price declined to ₹ 130 say, in April. *As per the principle of hedging, if a farmer/trader enters the futures contract as a seller/buyer, he has to come out of the futures contract as a buyer/seller in the cash market.* So, here, farmer tries to offset his futures contract by entering into opposite position as a buyer in the cash market. So, he will buy the same contract in the cash market (April) at ₹ 130/kg (he has to buy in the cash market within the contract period *i.e.*, before July). So, he will get a profit of ₹ 10 per kg and total profit is ₹ 10,000 (July futures sale: 140000 – April buying in cash market:130000 = ₹ 10000). This profit will be added to Ravi's initial margin account. His contract is settled. This example implies, if we expect the price of a commodity decrease over a given period of time, we can get profits by selling the futures contracts *i.e.*, in Going short. That means, the trader first sells a futures contract. If, as per the trader's expectations, the price decreases, profit can be realized by later buying a contract at the lower price in the cash market. The gain per unit will be the amount by which the selling (future) price is above the later purchase price (in cash market).

Case b. If the Futures Price goes Higher

As mentioned earlier, in the month of January, farmer, Ravi made an agreement with the buyer, Sri Ram to sell a futures contract on pepper *i.e.*, 1000 kgs at ₹ 140 per kg by July, at a total valuation of ₹ 1,40,000. Here, farmer entered into this futures contract as a seller to obtain protection from an expected price decrease in the future. But, the price was increased to ₹ 150 say, in April. *As per the principle of hedging, if a farmer/trader enters the futures contract as a seller/buyer, he has to come out of the futures contract as a buyer/seller in the cash market.* So, here, farmer tries to offset his futures contract by entering into opposite position as a buyer in the cash market. So, he will buy the same contract in the cash market (April) at ₹ 150/kg (he has to buy in the cash market within the contract period., *i.e.*, before July and to avoid excess loss may be due to further increase in price). So, he will get a loss of ₹ 10 per kg and total loss is ₹ 10,000 (July futures sale: 140000 – April Buying in cash market,150000 = -10000/-). This loss will be deducted from Ravi's initial margin account. His contract is settled. However, one mechanism is there to compensate this loss. He gains back this by selling the same commodity in the futures market (say, October) at high price, as increase in cash (April) price is accompanied by rise in futures price. If the prices in October cash market fall, the farmer, Ravi can enjoy profits.

Case c. If the Futures Price does not Change

Here, the futures price does not change in June. So, he tries to offset his contract by entering into opposite position as a buyer in cash market because, in order to exit from the futures market this offsetting is required. So, he will buy back the same contract at the same price of ₹ 140/kg (in April). So he does not get any profit or loss. His contract is settled. There is no change in initial margin account of the farmer, Ravi.

In the above cases, the contracts are settled by cash settlement because, farmer, Ravi offset the contract before the expiry date *i.e.* before July. In the cases given, the money difference is ₹ 10,000. Profit of ₹ 10,000 means ₹ 10,000 will be added to Ravi's initial margin account. Loss of ₹ 10,000 means ₹ 10,000 will be deducted from Ravi's initial margin account. So, in case of short hedge, if a farmer is correct in forecasting the direction and timing of the price change, the farmer can buy the same commodity in the cash market at low price thereby, making a profit. If the price of the commodity in the cash market increases, the trader will result in loss to farmer, Ravi. Because of leverage, losses as well as gains are larger than the initial margin deposit of ₹ 9000. In the above Case 'a' of Going Short hedge, a gain of ₹ 10000 is above the initial margin deposit of the trader *i.e.*, 9000/-, due to just ₹ 10 fall in price in April. So, here the leverage is working at the farmer's advantage. But, due to rise in price of pepper by ₹ 10 (Case 'b'), the loss is ₹ 10000 which is above the initial margin deposit. So, here the leverage is working at the farmer's disadvantage. It's the other edge of the sword. Note that, if at any time the loss on the trade had reduced funds in farmer's (Ravi's) initial margin account to below the maintenance margin requirement, the farmer will receive a margin call from the exchange or clearing house, to deposit the amount, so as to restore the traders account up to the initial margin requirement.

Example 3: Selling (Going Short) to Profit from an Expected Price Decrease

If we expect the price of a particular commodity to decrease over a given period of time, we can get profits by selling futures contracts *i.e.*, in Going short, the trader first sell a futures contract. If, as per the trader expectations, the price does decline, a profit can be realized by later purchasing the contract in the cash market at the lower price. The gain per unit will be the amount by which the purchase price (in cash market) is below the earlier selling (futures) price.

For example, assume it's now January. The July future price of commodity X is presently quoted at ₹ 1000/- per quintal and over the coming months, it is expected that, the price will decline. A farmer, Ravi decided to sell this futures contract in January for 10 quintals. So, the total valuation of contract is ₹ 10000/-. The farmer (Ravi) deposits an initial margin of ₹ 1,000 (10 per cent of total valuation of contract) and assume maintenance margin requirements is ₹ 600. Suppose, there is decline in price by April (cash market) to ₹ 950/-, and to get the profits, Ravi decides to buy the contract at that price in cash market. On the 10 quintals of commodity X, the profit will be ₹ 500/- and this is added to the initial margin account of farmer, Ravi. The same is explained below.

		Price per Quintal	*Value of 10 Quintals Contract*
January	Sell commodity X in futures market by July	1000	10000
April	Buy commodity X in cash market	950	9500
	Profit	50	500

Otherwise, assume that, in April, the commodity (cash market) price has risen to ₹ 1050/- and to minimize the loss due to further rise in price, farmer, Ravi will buy the same contract in the cash market. Since the futures contract is for 10 quintals, the trader gets a loss of Rs, 50 x 10 = 500/-. This ₹ 500 is deducted from the initial margin account of farmer, Ravi. (For simplicity, examples do not take into account commissions and other transaction costs.)

		Price per Quintal	*Value of 10 Quintals Contract*
January	Sell commodity X in futures market by July	1000	10000
April	Buy commodity X in cash market	1050	10500
	Loss	-50	-500

Note that, if at any time, the loss on the trade or open position had reduced funds in the farmer/trader initial margin account to below the maintenance margin requirement, the farmer/trader will receive a margin call to deposit the amount (maintenance margin), so as to restore the traders account upto the initial margin requirement. In this example, the initial margin is ₹ 1000 and maintenance margin level is ₹ 600. But, the loss incurred is ₹ 500. So, the farmer, Ravi gets a margin call from the exchange or clearing house to increase the initial margin upto ₹ 1000 by depositing the maintenance margin of atleast ₹ 500.

b. Going Long/Buying Hedge/Long Hedge

It refers to buying a futures contract. That means, in Going long, the buyer (trader) makes an agreement with the seller (farmer) and buy the futures contract (say, November, 2012) with the expectation that, prices will still rise in the future (May, 2013). So, in Going long, the trader will purchase the futures contract at low price with the expectation that price will rise in the future. This futures contract will get closed by selling equal number/quantity and same type of contract in the cash market. The following are the interesting examples for long hedge:

Example 1

A silversmith must secure a certain amount of silver in six months time for preparing ear rings and bracelets that have already been advertised in an upcoming catalog with specific prices. But, what if the price of silver goes up in the next six months? Because the prices of the earrings and bracelets are already set, the extra cost of the silver (due to rise in future price) can't be passed on to the retail buyer rather, it should be beared by the silversmith. So, the silversmith needs to hedge to minimize the risk against a possible price increase of silver in the future. Then, the silversmith would enter the futures market *i.e.,* makes an agreement with the seller (January 1st, 2012) and purchase a silver contract for settlement in next six months time (July 1st, 2012) at a price of 250/unit. Say, on 1st July, 2012, the price of silver in the cash market is ₹ 300/unit, *i.e.,* increased by ₹ 50/unit. So, the silversmith benefits from the futures contract and escape the price rise. Had the price of silver declined in July 1st, 2012, the silversmith will incur loss or would have been better off without the futures contract. But, since the silver market is highly volatile, this futures market will provide protection against price rise to the silversmith. To offset this futures contract, the silversmith must sell the same quantity in the cash market (since, he purchased contract in the futures market) and he will choose the period during which, the cash prices are low and sell the contract in the cash market.

Example 2

A processing firm wants to purchase potatoes from a farmer to manufacture chips in the month of July, 2012. Since the prices of potatoes tend to rise in July, the processor hedges with the farmer in the month of January, 2012 (even when the crop is in the field) to procure the produce at the future price (July, 2012). So, the processor will enter the futures market with the farmer in the month of January, 2012 and purchase potatoes contract for settlement in next six months time (July, 2012) at a price of 2000/- per quintal. But, in July, 2012, the actual price of potatoes in the cash market is ₹ 2050/- per quintal. So, the processor benefits from the futures contract and escape the higher price. Had the price of potatoes decreased in July, 2012, the processor will incur loss or would have been better off without the futures contract. But, since the potato market is highly volatile, this futures market will provide protection against price risk to the processor.

Example 3

We know, a person enters into futures contract as a buyer because, he wants to obtain protection against further rise in prices in the future. Assume that, a trader, Sri Ram entered into futures contract as a buyer in January, 2012 to purchase pepper of 1,000 kg at a price of ₹ 140/kg in the month of July from a farmer. So, this contract will expire in July. The total valuation of contract is ₹ 1.40 lakhs. Suppose, the initial margin is five per cent of total valuation of contract thereby, he will pay ₹ 9,000 as an initial margin. His contract is accepted by one farmer, Ravi. Then, three possible situations may arise on the trading floor *viz.,* the futures price may go lower or higher or do not change

Case a. If the Futures Price goes Higher

As mentioned earlier, in the month of January, 2012 a trader, Sri Ram made an agreement with the seller (farmer), Ravi to buy a futures contract on pepper *i.e.,* 1000 kgs at ₹ 140 per kg by July, at a total valuation of ₹ 1,40,000. Here, trader entered into this futures contract as a buyer to obtain protection from an expected price increase in the future. As he expected, the price increased to ₹ 150/kg say, in April. *As per the principle of hedging, if a trader enters the futures contract as a seller/buyer, he has to come out of the futures contract as a buyer/seller in the cash market.* So, here, trader tries to offset his futures contract by entering into opposite position as a seller in the cash market. So, he will sell the same contract in the cash market (April) at ₹ 150/kg (he has to sell in the cash market within the contract period, *i.e.,* before

July). So, he will get a profit of ₹ 10 per kg and total profit is ₹ 10,000 (April selling in cash market:150000 – July futures purchase: 140000 = 10000/-). This profit will be added to Sri Ram's (trader) initial margin account. His contract is settled. This example implies that, if we expect the price of a commodity increase over a given period of time, we can get profits by buying futures contracts *i.e.*, in Going long. That means, the trader first buys a futures contract. If, as per the trader's expectations, the price increases, profit can be realized by later selling that contract at the higher price in the cash market. The gain per unit will be the amount by which the selling price in cash market is above the earlier purchase (futures) price.

Case b. If the Futures Price goes Lower

As mentioned earlier, in the month of January, a trader, Sri Ram made an agreement with the seller (farmer), Ravi to buy a futures contract on pepper *i.e.*, 1000 kg at ₹ 140 per kg by July, at a total valuation of ₹ 1,40,000. Here, trader entered into this futures contract as a buyer to obtain protection from an expected price increase in the future. But, the price is declined to ₹ 130/kg say, in April. *As per the principle of hedging, if a trader enters the futures contract as a seller/buyer, he has to come out of the futures contract as a buyer/seller in the cash market.* So, here, trader tries to offset his futures contract by entering into opposite position as a seller in the cash market. So, he will sell the same contract in the cash market (April) at ₹ 130/kg (he has to sell in the cash market within the contract period, *i.e.*, before July). So, he will get a loss of ₹ 10 per kg and total loss is ₹ 10,000 (April selling in cash market:130000 – July futures purchase: 140000 = -10000/-). This loss will be deducted from Sri Ram's (trader) initial margin account. His contract is settled. One mechanism is there to compensate this loss. He gains back this by buying the same commodity in the futures market (say, October) at low price, as decrease in cash price is accompanied by decrease in futures price. If the prices in October cash market increases, the trader enjoys profits.

Case c. If the Futures Price does not Change

Here, the futures price does not change in July. So, the trader, Sri Ram tries to offset his contract by entering into opposite position as a seller because, in order to exit from the futures market this offsetting is required. Now, he tries to sell the contract at ₹ 140/kg worth *i.e.*, at the same price (in April). So, he does not get any profit or loss. His contract is settled. There is no change in initial margin account of the trader, Sri Ram.

In the above cases, the contracts are settled by cash settlement because, he offset the contract before the expiry date *i.e.* before July. In the cases given, the money difference is ₹ 10,000. Profit of ₹ 10,000 means ₹ 10,000 will be added to Sri Ram's initial margin account. Loss of ₹ 10,000 means ₹ 10,000 will be deducted from Sri Ram's initial margin account. So, in case of long hedge, if a trader is correct in forecasting the direction and timing of the price change, the futures contract can be sold later for the higher price (in cash market) thereby yielding a profit. If the price declines rather than increase, the trade will result in a loss. Because of leverage, losses as well as gains are larger than the initial margin deposit of ₹ 9000. In the above Case 'a' of Going long hedge, a gain of ₹ 10000 is above the initial margin deposit of the trader *i.e.*, 9000/-, due to just ₹ 10 rise in price in April. So, here the leverage is working at the trader's advantage. But, due to fall in price of pepper by ₹ 10, the loss is ₹ 10000 which is above the initial margin deposit. So, here the leverage is working at the trader's disadvantage. It's the other edge of the sword. Note that, if at any time, the loss on the trade had reduced funds in traders' initial margin account to below the maintenance margin requirement, the trader will receive a margin call from the exchange or clearing house, to deposit the amount, so as to restore the traders account up to the initial margin requirement.

Example 4: Buying (Going Long) to Profit from an Expected Price Increase

If we expect the price of a particular commodity to increase over a given period of time, we can get profits by buying futures contracts *i.e.*, in Going long, the trader first buys a futures contract. If, as per the trader's expectations, the price increases, profit can be realized by later selling the futures contract at the higher price (in cash market). The gain per unit will be the amount by which the selling price (in cash market) is above the earlier purchase (futures) price.

For example, assume it's now January. The July future price of commodity X is presently quoted at ₹ 1000/- per quintal and over the coming month, it is expected that, the price to increase. A trader, Sri Ram decided to buy futures contract in July for 10 quintals. So, the total valuation of contract is ₹ 10000/-. The trader (Sri Ram) deposits an initial margin of ₹ 1,000 (10 per cent of total valuation of contract) and assume Maintenance margin requirement is ₹ 600. Further, assume that, by April, the commodity cash price has risen to ₹ 1050/- and the trader decide to take the profit by selling in cash market. Since the futures contract is for 10 quintals, the trader gets a profit of Rs, 50 x 10 = 500/- and this is added to the initial margin account of trader, Sri Ram. For simplicity, this example does not take into account commissions and other transaction costs.

		Price per Quintal	*Value of 10 Quintals Contract*
January	Buy commodity X futures contract for July	1000	10000
April	Sell commodity X in cash market	1050	10500
	Profit	50	500

Suppose, instead of rise in the price, if there is decline in price by April to ₹ 950/-, and to avoid the possibility of further loss, the trader decide to sell the contract at that price in cash market. On the 10 quintals of commodity X, the loss will be ₹ 500/-. This ₹ 500 is deducted from the initial margin account of trader, Sri Ram.

		Price per Quintal	*Value of 10 Quintals Contract*
January	Buy commodity X futures contract for July	1000	10000
April	Sell commodity X in cash market	950	9500
	Loss	-50	-500

As mentioned earlier, if at any time the loss in the trade or open position had reduced funds in the trader's initial margin account to below the maintenance margin requirement, the trader (Sri Ram) will receive a margin call to deposit the amount (Maintenance margin), so as to restore the traders account up to the initial margin requirement. In this example, the initial margin is ₹ 1000 and maintenance margin level is ₹ 600. But, the loss incurred is ₹ 500. So, the trader, Sri Ram gets a margin call from the exchange or clearing house to increase the initial margin up to ₹ 1000 by depositing the maintenance margin of atleast ₹ 500.

The other examples of futures contract includes, a fast food chain would hedge against higher potato prices, a company in need of a loan in six months could hedge against rising interest rates in the future, a coffee processing firm could hedge against rising coffee bean prices next year etc. Whatever the hedging strategy, the common denominator is that, hedgers are willing to take up the opportunity to benefit from favorable price changes in order to achieve protection against unfavorable price changes. So, the futures contract provides a definite price certainty for both parties, which reduces the risks associated with price volatility. Buyers are thus, able to protect themselves against (hedge against) future higher prices and sellers are able to hedge against future lower prices.

5.2. Speculators

Speculators are the other important players of futures market, who, however, do not aim to minimize the risk, but rather aim for profit making duly taking price fluctuations of the commodity as an advantage. A trader, who trades or takes position without having exposure in the physical market, with the sole intention of earning profit from price movements is a speculator. Speculators always observe the price movements of commodities and if they are favourable to them, they participate in trade. They are prepared to assume or take the risks, which the hedgers are trying to avoid in the futures market. They provide depth and liquidity to the market. It would not be wrong to say that, in absence of speculators, the market will not be liquid and may even collapse. Thus, speculators tries for profit making duly taking the advantage of anticipating future price movements.

Speculation refers to sale/purchase of a product at a present price with the objective of purchase/ sale of the product at some future date at a favourable or higher price. But, the buying and selling activities of speculators need not be in opposite direction, unlike hedgers. So, the main objective of speculation is profit making in futures trading. Speculator purchases the commodity when the prices are low. He is not a regular buyer of the commodity. He will sell the commodity at a future date at a favourable (high) price and reap the profits based on price differences. Sometimes, he loses in this process. The following are the essentials of a speculator:

- ✰ He enters the trade at current (low) prices
- ✰ The transactions of speculators are completed on some future date
- ✰ Their main aim is profit making
- ✰ Except in few cases, the physical delivery of goods is neither taken nor given. Only the differences in prices is given/taken
- ✰ They are not regular buyers of the commodities and purchase the commodity only when the prices are low.

Speculation dampens/minimizes price fluctuations. Speculators buy at current prices when the prices are low, in the anticipation of rise in prices in the future. So, this encourages production and reduces consumption. Some speculators sell commodities at current prices in the anticipation of fall in prices in the future. This discourages production and encourages consumption. So, the sum total of both the effects of speculation results in minimizing the price fluctuations. Hedgers try to minimize their risk no matter, what they're investing in, while speculators accept the risks and therefore, maximize their profits. In the futures market, a speculator buying a contract low (at low price) in order to sell high in the future would most likely be buying that contract from a hedger selling a contract low in anticipation of declining prices in the future. Unlike the hedger, the speculator does not actually seek to own the commodity in question. Rather, he will enter the market seeking profits by offsetting rising and declining prices through the buying and selling of contracts (Table 5.96).

Speculators put their money at risk in the hope of profiting from an anticipated price change. Buying futures contracts with the hope of later being able to sell them at a still higher price (in cash market) is known as 'going long'. Conversely, selling futures contracts with the hope of being able to buy back identical and offsetting futures contracts at a lower price (in cash market) is known as 'going

Table 5.96: Strategies of hedgers and speculators in futures market.

Players	Reasons for BUYING futures contracts	Reasons for SELLING futures contracts
Hedgers	To lock in a price and thereby, obtain protection against rising prices in the future.	To lock in a price and thereby, obtain protection against declining prices in the future
Speculators	To profit from rising prices	To profit from declining prices

short'. An attraction of futures trading is that, it is equally as easy to profit from declining prices (by selling futures contract) as it is to profit from rising prices (by buying futures contract).

Speculators are essential participants in the futures market. This is so because, hedgers in futures market tries to shift risk, but it is, however, extremely difficult, for every hedger to find a hedger counterparty with matching requirements and willingness to bear the risk. The hedgers can shift the price risk, only if there are participants in the futures market, who are willing to accept the risk. Speculators are such participants who are willing to take risk of hedgers in the expectation of making profit (Table 5.96). Speculators, therefore, add liquidity to the futures market and assume market price risk. So, it is difficult to imagine a futures market functioning without speculators. It is important to note that, speculators are not gamblers in the futures market, since they do not create risk, but merely accept the risk, which already exists in the market. The speculators are the persons, who try to assimilate all the possible price-sensitive information, on the basis of which they can expect to make profit. The speculators, therefore, contribute in improving the efficiency of price discovery function of the futures market. This speculation is of two types: Legitimate speculation and Illegitimate speculation.

Legitimate Speculation

This speculation is also called as Speculation proper or Informed speculation. In this activity, speculators are active participants of futures trading in the sense that, they frequently purchase and sell the futures contracts and gain the profits through price differences. They accept the risk offered by hedgers and function with a profit motive in the trading. This form of speculation is considered healthier for the futures trading, as it adds liquidity to the futures markets. These speculators will not create artificial scarcity of commodities in the market.

Illegitimate Speculation

This speculation is also called as Speculation improper or Over speculation. In this form of speculation, speculators do not frequently purchase and sell futures contracts. Though their main objective is profit making as under Legitimate speculation, but these speculators gain profits through price differences by creating artificial scarcity of commodities. This form of speculation results in inflationary rise in prices of commodities and hence, not considered healthy on the part of futures trading. So, this form of speculation is altogether banned by the Government.

So, an informed speculation/speculation proper is good for the market. However over-speculation/speculation improper/illegitimate speculation needs to be curbed. In order to curb over-speculation (which may lead to distortion of prices), limits will be imposed on the open position held by speculators. Excessive speculation has long been criticized for wide price fluctuations and many times the commodity future trading has been suspended due to it. People speaking for it have strong view point that, speculator bears the risk of a hedger and if speculation vanishes out of futures market, the hedging will eventually vanish too. If speculation and hedging go out of futures market, then the markets will turn into conventional spot markets. Speculation increases participation of traders in the futures market and application of IT will make the things much more transparent. The people speaking against it say that, it is an organized gambling dominated by rich traders and they try to increase the prices to unrealistic levels. However, the problem starts, when persons who should act as hedgers start acting as speculators under the influence of news flow and greed.

Hedging vs Speculation

It is important that, the every commodity that we consume would likely cost more in the absence of commodities futures market. This is because of the intrinsic risks associated with the business and lacking the ability to shift risk. So, this makes the farmers or processors to purchase inputs or services at higher prices and hence, the consumer would have to pay higher prices for the commodities. This shifting of risk to someone who is willing to accept it, is called hedging. The person who is willing to accept that risk, so as to gain profits from the fluctuating price movements is called speculator and the process is called speculating/speculation. The cotton shirt, orange juice, cereal, coffee etc., we enjoy all would have been be priced many times higher without the participation of speculators in the futures market. Like in mandis, even in exchanges, the equilibrium prices are reached due to interaction of demand and supply forces and with the participation of large number of traders, we can realize a competitive price from futures trading.

Most of the times, it is noticed that, the same agent may act as a hedger at one instance and as a speculator at another instance. The general rule is that, hedgers form the two extreme ends of a chain of transactions, the intermittent roles are filled by a number of speculators, who may further be divided into a number of categories like arbitragers, day-traders, etc. So, in a futures market, the trade offs in risk is between the 'risk aversive hedgers' and the 'risk taking speculators' (Table 5.97). Both speculators and hedgers have their own demand and supply schedules and it is the interaction between these two market forces that fix prices in a future market, if they are given a free hand.

Though speculators and hedgers account for the demand and supply forces in the futures market, they are not the only entities that participate in futures trading. Clearing houses and brokers are the people, who also appear in the trading scenario. In every deal between two parties (seller and buyer),

Table 5.97: Hedging vis-à-vis speculation in a futures contract.

Item	*Hedging*	*Speculation*
Objective	To transfer price risk ie., hedging offers an element of protection against price risk	To gain more profits ie., speculation involves deliberately taking a risk on price movements, up or down, in the hope of obtaining a profit.
Availing the opportunity	The hedger gives up some opportunity in exchange for reduced risk.	The speculator acquires opportunity in exchange for taking on risk.
Activities of buying and selling	Hedgers manage the price risk of a commodity by entering into a counter balancing futures contract. The activities of buying and selling are always in opposite direction	Speculators are lured into the futures by the reason of the opportunity to realize a profit on their anticipation of a price change. So, the activities of buying and selling are not necessarily in opposite direction
Quantity of product purchased and sold	The quantity of product purchased and sold should be equal	The quantity of product purchased and sold need not be equal
Storage of goods	There is no storage of goods and only the differences in prices is paid	The storage of products may arise, as the speculator will not sell the commodity unless the price rises upto the level of his expectations
Attitude towards risk	Transfer price risk to others (Risk-aversive)	Accept price risk (Risk-takers)
Attitude towards price movements	Hedgers are concerned about securing a fixed price for the commodity	Speculators are concerned about possible price fluctuations of the commodity and try to reap the profits from the price differences.
Short (Selling)	Secure a price now to protect against future declining prices	✩ Secure a price now in anticipation of declining prices ✩ To profit from declining prices
Long (Buying)	Secure a price now to protect against future rising prices	✩ Secure a price now in anticipation of rising prices. ✩ To profit from rising prices

clearing house stand between the parties and take the stand opposite to what each party has taken. That is, the person taking short position, sells the commodity to the clearing house and the person taking the long position, buys the contract from the clearing house. Clearing house require the parties to deposit a margin, proportional to the value of the deal, so as to ensure the trading. Brokers are hired by traders parties to trade on their behalf in the market.

5.3. *Arbitragers*

Arbitragers are in the business to take advantage of a discrepancy between prices in two different markets. Arbitrage refers to the simultaneous purchase and sale of same commodity in two markets so that, the selling price in one market is higher than the buying price plus transaction cost in another

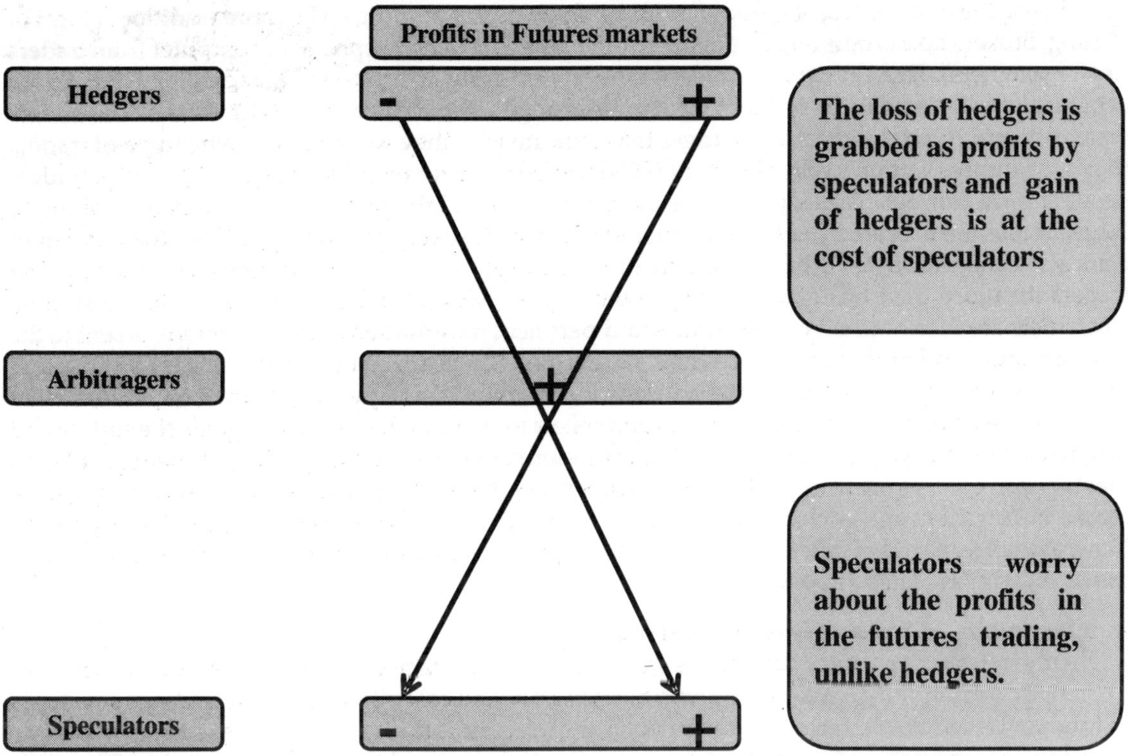

Figure 5.46: Profits in commodities futures trading.

market and thereby, resulting in risk-less profit to the arbitrager. So, arbitrager takes advantage of price discrepancy/differences.

The roles of hedgers, speculators and arbitragers as discussed above can be summarized through Figure 5.46. We know that, hedgers aim only at mitigating the price risk in the business. Any loss incurred by the hedger in the futures transactions is the gain realized by the speculator and *vice versa*. So, speculators aim for making profits in the futures trading, unlike hedgers and they take or accept the risk avoided by the hedgers. However, arbitragers also aim for making profits in the futures trading by closely following the differences in price movements between cash and futures markets and they operate in the futures till there is significant price difference between prices in cash and futures markets.

2.10. Mechanism of Trading in Futures Exchange

The futures market is a centralized market place for buyers and sellers from around the world, who meet and enter into futures contracts. There are two methods of trading in futures exchanges *viz.*, Open outcry trading and Electronic trading.

a. Open Outcry Trading

The largest organized futures exchanges in the world most notably, the CBOT, the Chicago Mercantile Exchange (CME), USA and London's International Financial Futures and Options Exchange

(LIFFE) use the open outcry method of trading. Open outcry trading is the more traditional form of trading. Brokers take orders (either bids to buy or offers to sell) by telephone or computer from traders (their customers). Those orders are then communicated orally to brokers in a trading pit. The pits are octagonal, multi-tiered areas on the floor of the exchange where traders conduct business. The traders wear different colored jackets and badges that indicate who they work for and what type of traders they are (Futures Commission Merchant (FCM) or local). It is called 'open outcry' because, traders shout and use various hand signals to relay information and the price at which they are willing to trade. Trades are executed (matches are made) when the traders agree on a price and the number of contracts either through verbal communication or simply some sort of motion such as a nod. The traders then turn their trade tickets over to their clerks who enter the transaction into the system. Customers are then notified of their trades and pertinent information about each trade is sent to the clearing house and brokerages. At the end of each day, the clearinghouse settles trades by ensuring that no discrepancy exists in the matched-trade information. This Open outcry method is an efficient means of 'price discovery', allowing buyers and sellers to arrive at the best prices given the supply and demand for a futures or options product. The explosion of electronic trading has prompted speculation that the open outcry system is quickly becoming obsolete. New advances in screen-based trading have forced both traders and exchanges to keep up with rapidly evolving technological developments. Now, exchanges like the CME bring in new technology to enhance the open outcry process and to keep the old system as viable as possible.

b. Electronic or Screen-Based Trading

Although the open outcry method is still the best way to trade highly active contracts on futures exchanges, electronic systems can improve the efficiency and cost effectiveness of trading some types of futures (Table 5.98). In recent years, the volume of electronic trades on futures exchanges has more than doubled, and it should continue to grow rapidly. An automated trade execution system (Electronic or screen-based trading) has three components *viz.*, computer terminals, where customer orders are keyed in and trade confirmations are received, a host computer that processes trades, and a network that links the terminals to the host computer. In electronic trading, customers (who have been pre-approved by a brokerage for electronic trading) send buy or sell orders directly from their computers to an electronic marketplace offered by the relevant exchange. There are no brokers involved in the process. Traders see the various bids and offers on their computers. The trade is executed by the traders lifting bids or hitting offers on their computer screens. The trading pit is, in essence, the trading screen and the electronic market participants replace the brokers standing in the pit. Electronic trading offers much greater insight into pricing because the top five current bids and offers are posted on the trading screen for all market participants to see. With electronic order-matching systems, the host computer matches bids with offers according to certain rules that determine an order's priority. Priority rules on most systems include price and time of entry. In some cases, priority rules may also include order size, type of order, and the identity of the customer who placed the order. In the simplest case, matching occurs when a trader places a buy order at a price equal to or higher than the price of an existing sell order for the same contract. The host computer automatically executes the order so that, trades are matched immediately. Trades are then cleared as they occur, as long as the host computer is linked to the clearing house. With advances in technology, electronic trading has gained momentum. Computers handle all trading activity - the software identifies matches of bids and offers and generally fills orders according to a first-in, first-out (FIFO) process. Dissemination of information is also faster on electronic trades. Trades made on CME® Globex®, for example, happen in milliseconds and are instantaneously broadcast to the public. In open outcry trading, however, it can take from a few seconds to minutes to execute a trade.

Table 5.98: Comparison between open outcry and electronic trading.

Item	*Open Outcry Trading*	*Electronic Trading*
Main suppliers of liquidity	Locals	Large institutions; market-making firms
Primary costs	Upkeep and staffing of trading floor; back-office tasks	Upgrading of software and hardware, telecommunications costs etc.
Information sources	Traders' observations of market activity	Order book; outsidenews sources
Operating efficiency	More time consuming, heavy investment on labour and more scope for errors	Less cost to the exchange, transparency, more speed of execution, anonymity, lower cost to trade
Limitations	Lack of precise trade records; lack of anonymity in trading	Manipulation of orders prior to entry

Electronic trading systems have become increasingly popular in the past decade and currently account for about one-fifth of volume on futures exchanges. Driving this rise in popularity is the potential for electronic trading to improve efficiency and lower costs. In addition, electronic trading systems make exchanges available to remote investors in real time, an important benefit in a period of increased cross-border trading.

2.11. How the Futures Market Works

For example, a producer of wheat may be trying to secure a selling price for next season's crop (after harvest), while a bread maker may be trying to secure a buying price for wheat to determine, how much bread can be made and at what profit. So, the farmer and the bread maker may enter into a futures contract on 1-6-2012 requiring the delivery of 50 kg of wheat grain by the farmer to the buyer (bread maker) at a price of ₹ 25/kg. It is this contract - and not the wheat grain that can then be bought and sold in the futures market. So, a futures contract is an agreement between two parties: a short position - the party who agrees to deliver/sell a commodity, and a long position - the party who agrees to receive/buy a commodity. In the above scenario, the farmer would be the holder of the short position (agreeing to sell), while the bread maker would be the holder of the long (agreeing to buy). Thus, every contract involves both the positions. In every futures contract, everything is specified: the quantity and quality of the commodity, the specific price per unit, and the date and method of delivery. The 'price' of a futures contract is represented by the agreed-upon price of the underlying commodity that will be delivered in the future. In the above scenario, the value of the contract for 50 kg wheat is ₹ 1250/- @ ₹ 25/kg.

Profit and Loss in Futures Trading - Cash Settlement

The profits and losses of a futures contract depends on the daily movements of the market for that contract and are calculated on a daily basis. Suppose, say the price for wheat increases to ₹ 30/kg on 2-6-2012 *i.e.*, the day after the farmer and bread maker enter into the futures contract at ₹ 25/kg. The farmer, as the holder of the short position, has lost ₹ 5/kg because, the selling price (in cash market *i.e.*, ₹ 30/kg on 2-6-2012) just increased from the future price (₹ 25/kg) at which he is obliged to sell his wheat. The bread maker, as the long position, has profited by ₹ 5/kg because, the price he is obliged to pay (₹ 25/kg) is less than, what the rest of the market is obliged to pay (₹ 30/kg on 2-6-2012 in the cash market). So, the change occurs, the farmer's account is debited by ₹ 250/- (₹ 5/kg x 50kg) and the bread maker's account is credited by ₹ 250/- (₹ 5/kg x 50kg). As the market moves every day, these kinds of adjustments are made accordingly. Unlike the stock market, futures positions are settled on a daily basis, which means that gains and losses from a day's trading are deducted or credited to a person's account each day. (In the stock market, the capital gains or losses from movements in price aren't realized, until the investor decides to sell the stock or cover his or her short position.) As the accounts of the parties in futures contracts are adjusted every day, most transactions in the futures market are settled in cash, as if the actual physical commodity is bought or sold in the cash market. Prices in the cash and futures market tend to move parallel to one another, and when a futures contract expires, the prices merge into one price. So, on the maturity date either party decides to close out their futures position, the contract will be settled. If the contract was settled at ₹ 30/kg, the farmer would lose ₹ 250/- on the futures contract and the bread maker would have made profit by ₹ 250/- on the contract. But, after the settlement of the futures contract, the bread maker still needs wheat (50 kgs) to make bread, so he will in actuality buy the wheat in the cash market (or from a wheat pool) for ₹ 30/kg (*i.e.*, he incurs ₹ 1500/- *i.e.*, ₹ 30/kg x 50kg) because, that is the price of wheat in the cash market (2-6-2012), when he closes out his contract. However, technically, the bread maker's futures profits of ₹ 250/- go towards his purchase, which means he still pays his locked-in price of ₹ 25/kg (₹ 1500 – ₹ 250 = ₹ 1250, which is equal to ₹25/kg x 50kg). So, the gain in the futures market is offset by the loss in the cash market and this is referred as Hedging. The farmer, after also closing out the contract, can sell his wheat in the cash market at ₹ 30/kg (2-6-2012), but because of his losses from the futures contract to the bread maker, the farmer still actually receives only ₹ 25/kg. In other words, the farmer's loss in the futures contract is offset by the higher selling price in the cash market and this is referred to as hedging. Instead of farmer and bread maker, if the two parties in the wheat futures contract are speculators, in such a case, the short (selling) speculator would simply have lost ₹ 250/- while the long (buying) speculator would have gained that amount ₹ 250/-. In other words, neither would have to go to the cash market to buy or sell the commodity after the contract expires.

It is important that, gains and losses on futures contracts are not only calculated on a daily basis, they are also credited or debited to each market participant's brokerage account on a daily basis. Thus, if a speculator were to have a ₹ 500 profit as the result of a day's price changes, that amount would immediately be credited to his or her account and, unless required for other purposes, could be withdrawn. On the other hand, if the day's price changes resulted in a ₹ 500 loss, the account would be debited for that amount. The process just described is known as daily cash settlement and it's an important feature of futures trading.

The above discussion infers that, two mechanisms are involved in futures trading *viz.*, Futures trading through physical settlement and Futures trading through cash settlement.

a. Futures Trading through Physical Settlement

Any member can enter into futures trading as a buyer or a seller. The member can trade on his/her own account or may enter through brokers. The person has to place an order with the broker. Then the broker will contact a brokerage house at the commodity exchange and relay the order.

Bidding is done to match the contract with others. After matching, the role of clearing house will start. All commodity exchanges use a clearing house to handle the book keeping of trading futures contracts. The clearing house is responsible for keeping records of all buyers and sellers by acting as a third party between on all buys and sells. In other words, the clearing house acts as a seller to all buyers and buyer to all sellers, once the original buyer's and seller's names and the prices are reported for the trade. After each day's trading, all exchange members must report their buys and sells to the clearing house.

Clearing bank clears or settles the contracts of both buyers and sellers in physical settlement as well as cash settlement. If the settlement takes place by delivering the commodities physically, then this mechanism is called futures trading through physical settlement. All contracts materializing into physical deliveries would settle in a period of 2-7 days after the expiry date. The exact settlement date would be specified in the settlement calendar released by the exchange.

b. Futures Trading through Cash Settlement

The participants entering into futures trading have an option to liquidate their outstanding position by entering into opposite contract, before maturity of the contract. If a person entered into futures as a buyer, he has to take the opposite position (seller) in cash market in order to exit from the futures contract before the expiry date. If he entered into futures as a seller, he has to take the opposite position (buyer) in the cash market in order to exit from the futures contract before the expiry date. After liquidating their positions, (by entering into an opposite contract) participants are not required to give/take deliveries. Then, his contract will be settled by payment of money difference rather than by delivering the commodity physically. If he is in gain, the money will be added to his initial margin account and if he is in loss, money will be deducted from his initial margin account. This process is called cash settlement. All contracts materializing in cash would be settled on the following day after liquidating the contract. Most of the futures contracts all over the world are performed through cash settlement rather than by physical settlement.

In trading mechanism through physical delivery of the product, the product itself is specifically defined (quality) and the month, when actual delivery or settlement is to occur. For example, a July futures contract implies, the delivery or settlement in July. It should be noted that, even in the case of delivery-type futures contracts, very few actually result in delivery. Not many speculators want to take or make delivery of the products. Rather, the vast majority of both speculators and hedgers choose to realize their gains or losses by buying or selling an offsetting futures contract prior to the delivery date.

2.12. Futures Basis

The basis reflects the relationship between cash price and futures price. Basis consists of carrying charges and costs of transportation for the commodity to an exchange approved delivery point. The basis is obtained by subtracting the futures price from the cash price. That is,

Basis = Cash Price - Futures Price

where, Cash Price is the cash price for a specific commodity at a given location and *Futures Price* is the relevant futures price for that commodity at an exchange. The basis can be a positive or negative number. A positive basis is said to be *'over'*, as the cash price is higher than the futures price. A negative basis is said to be *'under'*, as the cash price is lower than the futures price (Figure 5.47 and Table 5.99).

Example

Assume a farmer raises maize and cotton on his farm. On 1-1-2012, the local wholesaler purchased maize for ₹ 10/unit and a processor purchased cotton at ₹ 72/unit. On the same day (1-1-2012), the closing price of maize and cotton futures price at an exchange are ₹ 10.13/unit and ₹ 70.50/unit respectively. Now, if farmer wants to know his basis, he would simply take the cash price and subtract the futures price for each commodity.

	Item	*Maize*	*Cotton*
	Local Cash Price	₹ 10.00	₹ 72.00
Less	Futures Market Price	₹ 10.13	₹ 70.50
	Basis	–0.13	+1.50

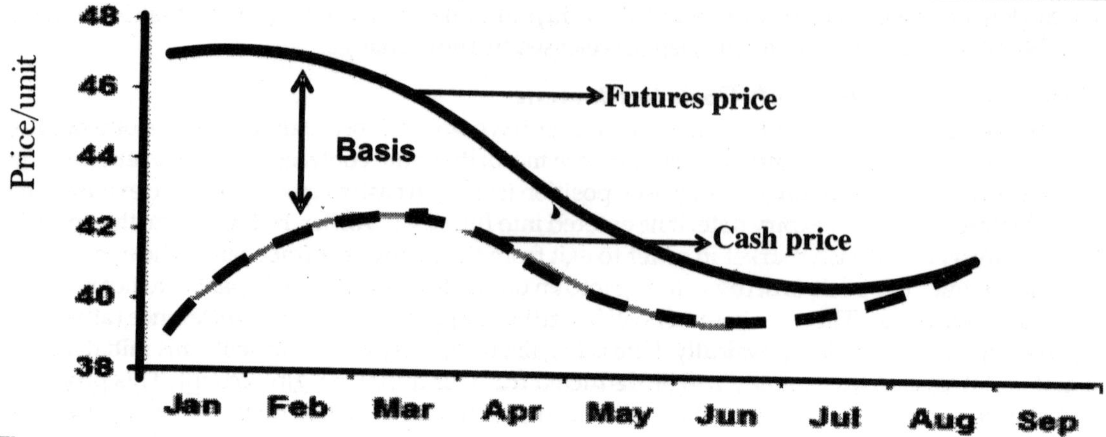

Figure 5.47: Cash price versus Futures price (August wheat market).

Table 5.99: Calculation of basis.

Spot (Cash) Price	₹ 42/unit
August Futures Price	₹ 47/unit
Basis	-5/unit (In market lingo, the basis is '$5 under August'.)

A negative value represents a cash price 'under' the futures price and a positive value represents a cash price 'over' the futures price. Figure 5.48 is used to describe basis movements. A basis that becomes more positive or less negative over time is said to narrow or strengthen. A basis that becomes less positive or more negative over time is said to widen or weaken.

Inferences Drawn from ' Basis'

For the above example, basis describes two separate relationships for maize and cotton. Therefore, these enterprises are separated in the discussion below.

Table 5.100: Use of basis in marketing strategies.

	High Price Difference	*Low Price Difference*
Strong basis	Sell cash	Sell cash or Hedge with futures (Long futures)
Weak basis	Hedge with futures (Short futures) or Delay cash sale	Store the produce

Maize

For maize grain, basis is typically used as an indication of current local demand. Weak basis (when cash price < futures price or current basis is less than normal basis) indicates that, the market doesn't want maize grain now, but the market may or may not want it later. But, strong basis (when cash price > futures price or current basis is more than normal basis) indicates the market wants the grain. So, basis is best used in deciding how to sell. Table 5.100 shows, what the farmer should do given certain market situations. For instance, assume the maize producer believes the maize price is high and basis is strong relative to historical patterns. So, sell the maize grain in the cash market, as there is little opportunity to better the current market price through storage and/or taking a futures position.

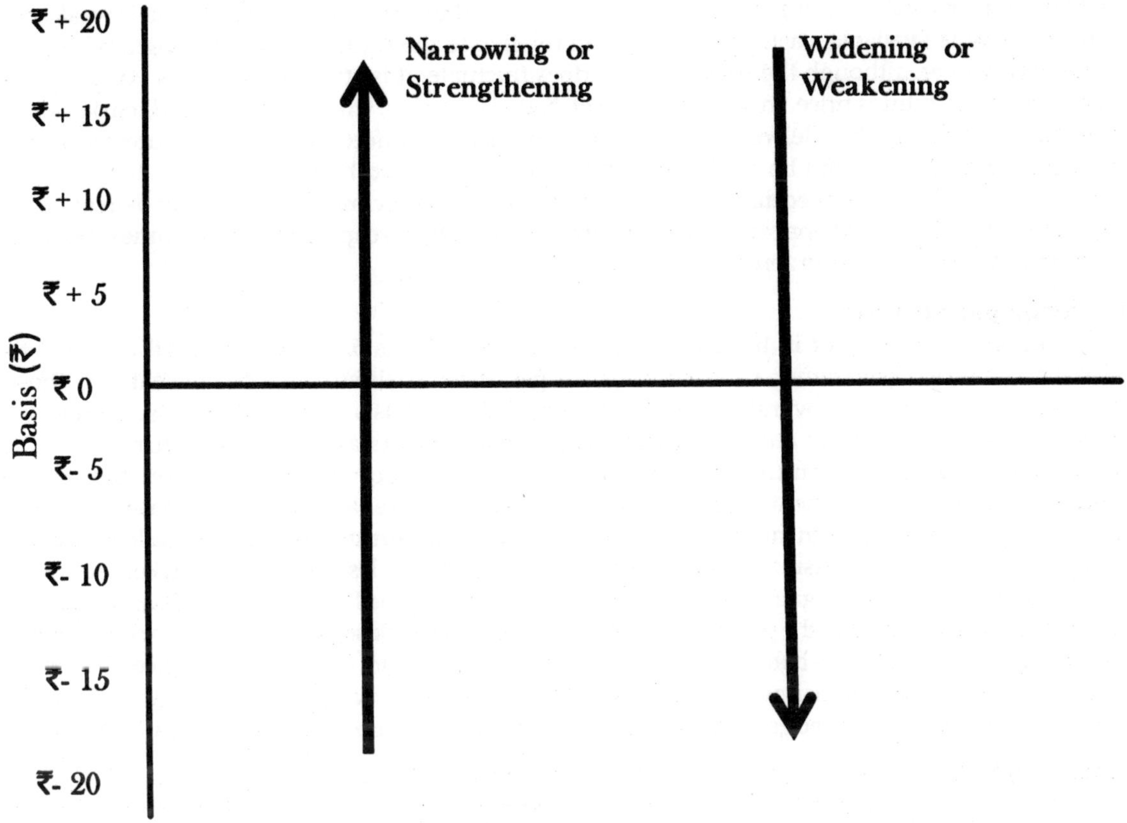

Figure 5.48: Basis – Strong basis and Weak basis.

Cotton

For cotton, basis is strong (+ve) and this implies, the market want cotton now. So, sell the produce in the cash market, as there is little opportunity to better the current market price through storage and/ or taking a futures position.

Thus, understanding the basis can help farmers and agri-business personnel in evaluating forward contracting and hedging decisions. It can be concluded from the Figure 5.45 that, if the current basis is more positive than the expected (average) basis for the period, it is often referred to as 'narrower' or 'stronger' than normal. For example, if the current basis is –₹ 0.10, when –₹ 0.25 is 'normal', the market is offering a strong or narrow basis. A basis of –₹ 0.40 when –₹ 0.25 is 'normal' would be a weak or wide basis. When basis is stronger or narrower than normal, the market is providing a financial incentive to make cash sales. When there is a wide or weak basis the market is discouraging cash sales and encouraging storage in the case of crops.

a. Strong or Weak Basis

The basis changes from time to time. If the basis gains in value (say from -20 to -5), we say the basis has strengthened. On the other hand, if basis drops in value (say from 20 to 5), we say the basis has weakened. Short term demand and supply situations are generally the main factors responsible for the change in the basis. If demand is strong and the available supply is small, cash prices could rise relative to futures price, causing the basis to strengthen. On the other hand, if the demand is weak and a large supply is available, cash prices could fall relative to the futures price, causing the basis to weaken. However, although the basis can and does fluctuate, it is still generally less volatile than either the cash or futures price. In a normal market, basis is positive, while in an inverted market basis is negative. As the month of delivery of product approaches, the difference between futures price and spot price will decrease and both the prices tend to converge and bring down basis to zero. The convergence of price occurs because, any non-zero basis, at a period close to the month of delivery, will provide the traders ample opportunities of arbitrage and as they reap these opportunities the basis will come closer and closer to zero.

b. Hedging and Basis

Basis is a crucial factor in hedging using the futures. Tables 5.101A and 5.101B are used to describe a gain or loss to either a short or long hedger, when basis strengthens or weakens. These tables further indicate the overall impacts of a change in the basis under different scenarios. By looking at basis historically and/or seasonally, it will be much easier to assess current market conditions in both cash and futures markets and the implications for basis. For the short hedger, the hedger gains from a strengthening basis. That is, the hedger realizes a cash price increase relative to the futures price and may gain more from the position taken in the futures market. For the long hedger, the hedger prefers for the basis to weaken. That is, the hedger pays less in the cash market relative to the futures market and may gain more from the position taken in the futures market. Basis risk arises because of uncertainty about the price difference when the hedge is closed out. Basis risk is considerably less than price risk, but basis behaviour can have a significant impact on the performance of a hedge. A stronger-than-expected basis will benefit a short hedger, while a weaker-than-expected basis works to the advantage of a long hedger. Let us discuss, Short hedge and Long hedge in-detail here under.

Table 5.101A: Direction and impact of basis movement for short and long hedges.

	Short Hedge	*Long Hedge*
If basis weakens (widening)	Basis loss (Unfavourable)	Basis gain (Favourable)
If basis strengthens (narrowing)	Basis gain (Favourable)	Basis loss (Unfavourable)

Table 5.101B: Impact of basis change on overall hedge.

Expected Basis	*Short Hedge*		*Long Hedge*	
	Basis Widens	*Basis Narrows*	*Basis Widens*	*Basis Narrows*
Negative	Loss	Profit	Profit	Loss
Positive	Profit	Loss	Loss	Profit

1. Short Hedge

As discussed earlier, short hedge is a hedging strategy used by farmers and processors to lock in the price of a commodity to be delivered sometime in the future. Hence, the short hedge is also known as output hedge. The short hedge involves taking up a short futures position, while owning the underlying commodity or commodity to be delivered. Should the underlying commodity price fall, the gain in the value of the short futures position will be able to offset the drop in revenue from the sale of the underlying. To understand this concept, let us go through the following example:

Assume that, a farmer is sowing crop X on 1-6-2012, which will be ready for harvest by September and to delivery in October. At the time of sowing the crop (1-6-2012), the local cash market price of this commodity X is ₹ 6.60/unit. The farmer knows from past years that, the Total Cost of Cultivation (TCOC) of crop X is about ₹ 6400. On 1-6-2012, the October futures price is at ₹ 6.70/unit and the farmer wishes to lock in this futures selling price. To do this, he enters a short hedge by selling October futures with a local trader. By, mid-September, the crop X is ready for harvesting and at that time, the local wholesale price (in the cash market) has dropped to ₹ 6.20/unit. Correspondingly, prices of October futures have also fallen and are now trading at ₹ 6.33/unit.

Loss in Cash Market

Total production of commodity X : 1000 units

TCOC = ₹ 6400

Gross returns = 1000 x 6.20 = ₹ 6200

Net Returns = 6200 – 6400 = - ₹ 200

So, for all his efforts, the farmer might have ended up with a loss of ₹ -200.

... is Offset by Gain in Futures Market.

Fortunately, he had hedge his output with a short position in October futures, which have since gained in value.

Value of commodity X futures sold in June = 6.70 x 1000 = ₹ 6700

Value of October commodity X futures purchased in September = 6.33 x 1000 = ₹ 6330 (offsetting in cash market)

Net Gain in Futures Market = 6700 - 6330 = ₹ 370

Overall profit = Gain in Futures Market - Loss in Cash Market = 370 - 200 = ₹ 170

Hence, with the short hedge in place, the farmer can still manage to make a profit of ₹ 170 despite falling commodity X prices.

The above short hedge is not perfect. In this example, while cash prices have fallen by ₹ 0.40/unit, futures prices have only dropped by ₹ 0.37/unit (Table 5.102) and so the short futures position have only managed to offset 92.5 per cent of the drop in price. This is due to the weakening of the basis.

Table 5.102: Short hedge to assess the profitability to farmer.

Item	*Cash*	*October Futures*	*Basis*
March	₹ 6.60	₹ 6.70	– ₹ 0.10
August	₹ 6.20	₹ 6.33	– ₹ 0.13
Net	– ₹ 0.40/unit	+ ₹ 0.37/unit	– ₹ 0.03 (Weakened by ₹ 0.03)

2. Long Hedge

The long hedge is a hedging strategy used by farmers and processors to lock in the price of a commodity to be purchased sometime in the future. Hence, the long hedge is also known as input hedge. The long hedge involves taking up a long futures position. Should the underlying commodity price rise, the gain in the value of the long futures position will be able to offset the increase in purchasing costs. To understand this concept, let us go through the following example:

In May, a flour manufacturer has linked a contract to supply wheat flour to a super market in September. Let's assume that, the total amount of wheat needed to produce the flour is 5000 units. Based on the agreed selling price for the flour, the flour maker calculated that, he must purchase wheat at ₹ 7.00/unit or less in order to breakeven. At that time (May), wheat is going for ₹ 6.60/unit at the local wholesale market (cash market), while September wheat futures are trading at ₹ 6.70/unit and the flour maker wishes to lock in this purchase price (₹ 6.70/unit), as he needs wheat flour in the month of August for beginning the processing of wheat. To do this, he enters a long hedge by buying September wheat futures.

In August, the processing of wheat should begin and hence, the flour maker needs to purchase the requisite quantity of wheat from the local wholesaler. However, the price of wheat have since gone up and at the local wholesaler, the price has risen to ₹ 7.20/unit. Correspondingly, prices of September wheat futures have also risen and are now trading at ₹ 7.27/unit.

Loss in Cash Market

Since the flour manufacturer's breakeven cost is ₹ 7.00/unit, but he has to purchase wheat at ₹ 7.20/unit in August cash market, he will lose ₹ 0.20/unit. So, at 5000 units, he will lose ₹ 1000 in the cash market. So, for all his efforts, the flour maker might have ended up with a loss of ₹ 1000.

... is Offset by Gain in Futures Market

Fortunately, he had hedge his input (wheat) with a long position in September wheat futures, which have since gained in value.

Value of September wheat futures purchased in May = 6.70 x 5000 = ₹ 33500

Value of September wheat futures sold in August = 7.27 x 5000 = ₹ 36350 (offsetting in cash market)

Net Gain in Futures Market = 36350 – 33500 = ₹ 2850

Overall profit = Gain in Futures Market - Loss in Cash Market = 2850 - 1000 = ₹ 1850

Hence, with the long hedge in place, the flour maker can still manage to make a profit of ₹ 1850 despite rising wheat prices.

This long hedge is also not perfect. In the above example, while cash prices have risen by ₹ 0.60/unit, futures prices have only gone up by ₹ 0.57/unit (Table 5.103) and so the long futures position have only managed to offset 95 per cent of the rise in price. This is due to the strengthening of the basis.

Table 5.103: Short hedge to assess the profitability to farmer.

Item	*Cash*	*September Futures*	*Basis*
May	₹ 6.60	₹ 6.70	– ₹ 0.10
August	₹ 7.20	₹ 7.27	– ₹ 0.07
Net	+ ₹ 0.60/unit	+ ₹ 0.57/unit	+ ₹ 0.03 (Strengthened by ₹ 0.03)

Importance of Futures Trading

Futures trading offers the following advantages:

- Futures trading is useful to farmers because, they can get an idea of the price likely to prevail at a future point of time and therefore, they can decide between various competing commodities, the best that suits them and guides them in cost-effective production. It also enables them to decide on the crop that they want to grow. Since there is transparency in prices, the farmers can decide when and where to sell, so as to maximize their profits. The farmers also benefit by the dissemination of the futures prices of the exchange traded products, as it improves their bargaining capacity. Further, farmers can thus, avoid taking unproductive decisions based on prices forecasted to them. These futures prices also guide the farmers in deciding between selling their produce now or storing and selling at a later date at better prices.

☆ Futures trading ensure the sellers towards improving the product quality. Since the contracts for commodities are standardized, it becomes essential for the farmers/sellers to ensure that the quality of the commodity is as specified in the contract. The advent of futures markets has enabled defining quality standards of different commodities. The quality testing and certification agencies contribute towards standardization and assessment of commodity quality.

☆ Under a multitude of market scenarios *viz.*, rising prices, falling prices or even stable prices, agricultural futures can help farmers and traders to achieve risk-management objectives. The farming community and traders can significantly reduce risk exposure with agricultural futures.

☆ Futures trading is considered as a highly leveraged investment. If the trader wants to own a futures contract, he only needs to invest a small fraction of the value of the contract (margin money). Most traders only invest about ten per cent of the contract's value in exchange for trading them. This way, traders may be able to trade larger amounts of commodities than if they ever bought the commodities in the cash market. For example, trading in pepper calls for only five per cent initial margin. Thus, if one pepper futures contract is valued at ₹ 9,00,000, the trader is expected to deposit an initial margin of only ₹ 45,000, so as to ensure futures trading. If the price of pepper goes up by even two per cent, the trader would make a profit of ₹ 22,500 on a deposit of ₹ 45,000 before the expiry of the contract. This is the benefit of leveraged trading transactions. With futures contracts, the trader trades (first buys) in the expectation of the price (rise) at a later date. This is possible with a margin deposit, which is usually between 5-10 per cent of the value of the commodity.

☆ Although futures trading involve certain commodities, the trader does not have to worry about how to take care of the products himself. Trading is done with the futures contract changing hands instead of the commodity itself. This makes it quite convenient, since the trader does not have to worry about, where to store and keep the commodities being traded for the meantime. So, futures trading is basically a paper investment.

☆ For a trader, it may be quicker to make money on futures trading. Aside from the leverage provided by futures, the markets tend to move more quickly as compared to cash markets. But, this can also work against the trader since, the quick pace of the market can also lead to quick losses for the trader for incorrect predictions on their positions.

☆ It provides the exporters an advance indication of the price likely to prevail and thereby, helps them in quoting a competitive price and secure export contract in a competitive market. Thus, the exporters can hedge their price risk and improve their competitiveness by making use of futures market. A majority of traders, who are involved in physical trade internationally intend to buy forwards. The purchases made from the physical market might expose them to the risk of price risk resulting to losses. The existence of futures market would allow the exporters to hedge their proposed purchase by temporarily substituting for actual purchase till the time is ripe to buy in physical market. In the absence of futures market it will be meticulous, time consuming and costly physical transactions.

- It helps the consumers about an idea of the price at which the commodity would be available at a future point of time. So, the consumers can do proper costing of commodities and minimizes their expenditure through making futures contracts. Predictable pricing and transparency is an added advantage.
- It aids in the process of proper price discovery. It is the mechanism by which a 'fair value price' is determined by the large number of participants in the futures exchange. This is the result of open outcry and electronic trading systems established on the exchanges. These prices get disseminated instantaneously to every participant, as they are made available on the centralized trading screen of the exchange.
- The larger, more frequent and more unforeseen price variability in a commodity, the greater is its price risk. This price risk is more common in spot or cash market. However, in futures market, the risk mitigation or 'hedging' mechanism guides the farmers and traders for an efficient and effective mechanism for management against price risks. For example, a farmer can lock in the price at which he would sell his produce, when it is ready for harvesting, so as to overcome the threat of future fall in price. Similarly, a trader or an exporter can 'hedge' his price risk on the trade contract by using the futures market, so as to overcome the risk of future rise in price.
- The futures market mechanism helps prevent rampant fluctuations in prices leading to price stabilization. It ensures in price stabilization in times of violent price fluctuations, dampening the peaks and lifting the valleys *i.e.*, the amplitude of price variation is reduced. So, it leads to integrated price structure throughout the country through achieving balance in supply and demand positions. As discussed earlier, efficient price discovery with the involvement of various participants helps in preventing seasonal price volatility. Further, the centralized trading platform gives access to several market players, unlike the bilateral forward market, which has only restricted access. So, this ensures the prevalent of more competitive price in the futures market with less volatility. This encourages competition and acts as a price barometer for farmers and traders.
- Low brokerage costs and less time-consuming negotiations helps in the reduction of transaction costs of agricultural commodities.
- As the clearing house acts as a counterparty for each buyer and seller *i.e.*, for each buyer, clearing house is the counterparty and for each seller, clearing house is the counterparty, it eliminates the counterparty risk in futures market transaction. This act of the clearing house is called 'novation' and this removal of the counter party risk is also facilitated by the standardized and centralized nature of all transactions. The clearing house also maintains various types of margins collected from the participants and this helps in guaranteeing the performance of future contracts thereby, bringing down the credit risk. Further, the management of price risk by the farmers and traders through futures contracts (through hedging) also improves farmer's income and ensure adequate investments in agriculture. This management of price risk is very crucial, as it also reduces credit risk from the other side. This reduction of credit risk through futures trading is very crucial for the developing countries like India. This reduction of price risk and credit risk enable the farmers to concentrate only on their core activity *i.e.*, production, without worrying about other types of possible risks. Moreover, as futures contracts are traded in commodity exchanges with

robust clearing and settlement systems, farmers are assured of trade and payment guarantee with no counterparty risk. In a nutshell, as the exchanges maintain neutrality, conflict of interest is always avoided.

- Futures markets are extremely transparent, in the sense that, the manipulation of prices of a commodity is extremely difficult due to globalization of economies thereby, providing for prices benchmarked across different countries and continents. For example, gold, silver, crude oil, natural gas, etc. are international commodities, whose prices in India are indicative of the global situation.
- The futures contracts being traded are considered very liquid. This means that, there are huge amounts of contracts being traded in the market on a daily basis. Orders can be placed quickly and they can be bought or sold in a similar fashion. There are always quite a good number of available buyers and sellers for the futures contracts, whatever commodity it might be. Further, the standardization of contracts enables participants to come out of their positions easily. The large number of market participants trading on an exchange further enhances liquidity.
- Buyers and sellers can avail of the bank finances for trading in commodities. Nationalized banks and private sector banks have come forward to offer credit facilities for futures trading. More and more banks are likely to fall in line looking at the huge potential that commodity futures markets offer in India.
- Commissions on futures markets tend to be smaller as compared to other trading markets. The commissions are usually paid after the position has ended.
- Futures contracts can be sold as easily as they are bought enabling a speculator to profit from falling markets as well as rising ones.
- Due to the liquidity of the futures market, the transaction costs are very competitive hence, usually minimal.
- The commodity exchanges act as the key source of price information. Besides the prices of futures contracts traded in the respective exchanges, these exchanges also provide useful information pertaining to spot prices of agricultural commodities from nearly 7,000 local mandis all over the country at frequent intervals. In fact, many mandis now begin functioning only after the commodity exchanges start trading, as both farmers and traders are interested in knowing current market conditions and future prices of commodities. Such price information is made available to the whole of India through various channels like TV, 'Kisan' call centres, display boards kept in mandis, warehouses, rural bank branches etc.
- Futures market is a powerful tool for breaking the market power of monopoly and collusive oligopoly traders. Since the trading is open to all the participants through an exchange by an open outcry or electronic trading methods, it ensures transparent determination of market prices in a competitive manner. Further, the active involvement of several corporate players like Cargill, Britannia, Reliance, Pantaloon and agencies such as Commodity Boards, MARKFED etc., in the commodity exchanges enhances the liquidity, competition, direct procurement from farmers, shortening the value chain etc. Further, physical proximity to the market is no longer an issue, farmers and traders can have the opportunity to sell anywhere in the country through online trading and this ensures transparent trading of commodities. This promotes competition and checks the market power of few trading firms.

- Commodity exchanges have developed alliances with other agencies to ensure that, farmers get scientific storage facilities. For example, NCDEX has also promoted a separate company – National Collateral Management Services Limited (NCMSL) to promote quality warehousing infrastructure. Similarly, MCX is taking its own initiatives for accreditation of existing warehouses.
- There are inter-linkages between physical commodity (spot) markets and commodity futures market. Commodity futures markets have been functioning in India, since 2002-03 and has made immense progress especially in the dissemination of futures prices of agricultural commodities using electronic telecommunications links with distant markets. These facilitated the farmers to access and enable them to plan their cropping pattern and managing the price risk. This dissemination of price information across different mandis ensures homogeneity of prices in the market economy. Above these, in future exchanges, the transacted commodities are highly standardized and this prompted the farmers and traders to transact the commodities as per the grades laid down.
- Above all, futures trading offers several advantages and benefits to the Indian economy in terms of business generation and growth in employment opportunities.

Reasons for Poor Participation of Indian Farmers in Futures Trading

In spite of the various benefits offered by futures trading, the use of these exchanges by the Indian farmers is very miniscule. The following are the important reasons:

- For opening trading accounts in the commodity exchanges, a PAN number is mandatory. Most of the farmers do not possess a PAN card.
- The exchanges also demand a proof of identity like electricity bills, phone bills, passport, voter cards and driving licences. But, it is not possible for many small farmers to produce any of these documents.
- Another major limitation is that, the minimum lot size specified by the commodity exchanges. The specifications are much higher compared to the normal quantity harvested by an average farmer. For example, the minimum lot size for wheat and chana in NCDEX is 10 m tonnes. But, the average yield of wheat harvested by small and medium Indian farmers is only around 1.6 m tonnes per acre. Hence, only a farmer who cultivates at least 7 acres of wheat will have adequate stocks to meet the minimum lot size requirement. On similar lines, since the average yield of chana is only 1.0 MT/acre, a farmer must cultivate minimum 10 acres of chana to be able to meet the minimum lot size stipulation.
- It is very difficult for small and medium farmers to understand the concepts like initial margin, maintenance margin, maintenance margin requirement etc. Further, if the clearing house or exchange calls the farmer to deposit the amount in the account to update his initial margin, the farmer may not have liquid cash in time to deposit the in his account. Further, in case of unfavourable price movements, margin calls could touch very high amounts.
- The number of delivery centres nominated by commodity exchanges is limited. Hence, if farmers want to make delivery at the expiry of the futures contracts, they may have to transport their produce over long distances. This involves logistical difficulties and higher costs.
- Issues like lack of warehouses at nearer locations for storing the produce till sales, grading, taxation etc. are also discouraging farmers from participating in commodity exchanges.

- ☆ Dissemination of price information is still not adequate. Hence, farmers will not show willingness to transact their produce through futures exchanges. Moreover, they are of the opinion that, these exchanges are more biased towards members rather than for farming community.
- ☆ Since majority of the farmers are largely illiterate, they could not follow the mechanisms of hedging and speculation trade. So, they develop false suspicions regarding the functioning of futures markets.

Ban on Futures Trading in India

There is a misconception that, futures trading leads to inflationary rise in prices in the economy because, some speculators goes for illegitimate speculation and create artificial scarcity of commodities. Some opine that, the traders purposefully quote higher future prices for the commodities. But, as per Economic Survey Report, 2007, in NCDEX, guar seed, chana and soya oil had the highest volumes of trade. Guar seed accounted for the largest share (11 per cent) trade, followed by chana (10 per cent). Clearly, the volumes of commodity futures in tur dal, urad dal, wheat and rice are insignificant and does not figure in the rising price scheme. To be more specific, for example, the wheat trading volumes on commodity exchanges were very low with only 20,000 m tonnes being traded in the country, which produces around 70 m tonnes in a year. So, we can strongly argue that the speculative activity in the futures exchanges is not the cause for fuelled inflation. However, it is quite undisputed fact that, futures trading has a direct impact on inflation. This is brought about by the presence of basis (the difference between spot and future prices). In usual cases, the futures prices are higher than spot prices (In support of this proposition, see market data for different commodities and different dates available at www.ncdex.org). The cause of this phenomenon is the presence of speculation and positive expectations. When futures prices are higher than spot prices for an agricultural commodity, the farmers would like to sell the commodity in future, bringing down the supply levels at present. This mechanism causes excess demand in the present time period and pushes up the price of the commodity in question. If the futures prices were almost always higher than the spot prices for majority of the agricultural commodities, it leads to rise in inflation in the economy and then, it is imperative on the part of the Government to impose ban on futures trading in agricultural commodities. However, the imposition of ban will really help in curtailing overall inflation or merely act as a stop gap measure in bringing down the prices of agricultural commodities alone can be ascertained, only after analyzing the actual causes of inflation. It is to be noted that, inflation in the economy is not only influenced by rise in prices of agricultural commodities alone, but even by other factors like RBI's (liberal) credit policy, increase in prices of non-agricultural commodities etc. For example, between 2004-05 and 2005-06, there has been acceleration in growth of broad money (M3) from 12.3 per cent to 17 per cent and this rapid growth cannot be attributed merely due to the presence of future trading, but due to liberal credit policy followed in the country. So, banning futures trading without analyzing the real causes will merely bring down price of agricultural commodities without causing a corresponding deflation for non-agricultural commodities. This will lead to a highly inequitable result, leaving the farmers at a worsened position. It is also essential to note that, the gains that the typical Indian farmer used to reap from futures trading are those that arise as a spill-over effect of futures trading and not from the participation of farmers in futures trading. This is because, the direct participation of Indian farmers in futures trading is highly insignificant due to factors mentioned earlier. This futures trading has been advocated as a tool to solve the market uncertainties by promising the farmers a fixed price for their future output. Given the market situation and volatility of prices of agricultural commodities in India, such an assurance is of extreme importance. Thus, futures trading in theory, has the potential to solve a number of problems in the agricultural sector. However, in the present scenario, the futures market is beyond the reach of a vast majority of Indian farmers especially small and marginal farmers. It is these farmers, who are the most affected by the market fluctuations and the monsoon failures. Hence, it is these farmers, who really need to hedge risks through futures trading. So, suitable strategies (Aggregator Models, collective marketing) should be formulated to transact the farmers' produce through futures trading.

So, the imposition of ban on futures trading in certain agricultural commodities is a mere stop gap measure to curb inflation. It certainly will affect the farmers adversely by lowering the price level of agricultural commodities unaccompanied by a corresponding lowering in prices of non-agricultural products. It is further to be noted that, commodity exchanges in India are in their nascent stage of development. There are numerous bottlenecks in the growth of this particular segment in India. These problems are not unknown to the Government and the FMC. This calls for a more focused and pragmatic approach from them in making the agricultural futures markets a vibrant segment for risk management. Further, India is one of the top producers of a large number of commodities and also has a long history of futures trading in agricultural commodities. The futures markets have made enormous progress in terms of technology, transparency and the trading activity. Interestingly, this has happened only after the Government protection was removed from a number of commodities and market forces were allowed to play their role. This should act as a major lesson for the policy makers in developing countries like India, that pricing and price risk management should be left to the market forces rather than trying to achieve these through administered price mechanisms.

6

Market Functionaries: Institutional Approach of Agricultural Marketing

In the earlier chapter, we conceptualized 'functional approach' of agricultural marketing. This functional approach attempts to answer the 'what' in the question 'who does what' in the marketing process. There is another method of market analysis, which looks at the involvement of the various agencies or intermediaries or institutions or players or middlemen in performing the marketing processes. So, this institutional approach focuses attention on the 'who', in the above question *i.e.,* 'who does what'. In this approach, the human element receives primary emphasis. The marketing institutions are the wide variety of business organizations or agencies involved in the purchase and sale of agricultural produce, as it is moved from farmer to the ultimate consumer. So, beginning from the agriculturists, down to the final consumer, one can find a long chain of different functionaries. These can be conveniently classified based on the place of working or executing the operations: This criterion classifies the functionaries as follows:

I. ***Functionaries working at the village level:*** They include, Farmers, Assemblers or Village merchants, Itinerant traders and Mashakhores.

II. ***Functionaries working at the mandi/market level:*** They include, Agent middlemen (Commission agents and Brokers), Merchant middlemen (Wholesalers and Retailers), Processors, Speculative middlemen and Facilitating middlemen.

I. Functionaries Working at the Village Level

1. Farmers

They are the cultivators, who offer their marketable surplus to the non-farming population. They constitute the first type of market functionaries operating at village level. They also own tractor, carts and other means of transportation to bring the produce to the market place. They also perform other marketing functions like storage, grading etc. In case of small farmers, they will sell the produce mostly in villages. But, large farmers will take the produce to the market yards and they maintain good contact with commission agents and traders. Frequent visits of the farmers to the market yard enable them to gain useful practical knowledge about the marketing practices.

2. Assemblers or Village Merchants

They are known by different names in different parts of India such as 'Beopari', 'Baniya', 'Sahukar', 'Paikars', 'Farias' etc. 'Assembler' is the first link between the farmer and other middlemen. He takes title and collects several smaller lots of produce from the farmers in the village by his own capital and combines them into a single lot at one location. To the extent, he arranges for or provides transport, the key function the assembler provides in addition to assembly. So, he is also called as 'Transporter'. The produce so assembled from different farmers in the village and supply the same to the nearby markets. They also act as liaison intermediaries between farmers and traders in wholesale markets or they themselves may sell the produce in assembly markets. They have small establishments in the villages. They purchase the produce of those, who have either taken finance from them or those who are not able to go to the distant markets. They also supply essential consumption goods to the farmers. They often visit nearby markets and keep in touch with the prevailing prices. They either sell the collected produce in the nearby market or retain it for sale at a later date in the village itself. Their main job is to collect the marketable surplus from villages and village markets and carry it to the assembly-markets or wholesale mandis in nearest towns. It is in this manner and through this agency, the produce is brought to the secondary and terminal markets. Before gathering produce from several farmers, they contact them, who are ready to harvest, in advance. Prices are negotiated based on quality and quantity of produce to be purchased. The price is based on the current average price. In such a case, the assemblers benefit from price differentiations after the produce was graded based on scientific standards in the assembly market. They operate in some cases with their own capital, but some are dependent on the finances offered by 'Arhatias' or 'Arhatdars' or large wholesale traders in assembling and distributing centres.

3. Itinerant Traders

They are also called as 'Beoparies'. They are petty merchants, who move between villages and purchase the produce for the farmers directly. They own or hire a bullock transport to move the produce to the nearby markets. They offer a lower price to the farmers than that ruling in the nearby market, as they take into consideration all factors such as transportation, market charges and their profit margin. They generally pay the farmers in 3-4 days after the produce has been disposed off in the market. They also act as a financer to poor farmers. These represent an alternative marketing channel for small holders. The itinerant traders perform a similar range of marketing functions to those undertaken by the assembler or village trader. Their only disadvantage is that, they are not generally as well known to small holders as the purchaser of the produce.

4. Mashakhores

They are small wholesalers or big retailers in the villages and they mainly deal in vegetables and fruits.

II. Functionaries at Mandi/Market Level

1. Agent Middlemen

These middleman acts as the representative of their clients. They do not take title to the produce *i.e.*, they do not purchase and own the produce, but they simply negotiate purchase and sale of commodity between the farmer and potential buyers. So, the agent middlemen don't own what they handle. That is, they may or may not take possession of the product, but they certainly never accept legal ownership of the product. In reality, these middlemen sell their services to their clients, not physical goods. These middlemen are basically hired by their clients (farmers or traders). They derive their income from the fees paid by their clients and they work for either a flat rate or percentage (of the selling price) commission. Their main stock in trade is their knowledge of market in which they participate. They use the knowledge in bringing together potential sellers and buyers on the same platform. Their services will be retained either by buyer or the seller, who do not have adequate bargaining power in the market transactions. There are two categories of agent middlemen *viz.*, Commission agents and Brokers.

a. Commission Agents

They are also called as Arhatiyas. They are the agent middlemen commonly seen in regulated markets and they render service on behalf of either farmer or buyer (processor or wholesaler). These commission agents have permanent establishment in the market yard and they will have physical control over the product. They will charge 'commission' for the services rendered by them. They will arrange the produce for sale, helps the client in negotiating about the price and other trading aspects with their counterparty, collect the price from buyers, deduct his expenses and service charges and the remaining balance will be paid to the client. Generally, in regulated markets, these commission agents will work on behalf of the farmers, as they are largely illiterate on the marketing aspects compared to the traders. In regulated markets, these commission agents are also called as 'A Class Traders'. These commission agents are of two types *viz.*, Kachha Arhatiyas and Pucca Arhatiyas.

- ✰ ***Kachha Arhatiyas:*** These commission agents render service on behalf of the farmers and sometimes they will provide loans to the farmers and put a condition that, the produce must be sold through him after harvest. They are small commission agents compared to Pucca arhatiyas. Their sphere of activity is purely local and they act mainly as middlemen or an intermediary between the primary producer (farmer) and buyer. His main business is to establish contact between farmer-seller and the buyer in the regulated market.
- ✰ ***Pucca Arhatiyas:*** These commission agents will render service on behalf of buyers or traders. In general, wholesalers and processors will approach them for purchasing the produce at remunerative price. They deal with a wide range of commodities and in large quantities compared to Kachha arhatiyas.

Functions of Commission Agents

The commission agents in RAMS perform the following functions.

- ✰ They will advance production loans to the farmers to an extent of 40-50 per cent of the expected value of the crop. So, they will act as bankers to the farmers. They charge 'commission' in addition to the normal rate of interest on the money they pay in advance.

- ☆ They will provide all advices to the farmers regarding purchase of inputs and sale of produce.
- ☆ They will provide empty bags/containers to the farmers to bring their produce to the market yards.
- ☆ They will have permanent establishment in the market yard and hence, they will provide accommodation to farmers and shelter to their animals.
- ☆ They will have owned godowns in the market yards thereby, they extend storage facility to the farmers.
- ☆ They will also provide transportation facilities to the farmers for bringing the produce to the market yards.
- ☆ They help the farmers in times of personal difficulties and personal needs.

b. Brokers

These agent middlemen are commonly seen in unregulated markets and they are totally absent in regulated markets because, 'open auction' is the common method of sale in the regulated markets. Unlike commission agents, brokers will not have any permanent establishment in the market and they will not take any physical control over the produce. They will simply move here and there in market yard and identify the potential buyers for the produce. Their main aim or job is to bring both buyer and seller on to the same platform for negotiating the price. They use their prior experience, skills, and connections to find farmers and potential buyers. They also use their market knowledge of current market value and produce availability to settle market price for the produce, which is acceptable to both farmer and buyer. For the service rendered by them, they will charge 'Brokerage' either from farmers or buyers. In some cases, they may even claim brokerage from both buyer and seller, depending upon the market situation and the services rendered. Thus, broker finds produce, finds a buyer and puts the two together. They ordinarily follow directions from their clients thereby, they enjoy little power over terms of sale or revenue collection. Further, they will not take any risk in the marketing transactions. So, commission agents enjoy more powers over physical handling of the produce, arrangement for terms of sale/purchase and collection of revenue from sale (Table 6.1). The brokerage is charged in two ways: a fixed price for selling each unit of produce and in terms of percentage of sales price. In the former case, the broker receives little incentive because, he gets the same regardless of price. But, in the latter case, the brokerage is in proportion of the sales price of the produce.

Thus, the role of commission agents and brokers is to facilitate distribution by bringing buyers and sellers together. Among these two, commission agents have close relations with particular farmers or processors or wholesalers and enjoy contract to sell on their behalf in return for a commission. Brokers, on the other hand, do not have intimate relations either with farmers or processors or wholesalers and they simply inform buyers of possible sellers and inform sellers of possible buyers. The clients use the services of a broker intermittently, since their supply of the produce into the market is intermittent. But, in case of commission agents, the farmers depend on them for loans right from the plough share point till transacting the produce through their shops in the market yards. So, commission agents normally represent buyers or sellers on an ongoing basis, whereas brokers are employed by clients for a short period of time.

2. Merchant Middlemen

These middlemen take title of the produce they handle *i.e.*, they will purchase the produce and own the produce. The buy and sell the produce for their own gain and derive their income from the margins arising from the sales (*i.e.*, difference between selling price and purchase price). Unlike other

Table 6.1: Differences between commission agents and brokers.

Item	*Commission Agents*	*Brokers*
Seen in	Regulated markets	Unregulated markets
Permanent establishment	Have permanent establishment in market yards	Do not have permanent establishment in market yards
Relationship with farmer and buyer	Enjoy close relationship	Less closer relationship compared to commission agent.
Physical control over the produce	Have physical control over the produce	Do not have any physical control over the produce
Fees	Commission is charged	Brokerage is charged
Services to farmers	Provide innumerable services to farmers like provision of finance for cultivating crops, arrange transport facilities, rest houses to farmers, gunny bags to farmers for transporting the produce, render advice to the farmers regarding purchase of inputs and sale of produce etc., besides guiding the client for efficient marketing of produce.	They do not provide such services and simply bring both buyer and seller on to the same platform for negotiating the price of the produce.
Representing buyers and sellers	Commission agents represent buyers and sellers on a permanent basis	Brokers represent buyers and sellers on a temporary basis.
Power over terms of sale	Commission agents enjoy more power in making arrangement for terms of sale/purchase and collection of revenue from transactions.	Brokers ordinarily follow directions from their clients thereby, they enjoy little power over terms of sale or revenue collection.

classes of middlemen, they hold uncertainty to a minimum *i.e.,* know what the buyfng and selling prices are going to be or should be. So, they are not risk takers. Merchant middlemen are of two types *viz.,* Wholesalers and Retailers.

a. Wholesalers

These are the merchant middleman, who purchase and sell the produce in large quantities. They buy and sell produce on their own account and thereby, ownership rights of the produce are transferred to them. They will purchase the produce either from farmers and other wholesalers and they will sell the produce to the processors or retailers. But, they will not sell the produce in significant quantities to the consumers. Thus, Merchant wholesalers are independent intermediaries that buy goods from farmers or processors or other wholesalers and sell to retailers and other Business to Business (B2B) customers. They sell the produce either in the same market or in other markets. They own godowns for the storage of the produce. They are a largely heterogeneous group of varying sizes and characteristics. One extreme of the more numerous groups of wholesalers are the local buyers or assemblers, who buy goods in producing areas directly from farmers and transport the same to distant markets. Another extreme of wholesalers is located in the large urban centres and they handle diversified (processed) products.

Wholesalers are sometimes referred as 'middle agents,' as they represent one of the important links in the supply chain along which, most goods pass on their way to the market place. As intermediaries between farmers and consumers of goods, wholesalers facilitate the transport, grading, storage and sale of produce ultimately destined for consumers. Wholesalers are extremely important in agri-business like in a variety of industries, as they are particularly vital to the operations of even for small retailers. They are successful, only if they are able to serve the needs of their consumers, who may be retailers or other wholesalers. Some of the important marketing functions provided by the wholesalers include:

- They assemble the produce from various areas
- They will store the produce during peak marketing season and offer the same in off-season
- They transport the produce to distant areas, where prices are high
- They will perform grading and standardization for the produce to fetch higher prices
- They will provide finance to the farmers to meet their production costs
- They will assess the demands of prospective buyers in the market.
- Provide producer's goods in an appropriate quantity for re-sale to buyers.
- Provide wider geographical access and diversity in obtaining goods.
- Ensure and maintain a quality dimension with the goods that are being obtained and resold.
- Provide cost-effectiveness by reducing the number of producer contacts needed (Figure 6.1).
- Provide ready access to a supply of goods.
- Assemble and arrange goods of a compatible nature from a number of producers for resale.
- Minimize buyer transportation costs by buying goods in larger quantities and distributing them in smaller amounts for resale.
- Maintain close association with producers to understand and appreciate consumerism in their production process.

b. Retailers

These are the merchant middlemen, who purchase the produce from wholesalers or processors and sell the same directly to the ultimate consumers. Thus, retailers engage in Business to Consumers (B2C) business. With the advent of internet, this B2C marketing is gaining rapid significance, as the firm is able to strike geographical barriers and achieve world-wide audience. So, a small retailer can enjoy same competitive advantage as their largest competitors (wholesalers) in the process of marketing. Retailers are often recognized as *'producer's personal representatives to the consumers'*. In the marketing channel, retailers are the closest to consumers and sell the requisite quantities of produce to the consumers (Table 6.2). They present the good to the consumer in the manner, location and form desired. They represent the most numerous types of agencies involved in the marketing process. In terms of undertaking marketing functions, their role in no easier compared to wholesalers. In fact, a retailer may have to do all the functions of marketing on his own *i.e.*, his job is complex. In the case of livestock, they may also carry out processing activities (*e.g.*, butchering). The retail sector includes a wide range of outlets such as merchants, dealers (say, fertilizers and pesticides dealers), departmental stores, supermarkets, smaller grocery stores etc.

Table 6.2: Differences between wholesalers and retailers.

Wholesalers	*Retailers*
They handle the commodities in large quantities	They handle the commodities in small quantities
They purchase the produce either from farmers or processors or other wholesalers and transact the produce in small quantities to retailers	They purchase the produce either from processors or wholesalers and transact the produce in requisite quantities to consumers
They usually deal with the raw produce or sometimes processed products, but not necessarily the form of product actually preferred by the consumers.	They deal with the form of product actually preferred by the consumers.
They do not deal with the consumers directly	They deal with the consumers directly
Wholesaler job is less complex compared to retailer, as he will not deal with the consumer's or end user's satisfaction directly. He can sell the produce to other wholesalers or processors or can export the produce outside.	Retailer's job is more complex compared to wholesaler, as his clientele is only consumer and he has to gain his confidence in making the business transactions.
He can enjoy profit margin over the large quantum of produce sold by him.	For retailer, profit margin on each unit of product from each consumer is important, as his transactions are in only requisite quantities, as desired by consumers.

Since, merchant middlemen take title to the produce, they assume certain risks and can suffer losses, if produce gets damaged or infested or stolen etc. At the same time, because they own the produce, they are free to develop their own marketing strategies including setting prices.

3. Processors

These middlemen carry on their business either on their own or on custom basis. Some processors employ agents to purchase raw material (agricultural produce) for them from the farmers or wholesalers in the market or producing areas, store the same and process it throughout the year on a continuous basis. Their role in marketing is to undertake some action on the products in order to change their form. Form changing is basically a marketing service. Processors may take active role in other institutional aspects of marketing. For example, they may act as own buying agents in the producing areas, wholesaling of finished products and promotion. Thus, processing is the only part of activities they get involved in. They also engage in advertising activity to create a demand for the processed products.

In the modern days of agri-business, processors are assuming crucial role, as the demand for processed products is increasing both in the domestic and international markets. They procure fresh agricultural produce from several sources. They may buy directly from farmers, assembly markets and wholesalers, and also from other processors. For example, to process fish into minced/ground fish, processors can buy fish fillet from other processors, who are engaged in filleting. Normally, small processors sell their products to retailers or even directly to consumers. Processors, who carry large volumes of products, usually sell their products to the wholesalers. In addition to processors who handle traditional products, several other processors operate cold storage facilities, which they either own or rent, for the purpose of exporting the processed perishable food products. These kinds of processors tend to trade large volumes and buy produce directly from farmers and assembly markets.

4. Speculative Middlemen

They are also called as Scalpers or Spreaders These middlemen take title to the product with a view to make a profit on it. They are not regular buyer or sellers of produce. They specialize in risk taking and work with a profit-motive from price movements. They buy at lower prices when arrivals are substantial in the market, stock the produce and sell in the off-season, when prices are high. So, they make a profit from the short run as well as long run price fluctuations. It is essential to note that, both buying and selling functions of speculative middlemen are not necessarily opposite to each other, unlike hedging. Thus, they are not interested in moving the produce down the marketing channel, but rather interested in short term price fluctuations. Thus, these middlemen derive their income from short term price fluctuations in produce they handle. Sometimes, they are also engaged in advertising activity to create demand for the products among traders and consumers.

The emergence and growth of speculative middlemen is due to the fact that, merchant middlemen are not willing to engage themselves in added risk involved in purchasing and storing of goods for longer period of time. Speculative middlemen play important role in marketing process in ensuring that, the commodities are made available from time to time. Their activities are desirable, especially if their expectations are met or true. Their speculating activities ensure more stable market prices for the produce.

5. Facilitating Middlemen

These middlemen are not engaged in sale or purchase of produce, but simply assist the marketing process. The process of marketing takes place even in their absence, but their presence ensures higher efficiency in the performance of marketing operations. These middlemen receive their income in the form of fees from those who use their services. The important facilitative middlemen include:

a. Hamalis

They perform loading and unloading of the produce, cleaning the produce, filling the same in gunny bags, stitching the gunny bags etc. They are considered as the *'hub of marketing wheel'*, as without their co-operation, marketing system will not function smoothly.

b. Weighmen

They perform weighment or measurement of produce and they receive their payment through commission agents in the regulated markets. Their function is so crucial in the market, as it influences the earnings of farmers.

c. Graders

They perform grading of produce based on the standards laid down.

d. Transport Agent

It facilitates movement of produce form one market to another.

e. Communication Agent

It helps in communication of marketing information such as market arrivals, prices, demand and supply information about the produce etc. The important agents include, Post and Telegraph, News papers, Radio, T.V., Internet etc.

f. Advertising Agent

It enable the prospective buyer to know about the quality of product and make purchase decisions. Thus, it aims at creating demand for the products among the traders and consumers. The important media include, news papers, radio, T.V., cinema slides, internet etc.

g. Auctioneers

They help in conducting the auction process in the market yard so that, effective communication takes place between buyers and sellers. The auctioneer is an employee of the regulated market. Neither the auctioneer nor his employer participates in buying or selling the commodity on their own account and they simply facilitate the buying and selling process. During the bidding process, the auctioneer invites bids for specific lots of produce and listens to the prices announced by various buyers. Later, in consultation with the farmer, the auctioneer invites the highest bidder to purchase the produce.

Importance of Market Functionaries

No matter how well a product meets the needs of consumers, without effective and efficient functioning of market intermediaries in the marketing channel, it is unlikely to succeed in the market place. Of course, in case of direct marketing, the absence of market intermediaries between farmer and consumers does not mean that, they are not required for marketing the produce. It all depends upon the nature of produce and the complexity of marketing functions that are to be executed before making the product ready for the consumer's satisfaction. For example, in case of vegetables and fruits, the direct marketing ensures 100 per cent benefits to the farmers. However, in marketing paddy, say by a small farmer, processor (rice miller) is an inevitable market functionary that must appear in between farmer and consumer. Further, the removal of market intermediaries from a marketing channel does not mean that, the entire savings are passed on to the farmers. Say, for example, if a farmer wishes to perform all the functions like packaging and packing, transportation, processing, storage, grading

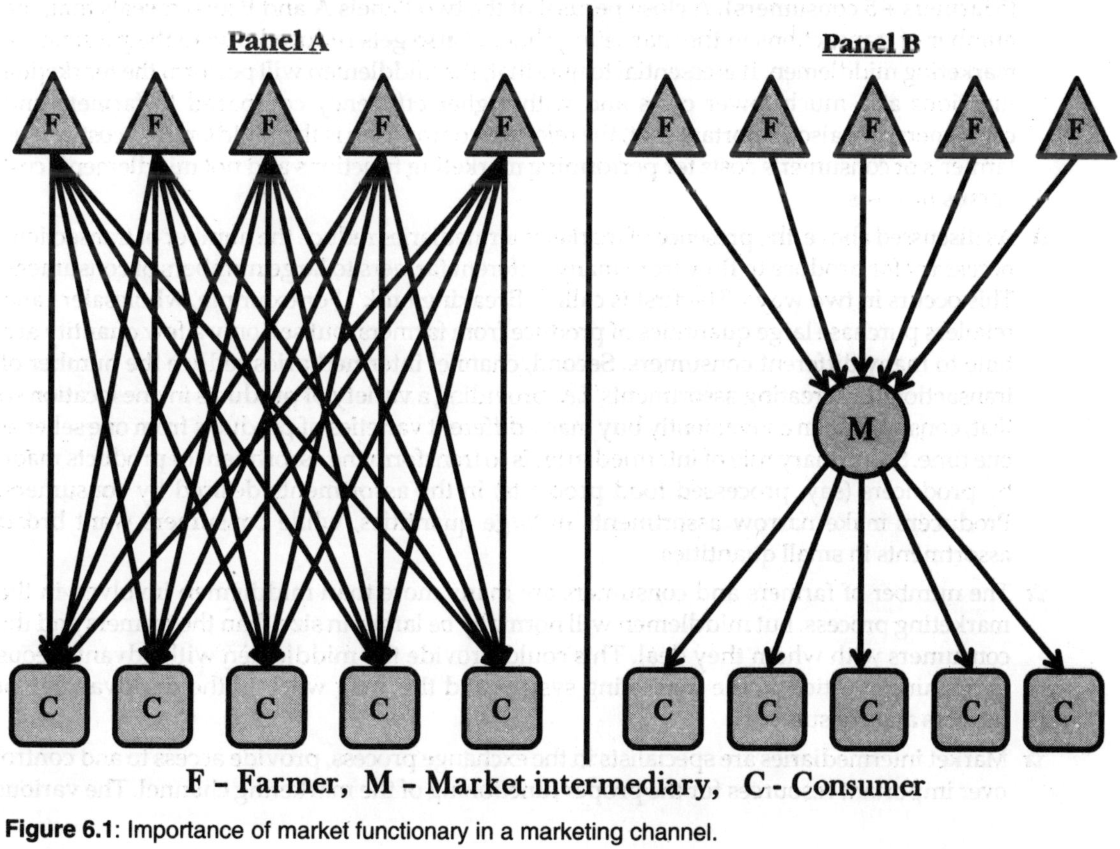

F – Farmer, M – Market intermediary, C - Consumer

Figure 6.1: Importance of market functionary in a marketing channel.

etc., he may be able to reduce marketing costs, but the inventory costs on the part of farmer may increase drastically and as a consequence, there are no savings and even the functions are performed less efficiently as compared to the presence of various intermediaries. The importance of market functionaries can be well-understood from the Figure 6.1.

The following are the aspects that justify the requirement of market functionaries in the marketing of agricultural produce:

- The presence of market functionaries improves exchange efficiency. There are certain costs associated with an exchange therefore, it is essential to reduce the number of transactions (exchanges).
- In the absence of market intermediaries (Panel A of the Figure 6.1), each farmer (F) sells directly to each consumer (C) and all of the marketing functions are performed by either farmers or consumers. That is, each farmer has to negotiate and exchange with each consumer (buyer) separately. So, the total number of contacts without intermediaries is 25 (5 farmers x 5 consumers). However, when there are specialized middlemen in between farmers and consumers, they perform the marketing functions, allowing farmers and consumers to specialize in other activities and sparing them search and transaction expenses. For example, with one intermediary (Panel B of the Figure 6.1), each seller (farmer (F)) negotiates with one intermediary (M) (as opposed to five buyers) and each buyer negotiates with one intermediary (M) (as opposed to five sellers). So, the total number of contacts with one intermediary is 10 (5 farmers + 5 consumers). A close perusal of the two Panels A and B also reveals that, the number of transactions in the marketing channel also gets reduced due to the presence of marketing middlemen. It is essential to note that, the middlemen will perform the marketing functions at a much lower costs and with higher efficiency compared to farmers and consumers. It is also important that, the relevant comparison is the middlemen's cost versus farmer's or consumer's costs for performing marketing functions and not middlemen's cost versus no costs.
- As discussed above, the presence of market intermediaries reduce the number of transactions necessary for produce to flow from many different farmers to large numbers of consumers. This occurs in two ways. The first is called 'Breaking bulk'. For example, wholesalers and retailers purchase large quantities of produce from farmers, but sell only a few quantity at a time to many different consumers. Second, channel intermediaries reduce the number of transactions by 'creating assortments' *i.e.,* providing a variety of products in one location so that, consumers can conveniently buy many different varieties of products from one seller at one time. So, primary role of intermediaries is to transform the assortment of products made by producers (say, processed food products) in the assortments desired by consumers. Producers make narrow assortments in large quantities, while consumers want broad assortments in small quantities.
- The number of farmers and consumers are many more than middlemen involved in the marketing process, but middlemen will normally be larger in size than the farmers and the consumers with whom they deal. This could provide the middlemen with advantageous bargaining position in the marketing system and this may work to the disadvantage of farmers and consumers.
- Market intermediaries are specialists in the exchange process, provide access to and control over important resources for the proper functioning of the marketing channel. The various

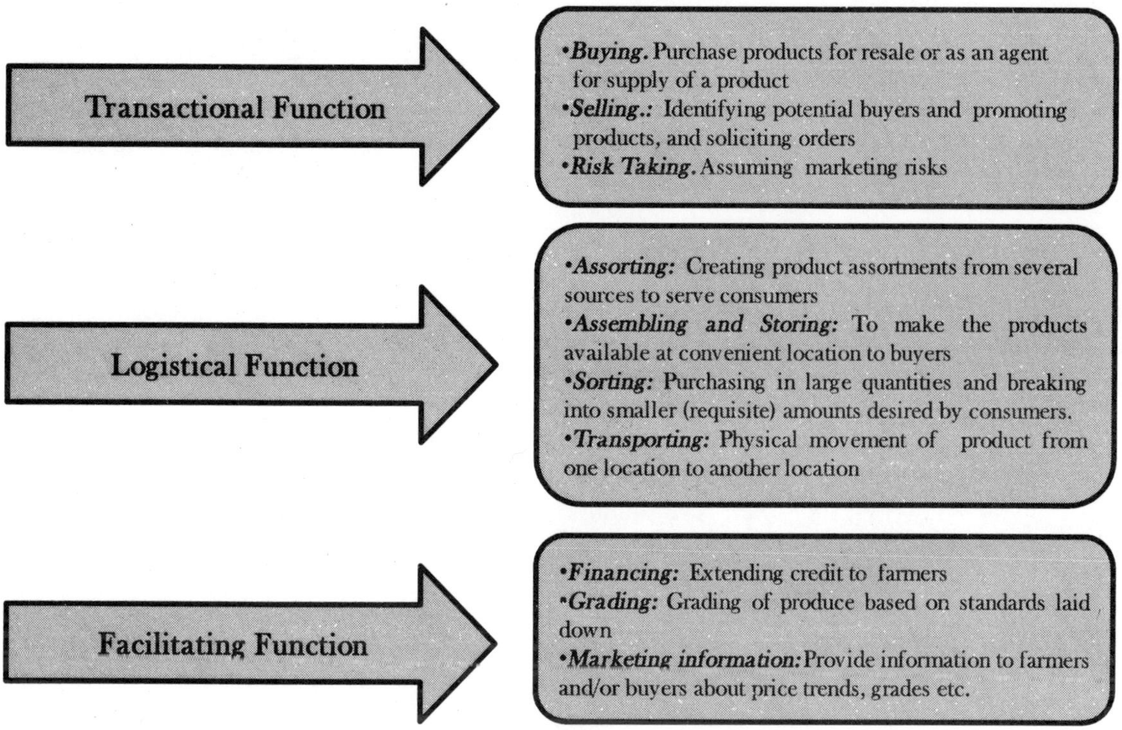

Figure 6.2: Functions performed by market intermediaries.

functions performed by market intermediaries can be conveniently categorized, as shown through Figure 6.2.

- The functions of different market functionaries highlight the concept of *'Division of labour'*.
- A well accepted maxim in agricultural marketing is that, *'you can eliminate the middlemen, but you can't eliminate their functions'*. This implies that, if market intermediaries are eliminated from the channel, then someone else would have to assume the tasks (either farmer or consumer). Thus, functions can be shifted and shared among channel members, but cannot be eliminated. However, if others share these functions, they are performed with less efficiency and they have to bear the burden of heavy inventory costs.
- A farmer will engage an intermediary, when he believes that, the intermediary can perform the function(s) more economically and efficiently than him.

However, in the modern days of agri-business, with the advent of internet, radical changes are noticed in distribution strategies. We know, in food processing industries, most products are mass-produced and in most cases, end users do not obtain products directly from manufacturers. However, with the advent of internet technology, the consumers can come in direct contact with the processors (through e-mail) and this facilitates direct transfer of products. So, the need for intermediaries that physically handle the products may become largely obsolete. Thus, many traditional intermediaries are already being eliminated from the distribution channels and this process of removal of intermediaries is termed as Disintermediation. This cut the unnecessary marketing costs and improves the efficiency of the marketing channel.

7

Present Agricultural Marketing System (PAMS) in India

In the earlier pages, we studied that, agricultural marketing involves a series of activities involved in moving the produce from the point of production to the point of consumption. It also refers to the study of all the activities, agencies and policies involved in the procurement of farm inputs by the farmers and the movement of agricultural produce from the farmers to the ultimate consumers. From the producer point of view, it is important to know, whether the prices prevailing in the market enable him to continue to produce or not, and what he should produce and where and at what time he should sell it. A consumer looks at marketing from the point of view of commodities and the prices at which they are offered. Middlemen try to increase their profit margin by discharging various marketing functions.

We know, agriculture is fundamentally different from manufacturing industry (Table 2.1) and marketing of agricultural produce is beset with a number of complex problems. The farmer is therefore, generally handicapped in efficient marketing of produce. That is, he has to sell his produce at an unfavourable place at an unfavorable time and on unfavorable terms. However, before going into these details, it is essential analyze the features of an 'ideal' marketing system. Though this form of perfection is not seen in the market economy, it is essential to discuss the same considering the following grounds:

- The analysis of an ideal marketing system serves as a backdrop to locate the loopholes in the PAMS.
- Studying the ideal marketing system assuming all perfections in a market guides us to clearly understand the complicated imperfections involved in the actual market situation.
- The analysis of imperfections in PAMS helps us to study the extent of imperfections and degree of inefficiency as compared to an ideal market system.

a. Features of the 'Ideal' Marketing System

An ideal marketing system is one that maximizes the long run welfare of society. That means, an ideal marketing system ensures efficiency in the entire marketing process of a commodity. That is, in performing each and every marketing function, efficiency is to be increased and its cumulative maximum of all the marketing functions implies an ideal marketing system. For example, in performing physical functions like packaging and packing, transportation, processing and storage, efficiency will be achieved by carrying all these functions such that, highest income is achieved per unit of cost incurred on them. Similarly, an ideal marketing system must allocate agricultural products over time, space and form to intermediaries and consumers in such a way that, no other method of allocation would make them better off. To achieve this condition, prices realized in the marketing system must be efficient and must equal to the marginal cost (M_C) of production and marginal consumer utility. The following are the important requisites to achieve an ideal marketing system:

- It ensure highest possible FSCR *i.e.*, to reduce the price difference between the producer and ultimate consumer to the maximum possible
- It makes available all types of products of farm origin to consumers at reasonable prices without impairing on the quality of the produce.
- There should be free interaction between demand and supply forces in the determination of market prices of produce. No Government interference should be there in fixing the prices. If the Government intervenes the market in terms of imposing restrictions regarding the movement of agricultural produce, restrictions on the quantity to be processed or on the construction of processing plant, price supports, rationing, price ceiling, entry of persons in the trade etc., inefficiency in the market system creeps in and commodities pass into the black market. They are not then easily available at the fair prices.
- All the marketing functions must be carried out on scientific basis, so as to ensure both physical and pricing efficiencies.
- The marketing system should be dynamic such that, it is consistent with the tastes, preferences and income levels of the consumers.
- The marketing system must show positive influences on strengthening agri-business sector and overall development of the rural sector.
- It must offer livelihood security to vast majority of rural people involved in marketing and agri-business activities.
- Prices of produce must be commensurate with its quality (grades) and all types of grades of produce must be available in the market, so as to meet the needs of different categories of consumers.
- Farmers and other stakeholders must have ready-made information about the prices and quality of produce available in the market so that, they have options regarding when and where to sell their produce.

- The marketing system must serve the basic objectives of economic development of the country namely, output growth, price stability and poverty alleviation.
- It must ensure optimization of input use and output produced.
- It must guarantees better prices to farmers for their farm produce and induce them to invest their surpluses in the purchase of modern inputs so that productivity may increase. This again results in increase in the marketed surplus and incomes of the farmers.
- It must widen the market for products by taking them to remote consumers of the country to areas far away from the production point. For example, paddy produced in Andhra Pradesh and Tamil Nadu is also sold in remote tribal areas. This widening of the market helps in increasing the demand on a continuous basis and thereby guarantees a higher income to the farmer.
- It must help the farmers in the adoption of new scientific and technical knowledge both in the areas of production and marketing.
- It must aim at diminishing the poverty of agricultural population, reducing consumer food prices, earning more foreign exchange and eliminating economic waste.
- Ideal marketing system remedies the imbalance in the supply of the surplus produce to the deficit areas.
- It ensures remunerative prices to the farmers and a reduction in the Mc and MMs, so as to provide commodities to the consumers at reasonable prices and promote the movement of surpluses for economic development.
- It must generate employment in the rural economy and this is necessary in the light of the faster falling share of agriculture in the GDP as compared to the slower falling rate of people's dependence on agriculture.
- It must change the crop rotation in favour of more nutritious, more valuable and environmentally sustainable agricultural crops.
- It must increase the net income of the farm sector to make it self-sustainable and attractive.

Any deviation from the above essentials of an ideal marketing system leads to inefficiencies in marketing system. The PAMS in India is a good example for these inefficiencies, as it shows a number of deviations from the above essentials discussed under ideal marketing system. So, it infers that, the PAMS in India suffer from several defects or constraints or limitations and they are discussed here under.

b. Defects in PAMS in India

They include:

- ***Illiteracy and ignorance of the farmers:*** Majority of the farmers are largely illiterate and hence, they are ignorant of modern marketing techniques and methods in transacting the produce. Moreover, the farmers are not aware of current prices of produce in different markets and also about demand for their produce in the market.
- ***Prior indebtedness of the farmers:*** Generally, the farmers will take the loans either from the financial institutions or from private money lenders for crop cultivation purposes. But, they mostly avail loans from private money lenders even at high rates of interest because, they will provide loans both in terms of adequacy and in right time. So, immediately after

harvesting the produce, the farmers will be forced to sell their produce in the villages itself to clear off their prior debts. Sometimes, even the farmers will resort to distress sales for clearing their prior debts and for getting liquid cash to meet their family living expenses and to start the next crop season.

☆ ***Post-harvest immediate sales by farmers:*** It is estimated that, in India, on an average, nearly 65 per cent of the produce is sold in the villages itself. About 60 to 80 per cent of the food grains are still marketed in the first quarter of the harvest season. Majority of the farmers, especially small and marginal farmers resort to village sales and even distress sales of produce due to the following reasons:

- Immediate requirement of liquid cash to meet pressing family needs
- To clear prior debts
- To pay wages to agricultural labour
- Lack of proper transportation and storage facilities
- High perishability of agricultural produce
- Fear of loss of produce by fire, theft or other uncertainties
- Low risk bearing ability of the farmers.
- Difficult to carry the produce in bullock carts to the markets, which are often situated at long distances.
- Small quantity of marketable surplus due to sub-division and fragmented land holdings.
- Farmers may purposefully avoid selling in distant (big) markets, as they lack adequate knowledge about prevailing market practices, possibility of cheating them by traders etc.

☆ ***Low marketable surplus:*** At farmer's level, the marketable surplus is low because of sub-divided and fragmented land holdings. As a result, the farmer could not sell much of produce in distant markets and hence, he derives low income from the village market.

☆ ***Agriculture the unorganized sector:*** Unlike the industrial sector, the agriculture sector comprising of millions of scattered land holdings remains unorganized and is not conducive to collective bargaining. This is even more applicable to small farmers, who lack bargaining capacity because of low marketable surplus. Small farmers, being generally less educated, are often serviced by agencies which exploit their lack of knowledge. There is need for small farmers to get organized. However, sometime small farmers are in competition with each other which prevents their coming together for a common interest. Often caste loyalties also prevent such unions.

☆ ***Lack of farmers organization or CMSs:*** Due to lack of farmers organization and CMSs, the farmers will have only less amount of marketable surplus. Further, they have to market their produce individually. Because of this, they have low bargaining power and they have to deal with traders having strong organization. They cannot therefore, insist on a competitive price for their produce and they silently watch as the open auction takes place. The following are the important reasons for lack of organization among the farmers:

- Caste feeling among the farmers
- Locational disadvantages and difficulty in bringing them under one organization.

- Difference in the size of holdings and the surplus available with the farmers.
- Marketing aspect is not given due importance by the farmers, because of their ignorance.

☆ ***Drastic increase in cost of production of commodities:*** On one side, the cost of cultivation of crops is increasing due to increase in prices of inputs, but on the other side, the yield is more or less stable. As a result, the cost of production of commodities is increasing significantly thereby, adversely affecting both domestic and export competitiveness of commodities. Further, the input prices are too high in relation to the output prices. The output price fluctuations are excessive and this in addition to high and unjustified marketing levies escalates the farmers' problems.

☆ ***Heavy village sales of agricultural commodities:*** A majority of the farmers in India sell a large part of their produce in the villages itself, which result in low returns for their produce.The village sales is 20 to 60 per cent in the food grains, 35 to 80 per cent in cash crops and 80 to 90 per cent in perishables. The following factors are responsible for these village sales:

- Prior indebtedness to local money lenders or traders or landlords. So, the farmers are often forced either to enter into advance sale contract or sell the produce to them at low prices.
- Difficult to transport the produce to distant markets.
- Small size of marketable surplus
- Perishability of the produce and lack of scientific storage facilities.
- Lack of awareness of farmers about modern marketing techniques.

☆ ***Technological advances in agricultural production:*** The recent technological advances in agricultural production also aggravated the problems in agricultural marketing. For example, by using paddy harvesters, paddy grains with high moisture content will be harvested, by mechanical picking of cotton kapas, the cotton kapas is mixed with trash, by using potato diggers injuries are noticed on potatoes etc. These problems led to decline in prices of the commodities.

☆ ***Poor quality of produce:*** The produce offered by the farmers in the market is generally of poor quality due to absence of grading, pests and diseases infestation, deliberate adulteration etc. For example, cotton, potatoes, grapes etc., do not enjoy good reputation in the international market on account of SPS standards imposed by the importing countries. As such, the cash returns to the farmers and exporters are low.

☆ ***Inadequate market finance:*** As discussed above, most of the farmers resort to village sales and even distress sales of agricultural produce. So, to safeguard the farmers from such forced sales, adequate market finance (pledge loans) is to be provided by the Government so that, they can wait for better times and better prices. Even though, such credit facility is available, it is not adequate and not even timely (because of stringent procedural formalities), thereby, the farmers are forced to take loans from private money lenders and commission agents, while agreeing to pledge their produce to them at less than market prices. These non-institutional sources are extremely expensive sources of funds and often lock the farmer-borrower into a never-ending cycle of debt. The non-institutional sources provide marketing loans at a high rate of interest to the farmers. This naturally leads to a rise in the cost of marketing with the ultimate result that, the share of the price received by the farmer in the consumer's rupee is correspondingly reduced. The immediate requirements of liquid cash compel the farmers to go for distress sales. It is to be noted that, the provision of market finance in terms of adequacy, in right time and the flexibility in its repayment makes the private finance more attractive to the farmers despite high rate of interest. Though Government has taken several measures to provide cheap institutional market finance, a large majority of farmers are still dependent on private money lenders and commission agents in obtaining credit, because of the said advantages. It may be noted that, the activities of private money lenders and commission agents through money lending curtail the freedom of the farmer-sellers to dispose off their produce profitably in market yards and make them permanently indebted. Thus, the inadequate access to institutional market finance leads to limited bargaining power on the part of the farmers and forces them to sell their produce just after harvest, when the prices are low. This linked to asymmetric market and price information also hinders small and marginal farmers from realizing remunerative prices for their produce. So, the Government has to realize the gravity of the situation and take effective steps to release the poor and innocent farmers from the clutches of the private money lenders and commission agents.

- ☆ ***Poor credit flow to agricultural sector:*** Poor credit flow has been one of the major bottlenecks in the country's agricultural system. Most of the lending to the agriculture sector is short term, and is in the form of crop loans. Farmers are expected to buy all the required inputs to raise a crop with the credit made available to them and then return all loans soon after harvest. As discussed above, post-harvest credit from banks (market finance) is usually not available in adequate quantity. Given the importance of a good credit history, most farmers, especially small and marginal farmers, take the produce to market soon after harvest. Most farmers face the same objective function and produce is brought to market simultaneously by several farmers. As a result of the temporal supply conditions, prices will fall. Often such fall in prices leave little surplus for the investment in next season's crop.
- ☆ ***Inadequate access to markets:*** This is an area, where farmers are handicapped in general, and small farmers in particular. It is generally perceived that, agricultural marketing begins with post-harvest operations. Indeed with the crop standing in the field, marketing affects field operations too. It influences harvest timings, maintaining market preferred quality, value addition and treatment of produce from marketing angle, packaging preferences, quality standards, minimizing transport cost as well as transport losses, use of cold storage facilities, marketing mechanism etc. Knowledge and adoption of these practices are very important for farmers to market their produce to earn reasonable profit. Apart from farmers, extension functionaries also need to undergo training on market-related skills. However, the lack of these practices/skills are rampant among both farmers and extension functionaries.
- ☆ ***More number of middlemen:*** Between farmer and consumer, more number of middlemen are involved in transacting the produce. As a result, the FSCR will be low. The important intermediaries in the agricultural marketing system are village traders, kachha arhatiyas, pucca arhatiyas, brokers, processors, wholesalers, retailers etc. Each of these intermediaries take their own margin and thereby, diminish the returns of the farmer. Since, there is no restriction from the Government regarding the entry of market middlemen in transacting the agricultural produce, there may be any number of middlemen between farmer and consumer. So, with increase in the number of market middlemen in the marketing channel, the Mc and

MMs go up. Hence, farmers receive low price, while consumers pay high price for an equivalent quantity of the commodity.

☆ ***Malpractices by middlemen:*** As the farmers are largely illiterate and ignorant of modern marketing techniques, they have to depend upon commission agent and brokers for selling their produce. These middlemen as well as merchant middlemen take advantage of the ignorance and illiteracy of the farmers and use unfair means to cheat them by imposing unauthorized market charges, drawing large number of samples from the produce, improper weighment of produce, quoting low price for the produce by saying that the produce is of inferior quality etc. There are all kinds of arbitrary deductions for religious and charitable purposes. The burden falls entirely on the farmer and he has no effective means of protest against such practices. So, due to improper market structure, traders or middlemen have become all powerful. They have bent the rules in such a way that, it is possible for them to cheat and get away. Moreover, the unorganized farmers and the market machinery are no match to the powerful traders legally. Even in APMCs, middlemen resort to several malpractices.

☆ ***Multiplicity of market charges:*** As soon as the farmer enters into market, he will approach either a commission agent or broker for marketing his produce. They will charge commission or brokerage respectively for the services rendered by them. Besides these charges, the farmers have to pay for hamalis for unloading the produce in market yard, for weighment of produce, for cleaning produce, for fixing grade to produce etc. APMCs also will collect fees from the farmers towards maintaining charity fund for keeping the market yard in hygienic condition, for maintaining farmer's rest house, providing shelter to farmers animals etc. The rates of these charges also vary from market to market. The other charges like octroi duties (it is the tax paid by the heavy vehicles like trucks carrying goods and the State Government levies the octroi charges, when the product enters the state), toll taxes etc., are also paid by the farmer, while transporting produce to the market.

☆ ***Lack of proper transportation facilities:*** The transportation system, which consists of roads network, railways, air transport, ports and shipping services is not only inadequate, but also very costly for the transportation of the bulky agricultural produce from one place to another. Only a small number of villages are joined by railways and pucca roads to mandies. So, majority of the farmers transport the produce through traditional slow moving bullock carts. But, this mode is unfit to transport the produce to distant markets and totally unfit for highly perishable produce. So, the farmers have to dump their produce in nearby markets, even if the prices obtained in these markets are considerably low. Thus, due to lack of proper transportation facilities, the farmers are selling the produce in villages itself and they could not transport the same to the distant markets, where the prices are high. Thus, the transportation costs on the part of the farmer are very high and nearly 30 per cent of the total value of the produce or CPP is due to transportation costs only. Moreover, road conditions from remote village to mandis are often extremely poor and defective. These conditions not only add to the loss of produce during transportation and aggravate the strain on bullocks and other pack animals, but also lead to the multiplication of small dealers and intermediaries. They also restrict market by hindering cheap and rapid movement of agricultural produce.

☆ ***Lack of processing facilities:*** Due to lack of adequate and scientific processing facilities, the farmers prefer to sell fresh produce only. But, the value of fresh produce is low when compared to processed products. Moreover, in the international market, processed products are more preferred over fresh produce.

☆ ***Lack of scientific storage facilities:*** Due to lack of scientific storage facilities, the farmers are subjected to distress sales of agricultural produce. Further, they are compelled to store their produce in kachha storage structures. The indigenous methods of storage adopted in the villages as well as in other markets do not adequately protect the produce from dampness, pests and diseases, extreme high and low temperatures, rodents, birds etc. The losses due to inadequate storage have been estimated to range from 1.5 to 2 per cent (Food grains Investigation Committee). A recent estimate puts the loss at from 5 to 15 per cent by weight of the production and it is due to defective storage. Even at five per cent level, the loss of cereals, millets, spices, oilseeds, jute, cotton, tobacco would come to over ₹ 4,000 million every year in India. Further, the networks of CWC and SWCs are limited to major market centers only and they are not extended up to the villages. Due to this reason, supply in the village market increases substantially and the farmers are not able to get a fair price for their produce. This also results in high prices of produce during off season.

☆ ***Lack of proper standardization and grading facilities:*** Due to lack of proper standardization and grading facilities, the farmers are selling the produce as a single lot (Dara method of sale), which generally fetches low income to farmers. The quality consciousness among farmers is low and this led to decline in export potential of major agricultural commodities from India. Further, the graders employed in the market yards are not adequately trained thereby, the standards and grades fixed by them do not show much favourable impact on pricing of commodities. A large number of farmers have little knowledge about grading and they are even penalized by traders for the existence of a small percentage of poor quality produce in the lot.

☆ ***Lack of proper communication and marketing information facilities:*** Due to lack of proper communication and market information facilities, it is often not possible for the farmers to obtain information on exact market prices prevailing in different markets. So, they accept, whatever price the traders offer to them in the local markets. Generally, the farmers depend upon oral information about market conditions, market arrivals, demand conditions, ruling prices and market trends etc., from co-farmers or commission agent or their neighbours, which may be manipulated by them for vested interests. They have practically no contacts with the reliable sources working in APMCs. Even in cases, where information on prices is available, they are not comparable on account of the lack of standard grades acceptable to the whole country, inaccuracy of information supplied by various agencies concerned and variations in weights and measures used in several markets in the absence of standardization of weights and measures. Further, the price quotations provided various sources are sometimes not reliable and suffer a great time-lag. The traders exploit the same and generally offer less than the price quoted even by the Government news media.

☆ ***Lack of proper weighment facilities in market yards:*** Due to lack of proper weighment facilities in the market yard, the farmers are often cheated several times by marketing middlemen. As a result, the farmer gets low income from the produce.

☆ ***Multiplicity of Weights and Measures:*** Till recently, there had been an absurd multiplicity of weights and measures in India. This multiplicity of weights and measures offers bright scope for the marketing middlemen to cheat innocent and ignorant farmers. They further make the consumer to pay an unnecessarily high price for produce. Weights made of sticks, stones and bits of old iron are common features in the markets and villages.

- ☆ ***Adulteration:*** Adulteration of produce ensures benefits to the traders in two ways *viz.*, a fixed deduction in actual produce is done to add impurities and the cost of added impurities is far less than the value of produce deducted. It is estimated that, in case of wholesale markets, a trader offering the produce containing five per cent impurities fetches the same price as the producer offering cleaner produce containing only one per cent impurities. Various devices of adulteration are in vogue. For example, tomato sauce is often only a mashed pumpkin with a small percentage of tomato, vinegar in some cases has been found to be acetic acid, small chippings of white stone have been found in rice bags in some localities, common salt packets contains large-quantities of white chalk, turmeric powder is adulterated with lead chromate which has a deep yellow colour, lead oxide is added to red chillies to brighten the colour and add weight and so on.
- ☆ ***Variations in market taxes and fees:*** There is a considerable variation in the structure of taxes and fees on the agricultural produce in various States, which also distorts the operation of the domestic market, gives wrong signals to the farmers and leads to considerable fallacy in the efficiency of the operation of private trade vis-à-vis farmers co-operatives and public agencies.

Thus, the PAMS in India is not well-organized and the farmers have to depend largely on the middlemen for the disposal of their produce thereby, traders have no hesitation in taking advantage of the farmer's dependency upon them. The real evils are the lack of adequate infrastructure in the market yards coupled with the tendency of these intermediaries to exploit the ignorance and helplessness of the farmers to increase their own profit. Market infrastructure in the country is more regulatory than facilitative. It provides no help in direct and free marketing, organized retailing, smooth raw material supplies to agro-processing firms, competitive trading, information exchange and adoption of innovative marketing technologies. The linkage between spot markets and future markets is poor due to domination of speculators. Marketing support provided by Government is more reactive than proactive. As a result, most farmers dispose off their produce through middlemen/ traders and do not get a reasonable price for their produce. The AIRCSC (1954) described this situation as follows: *'while standards of marketing have improved, in most of the relatively few regulated markets which have been established, a number of malpractices still exist even in them since personnel and enforcement are two great problems, not always sufficiently attended to, much less solved'. Sometimes, the malpractices take fresh lease of unauthorized life just outside the market, for the private interests are strong, the advantages of evading strict regulation are many and the producer is in no position to seek eventual advantage and protection from law at the cost of the immediate disadvantage involved in the loss of powerful customers, which are also sources of credit and finance. Moreover, there is a very great lacuna that no control at all is exercised over village sales, in which the primary producer is literally, legally and in practice at the mercy of the village trader'*. Thus, farmers are not conversant with the complexities of the marketing system, which is becoming more and more complicated. They are handicapped by several disabilities as sellers. They sell their produce at an unfavorable place, time and price.

c. Suggestions to Improve PAMS in India

We know, the agriculture sector needs well-functioning markets to drive growth, employment and economic prosperity in rural areas of India. But, the above discussion infers that, there are several challenges involved in marketing of agricultural produce. In executing each and every function of the marketing process, there are several defects. Further, there are multiple channels of distribution that eats away the pockets of both farmers and consumers. The Government funding of farmers is still at nascent stage and most of the small and marginal farmers still depend on the local money lenders, who charge high rates of interest. There are several loopholes in the present legislation and hence, the traders are exploiting the farmers at every stage of marketing. This situation turned much complicated with the globalization and trade liberalization of agricultural marketing system and now, it is not possible to view the PAMS in India in isolation. Keeping in view of these, the following policy measures need to be implemented for revamping the PAMS in India:

- ***Agricultural Marketing Extension, Training and Research:*** Agricultural marketing is witnessing major changes owing to globalization and liberalization of markets. In this context, agriculture has to be market driven, more cost effective, competitive, innovative and responsive to high tech and IT applications. Training and extension systems in agricultural marketing will have to sensitize and orient the beneficiaries to respond to these challenges. It is necessary to build capacity of each of the beneficiary group namely, the farmers, market functionaries and other officials involved in the agricultural marketing activities. Knowledge has to be imparted at the grass root level in areas such as, market driven production program, the post harvest management of agricultural and horticultural crops, availability of market finance, information on facilities for quality assurance and standards, grading, packaging, storage, transportation, contract farming, direct marketing, alternative markets including Forward and Futures markets, online market information system etc. Training and education modules will have to be prepared in these areas for reaching the region specific farmers in vernacular languages. The main objective of imparting training to marketing functionaries and stake holders should be to create an ambiance of Good Marketing Practices in the country to promote the interests of farmers as well as consumers. In 1962, DMI has taken initiative by establishing an 'Extension cell' through which trainings will be given to the farmers in the field of agricultural marketing. At present, State Agricultural Marketing Departments, SAMBs, CWC, SWCs, SAUs, line departments of agriculture are playing vital role in providing training to the farmers. These extension strategies make the farmers to realize the importance of adopting new techniques of production and in making efforts for getting more income and improving their standards of living. As a consequence, the cropping pattern is no longer dictated by what they need for personal consumption, but what is responsive to the market in terms of prices received by them. So, suitable policies, regulatory frameworks and incentives are essential for designing trainings and studies for marketing reforms to provide level playing field across different size groups of farmers in the context of globalization. In order to strengthen and develop agricultural marketing system in the country, efforts in the area of training and extension have to be made at three levels. At policy level, it is necessary to formulate an effective policy on agricultural marketing under which, various components of marketing programmes and activities can be integrated and coordinated. At managerial level, the managerial and technical capabilities of those technical institutions involved in the implementation of the marketing policy needs to be improved to enable the work to be carried out more efficiently and economically. At farm level, marketing policies and programmes should assist farmers in marketing their produce, purchase their inputs and raise their incomes. At farm level, however, the effectiveness of Government policies and programmes may often loses its impact, especially for those farmers whose marketable surplus is too small and not oriented to marketing requirements.

  In India, although at village and farm levels, agricultural production extension services exist, but marketing extension work designed to benefit farmers and other market functionaries does not exist. Marketing extension has not been accorded due importance it deserves in the light of shifting agricultural policies towards sound marketing system. Consequently, a number of States do not have even separate marketing extension cell to undertake regular extension activities. In the absence of emphasis on marketing extension, technology transfer in the field of agricultural marketing has been sluggish. As the days of the mass production and mass marketing are now being replaced by customer based or market driven strategies, an effective marketing extension service is need of the hour. This has added significance in the light of post-WTO scenario. If the Indian farmers have to be competitive both in domestic and international markets, the marketing extension would be an effective instrument to safeguard farmers' interest through proper education and guidance on regular basis. The marketing extension service to assist small and marginal farmers in solving the problems faced in marketing their produce is therefore, a *sine-qua-non* in the free trade environment. The marketing extension envisages advise on product planning, marketing news and information, securing markets for farmers, advise on alternate marketing, advise on improved marketing practices, advise on establishing and operating markets etc.

Considering the limited reach of public extension service, it is felt that privatization of extension services with appropriate financial backup from the public sector is considered more appropriate and practical. Privatization of extension activities would facilitate tailor made extension services beneficial to both farmers as well as entrepreneurs. The NGOs, co-operatives, trade associations, private firms and corporate bodies should be allowed and encouraged to undertake marketing extension. To facilitate private agencies to undertake marketing extension programmes on regular basis for the country as a whole, a 24 hours TV Kisan Channel on Doordarshan is necessary. There are 75 channels operated by Doordarshan, out of this one channel should be devoted exclusively to the farmers service. The Kisan Channel would be best visual media to educate farmers by both public as well as private agencies in agricultural production as well as agricultural marketing.

A substantial training efforts will be needed to upgrade technical, managerial and organizational skills and knowledge of field and marketing personnel. The major thrust areas of the training in the changing scenario should be on the following lines:

- Agri-business management
- WTO and its implications
- Post-harvest management
- Grading, Standardization and Quality assurance
- IT
- Emerging areas in agricultural marketing
- Organic Food Marketing
- Market led extension
- Post-graduation courses in SAUs
- Entrepreneurship programmes for rural Agri-clinics.
- Training of Trainers
- Legal Reforms
- Pledge Financing
- Warehousing

- Packaging
- Storage and Cool Chain
- Direct Marketing
- Contract Farming
- SPS Measures
- Forward and Futures Market

In view of the changes in agriculture in general and agricultural marketing in particular, there is need to orient the faculty of the national level nodal institutes. Until and unless the faculty is exposed to the complex international problems and ground realities of Indian fields, it is not fair to expect solution of all problems from them. At the same time, they should be trained involving practical problems of farmers in the fields and marketing of agricultural produce. The skills should be developed in management, planning and leadership.

☆ ***Loans from financial institutions:*** Farmers should be encouraged to avail marketing loans as far as possible from the financial institutions at lower rates of interest so that, they can take adequate time for efficient marketing of produce. Further, from the Government side, the disbursement of loans towards market finance should be enhanced for efficient marketing of produce by the farmers. On the other side, the farmers should also be educated for availing market finance facility instead of resorting to forced sales in the villages. The pledge loans given for a period up to six months to the farmers availing crop loans up to ₹ One lakh are reckoned as priority sector advance. Banks are financing wholesalers and other traders, but they do not fall under agricultural finance. Loans for construction / running storage facilities are already included under priority lending. It is suggested that, the existing limit of priority sector advance should be revised upward suitably taking into consideration innovations in agricultural sector in the post-WTO regime. The present credit flow to the agricultural marketing sector is very meager. The banking environment and lending policies and programmes for financing is not found conducive for the increased capital needs of agricultural marketing. Therefore, Government is requested to design full-fledged agricultural marketing credit policy, considering the requirement of increased production, market innovations, technologies and socio-economic changes with specific reference to post-WTO regime.

☆ ***Establishment of Co-operative Farming Societies (CFSs) and CMSs:*** Government should promote CFSs and CMSs so that, the farmers can increase their agricultural production and also enjoy the benefits of sale of large scale marketable surplus. This is because, with the commercialization of agriculture, efficient marketing is as necessary as scientific agricultural production and these two must go side by side. The income of the farmer mainly depends on his ability in transacting the produce at a fair market price. Even if more output is produced through improved production methods, the farmer would not gain much, if there is no proper arrangement for the marketing of his produce, as the benefits of better farming would probably be reaped by middlemen intervening between him and the ultimate consumer. So, both CFSs and CMSs must be strengthened to ensure the farmers realize the potential benefits of large scale production and large scale marketable surplus respectively.

☆ ***Strict supervision by APMC officials:*** In APMCs, strict supervision should be done by the Officials regarding transactions of produce in the market, method of sale, proper weighment of produce etc., such that, the malpractices resorted by traders and other intermediaries can be effectively checked.

☆ ***Checking the multiplicity of market charges***: The market charges incurred by the farmers in the market yards should be effectively checked by the strict supervision of APMC Officials. Co-operative marketing and direct marketing of produce should be promoted in the market yards, to check the multiplicity of charges in the markets so that, the FSCR can be increased.

☆ ***Promoting cost effective production and marketing technologies***: The Government should disseminate cost-effective production technologies among the farmers so that, the competitiveness of commodities will be increased both in domestic and international markets. This concept is gaining more significance especially in the context of WTA. Similarly, cost effective marketing technologies should also be promoted so that, the transaction costs can be greatly reduced, which indirectly increases the income to the farmers.

☆ ***Provision of scientific storage facilities***: Government should provide storage facilities to farmers on scientific lines so that, the quality of produce can be maintained for a longer period. The CWC and SWCs must establish their network of godowns up to the village level. Village Co-operative societies are given necessary technical and financial help for construction of godowns.

☆ ***Quality control:*** Quality control issues in food products through promoting HACCP system, code of practices, good hygienic practices, GMP etc., need to be undertaken. The requisite quality and food safety infrastructure is to be established on war-footing basis either with public or public-private modes. DMI should be made as a Nodal agency for promoting GAP, as they ensure cost-effective production and increases the competitiveness of commodities both in domestic and international markets.

☆ ***Improvement in transportation facilities***: All agencies involved in agricultural marketing are affected by the shortage of transportation facilities. The APMCs are under no statutory obligation to provide requisite transportation facilities to farmers from villages to market yards. So, the Government should play a vital role in improving the transportation facilities from remote villages to distant markets for efficient marketing of produce. Even though, it is not the statutory responsibility on the part of APMCs to provide transportation facilities to farmers, but in view of the significance, they should also assume this function. This increases the arrivals of the produce into the market yards and checks the village sales effectively. To improve the efficiency of transportation system in the country the Government has passed Motor Vehicles Act, 1939. Similarly, 'Quarter parcel rate' was also introduced by the Government for transporting the commodities at lower freight charges through the locomotives. Further, the Government introduced refrigerated vans for perishable commodities like fruits, vegetables, fish, milk etc. The following suggestions must be looked into for effective improvements in transportation system that reduce transportation costs:

- The container in the transportation facility must be fully utilized to reduce per quintal cost of transportation.
- The transportation cost per quintal can be reduced by fixing the rate of transportation for different means.
- Use of proper types of wagons to reduce spoilage, damage, breakage or pilferage.
- There should be reduction in the barriers to inter-state movement of produce.

☆ ***Strengthening of processing facilities:*** Government should strengthen the processing facilities of agricultural produce, as processed products fetch more income to the farmers. The existing processing facilities for rice milling, flour milling, pulses milling, oil extraction, cotton ginning etc., are inadequate and need to be augmented and modernized to meet the growing demand for quality products in domestic as well as for export markets. The processing facilities also need to be augmented for processing the perishable products keeping in view of their demand especially in the export market. So, there is a need to enhance the capacity of agro-processing sector. This will not only help in stabilizing the prices realized by farmers, but also in creating employment in rural areas. The food-processing sector alone provides tremendous potential in this area. For attracting private initiative and investment in food processing, the Government of India through MOFPI and NHB have already formulated several schemes of assistance. A ten year tax holiday has been announced. However, the State Governments should also come forward and grant relief in terms of sales tax and other local taxes on processed products. Cheaper processed products will expand demand for such products. At present, value addition is estimated at only seven per cent and processing is only two per cent of the total production. Within next ten years, there is a need to increase value addition to 35 per cent and processing to atleast 10 per cent. Quality control and standardization will be extremely important in this endeavour. The Central Government should establish or encourage a network of food analysis laboratories in the country. This will also be necessary to face competition from imported processed products. The investment potential in value addition and food processing is quite large. If conducive policy environment and incentive frame work is created, private sector can be attracted to make investment of this magnitude.

Further, there is need for small-scale farm level (local level) processing and value addition for agricultural produce particularly at primary level, *viz.,* fruit and pulp making, pickle making, preservation of vegetables, jam and marmalade. Value-adding techniques provide higher income to the small and marginal farmers. This will need to be supported by access to market information and linkage with potential markets, where the processed products could be sold. Agro-processing units in semi-urban centers would create employment generation and also increase demand for agricultural produce. This will ensure reasonable remuneration and increase in the income of small and marginal farmers. Lesson learnt from the White Revolution is itself a guiding beacon for commercialization in areas of dairying and animal husbandry. 'Operation Flood' provided the example of large-scale modern milk processing fed by a well-organized milk shed, procuring milk from a large number of small producers in small amounts.

☆ ***Strengthening of grading and standardization facilities:*** The Government should strengthen grading and standardization facilities in the market yards, as graded produce will fetch more income to farmers when compared to the produce sold as a single lot. To promote grading and standardization of commodities in the country, the Government established several grading stations in the country. The first grading station was established at Kodur in Madras state in 1938 for grading Sathgudi oranges and at Guntur in Andhra Pradesh in 1956, for grading ghee. The improvement in grading and standardization of agricultural produce ensures several benefits and they are already discussed under Section III of Chapter 5.

In view of the globalization and trade liberalization, grading has to bestrengthened on war-footing basis and for this, the following are the key suggestions:

- Promote establishment of grading units at village level through private sector participation.
- Expand the network of State Agmark laboratories covering all the major markets in each state.
- Grading and quality control must be made compulsory for each commodity, so as to reduce further chances of adulteration.
- Consumers and traders should be explained about the importance of Agmark grading.
- Graded products should be advertized frequently, so as to make them popular among consumers.
- DMI should conduct detailed techno-economic feasibility studies to promote and strengthen grading facilities at farmer's level.
- Presently different agencies like FCI, NAFED, and State Co-operative Marketing Federations, Civil supplies Department, CWC and SWCs fix their own standards and grades for the commodities and this is creating lot of confusion among the farmer–sellers and consumers. To avoid this, it is suggested that, Agmark standards formulated by DMI should be adopted and popularized by all these organizations.
- Consumers Service Centers equipped with consumer service laboratories be set up in the States to facilitate the consumers to lodge complaints and re-dressal of their grievances in respect of Agmark products. The samples brought by the consumers at these centers should be quickly got analyzed for appropriate action by the competent authorities.
- Necessary trainings should be imparted to the graders so that, the standards and grades fixed by them show favourable impact on pricing of commodities. To ensure this, Training Centers for Grading should be established in various States to meet the grading requirements of special commodities produced in different parts of India.
- There is need to promote proper packaging after grading so that, further chances of adulteration may not be there.

☆ ***Strengthening market information and communication facilities***: Marketing news and information is vital for taking production and marketing decisions. The domestic demand for food grains may be stable, unlike for cash crops like sugarcane turmeric, tobacco etc. So, information on demand pattern, price behaviour, better methods of production, handling and packing etc., are to be made available to the farmers and traders for taking appropriate decisions. Noticing this importance, the Government of India in collaboration with the various State Governments helped to setup an 'All India Market News Service' in the II FYP. This mainly benefited the farmers, as initially almost all the news relayed from these centers was agricultural in nature and market oriented. Since then, the Government has given a lot of importance for improving the quality of agricultural marketing news services. In the era of trade liberalization, there is much emphasis in strengthening the market information and communication facilities, especially with reference to price of commodities both in domestic and international markets. This guides the farmers and traders for transporting the produce to the areas, where markets prices are high. Government is playing important role in providing markets information through newspapers, radios, T.V, Commodity reports, Special bulletins, journals of agriculture etc. APMCs, SAMBs, CWC, SWCs, etc., are also playing important role in conducting market surveys and in dissemination of market information with reference to market arrivals, valuation and price of commodities in major markets. With the advent of IT, agricultural produce marketing requires connectivity between the markets and exporters or farmers or traders, through wide area network (WAN) of National and International linkages in order to provide day-to-day information with regard to commodity arrivals and prices, to provide links for online international market information, to provide export-related documentation, to inform about the latest research in agricultural marketing, information related to packaging/storage etc., and to provide linkage/connectivity with the APEDA, NIAM, NHB, DMI, IIP, SAMB and SAUs.

- ☆ ***E-Catalogue for Commodity Profiles***: In the context of export competitiveness, each and every commodity needs to the publicized highlighting its characteristics on nutritional values, quality standards, seasonally, quantity for supply and prices etc. A brief commercial profile of the commodity would help the buyer in making comparative analysis on account of cost and margins. Therefore, it is necessary that, each commodity have specific commercial profiles giving the details as mentioned above. These profiles should be transmitted to international markets through 'Web Pages'. Presently, AGMARKNET is providing commodity profiles for major agricultural commodities (18) highlighting (for each crop) origin of crop, importance of crop, major producing countries in the world, major producing states in India, zone-wise major commercial varieties, characteristics of international qualities of mustard-rapeseed, post-harvest losses, harvesting care, post-harvest equipments, grading, grade specifications, adulterants and toxins, grading at producers' level, packaging, transportation, storage, major storage pests and their control measures, storage structures, storage facilities, producer's storage, rural godowns, mandi godowns, CWC and SWCs warehouses, co-operative storage, pledge finance system, marketing practices and constraints, assembling, major assembling markets, market arrivals, inter-state movements, export and import, SPS requirements, export procedures, marketing constraints, marketing channels, Mc and MMs, marketing information and extension, alternative systems of marketing (Direct marketing, Contract marketing, Co-operative marketing, Forward and future markets), marketing related schemes of Government/Public Sector organizations, institutional credit facilities etc.
- ☆ ***Strict weighment of produce***: Government should enforce strict weighment of agricultural produce, as proper weighment fetches more income to the farmers. The Royal Commission on Agriculture (1928) stressed the importance of implementation of Standard Weights and Measures in the markets. Accordingly, in year 1934, 'Provincial marketing depots' were started in the country regarding the use of Standard Weights and Measures in the markets. In the 1935, Weights and Measures Department was established to supervise the strict implementation of standard weights and measures in all the markets in the country. In year 1939, Standards of Weight Act was passed by the Government and introduced 'tola' and 'ton' as standards in the country (1Tola=11.667 grams), (1Ton=10^6 grams) and (1 Ton=85714.04 Tolas). But, each individual State has used different multiples of Tolas and again this led to confusion in the markets in the country. Again in year 1948, Weight and Measures Act was passed regarding the strict implementation of standards of weights and measures in all the markets in the country. In the year 1956, the Metric Weights Act was passed introducing grams and litres as standard weights and measures in the country. This was again incorporated in the Standards of Weights and Measures Act in 1958, prescribed

compulsory use of metric system of weights and measures in the country (*i.e.*, in terms of 'grams' and 'litres'). By this,

- Farmers, who are largely illiterate can easily understand weights and measures of the produce.
- The malpractices of traders were effectively checked.
- Market information on weights and measures can be effectively communicated or disseminated.

'Standards of Weights and Measures (Packed goods) Act' was enacted in 1977 to ensure packing of goods of the correct weights in the packages already existing. To enforce these provisions of standard weights and measures in the market yards, APMCs employ licensed weighmen. In some markets, a weighbridge has been installed. This safeguards the interest of parties against cheating by false or under-weight.

☆ ***Strengthening of regulated markets:*** Government must strengthen the regulated markets in the country, as these markets aims at the elimination of unhealthy and unscrupulous practices in the market yards so that, the farmers' interests can be safeguarded in transacting their produce. These markets regulate the market practices, systematize the Mc, provide scientific storage facilities, proper weighment of produce, grading and standardization facilities, processing facilities, ensure proper method of sale, effective supervision of sales, prompt payment of sales proceeds, market information facilities etc., to the farmers. The management of such markets is done by a APMC, which has nominees of the State Government, local bodies, arhatiyas and farmers. These Committees are appointed by the Government for a specified period of time. The Committee is responsible for the licensing of arhatiyas and weighmen. It is nested with powers to punish anyone, who is found guilty of dishonest and fraudulent practices. No doubt, these regulated markets have helped in mitigating the market handicaps of farmers or traders at the wholesale assembling level. But, the rural periodic markets in general, and the tribal markets in particular, remained out of its developmental ambit.

☆ ***Improvement in market infrastructure:*** In order to provide dynamism and efficiency into the marketing system, large investments are required for the development of post-harvest and cold-chain infrastructure nearer to the farmers' fields. A major portion of this investment is expected from the private sector, for which an appropriate regulatory and policy environment is necessary. Also, enabling policies need to be put in place to encourage the procurement of agricultural commodities directly from farmers' fields and to establish effective linkage between the farm production and the retail chain and FPI. Accordingly, the State Governments were requested to suitably amend their respective APMC Acts for deregulation of the marketing system in India, to promote investment in marketing infrastructure thereby, motivating the corporate sector to undertake direct marketing and to facilitate a national integrated market. The strengthening of market infrastructure is important not only for the performance of various marketing functions and expansion of the size of the market, but also for transfer of appropriate price signals leading to improved marketing efficiency. So, private sector investment in the market infrastructure development may be encouraged by modifying various procedures backed up by package of incentives. Nevertheless, for providing infrastructure in remote and difficult areas, the public sector would need to continue to play an important role. To improve the post-harvest handling of agricultural produce, the Government of India launched a Central Sector Scheme 'Development/

Strengthening of Agricultural Marketing Infrastructure, Grading and Standardization' on 20th October, 2004 with a view to induce large investments from private and co-operative sectors by providing credit linked back-ended subsidy on capital investment for setting up/ strengthening of agricultural markets, marketing infrastructure and support services such as grading, standardization and quality certification. Assistance will be available to individuals, group of farmers, NGOs, SHGs, companies, corporations, co-operatives, local bodies, APMCs etc. This scheme will be implemented through NABARD and NCDC in those States, which amend their APMC Acts, wherever required, to allow direct marketing and contract farming and to permit setting up of markets in private and co-operative sectors. The scheme would provide increased access to small and marginal farmers to agricultural marketing infrastructure for better price realization and to open greater market opportunities to growth, employment and poverty reduction.

The changing trade environment in the wake of LPG phase and increased agricultural production and marketable surplus, further emphasizes the pivotal role to be played by the infrastructure in agricultural development. It is concluded by the Inter Ministerial Task Force on marketing reforms that, the huge investments on agricultural marketing infrastructure may pose huge burden on the Government's exchequer and they opined that, the need of the hour is to mobilize private capital to this sector. With the involvement of private sector, besides availability of private capital, there will be optimum utilization of resources with private management expertise and sharing of risks between the Private and Public sectors. This signifies the importance of introducing appropriate models of PPP ventures in the sector. These mutually beneficial PPP ventures have the potential to play a proactive role in infrastructure development in the sector through sharing of various rights and risks between the partners. It is against this backdrop that, the XI plan target for investment in agricultural marketing infrastructure is envisaged at ₹ 64312 crores with ₹ 30625 crores to be mobilized from private sector.

☆ ***Market research:*** The Government should give more importance for conducting market research through analyzing marketable surplus, minimum prices and maximum prices, modal prices, forecasting the price trends etc., so that, the farmers can plan their crop production programmes and also for efficient marketing of the produce. DMI is playing a vital role in publishing a quarterly journal (Agricultural Marketing) and also publishing commodity survey reports through which the market research data pertaining to both domestic and international markets are provided. These survey reports have brought out the various problems connected with the marketing of agricultural produce and have made suggestions for their removal. Besides DMI, SAMBs, CWC and SWCs, SAUs are also conducting commodity-specific studies and disseminating useful market information to the farmers and other stakeholders. Following are the other important areas for potential research on marketing:

- Marketing technology adaptable to Indian conditions like labour intensive technology without compromising quality
- Dynamic studies of comparative advantage of commodities of India
- Trends in prices of agricultural commodities
- SCPP of Agricultural markets
- Study on cost and margins of agricultural crops

☆ ***Export Oriented Agri-Zones (EOAZ):*** Strengthening of EOAZ for promoting export of agricultural produce by increasing the area under exportable crops (based on the principle of comparative advantage) and also providing necessary post-harvest management and other infrastructure required and information on prices prevailing at international markets is essential as an integrated approach. These should be promoted by providing institutional and physical infrastructure, as per the needs of the specific commodity. In some of EOAZs, there is also a need to establish, what is called Food Parks. In these parks, some common facilities like electricity and warehouse should be created with Central Government assistance, which will help in attracting investment by the private sector and the State Government. While most of the investment should be made by the private entrepreneurs as a way of incentive, Government should invest in common facilities and quality certification. In identification of EOAZs and Food Parks, the Government of India should take an active role, rather than leaving it to the State Governments.

☆ ***Risk management and e-trading:*** They have to be popularized by educating various stakeholders of the agricultural marketing system. For this, a separate resource center should be established in SAUs, NIAM, SAMBs etc. Forward and futures contracts enable price discovery. The price discovery function allows important economic decisions to be made as to which commodities to produce; how to produce the commodities, how much of each commodity to be produced, how much to sell and at what prices, how much to store and for how long, what commodities to consume and how much to consume and at what prices etc. The above questions and answers typically constitute the marketing functions of an economy. After committing to the AoA under the WTO, the need for effective price risk management has become imperative to protect the agricultural sector from price volatility. The active functioning of futures markets enables the farmers and other participants to ensure price risk management. The new economic liberalization policy has strengthened the process of commercialization of agricultural sector. As a result, several new agro-based industries including traditional ones are being set up. The performance of these industries depends heavily on the prices of raw materials (agricultural commodities), as they constitute a major share in their costs. High volatility in prices makes agro-based industries prone to high price risks. Therefore, agro-based industries need a risk management mechanism to safeguard their business from price risk and to plan their production and sales economically.

☆ ***Strengthening linkage between spot and futures markets:*** The regulated markets, which are spot markets, have not been able to link themselves with the forward and futures markets to receive price signals. The futures trading in agricultural commodities is also regulated by the Government. The linkage between the spot and futures markets seems to be poor due to domination of the speculators. Government has, to continue its efforts to strengthen and institutionalize commodity exchanges to instill confidence and awareness among market players.

☆ ***Applications of IT***: The use of IT in agri-marketing has become indispensable. Therefore, encouragement may be provided to generate and host useful portals, websites, databases, information packages and other softwares, generic as well as customize on agricultural marketing. Info kiosks may be encouraged to be set up in the markets, spot as well as futures, farmers organizations, associations of traders and other functionaries for exploiting the opportunities offered by information revolution like:

- online demand of different products
- product specifications with regard to quality, pack size, packaging material, quantity and the time frame of supply
- transport cost involved and the marketing charges likely to be incurred in the market where the goods are to be delivered
- facilities available to the farmer in the buying market
- re-handling of the produce, if necessary, in the supplying market to suit to the requirement of the buyer market
- rules and regulations of the destination market, if it is located outside the State at distant place, and other specific information as may be conducive for the seller to transact the business with the purchasers
- legal provisions related to storage, transportation, phyto-sanitary requirements etc.

☆ ***Promoting forward and futures markets***: Forward and futures markets may well be regarded as direct and alternative marketing system. In the context of trade liberalization process in India, interest in futures markets has been revived for their price risk management and price discovery roles, as well as handling the situation of plenty's and scarcities. Considering the importance of these forward and futures markets, it is suggested that, more and more commodities be added to facilitate competitive and free marketing system. In the process of the economic liberalization and signing of the WTA, Government has given a thrust to enhance international trade in raw agricultural commodities and their value added products. This would expose the domestic market to international price risk. Since exporters generally operate their business on a thin margin, an efficient price risk management mechanism would be essential in the new environment. Withdrawal of Government's intervention from commodity sector ensures only demand and supply forces will determine market price in the new economic environment. It is also argued that, an efficient price management mechanism for agricultural commodities encourages commercial banks to finance the commodity sector and attract FDI in the agricultural sector. Therefore, in the new economic environment, futures markets are expected to play a vital role in the agricultural commodity sector.

☆ ***Government marketing support programme:*** We know, the production in agriculture is uncertain and so, the prices of agricultural commodities. The prices rise sharply at times of poor harvest and fall drastically during bumper harvests. These two situations adversely affect the consumers and farmers respectively. To safeguard the interests of farmers and consumers, the Government has introduced MSP scheme for 25 agricultural commodities (Table 8.31), which together accounts for 82 per cent of the gross cropped area and 75 per cent of the total value of crop output in the country. However, the implementation of MSP scheme today is subjected severe criticism, as very frequently, the farmers are agitating for its hike and to be realistic in its fixation to selected commodities considering the rise in input costs at the farmers' level. The agencies involved in procurement of produce at the farmers' level (at times of distress sales) do not start operations in time. They do not make payments in time. They do not have sufficient arrangement for weighing, bagging, storage, documentation and payment of purchases. Still, this form of intervention by the Government in food grains especially rice and wheat, (in the form of price support, buffer stocking and PDS) has helped the farmers in overcoming distress sales, achieving self sufficiency in staple food and in improving economic access of masses to food. It is therefore, recommended that, continuation of this Scheme and food management system must be strengthened as essential component of National Food Security. Further, the agencies involved in this food management should be encouraged for optimum efficiency and diligence. However, besides food grains, there are several other commodities like fruits and vegetables, which are of considerable economic significance in various agro-climatic regions of the country. The fluctuations in their prices are considerably more. So, to protect the farmers against wide fluctuations in their prices, Government of India launched Market Intervention Scheme (MIS). The MIS for these commodities (fruits and vegetables) have not been included under the MSP policy, as they occupy small proportion of the gross cropped area in the country. At the regional level, however, these crops are of considerable importance. For these crops, the need for price support does not arise every year in all the regions. However, once in two or three years owing to a good harvest, the prices crash and as a consequence, the tempo of increasing their production gets thwarted. The MIS of the Government of India is intended to provide price support in such cases. This policy of MIS also has proved a boon to the farmers in overcoming distress sales of horticultural produce. However, this instrument has not been used at a required scale. The crops for which this type of support is best suited include onion, potato, chillies, cumin, coriander, fenugreek, turmeric, garlic etc. Keeping in view the flexibility in MIS as compared to that in the policy of MSP, there is a need for effective use of this instrument so long as the inter-year fluctuations in output continue to be wide. The distinguishing features of MSP and MIS are shown through Table 7.1.

Table 7.1: Differences between MSP scheme and MIS.

Item	*MSP*	*MIS*
Number of commodities	25 (Fixed number)	Number of commodities not fixed
Announcement of support	Regularly for kharif and rabi crops	Not regularly announced
Recommending agency	CACP, but announced by Government of India	Decided jointly by union andindividual state government
Quantum of supportPurchases	Entire quantity is to be procured, when open market prices fall below MSP.	Pre decided limited quantities.
Implementation of the Scheme	Throughout the country	Specified limited markets
Time of operation of the Scheme	Throughout the year	Specified period
If the Government incurs losses	Central Government	Equally shared betweenstate and union government
Infrastructure required for the implementation of Scheme	Large scale	Limited scale

Source: Economic Survey, 2000-01.

☆ ***Changes in forms of market intervention:*** The objectives and forms of intervention in the marketing system have undergone a substantial change from mid sixties (Figure 7.1), when the country opted for a package of interventions for protecting and reconciling the interests of farmers, consumers and other agencies. One of the important instruments used by the Government to intervene in produce markets consists of fixation and announcement of administered prices and arrangements for their implementation. The administered price

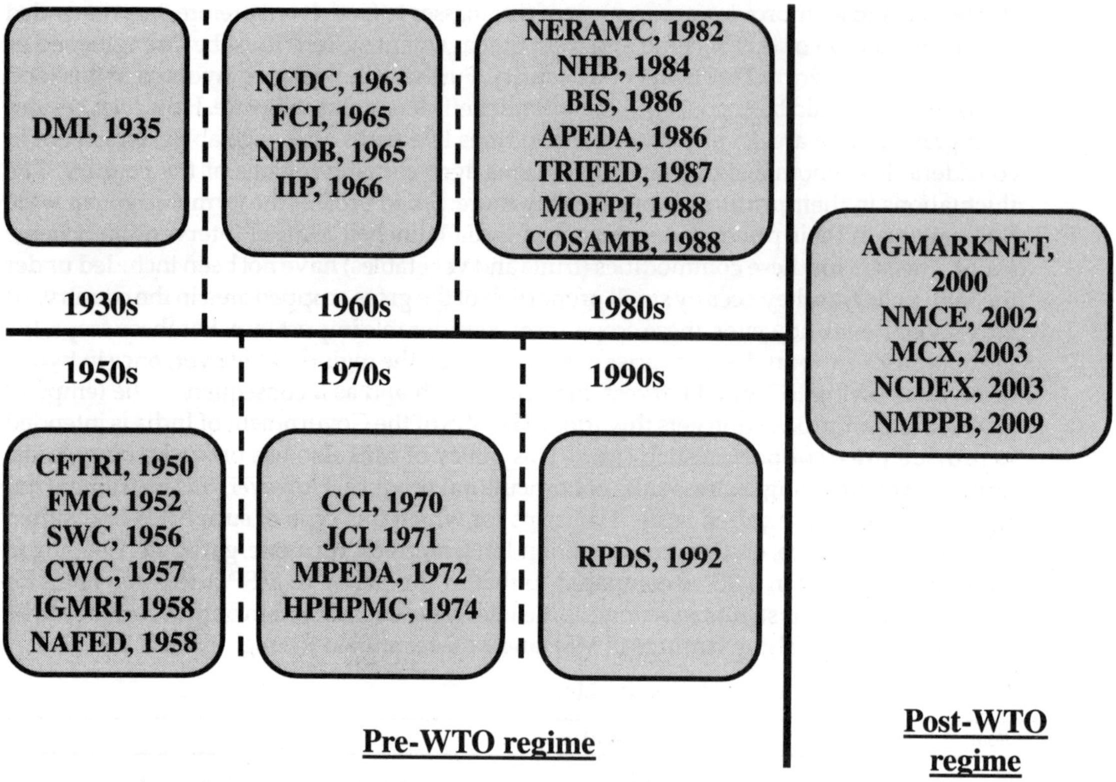

Figure 7.1: Different forms of Government intervention in agricultural marketing in India.

regime currently in vogue includes MSP for 24 commodities including seven cereals (paddy, wheat, barley, jowar, bajra, maize and ragi); five pulses (gram, arhar/tur, moong, urad and lentil); eight oilseeds (groundnut, rapeseed/mustard, toria, soyabean, sunflower seed, sesamum, safflower seed and nigerseed); copra, raw cotton, raw jute and Virginia Flu Cured (VFC) tobacco; Statutory MSP (SMP) for sugarcane; Levy prices for rice and sugar; and Central Issue Prices (CIP) for rice, wheat and coarse cereals for sale under PDS. Direct entry of public agencies in the marketing has influence on SCPP. Maintenance of buffer stocks of rice and wheat, distribution of cereals and sugar at prices lower than market prices (issue prices) and open market operations by public agencies cast their influences on market.

☆ ***Need for Legal Reforms:*** The nature of legal framework within which agricultural markets operate has a fundamental effect on the functioning of the agricultural marketing system. Legal reforms can play an important role in making the marketing system more effective and efficient by removing unnecessary restrictions and by establishing a sound framework to reduce uncertainty of the market. The volatility of commodity prices has been one of the main justifications for State intervention to control or direct the functioning of agricultural markets. Apart from the Agricultural Produce Markets Acts, activities of market functionaries are regulated by several other legal instruments promulgated by the Central Government and the States. The important ones are:

- The Indian Sale of Goods Act 1930.
- Agricultural Produce (Grading and Marking) Act, 1937, 1986
- Drugs and Cosmetics Act 1940.
- Vegetable Oil Products (Control) Order, 1947.
- The Emblems and Names (Prevention of Improper use) Act 1950.
- PFA Act, 1954, 1964, 1976, 1986
- The Essential Commodities Act, 1955
- The Sugar Control Order 1956.
- The Export (Quality Control and Inspection) Act 1963.
- Fruit Products Order (FPO), 1965, 1997.
- Solvent Extracted Oil, Deoiled Meal and Edible Oil (Control) Order, 1967
- Monopolies and Restrictive Trade Act, 1969.
- Meat Food Products Order, 1973
- Vegetable Oil Products (Standards of Quality) Order, 1975.
- The Packaged Commodities Order 1975.
- Standards of Weights and Measures Act, 1976
- Pulses, Edible Oilseeds and Edible Oils (Storage Control) Order, 1977
- Consumer Protection Act, 1986
- Cotton (Control) Order, 1986
- Milk and Milk Products Order, 1992
- Bureau of Indian Standards Act 1986.
- Jute and Jute Textiles (Control) Order, 2000.

In view of these multiple legal instruments, a review is required in respect of all Acts, which regulate markets in terms of licensing of functionaries, commodities traded, controls on packaging and labeling, laws affecting market place, laws affecting supply including controls on movement of produce and volume of commodities traded etc. So, a Task Force is to be set up under the Ministry of Agriculture, Department of Agriculture and Co-operation to undertake a review of all marketing legislations and suggest introduction of necessary legal reforms to promote free and fair marketing system for agricultural and allied products.

- ***Alternative Marketing Systems:*** Role of Government in managing markets is on the decline world-wide. The present system being not amenable to easy transformation, the only way to modernize marketing is to set up an 'alternative marketing system' that may operate parallel to, and in addition to present mandies. The purpose of the alternate marketing systems is to establish modern efficient trade practices as a catalyst for changes in the market towards improved transparency and efficiency. The different forms of alternative marketing could be:
  - Direct marketing by farmers;
  - Marketing through Farmers Interest Groups;
  - Setting up of Terminal markets by the private sector;

- Forward markets and futures exchanges;
- Electronic trading – e-commerce;
- Setting up mega markets covering all marketing functions for domestic as well as export trade; and
- Introduction of NWR system.

The Government should recognize direct marketing as one of the alternative marketing structure that sustains incentives for quality and enhanced productivity, reduce distribution losses, improving farmer incomes with improved technology support and methods. Direct marketing thus, enables farmers and buyers to economize on transportation costs and to improve price realization considerably. The direct markets will operate outside the purview of the Agricultural produce Marketing Act. Direct marketing through SHGs or informal groups, NGOs, co-operatives, farmers associations, companies, partnership, joint ventures may be encouraged by the Government to various policy back up and programmes. As another form of direct marketing, the unemployed youth could be involved in procurement of orders and supply of graded and packed products to different city dwellers. The youth could be trained in marketing practices of procuring products and supplying them. The financial assistance from the public sector to such ventures would generate entrepreneurship and provide profitable employment to the younger generation. They could be given assistance in the form of working capital to start the enterprise along with the necessary agri-business training. The Government's role should be that of a facilitator rather than that of having control over the management of the markets. The similar logic holds good for consumer organizations also, who can procure the produce directly from farmers and distribute to the consumers commensurate with their purchasing powers. The Government should support these organizations with schemes like providing back ended incentives for refrigerated as well general transport, setting up of grading and packing houses, credit at low interest etc., till they become financially self-sufficient on commercial lines. At present, these direct markets are being run at the expenses of State exchequer as a promotional measure, to inculcate habit of marketing without help of middlemen by the small and marginal farmers of fruit and vegetables. Considering the vastness of the country, more and more such markets need to come up in the organized sector with private investment so that, they can be developed in tune with the market requirements with backward and forward linkages. A common code of conduct and modalities with regard to ownership, operation and need based infrastructure will have to be prepared and circulated to spread the concept of direct marketing by the farmers. Besides Direct marketing, Electronic trading or e-Commerce is another more appropriate form of alternative system of marketing. Every APMC should be provided with facility of electronic trading by setting up a special kiosk for the purpose.

☆ ***Professionalizing the agricultural marketing system:*** The States should be incentivized to professionalize the management of the market yards through outsourcing operation and management, promoting modernization in PPP mode and privatizing some of these markets, wherever feasible. The modernization may consist of the infrastructure, which can add value to the farmers' produce and increase the price realization by the farmers. A transparent electronic auction system, electronic price display mechanism, bulk weighing and bulk handling infrastructure, computerized operations of the market, appropriate storage infrastructure, appropriate cleaning, grading, sorting and packing infrastructure wherever required should become mandatory infrastructure in each of such markets, with suitable modifications to suit the local requirements. Further, there is an imperative need to make Government administered marketing organizations administratively viable and managerially competent in keeping with liberalized trade atmosphere. APMCs including sub-yards, should be headed by professionals. Existing Secretaries need to be trained in professional management of the markets. The functions of APMCs and SAMBs may have to be remodeled for this purpose.

So, to make the PAMS in India more vibrant and competitive in the context of globalization and trade liberalization, Government of India has to implement the above suggested reforms, so as to attain the goals and objectives of an 'ideal marketing system'. These reforms will certainly make the farmers to get the best possible returns through minimizing the price difference between them and ultimate consumer. From the consumers' side, these reforms ensure the availability of all farm products to them at reasonable (affordable) prices without impairing on the quality of the produce. The suggestions discussed earlier under each marketing function (Chapter 5) must also be looked into to make the marketing system 'ideal' considering the interests of different agencies and stakeholders involved in the marketing process. We know that, agricultural marketing has been undergoing tremendous changes with the advent of LPG phase since the past one-and-half decades. Consequent upon the establishment of WTO, the agricultural sector in all the member countries of WTO has been thrown open to the vagaries of international economic transactions. The agricultural sector, especially in developing countries like India, has become extremely susceptible to changes in the global economic fluctuations. With the opening up of overseas markets for fresh as well as processed agricultural products, the need for global quality standards in products, SPS norms, changing marketing practices etc., has become more urgent. In view of this, the conventional farm products are gradually giving way to new and internationally demanded processed farm products. So, the age-old agricultural marketing systems are being replaced by new marketing systems as exemplified by the emergence of chain stores, modern food retailing and corporatization of supply chain in marketing. With such a fast changing agricultural marketing scenario, the farmers not only have a great opportunity to cash in on the advantages thrown up by modern marketing systems, but also face a great challenge in that their production and exports have become more vulnerable to the risks and uncertainties of globalization. Considering these dynamic changes, both central and local governments must look into the policy suggestions discussed above in designing the marketing system on the ideal grounds. To assure the benefits offered by an 'ideal marketing system' to the farmers, Government intervene in the marketing mechanism of agricultural produce and the various forms of interventions are discussed in the ensuing Chapter 8.